Second International Symposium Recycling of Metals and Engineered Materials

Edited by

Jan H. L. van Linden
Alcoa Laboratories
Alcoa Center, Pennsylvania 15069

Donald L. Stewart, Jr.
Alcoa Laboratories
Alcoa Center, Pennsylvania 15069

Yogeshwar Sahai
Ohio State University
Columbus, Ohio 43210

A Publication of

TMS
Minerals • Metals • Materials

A Publication of The Minerals, Metals & Materials Society
420 Commonwealth Drive
Warrendale, Pennsylvania 15086
(412) 776-9024

Printed in the United States of America
Library of Congress Catalog Number 90-62250
ISBN Number 0-87339-154-3

Preface

The first International Symposium on Recycle and Secondary Recovery of Metals was held in 1985 in Fort Lauderdale, Florida as the Fall Extractive and Process Metallurgy Meeting, sponsored by the Physical Chemistry Committee of The Metallurgical Society of AIME. The fact that the 500 copies of the Proceedings of that meeting are completely sold out speaks well for the success and the quality of that symposium.

Much has changed since the first Recycling Symposium: TMS has added Materials and Minerals to its name and field of endeavor, shortages of landfill space and resources have catapulted recycling into the role of "premier savior" of our standard of living and finally, the growing public awareness of the long-term effects of some industrial waste products has led to increased scrutiny of the entire industrial waste stream which, in turn, has stimulated recycling activity.

To reflect these changes the organizers of the second Recycling Symposium have actively solicited contributions pertaining to materials other than metals and, furthermore, included in the meeting a panel discussion on the technological, economic as well as regulatory aspects of recycling.

The Editors and Organizing Committee would like to thank the authors for providing the valuable information contained in this book. In addition, we would like to thank the members of the fledgling Joint Recycling Committee of the Extractive Processing and Light Metals Divisions of TMS for their many good suggestions. Finally, we would like to thank Barbara Kamperman and Janet Urbas from TMS for their organizational guidance.

Jan H.L. van Linden
Alcoa Laboratories
Alcoa Center, Pennsylvania 15069

Donald L. Stewart, Jr.
Alcoa Laboratories
Alcoa Center, Pennsylvania 15069

Yogeshwar Sahai
Ohio State University
Columbus, Ohio 43210

Table of Contents

USED BEVERAGE CONTAINERS

FERROUS METALS

LEAD - ZINC - TIN

ENVIRONMENTAL/HEALTH

ALUMINUM: BYPRODUCT RECOVERY

OTHER METALS A

OTHER METALS B

MIXED MATERIALS

PLENARY LECTURES

Session Chairman:
Jan H. L. van Linden

RECENT PROGRESS ON FERROUS SCRAP PRETREATMENT

Alan W. Cramb and Richard J. Fruehan
Department of Materials Science and Metallurgical Engineering
Carnegie Mellon University
Pittsburgh, PA 15213

The potential build-up of residual elements in steel is a cause for concern within both the steel and the scrap industries. High levels of elements such as copper, nickel, molybdenum, tin and chromium are thought to be deleterious to product consistency and have also been shown to cause problems during casting and subsequent processing.[1] The absolute level at which a residual element becomes deleterious to product quality is dependent upon the application; however, it seems reasonable to assume that, at some time in the future, prudent background levels of residual elements will be exceeded, leading to scrap sorting or pretreatment as a routine operation before charging into an electric arc furnace.

In addition to the residual problem, the growth in recycled automotive galvanized sheet scrap has resulted in increasing zinc content in the exhaust dusts of the steelmaking furnaces. This dust is considered hazardous by the EPA and disposal is expensive as it is classified as a hazardous waste. Thus zinc must also be removed from scrap before it is charged into the steelmaking furnace.

The purpose of this article will be to review scrap pretreatment for the removal of copper and zinc from scrap. Some recent efforts at Carnegie Mellon University to develop a method of scrap pretreatment which would allow solid copper to be separated from solid scrap will be highlighted.

Second International Symposium -
Recycling of Metals and Engineered Materials
Edited by Jan H. L. van Linden,
Donald L. Stewart, Jr., and Yogeshwar Sahai
The Minerals, Metals & Materials Society, 1990

The Copper Problem

The United States has a tremendous ferrous scrap surplus. To produce steel from scrap takes less than one third of the energy that it does when starting with ore. In addition, many of the major environmental problems are eliminated, in particular those associated with coke making. However, much of the scrap cannot be used because it contains elements, such as copper and zinc, which are detrimental to the steelmaking process or the properties of the steel produced. If these unwanted elements can be removed, considerably more steel can be recycled and other environmental problems related to steelmaking reduced.

Background

The copper residual problem is well recognized in both the solidification and deformation processing of liquid steel. Gross defects have been found during the casting, rolling and welding of steels with significant copper residuals[2-5], while variations in product properties have been correlated to copper variations in the alloy[6].

Generally copper residual contents are limited to less than 0.35%[1] in most alloy steels with the exception of weathering steels (0.5%) and steels exposed to nuclear radiation (0.1%); however, copper contents of less than 0.25% have been shown to cause fine intergranular cracks on bloom castings[3]. The maximum allowable copper levels for various grades of steel are given in Table I. In general the copper levels in steels produced by integrated producers of flat products are controlled to lower levels than those produced via electric arc furnaces.

Table I: Maximum Allowable Copper Levels

Application	%Cu
Deep Drawing Steel	0.06%
Tin Plate (critical)	0.06%
Drawing Quality	0.10%
Steel Forgings	0.35%
Irradiated Steels	0.10%
Bar Products	0.35%

Casting and rolling defects are caused by embrittlement due to the formation of a liquid copper film which forms when the iron is preferentially oxidized during the formation of scale[3,7]. This preferential oxidation causes an enrichment in copper and other residuals such as nickel, tin and antimony in the grain boundaries. Copper rich phases found in cracks on bloom surfaces at Round Oak Steel Works in England were found to contain 85% Cu, 5% Fe, 5%Sn, 4% Ni and 1% Sb which would be molten

[1]supplementary requirement S1 of ASTM Specification for Steel Forgings, General Requirements (A 788)

below 1100°C (Figure 1[8]) and give rise to classical 'hot shortness' due to penetration of the liquid film along the grain boundary. This phenomenon of liquid copper film embrittlement of steels in an oxidizing atmosphere is well documented, was originally proposed by Pfeil[9] and was discussed by a number of authors during the sixties[10 - 13]. One suggested method of eliminating this defect is to ensure that a solid intermetallic compound of copper and nickel is formed by ensuring that the copper to nickel ratio in the steel is 3 : 1 or larger[9]; however, at higher copper levels (0.35%) substantial quantities of nickel (0.12%) would have to be intentionally added to ensure avoidance of the defect. A listing of the factors which can favorably affect the hot workability of copper containing steels is given in Table II.

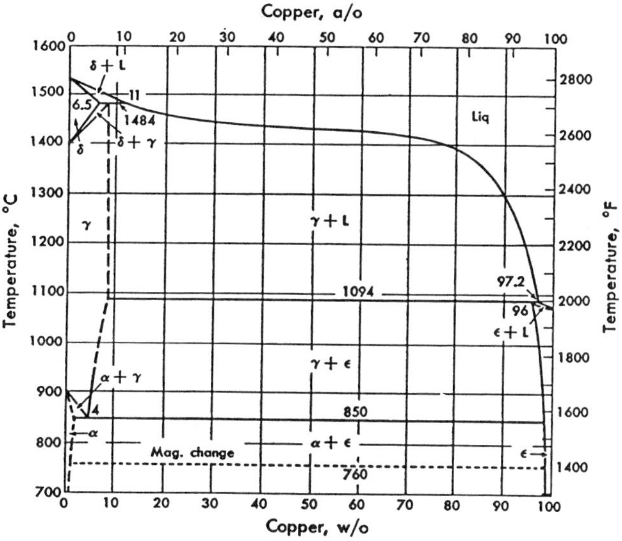

Figure 1: Phase Diagram of Fe-Cu system[8].

Table II: Factors Favorably Affecting the Hot Workability of Copper Containing Steels, after B. F. Glasgal, [1]

- Low Copper Residuals
- Fast Heating Rates
- Reduced Time at Temperature
- Limited Scale Formation
- Nickel:Copper Ratio greater than 1:3

Another copper induced crack which is very common in continuously cast slabs and blooms is the star crack. This crack is caused by localized surface concentrations of copper which are due to wear of the

Table V: Decrease In Average Copper Values Due to Scrap Preparation, After Pflaum [21]

Type	Before	After
Shredded	0.23	0.16
#1 Heavy Melt	0.24	0.19
#2 Heavy Melt	0.46	0.37
Turnings	0.29	0.18
#2 Bundles	0.51	0.42

technique with sulphide slagging to develop a new technique capable of the removal of solid copper from solid scrap at temperatures lower than the melting point of copper (1083°C). Fruehan's process, which operates at 1000°C or below, is based on an analysis of the chemistry of the system. For example, the equilibrium constant of equation 1 is as follows:

$$K = \frac{a_{Fe}\ a_{Cu_2S}}{a_{Cu}^2 a_{FeS}} \qquad [3]$$

and,

$$\frac{a_{Fe}\ a_{Cu_2S}}{a_{Cu}^2 a_{FeS}} = exp\ [\ -\ \frac{\Delta G^o}{R\ T}\ ] \qquad [4]$$

where ΔG^o is the standard free energy of reaction 1. In Fruehan's process solid iron is precipitated upon reaction.

In the above equation ΔG^o is negative (the reaction is spontaneous), and the value for K is approximately 2 at 1000°C[16]. Thus to maximize the amount of copper in the matte one needs a low starting activity of copper sulphide in the matte, a high iron sulphide activity in the matte and a high copper activity. The activity of solid iron is 1.

Thus the starting matte composition should be high in iron sulphide and low in copper sulphide. In addition the matte must be liquid and fluid at treatment temperatures, and sodium sulphide is normally added as the matte's liquidus can be lowered to 700°C by such additions[22] (Figure 2). Sodium sulphide also decreases the activity of copper sulphide in the matte, thus improving the thermodynamics of copper removal. A normal matte starting composition is 82% FeS - 18% Na_2S.

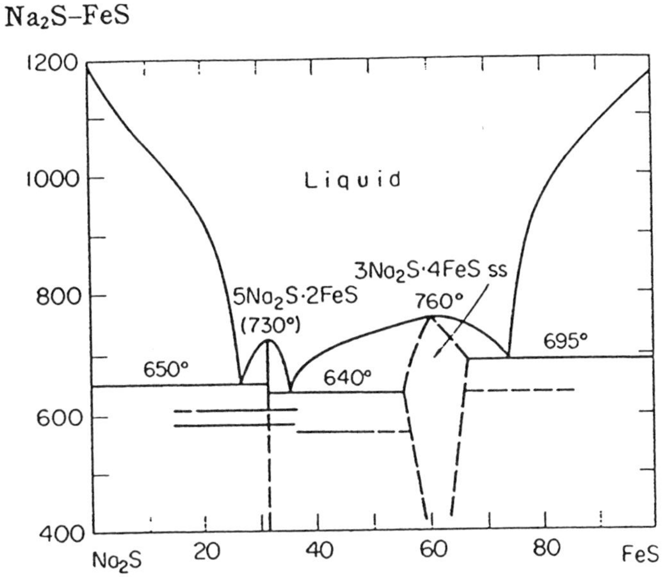

Figure 2: Phase Diagram of Fe-S-Na_2S system.

Melting scrap puts copper into solution in liquid iron and significantly reduces its activity when referred to pure liquid copper. Even in carbon saturated iron, where the copper activity coefficient is high, the activity of copper in solution in liquid iron is reduced by a factor of 25 - 50 as compared with pure copper. Due to the low copper activity in the melt the partition of copper between the matte and the liquid steel is significantly lowered and trials at the Bureau of Mines[22] in a 1.5 ton pilot operation could only remove 40% of the copper from the metal using 200 lb of an FeS-Na_2S mixture per ton of carbon saturated iron. The activity of copper as a pure solid is by definition unity. Thus it is more thermodynamically feasible to remove solid copper from solid scrap than to remove copper from scrap which has been remelted.

The iron sulphide - sodium sulphide matte reacts with solid copper, displaces solid iron and forms a iron sulphide - sodium sulphide - copper sulphide matte which is liquid and fluid at temperature (Figure 3). This matte can then be drained from the scrap.

Initial trials of the concept were carried out by Jimbo et al.[16] on a laboratory scale at Carnegie Mellon University using a matte containing 19 to 25% Na_2S. Simulated scrap consisting of pieces of iron and copper were treated in a rotating crucible with FeS-Na_2S fluxes (Figure 4). Approximately 100 grams of simulated scrap containing from 0.4% to 1% Cu was treated. over 90% of the copper could be removed at 1000 C and nearly 80% at 800 C. Initial trials indicated that only 4 kg of matte per metric ton of scrap would be necessary to treat steel containing up to 1% copper.

Process Benefits

Due to the success of the small scale tests in proving that Fruehan's concept had potential, a detailed comparison of solid and liquid scrap treatment was made. The advantages of the solid scrap treatment

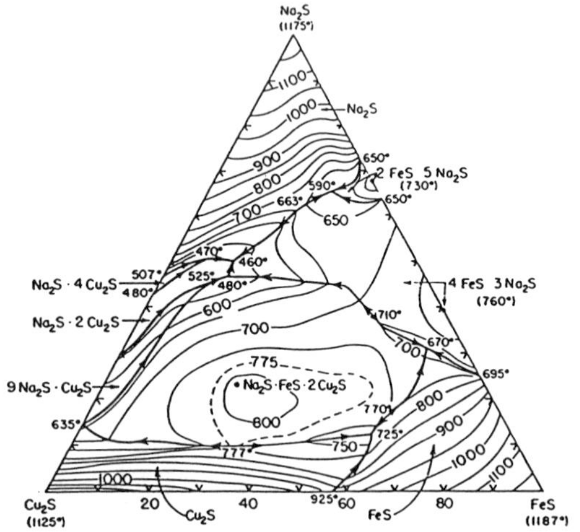

Figure 3: Phase Diagram of the FeS-Na_2S-Cu_2S system.

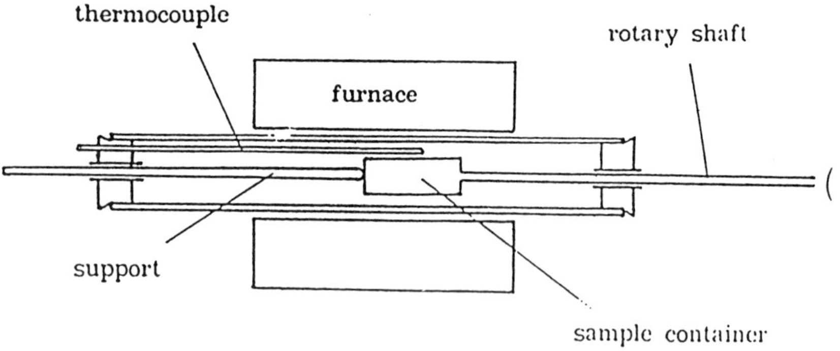

Figure 4: Schematic of Rotary Mixer Used in Initial Tests, after Jimbo[16]

process were as follows:

- Lower Matte Weight. To reduce copper from 0.4 to 0.1% in the scrap less than 10 kg of matte would be thermodynamically necessary via the solid process, while 120 - 200 kg would be needed for liquid processing.

- One Step Process. In the liquid treatment the slag would have to be removed, the matte added and the steel carburized before treatment and decarburized after treatment. In the solid process the matte would mixed with the scrap, and, after reaction, drained.
- Lower Na_2S Loss. Evaporation losses from the matte are much lower at lower temperatures.
- Increased Matte Copper Content. Matte copper content could be as high as 30% in the solid process; therefore, recovery of copper from the matte may be economically feasible.
- Decreased Resulphurization of Metal. In the liquid treatment there will be a significant sulphur pick-up by the liquid steel which would necessitate extensive desulphurization after processing. In the solid scrap process sulphur pick-up would be determined by the amount of matte which did not drain from the scrap.
- Process Flexibility. The solid scrap treatment process could be a stand alone unit off-site from the remelter or an on-site unit supplying preheated low copper scrap to an electric arc furnace. The liquid process would require an advanced ladle metallurgy station, equipped with reheat, slag skimming and injection facilities. The solid treatment would not affect process cycle time after melting, while the liquid treatment stage would have a considerable impact on productivity.

In addition, an economic assessment of the process was made which suggested that the energy cost of treatment would be approximately $5.00 per ton, and the reagent cost approximately $2.00 per ton. If the hot scrap is charged directly into the electric furnace there would be a decrease in electrical energy of approximately 250 kwh\ton and corresponding decreases in heat time and refractory wear which should lead to increased productivity and lower operating costs. Although it is difficult to exactly calculate total cost savings due to variations in raw material prices, we estimate that it may be possible, if there is a $40.00 per ton differential between high copper scrap and low copper scrap, that cost savings of approximately $20.00 per ton for direct electric arc furnace charging and $ 10.00 per ton for a stand alone processor may be possible.

Practical Concerns

The initial small scale experimentation resulted in a number of questions concerning this new process:

- What was the rate controlling step in the reaction?
- Would the matte spread and wet scrap?
- Could the matte be effectively drained from the scrap?
- Could the process be carried out in air?
- Would it be possible to treat electrical motors with this process?

To answer the above questions a number of experiments were carried out before large scale, industrial tests were initiated.

1. Rate Controlling Step.

It was assumed that liquid phase mass transfer would be the rate controlling step in the reaction. Thus, a rotating cylinder experiment was attempted under conditions where the mass transfer was well documented. Typical experimental results from the work of Y. Zhang[2] are given in Figures 5 and 6, where the increase of matte copper sulphide concentration can be seen as a function of time and rotation speed (Figure 4) with the attendant decreases of sodium sulphide and iron sulphide (Figure 5). At higher rotation speeds the reaction was found to be controlled by mass transfer; however, at low rotation speeds the matte formed a viscous layer which impeded mass transfer. Thus it is important to carry out the matte process in an environment, such as rotating kiln, which stimulates matte movement.

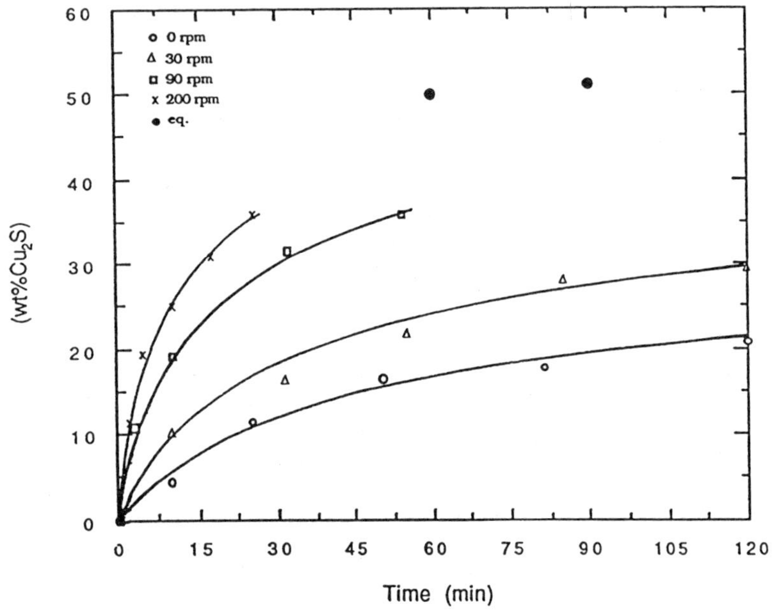

Figure 5: Concentration of Copper Sulphide in Matte as a Function of Time and Rotation Speed.

2. Matte wettability.

Experiments were carried out using an X-Ray Radiographic Technique to determine the contact angle between the matte and iron, copper and an alumina refractory. The results of this study indicated that the matte wet and spread upon iron, copper and the alumina refractory. Contact angles in all experiments were less than 5 degrees.

3. Matte Drainage.

To determine the efficiency of matte drainage L angles of steel (40 x 40 x 3 mm), steel plate (40 x 2 mm) and a copper sheet (40 x 2) were dipped into an 18% Na_2S - 82% FeS bath at temperatures of 800, 900 and 1000 C. The samples were emersed for 3 to 5 minutes before being withdrawn from the bath and air cooled.

---

[2]Y. Zhang, A. W. Cramb, and R. J. Fruehan: unpublished research, CISR, Carnegie Mellon University, 1990

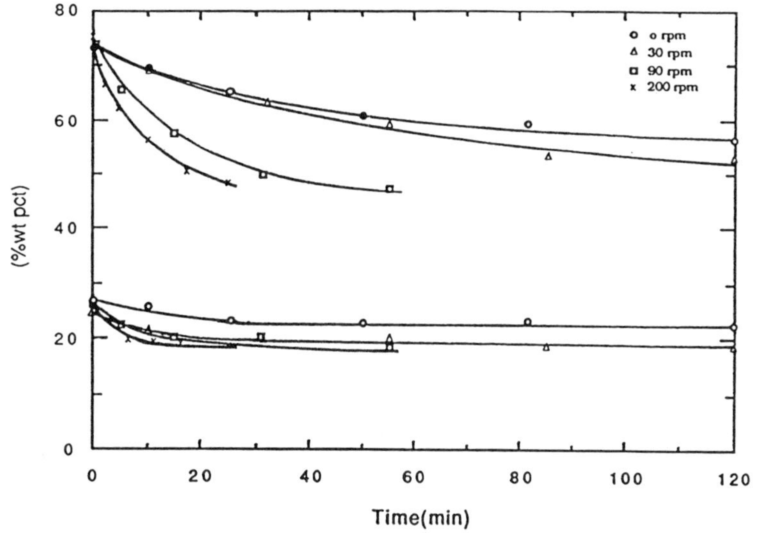

Figure 6: Concentration of Sodium Sulphide and Iron Sulphide in Matte as a Function of Time and Rotation Speed.

The copper sheet was completely dissolved at all temperatures. Results of the drainage experiments are given in Figures 7 and 8, where the ΔM% is the weight gain upon removal from the matte. In both cases the amount of matte adhering to the scrap was minimal at 1000°C; however, significant amounts of matte remained attached to the scrap at lower temperatures.

4. Process Feasibility in Air.

The above dip tests were carried out in air with no apparent deleterious effects; however, the time at temperature was small. Other experiments, carried out over longer time periods, indicated that if the copper was allowed to significantly oxidize, the process efficiency greatly decreased. In addition, if the iron was allowed to significantly oxidize, iron oxide would be picked up in the matte.

5. Removal of Copper from Electrical Motors.

To determine the efficiency of drainage from a copper wound motor, a large bath of matte (40 kg) was melted in an induction furnace using a carbon crucible. The motor was lowered into the matte and slowly rotated. The matte temperature was 1000 C during the experiment. Initially there was a large quantity of smoke as the plastic coverings, etc., on the copper windings burned off. The motor was rotated for 20 minutes and then removed. All copper which was emersed in the matte was removed from the motor.

Larger Scale Trials

In view of the process potential larger scale experiments were proposed to and accepted by the DOE[24]. A rotary kiln was purchased from CORECO, and set-up in the pilot facility of USX Corporation

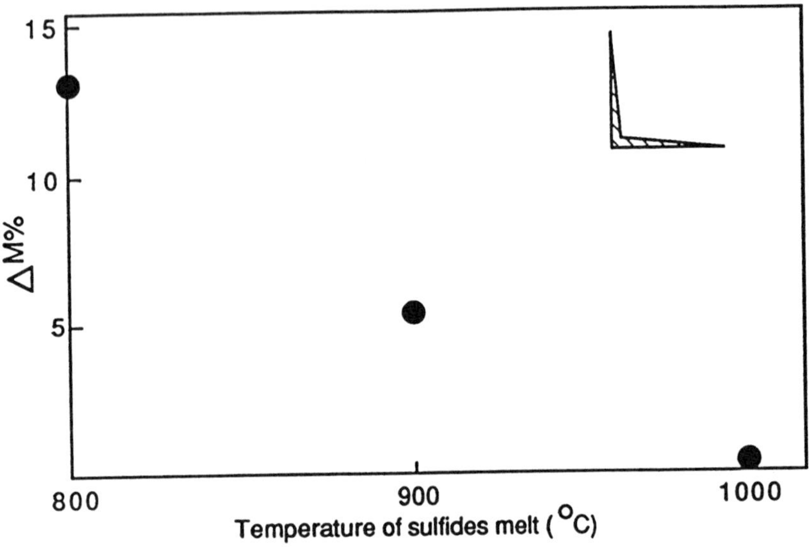

Figure 7: Matte Adherence Trial for Steel L Angle.

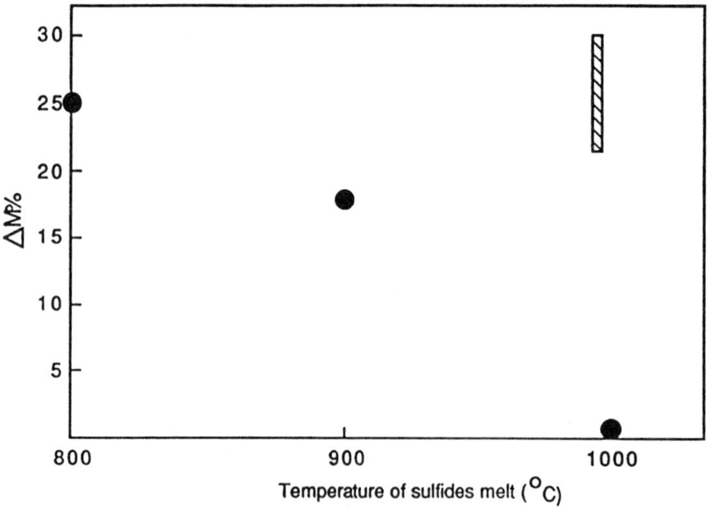

Figure 8: Matte Adherence Trial for Steel Plate.

in Universal, Pa. A schematic of the kiln is given in Figure 9.

To date the kiln has been installed and initial trials conducted. Preliminary results indicate that the process can be used on a larger scale to remove solid copper from scrap. Up to 100 kg of scrap has

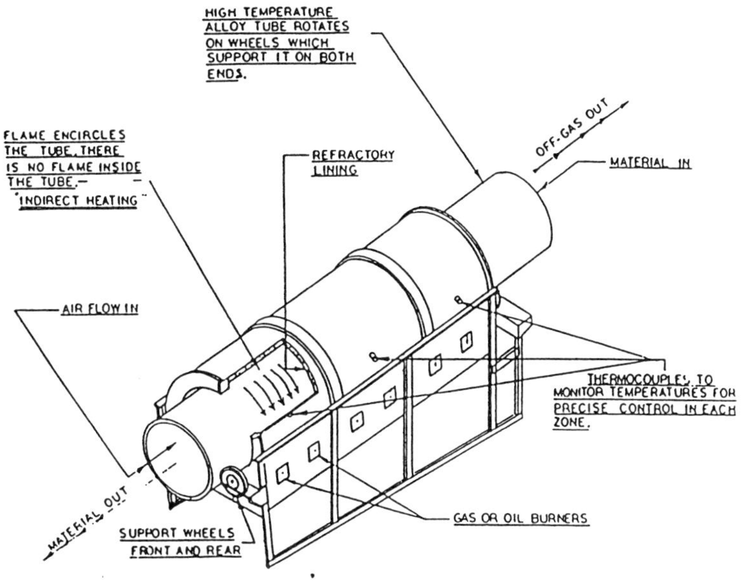

Figure 9: A Schematic of the CORECO Rotary Kiln.

been treated. Copper tubing and wire have been used to simulate the solid copper content of scrap. The process was operated at 1000°C and up to 4 kg of copper was removed from solid scrap in less than 10 minutes. At 1000°C there have been no problems in draining the matte from the kiln and copper removal efficiencies are approximately 90%. An example of the increase in copper sulphide content of the matte during a large scale trial is given in Figure 10. The fast rate of copper pick-up can be seen (4 kg of copper in less than 10 minutes). The copper sulphide content decreases as a function of time due to increasing matte FeO content as a consequence of oxidation of the scrap; therefore, it is best to carry out these experiments in a non-oxidizing atmosphere.

The Zinc Problem

The melting point of zinc under 1 atmosphere pressure is 419.58°C and its boiling point is 907°C. Thus, in any steelmaking process where temperatures can rise to 1600°C, zinc will vaporize, potentially oxidize and be found in the exhaust dust of the process. Electric furnace dusts are considered hazardous by the Environmental Protection Agency[25,26], due to the leachability of the toxic tramp elements of lead, cadmium and chromium and disposal of this dust is expensive. This has lead to numerous attempts to remove the zinc from the scrap before it is charged into the steelmaking vessel[26] in order to produce a material which can be more easily and profitably recycled.

The problem is being exacerbated by the growth in the amount of automotive galvanized sheet steel scrap which was estimated[25] to increase to 1.4 million net tons by 1990 by CMP. Automotive scrap is estimated to be only 10% of the galvanized scrap production, with the remainder joining the scrap problem in future years due to obsolescence of the galvanized part.

Residual Copper in Annealed and Postweld Heat Treated 2-1/4Cr - 1Mo Steel." , Residual and Unspecified Elements in Steel, ASTM STP 1042, A. S. Melilli and E. G. Nisbet, Eds., American Society for Testing and Materials, Philadelphia, 1989, pp202 - 231.

7. J. A. McNichol:"Some Aspects of the Effect of Copper in Cast Iron and Steel", The Australian Engineer, October 7, 1953, pp54 - 60.

8. O. Kubaschewski: Iron-Binary Phase Diagram, Springer-Verlag, 1982, p35.

9. L. B. Pfeil: Journal of the Iron and Steel Institute, 1929, Vol. 119, pp501 - 547.

10. D. A Melford: "Surface Hot Shortness in Mild Steel", Journal of the Iron and Steel Institute, 1962, Vol 200, pp290 - 299.

11. A. D. Nicholson and J. D. Murray: "Surface Hot Shortness in Low Carbon Steel", Journal of the Iron and Steel Institute, 1965, Vol. 203, pp1007 - 1018

12. D. A. Melford: "Influence of Antimony and Arsenic on Surface Hot Shortness in Copper Containing Mild Steels", Journal of the Iron And Steel Institute, 1966, Vol. 204, pp495 - 496.

13. W. J. M. Salter: Journal of the Iron and Steel Institute, 1966, Vol 204, pp478 - 488.

14. R. Rosegger: Radex Rdsch., 1964, No. 5, p277.

15. K. G. Speith and A. Bungeroth: Stahl und Eisen, 84, 1964, p1297.

16. I. Jimbo, M. S. Sulsky and R. J. Fruehan: Iron and Steelmaker, August 1988, pp20 - 23.

17. J. F. Jordan: US Patent 2,512,578, June 20, 1950.

18. N. Kopylov and S. Novoselov: Zh. Neorgan Khim., 9 (8), 1964, p1919.

19. W. Oelsen, E. Schuermann and G. Heinrichs: Archiv fur das Eisenhuttenwessen, 30, 1959, pp649 - 654.

20. J. Peace and D. Engledow: "Developments of Scrap Beneficiation by the British Steel Corporation", Ironmaking and Steelmaking, 1987, Vol.14 , No.5, pp248 - 252

21. D. A. Pflaum: "Residual Problems and the Scrap Industry", Residual and Unspecified Elements in Steel, ASTM STP 1042, 1989, pp11 - 25.

22. V. G. Leak and M. Fine: Bureau of Mines Report of Investigation 7809.

23. N. Kopylov and S. Novoselev: Zh. Neorgan. Khim., 9(8), 1964, p1042.

24. R. J. Fruehan and A. W. Cramb: "Removal of Copper from Scrap", A proposal to the DOE, December 14, 1987.

25. R. J. Fruehan: "Scrap in Iron and Steelmaking", Recycle and Secondary Recovery of Metals, P. R. Taylor, H. Y. Sohn and N. Jarrett, eds., TMS, 1985, pp13 - 34.

26. O. F. Angeles and Edward F. Petras: "Dezincing of Galvanized Steel Sheet Scrap" CMP Report No. 86-1, May 1986.

27. H. D. Reitere and W. Morth, European Patent No. 73186, March 2, 1983.

28. Belgian Patent No. 892, 897, October 19, 1982.

29. J. Frenay, J. Hissel and S. Ferlay: "Recovery of Lead and Zinc from Electric Steelmaking Furnace Dusts by the Cebedeau Process", Recycle and Secondary Recovery of Metals, TMS 1985, pp195 - 202.

30. R. J. Fruehan: "Plasma Processes for Metals Production; A Scoping Study", CMP Report No. 85-3, 1985.

31. T. Knoepke: "Initial Results from the Voest-Alpine Plasma Furnace", Iron and Steel Engineer, May 1985, pp23 - 26.

32. J. A. Dalessandro and D. L. McBride: "Superheating Blast Furnace Iron with Channel

Induction Furnaces", Iron and Steel Engineer, February 1985, pp29 - 36.

33. Japanese Patent No. 57/85916, 1982.

34. J. A. Vallomy: " Continuous Steelmaking Via Electric Furnace, The CONSTEEL Process", AIME Electric Furnace Conference, 1984.

35. J. E. Bonestell and R. Weber: "EOF (Energy Optimization Furnace) Steelmaking ", Iron and Steel Engineer, October 1985, pp16 - 22.

36. D. R. MacRae: "Electric Arc Furnace Dust - Disposal, Recycle and Recovery" CMP Report No. 85-2, May 1985.

A SECONDARY'S VIEW OF RECYCLING

JOSEPH S. VILAND

WABASH ALLOYS

Abstract

This paper will present an overview of the secondary aluminum industry which produces foundry alloys. The perspective will be primarily from an American producer's point of view; however, comparisons will be made to the European approach which generally takes a different path.

The secondary aluminum industry has been "mining" an aluminum resource since the advent of the primary aluminum industry, but developed as a significant industry during World War II. Processing equipment and improved furnace technology have increased the efficiency of the process of recovery significantly, making what were considered economically poor sources of scrap, recoverable. With stronger emphasis on recycling, the future looks bright; however, the handling of the industries' wastes and increased environmental controls provide a challenge to the industry.

History

In the early days of aluminum production, remelting of scrap was conducted by the primary aluminum industry. With the rapid expansion of the primary industry in the 1930's and its subsequent explosion in the war years, the reuse of aluminum was delegated to an industry which was largely the extension of the "smelting industry" which remelted copper, brass, lead, tin, and zinc.

The secondary aluminum industry developed, as many industries, with plants based near the source of raw materials and were, in most cases, small by current standards. Furnace capacities of 40,000 to 60,000 pounds were the norm whereas today we use furnaces of 200,000 pound capacity. Materials were shipped by rail for the most part because we didn't have the vast interstate highway system that allows us to move material from the West Coast to the East Coast in a matter of days.

I've often wondered why we were given the name the "secondary aluminum industry" rather than something more esoteric. I suppose it is because we weren't primary so by definition we were secondary. However, it's much nicer being called the recycling industry.

Second International Symposium -
Recycling of Metals and Engineered Materials
Edited by Jan H. L. van Linden,
Donald L. Stewart, Jr., and Yogeshwar Sahai
The Minerals, Metals & Materials Society, 1990

The secondary recycling industry primarily produces foundry alloys and, to a small degree, deoxidation products for the steel industry. It does not recycle cans, although in the infancy of the can recycling development, we did remelt most of the cans produced. From a metallurgical point of view, cans are not a good source of metal due to the high magnesium and manganese levels.

Manganese contributes to sludging in casting alloys so it has to be diluted. Magnesium limits on most die casting alloys are 0.1% and has to be removed from the alloy, generally by chlorination. Therefore, we are pleased that technology was developed to recycle cans back into can stock.

The growth of the secondary industry is shown in Figure 1. To put this in perspective, the industry with over two billion pounds of shipments in 1988 was the equivalent of eight modern 250 million pound/year primary smelters. To produce the volume of alloy currently shipped, a total of over 3 million pounds of scrap must be processed each day.

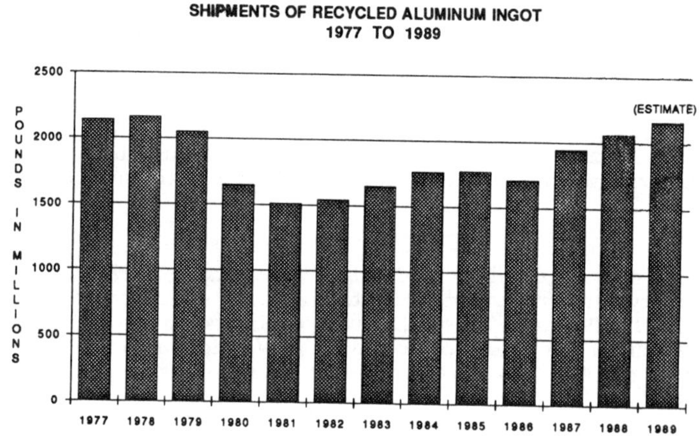

Figure 1: Shipments of Recycled Aluminum Specification Ingot

The impact of recycling is shown on Figure 2 which compares the secondary industry's recycling and can recycling to primary aluminum production.

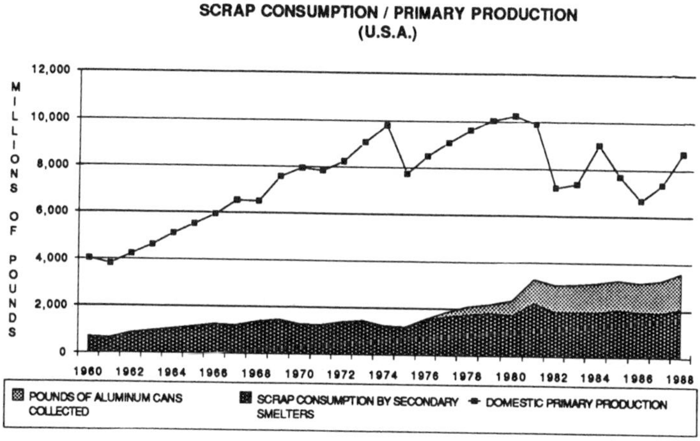

Figure 2: Secondary Aluminum vs. Primary Aluminum

Markets

The markets that are served are varied but fall into the following end use categories:

Direct Automotive	-	22%
Automotive Related	-	44%
Small Engine	-	8%
Appliance	-	7%
Other	-	19%

Automotive uses are primarily for transmissions, intake manifolds, and a variety of smaller parts and housings. Small engine uses are for lawn mowers, motor boats, etc., and appliance uses are principally motors and drive housings. The balance of parts range from barbeques to weedwackers.

Of the products produced, 75% are die cast, 15% are permanent mold cast and 10% sand cast.

Future applications which are projected to provide market growth are aluminum automobile engine blocks. A trend has started and by the year 2000 the cast iron block could be a "dinosaur." If all blocks were made of aluminum at 30 pounds each, a 300,000,000 pound annual market would result.

Raw Material

Our raw material source is primarily industrial scrap generated during manufacturing processes. This material is also called new scrap and is purchased directly from the manufacturer, i.e., machining chips from automobile producers and machine shops, and skeletons from stamping plants. New scrap is also purchased from scrap dealers that have consolidated the scrap material from many manufacturers into truckload quantities for delivery to our facilities.

Another source of our raw material is old scrap. This scrap is purchased from dealers who have obtained the material from scrap collectors (scavengers). The scrap may still be in its original form or it may have already been melted in sweat furnaces and formed into sows or ingot. These sows and ingot are referred to as recycled secondary ingot or R.S.I. Sweat furnaces are used to separate iron inserts from old cast products since iron, in high concentrations, is detrimental to mechanical properties of most aluminum alloys.

Table I in the appendix lists the various classifications of scrap that are used. In addition to scrap, silicon additions of approximately 6% are made to casting alloys, to provide fluidity. A network of over 1500 dealers delivers 2 to 3 million pounds of scrap/day to feed the secondary industry.

Processing

Figure 3 is a flow sheet of a typical smelting operation. There are two general types of scrap: furnace-ready or scrap requiring processing. The steps are quite simple and include crushing, drying, melting, refining, casting and reprocessing of dross.

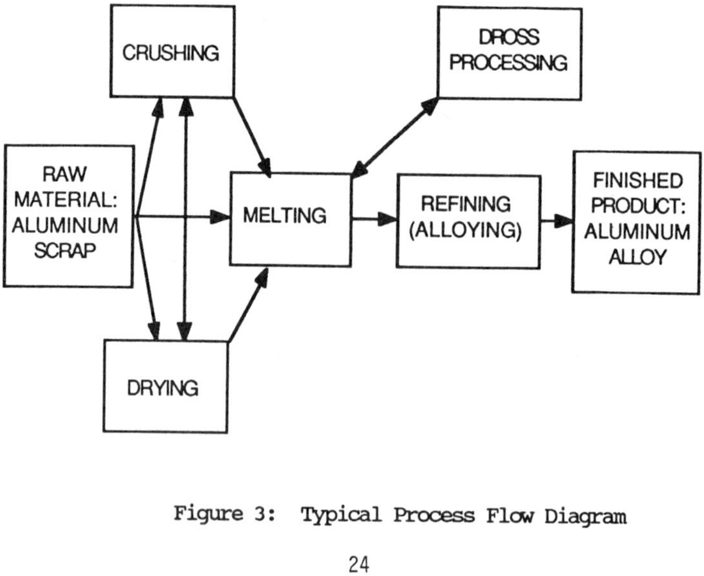

Figure 3: Typical Process Flow Diagram

Shredding/Crushing

Figure 4 is a typical shredding operation. The shredders and/or crushers are used for size reduction, but just as importantly for iron removal. It is not unusual to have several sets of magnets at the exit end of the crusher.

Figure 4: Scrap Shredder

Figure 5: Rotary Dryer

Dryers

Figure 5 is a typical dryer which resembles a rotary kiln with an afterburner and baghouse for pollution control. The dryer's function is to remove contamination, such as cutting oils, plastic and other organics for two primary reasons: to minimize air pollution from the melting furnaces and to minimize the amount of oxidation while melting.

Melting

In North America melting is performed almost exclusively in reverberatory furnaces (Figure 6), although some types of scrap are melted in rotary salt furnaces. The furnaces range in size from 40,000 pound to 220,000 pound capacity. They have an open charge well with the closed hearth used as the heating chamber with high velocity burners. While original designs used convection to transfer heat, molten metal pumps are used to circulate the molten metal through the charge well to improve thermal efficiency. Mechanical skimmers, as shown in Figure 7, are used to skim dross on many furnaces, replacing hand skimming which had been the previous practice.

Figure 6: Large Reverberatory Furnace
Approximately 200,000 Pounds Capacity

Figure 7: Mechanized Dross Skimmer

In Europe, melting is performed almost exclusively in rotary salt furnaces of up to 50 tons capacity. Rotary furnaces are believed to reduce melt losses due to the protection of the melting flux, however, have limited alloying flexibility and are used principally as melters. Molten metal is transferred to reverberatory-type alloying/holding furnaces or tapped into sows for further processing. A problem that has recently developed with rotary furnaces is the disposal of salt residues, especially in Europe. This effect has negatively impacted the economics of rotary furnaces.

Casting

Ingots are the principal product of the secondary industry, although sows are also cast. It is of interest, however, that in North America approximately 500 million pounds/year of the product produced is shipped in the liquid form in crucibles on trucks as shown in Figure 8.

Figure 8: Liquid Metal Transporter

Dross Processing

As a further attempt to recover the last bit of aluminum from the recycling process, dross generated during melting is processed to recover the approximately 20% of contained aluminum. Drosses are processed using two principal methods: (1) physical separation of aluminum from dross residue (salts and aluminum oxides) using hammermills, or (2) melting in rotary salt furnaces.

Metallurgical Considerations

In general, the die casting alloys can use the poorer grades of scrap due to higher specification limits on iron, manganese, copper, zinc and chromium. When corrosion resistance is required, such as for parts in outboard motors, copper limits are greatly reduced. In the permanent mold and sand casting processes, iron levels are reduced substantially in order to improve ductility. Therefore, those alloys require cleaner, higher quality scrap and are more expensive than the die cast alloys.

The major difference between wrought alloys and casting alloys, however, is the addition of 5 to 10% silicon for fluidity purposes. Die casting alloys are almost never heat-treated since they depend on a rapid solidification rate to develop fine grain size for their strength. Sand casting and permanent mold castings are generally heat treated for strength and/or hardness which also improves machinability.

Future Challenges

As was mentioned earlier, the secondary aluminum industry now is part of the recycling industry. Recycling has come of age and will become a preferred method of generating new materials. The pressures that have pushed recycling into a preferred method of material generation are those created by the solid waste landfill crises in our country and a general awareness of the environmental damage that has and can occur. You have heard, and will continue to hear, the arguments for and against the effects of fossil-fueled power versus nuclear, of the automobile versus mass transit, of chemicals versus organic gardening methods, etc. The point is that primary material facilities could find it more difficult to obtain permits than secondary industries, and the cost of handling wastes generated by primary industries will probably increase. Therefore, the secondary industry should continue to grow.

However, the secondary industry is not without a few problems. While we are recyclers of materials and reduce the overall waste generated, we do generate wastes of our own--primarily dross residues and baghouse dusts. In Figure 9, a graphic illustration of the material balance of a typical smelter is shown. For every 1 million pounds of scrap processed, 760,000 pounds of secondary aluminum is produced, 240,000 pounds of dross residues, and 3,000 pounds of baghouse dusts are generated.

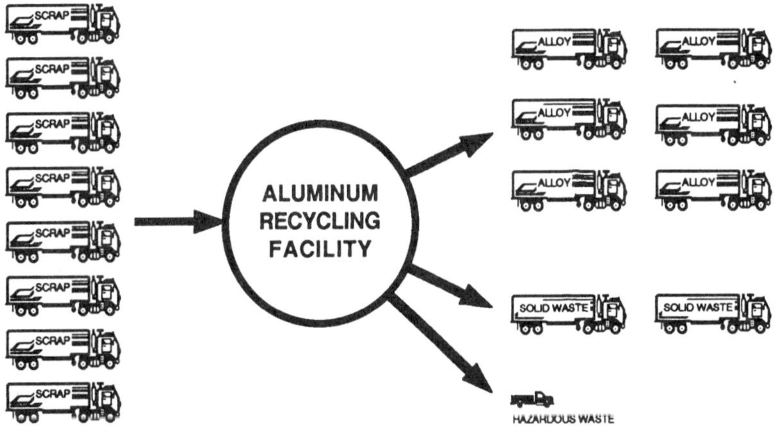

Figure 9: Material Balance

The dross residues are not hazardous but contain salts and are generally disposed of in solid waste landfills. As landfill space becomes more scarce, the product will have to be disposed of in mono-fills or recycled to recover the salts for reuse as fluxes, and aluminum oxides for use in hot topping compounds for the steel industry.

In Europe, the land disposal of salt slag residue has been forbidden, especially in Germany and Italy. Smelters are accumulating mountains of salt slag until salt recycling plants are installed. With the cost associated with land disposal projected to increase significantly in North America, the economics of recycling salt residues will become more favorable. New landfills for salt slag disposal require clay and/or synthetic liners, monitoring wells, and other high cost elements which make landfill costs significantly higher than those experienced in existing landfills.

Salt reclamation systems are not new. Many have been built in the past 20 years but just as many of them have become corrosion test beds due to the corrosive nature of salt solutions. In order to withstand the environment developed in the salt recovery process, expensive material such as titanium have to be used which stymied the adopting of the technology as long as inexpensive landfill opportunities existed.

Baghouse dusts often contain levels of lead and cadmium above the limits, as measured in the extraction procedure toxicity test (Ep Tox), because we are concentrating, in a small volume, contaminants spread over a large amount of scrap. In many cases, provisions have to be made to dispose of those products in hazardous waste landfills. However, it is currently very difficult to obtain a permit for siting a hazardous waste landfill due to the NIMBY attitudes present in the U.S. Efforts are being made to develop treatments to render these wastes not hazardous.

There are a variety of environmental issues that face us in the future: water runoff, PM-10, etc. These and other issues make the secondary aluminum industry one which will be making the transition from a simple business where you melted scrap into an ingot and dumped the dross in the back of the lot, to an industry with greater sophistication in processing poorer grades of scrap more efficiently and disposing of generated wastes in a regulated manner.

Summary

The recycling industry, from a secondary aluminum producer's point of view, is encouraging because as recycling gains greater acceptance, more raw material will be available. Aluminum can be recycled and reused on an economic basis--not due to legislation. We do not require subsidies and great market development efforts, such as will be required for recycled plastics. The greatest challenges are on the environmental side of our business, but we are confident that we can address those challenges.

When all is said and done, even though we generate some wastes, in the overall equation, we are waste minimizers. We truly are "the good guys."

Table I Scrap Categories

1. New Pure Aluminum Clippings
2. New Pure Aluminum Wire and Cable
3. Old Pure Aluminum Wire and Cable
4. Segregated New Aluminum Clippings
5. Mixed New Aluminum Clippings
6. Mixed Low Copper Aluminum Clippings
7. Mixed Old Alloy Sheet Aluminum
8. Scrap Sheet and Sheet Utensil Aluminum
9. Segregated New Aluminum Castings, Forgings, and Extrusions
10. Mixed New Aluminum Forgings
11. Mixed New Aluminum Castings
12. Aluminum Auto Castings
13. Aluminum Airplane Castings
14. Mixed Aluminum Castings
15. Aluminum Pistons
16. Wrecked Airplane Sheet and/or Breakage Aluminum
17. New Aluminum Foil
18. Old Aluminum Foil
19. All Other Aluminum Base Foils including Etched Foil, Radar Foil & Chaff
20. Segregated Aluminum Borings and Turnings
21. Mixed Aluminum Borings and Turnings
22. Sweated Aluminum
23. Aluminum Grindings
24. Aluminum Drosses, Spatters, Spillings, Skimmings and Sweepings
25. Aluminum Hair Wire
26. Aluminum Wire Screen
27. Coated Scrap
28. New Aluminum Can Stock
29. Old Can Stock
30. Painted Siding
31. Aluminum Copper Radiators
32. Containers of All Types (Oil, Food, Beverage, Aerosol)
33. Automobile Shredder Scrap
34. Items Not Covered Specifically by Above Classifications

NEW TECHNOLOGY DEVELOPMENT IN PLASTICS RECYCLING

David B. Spencer, Sc.D.

wTe Corporation

Abstract

This paper discusses the various aspects of the integrated systems -- collection, separation, refining, marketing and end use -- being developed today which will lead to the effective and efficient recycling of post-consumer recyclables including metals, plastics and paper. The focus is on the development of separation technologies using a large number of mining and mineral benefication unit operations for mixed plastics which consider both collection and processing requirements, as well as the requirements for refining and marketing the recovered materials as high grade polymer substitutes to off-grade or prime polymers.

Second International Symposium -
Recycling of Metals and Engineered Materials
Edited by Jan H. L. van Linden,
Donald L. Stewart, Jr., and Yogeshwar Sahai
The Minerals, Metals & Materials Society, 1990

Introduction

Everything that we produce with our industries and all the goods and services that we purchase with our ever increasing disposable income -- our entire gross national product -- will be discarded at the end of their useful life. Our nation's focus since the early 1900s has been with production. The industrial revolution was necessary to meet the needs of our consumer society. Now our focus needs to be on what to do with all those goods that we produced but no longer want. The result is 158 million tons per year of municipal solid waste which is in addition to the millions of tons of industrialized and mining wastes. Solid waste has become big business. Our trash, our beverage bottles, our grocery bags, our automobiles, our carpets, our skyscrapers -- everything will eventually have to be re-used, recycled, burned or landfilled. It is no wonder that the waste industry is growing so fast. We seem to be living so well as a nation. Our waste quantities and disposal patterns are a consequence of success.

Just as we have so successfully applied our advanced technology to the production of raw materials, we must now apply that same technology and innovation to the disposal or recovery of these goods. The materials which we have produced have a tremendous amount of intrinsic "value-added" in them. We spend so much time and effort making plastic polymers from petroleum by-products, or pure metals and alloys from "ore". To discard these valuable materials after they have been produced seems wasteful.

The machines which have been used in mining and metallurgy to produce and separate high grade, pure materials from vast bodies of ore must now be applied to separation of the same materials from a rich new ore -- termed "urban ore" by the United States Bureau of Mines.

The materials to be separated are new but the technology is old. In processing trash we employ air classifiers, mineral jigs, cyclone separators, rod mills, flotation cells, eddy current separators, magnetic separators, optical sorting devices, heavy media, and many other separation devices familiar to us in our metallurgy and mining experience. The bag of tricks is still the same; the feedstock is not.

Statement of Problem

Today in the United States we generate 160 million tons per year of municipal solid waste. Of that quantity, about 85% is now landfilled. However, landfills are closing at an ever increasing rate. Further, the cost of landfilling is becoming quite expensive as the regulatory pressures and environmental controls become more stringent. Leachate collection, methane gas monitoring, gas collection (or flaring), multiple natural and synthetic liners add to the cost of landfills. Landfills will continue for a long time to be the disposal method of last resort.

Today, 15% of our waste is either recycled or burned. About 10% is recycled -- most of which is paper. As you may be aware, mixed newspaper markets have saturated. The value of mixed newspaper is now either very low or non-existent. Although a high percentage of aluminum is recycled, it represents a small portion, less than 1%, of the overall waste stream. Another 5% is burned -- mostly in waste to energy plants which recover steam, electricity, hot water or chilled water from waste combustion. The amount of waste combustion is presently growing rapidly due to the fact that many plants permitted two to four years ago are soon coming on line. However, future growth will be slowed due to regulatory concerns regarding emissions from waste to energy plants.

The U.S. Environmental Protection Agency has established an integrated systems approach as the ideal solution for dealing with our wastes. This integrated approach involves a combination of (1) waste reduction, (2) recycling (including composting), (3) thermal reduction (incineration), and (4) landfilling. A hierarchy, or priority, has been established for these alternatives namely in the order given. While a sensible approach would involve the simultaneous implementation of all of these approaches at the same time, public pressure has been such that communities are virtually politically unable to implement an incineration solution until implementation of the recycling program has been exhausted.

To get industry to wake up to the urgent need for recycling and the extent of the disposal problem, some communities have threatened to ban the sale or use of materials which cannot be recycled. Polystyrene packaging has been the most criticized -- even though it makes up less than 0.25% of the waste stream. This has been a shock to industry leaders who have always been production oriented and now find themselves on the defensive to protect a material which they believed nearly indispensable to the consumer.

We make a great mistake when we attribute "excessive logic" to the legislative or consumer advocacy process. A great deal of attention is now being focused on the "green Isle" of our supermarket shelves. Even cars will be coming under pressure for the recyclability of the materials from which they are made.

Recycling seems to be the slogan for today. Consumers are willing to make preference choices in the goods they purchase not based upon the criteria of the past such as price, safety, and convenience, but now based on a new criteria -- recyclability.

Recyclable Materials

Materials That Can Be Recycled

At one time, the materials which could be recycled would have been purely a function of the demand for those materials in the market place. This is no longer the case. In the recycling business of the 1990s, avoided disposal cost can be as important or some times even more important than market demand.

For example, a project in Columbus, Ohio employs large automobile shredders to produce a "refuse-derived fuel" which is then fired in boilers to generate electricity. After the shredders size reduce the incoming waste, magnets are employed to remove ferrous materials. Because of both putrescible contamination and metallic tin and copper, this ferrous concentrate is normally not marketable. Our company installed its own proprietary technology to recover the ferrous metals in a salable form from this magnetic product. The investment was not so much motivated by the sales price for the finished product as it was motivated by the avoided landfill disposal costs which exceeded the value of the finished product.

Some communities on Long Island haul combustion residues to western Ohio to dispose of it at a cost of $140 per ton. Disposal costs in New England of $80 per ton are quite common. These costs exceed the value of a ton of ferrous metals recovered from trash or a ton of glass recovered from a curbside collection program.

The economic pressure on the plastic industry to recycle polystyrene has been enormous during the past year. This particular plastic has come under tremendous public pressure to ban its manufacture and use unless it can be recycled. The entire billion dollar annual market will be threatened if the infrastructure is not placed in service to collect, separate and recycle polystyrene. This has resulted in seven major oil and plastic manufacturers banding together to form an industry group, the National Polystyrene Recycling Company, to establish nationwide recycling facilities. Mobil, Dow, Amoco, Arco, Huntsman, Fina, and Polystar have all made investments to build and operate five to eight facilities nationwide to recycle polystyrene. The success of their product lines may depend very strongly upon the success of these recycling efforts.

Similarly, the very existence of any other material may be threatened if it is not made recyclable. Thus, when the question of what materials are recyclable is asked, the answer may be "any material."

Generally, the major consumer items in our grocery store which are presently recyclable include: aluminum, ferrous (tin) cans, glass (all three colors), paper and corrugated, and plastics (including HDPE, PET, PS, PVC, LDPE and PP). In the future, I would expect that film plastics will also be recycled.

Problems of Collection

One of the greatest hindrances to recycling is the cost of collection of the recyclables. Reports from the State of Rhode Island, where the entire state has begun to recycle, are that the costs range from $80 to $150 per ton. Collection of plastics can be particularly expensive. It is akin to collecting balloons at a circus. It doesn't take much time for the truck to fill up. Although the plastic industry has made a point that plastic makes up only about 7% of the waste stream, it is clearly a very large part of the waste stream by volume.

The methods of collection are continuously changing. Communities have not really figured out how to best accomplish the overall system for recycling of plastics from collection at the home through reuse at the point of manufacture.

Size reduction and granulation have been considered to achieve densification during the collection process. However, there can be more difficulty in separating the various components once they are shredded or compacted. It is clear that there is an interrelationship between the methods of collection and the methods of separation to be used after collection. A system which separates HDPE milk bottles from PET bottles by size, may be rendered ineffective if the HDPE bottles were granulated or flattened before separation. Separation of glass bottles by color will be difficult if the glass is compacted and broken during collection.

Problems of Contamination

The value of a recycled product is heavily dependent upon its purity. While this may be obvious to those in the metallurgy and materials science areas, it does not seem obvious to most waste collectors and community groups. For example, glass has recently come under criticism because of the rejection of most deliveries. Ceramics and dirt which is often delivered with broken glass forms "stones" in the finished glass product. These "stones" are stress concentrators in the finished bottles causing them to break when pressurized. The level of impurities by specification can be on the order of 1 particle per half-million. However, a garbage collector usually does not deal with specifications or low impurity levels. The attitude is that if it looks like it is mostly glass, then it must be glass. If the glass manufacturers cannot use it in the form they recycle it, then by definition, glass must not be recyclable -- or there are no markets. They often see little relationship between the impurity levels of the products they are trying to recycle and the recyclability of the material.

The same is true of other recyclable materials. Plastics must often be at least 99.9% pure if they are to be reused as substitute polymers to virgin or offgrade materials. While contamination is not a problem to plastic wood materials, it can be critical for other applications. Further, there are many grades of plastics as there are with metals. Composition and purity are critical to recyclability, but unlike traditional sources of recyclable materials, the homeowner and the waste collector are often very insensitive to these issues.

Part of the job of the processor is to make a pure, uncontaminated output product in spite of the contamination of the incoming material. Recently, a plastic manufacturer, after inspecting a shipment of HDPE milk bottles delivered to our plant, said he would reject any load containing as much contamination as we had received. He was unaware of the time and price spent to get the material received. If we rejected contaminated loads, we would have no feedstock at all for our plant.

USE OF VAPOR PHASE COMPLEXATION FOR THE EXTRACTION

OF ANTIMONY FROM SULFIDIC WASTES

B. R. Westphal and J. P. Hager

Department of Metallurgical and Materials Engineering
Colorado School of Mines
Golden, Colorado 80401

Abstract

The extraction of antimony from sulfidic wastes has been investigated. Using a special double-zone transpiration reactor the vapor transport chemistry of antimony sulfide has been established. By varying the sulfur potential in the carrier gas, though the use of an H_2S/H_2 mixture, it was determined that two transport mechanisms are possible:

$$Sb_2S_3(s,l) = Sb_2S_2(g) + \frac{1}{2} S_2(g)$$

$$Sb_2S_3(s,l) = Sb_2S_3(g)$$

Moreover, if arsenic sulfide vapor is present in the carrier gas the transport rate of antimony sulfide can be enhanced by several hundred percent due to the formation of the vapor complex $AsSb_3S_6(g)$:

$$6\ Sb_2S_3(s) + As_4S_6(g) = 4\ AsSb_3S_6(g)$$

The Gibbs free energies have been determined for the vapor transport reactions. The results have been used to construct vapor pressure/stability diagrams for the Sb-S-O system and to identify process conditions for the treatment of sulfidic wastes containing arsenic and antimony. Recommendations are also presented on the selective deposition of arsenic and antimony from the discharge gas stream of a transport reactor.

Second International Symposium -
Recycling of Metals and Engineered Materials
Edited by Jan H. L. van Linden,
Donald L. Stewart, Jr., and Yogeshwar Sahai
The Minerals, Metals & Materials Society, 1990

Introduction

The use of vapor phase complexation for the extraction of antimony has been investigated by Li and Hager [1] for oxidic materials. It was shown that in the presence of arsenic oxide vapor (As_4O_6), the vapor pressure of $Sb_2O_3(s)$ could be increased by a factor of 100 or more. It was determined through a series of transpiration experiments, where the carrier gas was doped with $As_4O_6(g)$, that a series of vapor complex species were formed that dramatically increased the apparent vapor pressure of the $Sb_2O_3(s)$. The thermodynamic properties of the following complex formation reactions were determined, from which the increase in the vapor transport of antimony oxide could be calculated for different processing conditions:

$$0.75\ As_4O_6(g) + 0.5\ Sb_2O_3(s) = As_3SbO_6(g) \qquad (1)$$

$$0.50\ As_4O_6(g) + 1.0\ Sb_2O_3(s) = As_2Sb_2O_6(g) \qquad (2)$$

$$0.25\ As_4O_6(g) + 1.5\ Sb_2O_3(s) = AsSb_3O_6(g) \qquad (3)$$

Process chemistry was also developed for the selective deposition and separation of the antimony oxide and arsenic oxide from the vapor complexes through the addition of oxygen to the product gas mixture containing the complexes. The use of the results for the recovery and separation of arsenic and antimony from oxidic materials is described in U.S. Patent 4,808,221 [2].

The work of Mauser [3] suggests that similar extraction chemistry may be possible for the treatment of sulfidic materials. Under certain conditions, the measured weight loss of mixtures of As_2S_3 and Sb_2S_3, under free evaporation conditions, was found to be greater than the calculated weight loss based on the measurements on the individual sulfides. Furthermore, direct-inlet-probe mass spectrometric measurements provided semi-quantitative results that indicated that the vapor complex species As_3SbS_6, $As_2Sb_2S_6$, and $AsSb_3S_6$ are formed in the presence of $As_4S_6(g)$. Thus, it would appear that a similar set of reactions exists that will enhance the vapor transport of antimony from sulfidic materials:

$$0.75\ As_4S_6(g) + 0.5\ Sb_2S_3(s) = As_3SbS_6(g) \qquad (4)$$

$$0.50\ As_4S_6(g) + 1.0\ Sb_2S_3(s) = As_2Sb_2S_6(g) \qquad (5)$$

$$0.25\ As_4S_6(g) + 1.5\ Sb_2S_3(s) = AsSb_3S_6(g) \qquad (6)$$

The purpose of the present study was to investigate, under equilibrium conditions, the formation of any As-Sb-S vapor complexes and to establish their thermodynamic properties. In this way, the operating conditions could be identified for a new extraction process that would utilize the formation of the vapor complexes.

The same experimental procedure was selected as used in the previous study [1]; namely, to dope a carrier stream with $As_4S_6(g)$ and to measure the enhanced vapor pressure of $Sb_2S_3(s)$. However, a review of the literature on antimony sulfide vaporization chemistry [4-18] revealed that considerable uncertainty exists regarding both the principal Sb-S vapor species and the vapor pressure as a function of temperature. The vapor pressure/stability diagram shown in Figure 1 indicates the complexity of species and the influence of the sulfur potential, as represented by the logarithm of the partial pressure ratio of H_2S to H_2, on the vapor pressure of the various species. This diagram was calculated from the thermodynamic data given in the recent U.S. Bureau of Mines compilation on sulfides [6], and the line labeled P_T^E is the effective total vapor pressure as defined by Kellogg [19]. Thus, the experimental plan was altered to first measure the vapor pressure of Sb_2Sb_3(s,l), and identify the principal vapor species, and then measure the increase in the "baseline" vapor pressure of antimony trisulfide in the presence of arsenic sulfide vapor. In view of the importance of controlling the sulfur potential in the system, mixtures of H_2S + H_2 were used as the carrier gas in the transpiration apparatus.

Experimental

The transpiration method is well suited for determining the molecularity of the transport species and measuring their vapor pressures. For antimony trisulfide, the following reactions can be written where a H_2S + H_2 mixture is used as the carrier gas:

$$Sb_2S_3(s,l) = Sb_2S_3(g) \qquad (7)$$

$$Sb_2S_3(s,l) + H_2(g) = Sb_2S_2(g) + H_2(g) \qquad (8)$$

$$Sb_2S_3(s,l) + H_2(g) = 2\ SbS(g) + H_2S(g) \qquad (9)$$

$$3\ Sb_2S_3(s,l) + 5\ H_2(g) = 2\ Sb_3S_2(g) + 5\ H_2S(g) \qquad (10)$$

$$Sb_2S_3(s,l) + H_2S(g) = Sb_2S_4(g) + H_2(g) \qquad (11)$$

These reactions represent the formation of the major species shown in Figure 1. The equilibrium constant expression can be written for each reaction and rearranged to give equations for LOG(P[complex]) as a function of LOG(P[H_2S]/P[H_2]). These equations are given in Table I for the above set of reactions. Thus, if one of the five Sb_XS_Y species is the principal vapor species under the experimental conditions, a plot of the logarithm of the measured vapor pressure (at a constant temperature) as a function of the logarithm of the P(H_2S)/P(H_2) ratio will give the slope indicated in Table I. This provides the basis for the identification of the species. A similar approach can be used to identify the As-Sb-S complexes by varying the partial pressure of the arsenic sulfide species in the carrier gas.

Transpiration Measurements

Apparatus. A schematic view of the transpiration apparatus is given in Figure 2. The reactor tube was constructed of mullite and the sample carrier tube of quartz. Two 4.5" slots were provided in the carrier tube to hold the alumina sample boats. Each slot provided for a set of four sample boats. The cross-sectional area between the sample carrier tube and the reactor tube was minimized to prevent any back diffusion of the carrier gas during a run. The sample carrier tube was inserted in and withdrawn from the heated furnaces by

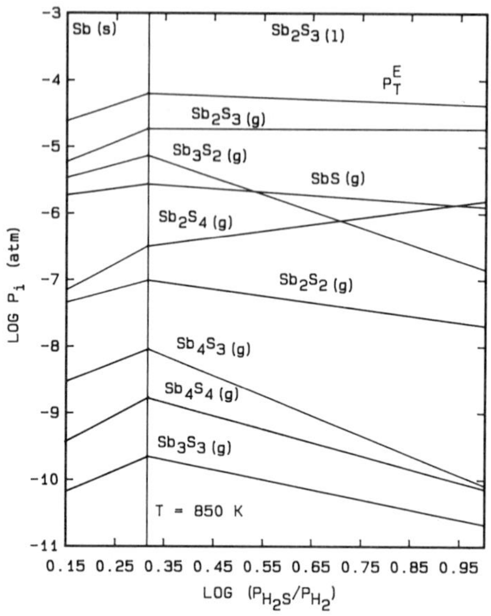

Figure 1 - Vapor pressure/stability diagram for the Sb-S system at 850K.

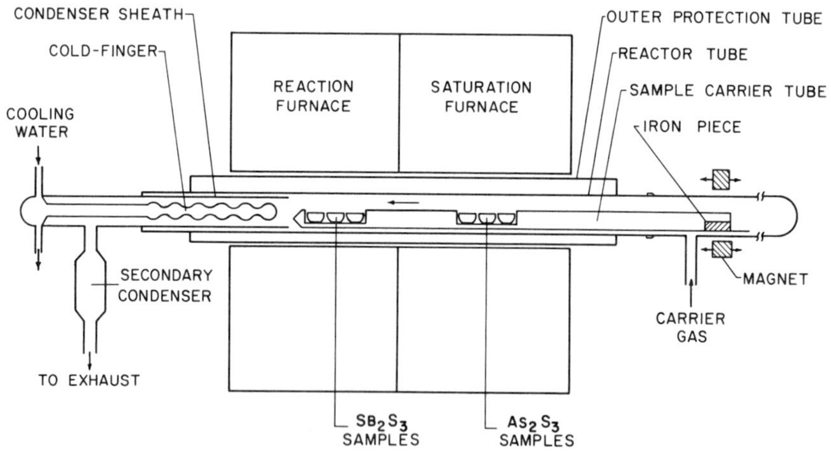

Figure 2 - Schematic diagram of the transpiration apparatus.

TABLE I - Vapor Pressure Equations for Different Sb_XS_Y Vapor Species

Species	Equation	Slope at Constant Temperature
Sb_2S_3	LOG P(Sb_2S_3) = LOG K7	0
Sb_2S_2	LOG P(Sb_2S_2) = - LOG [P(H_2S)/P(H_2)] + LOG K8	-1.0
SbS	LOG P(SbS) = -0.5 LOG [P(H_2S)/P(H_2)] + 0.5 LOG K9	-0.5
Sb_3S_2	LOG P(Sb_3S_2) = -2.5 LOG [(P(H_2S)/P(H_2)] + 0.5 LOG K10	-2.5
Sb_2S_4	LOG P(Sb_2S_4) = LOG [P(H_2S)/P(H_2)] + LOG K11	+1.0

use of a magnetic coupling arrangement. Two Marshall resistance furnaces (multiple tap) placed end to end provided for the proper temperature profile along the reactor tube.

The carrier gas consisted of a mixture of purified hydrogen and purified hydrogen sulfide. The gas composition and flow rate were controlled by metering the flow of each gas stream with precision capillary flowmeters. The carrier gas was introduced at the inlet end of the reactor tube. The gas stream could be first saturated with As_4S_6(g) before being equilibrated with the Sb_2S_3(s) samples. For those experiments in which only the volatilization of Sb_2S_3(s,l) was to be studied, the sample boats in the saturation zone were left empty. After the gas stream equilibrated with the Sb_2S_3(s,l), it passed through the condenser assembly. The condenser assembly was inserted at the outlet end of the reactor tube and consisted of an outer quartz sheath and an inner water-cooled cold finger. To minimize the danger of H_2S exposure, the waste gas stream was directed to a gas burner that mixed natural gas with the waste gas before introduction to the burner head. The combusted gases were then removed with an exhaust hood.

The temperature profile along the length of the reactor tube was determined before each experiment using a calibrated type K thermocouple assembly. The assembly was inserted in the reactor tube and allowed to equilibrate from 12-14 hours before the profile was determined. A second thermocouple was placed on the outside of the reactor tube and correlated with the temperature measured on the inside of the reactor tube. In this way, the sample temperature could be monitored during a run. The total uncertainty in measuring the sample temperature was estimated to be less than ± 2 degrees Kelvin.

Procedure. Samples of high purity As_2S_3 and/or Sb_2S_3 were loaded into the sample boats and the boats then weighed on a 5-decimal place analytical balance. The boats were loaded into the sample carrier tube and the system sealed and evacuated. The system was then backfilled and flushed with the carrier gas, and the sample carrier tube inserted into the reactor tube under a flow of the carrier gas. After an equilibration period of 5-7 hours, the sample carrier tube was withdrawn. The sample boats were reweighed and the condenser assembly removed and cleaned.

The mass transfer of arsenic and antimony sulfide was based on the weight change of the sulfide samples. However, samples of As_2S_3 and Sb_2S_3 were frequently analyzed by X-ray powder diffraction to verify that no phase changes had occurred during the run. Further, it was important to verify that the Sb_2S_3 samples did not become contaminated with arsenic sulfide during the complexation experiments.

To be certain that the results would not be dependent on the nature of the Sb_2S_3 sample that was used, vapor pressure measurements were made on several different samples. Excellent reproducibility was obtained with both Johnson Matthey Puratronic grade Sb_2S_3 (99.999%) and GFS Chemicals Sb_2S_3 (99.999%). However, there was a noticeable lack of reproducibility with Alfa Products Ultrapure grade Sb_2S_3, and sample fusion was observed at temperatures below the established melting point of antimony trisulfide. This material was not used in the reported vapor pressure measurements. For the complexation runs, GFS Chemicals As_2S_3 (99.999%) was used.

To confirm that equilibrium was achieved in the reaction zone, experiments were conducted at different flow rates of the carrier gas. The run time was varied, the ratio of H_2S/H_2 was varied, and for the complexation runs the percent of $As_4S_6(g)$ added to the carrier gas was varied.

Data Analysis for $Sb_2S_3(s,l)$ Volatilization Experiments. As indicated by reactions (7) to (11), a number of Sb_XS_Y gaseous species may be formed. However, as shown in Figure 1, there may be a predominate species within a given range of the H_2S/H_2 partial pressure ratio at a fixed temperature. The data analysis method presented below is based on selecting a given $Sb_XS_Y(g)$ species and calculating the equilibrium vapor pressure on that basis. In this way, the partial pressure can be calculated for a number of species and the results plotted as a function of the H_2S/H_2 partial pressure ratio. A comparison of the observed slopes with those given in Table I should then provide for the identification of the predominate species. If more than one species exists with similar vapor pressures, a variation of the method presented below is required in which the moles of vaporized antimony need to be distributed between the various species. As will be seen in the Results section, there appears to be predominant $Sb_XS_Y(g)$ species within given ranges of the H_2S/H_2 partial pressure ratio.

The total moles of atomic hydrogen, sulfur, and antimony in the equilibrated gas phase during an experiment can be calculated as follows:

$$N_H = 2 \cdot M(H_2S) + 2 \cdot M(H_2) \quad (12)$$

$$N_S = 3 \cdot M(Sb_2S_3) + M(H_2S) \quad (13)$$

$$N_{Sb} = 2 \cdot M(Sb_2S_3) \quad (14)$$

where:

- $M(H_2S)$ = moles of H_2S introduced in the carrier gas;
- $M(H_2)$ = moles of H_2 introduced in the carrier gas; and
- $M(Sb_2S_3)$ = moles of Sb_2S_3 vaporized as determined by the weight loss of the Sb_2S_3 samples.

The dissociation of the H_2S in the carrier gas can be written as:

$$H_2S(g) = H_2(g) + 0.5\ S_2(g) \quad (15)$$

and the equilibrium constant for the reaction written as:

$$K15 = N(H_2) \cdot N(S_2)^{1/2} \cdot P_T^{1/2} \,/\, \left(N(H_2S) \cdot N_T^{1/2} \right) \quad (16)$$

where:

- K15 = equilibrium constant [6];
- $N(H_2)$ = moles of H_2 at equilibrium;
- $H(S_2)$ = moles of S_2 at equilibrium;
- $N(H_2S)$ = moles of H_2S at equilibrium;
- P_T = total pressure in atmospheres; and
- N_T = total moles of gaseous species at equilibrium.

The mass balance equations can now be rewritten in terms of the moles of species in the equilibrated gas mixture:

$$N_H = 2 \cdot N(H_2S) + 2 \cdot N(H_2) \tag{17}$$

$$N_S = N(H_2S) + 2 \cdot N(S_2) + Y \cdot N(Sb_XS_Y) \tag{18}$$

$$N_{Sb} = X \cdot N(Sb_XS_Y) \tag{19}$$

where:

$N(Sb_XS_Y)$ = moles of Sb_XS_Y at equilibrium.

Also, the total moles of equilibrium gas species can be taken as the sum of the mole numbers of the individual species:

$$N_T = N(H_2S) + N(H_2) + N(S_2) + N(Sb_XS_Y) \tag{20}$$

Equations (16) to (20) represent five equations in five unknowns. This set of non-linear equations was solved using the geometric-arithmetic inequality algorithm developed by Baker [20]. Once the values of the mole numbers were calculated, the partial pressures of the various gaseous species were determined[1]:

$$P_i = N(i) \cdot P_T / N_T \tag{21}$$

Data Analysis for Complex Formation Experiments. The proposed complex formation reactions are given as reactions (4) to (6). A plot of the logarithm of the partial pressure of the complex as a function of the logarithm of the H_2S/H_2 partial pressure ratio should indicate which species is formed:

Species	Slope of LOG P(complex) versus LOG P(As_4S_6) Plot
As_3SbS_6	0.75
$As_2Sb_2S_6$	0.50
$AsSb_3S_6$	0.25

However, the calculation of the partial pressure of the complex species is complicated by the presence of other antimony-containing vapor species; in particular the antimony-sulfur species. This problem was resolved by using a sufficiently large value of the H_2S/H_2 partial pressure ratio that a single antimony-sulfur vapor species would be predominant. As can be seen from the Results section, the use of a H_2S/H_2 partial pressure ratio greater than 6.3:1 is sufficient to assure that $Sb_2S_3(g)$ will be the principal $Sb_XS_Y(g)$ species over the temperature range of interest. A ratio of 12.6:1 was used for all of the complexation experiments to guarantee that the following reaction would define the antimony trisulfide vapor pressure:

1 - Although pressures are reported in atmospheres, they can be converted to Pascals by using the relationship 1 atm = 101.325 kPa.

$$Sb_2S_3(s) = Sb_2S_3(g) \tag{7}$$

Again, the method of analysis presented below is based on the formation of a single vapor complex species. Another method of analysis would be required if two or more complexes are formed with similar vapor pressures.

The total moles of atomic hydrogen, sulfur, antimony, and arsenic were first calculated:

$$N_H = 2 \cdot M(H_2S) + 2 \cdot M(H_2) \tag{22}$$

$$N_S = 3 \cdot M(Sb_2S_3) + 3 \cdot M(As_2S_3) + M(H_2S) \tag{23}$$

$$N_{Sb} = 2 \cdot M(Sb_2S_3) \tag{24}$$

$$N_{As} = 2 \cdot M(As_2S_3) \tag{25}$$

where:

$M(As_2S_3)$ = moles of As_2S_3 vaporized as determined by the weight loss of the As_2S_3 samples.

Further, the dissociation of $H_2S(g)$ into $H_2(g)$ and $S_2(g)$ provides the equation:

$$K15 = N(H_2) \cdot N(S_2)^{1/2} \cdot P_T^{1/2} / \left(N(H_2S) \cdot N_T^{1/2}\right) \tag{26}$$

The equilibrium constant for reaction (7) can be written as follows:

$$K7 = N(Sb_3S_3) \cdot P_T / N_T \tag{27}$$

where:

K7 = equilibrium constant for reaction (7) as determined from the $Sb_2S_3(s,l)$ vapor pressure experiments.

The mass balance equations, in terms of the equilibrium mole numbers, now become:

$$N(H) = 2 \cdot N(H_2S) + 2 \cdot N(H_2) \tag{28}$$

$$N(S) = N(H_2S) + 2 \cdot N(S_2) + 3 \cdot N(Sb_2S_3)$$

$$+ 6 \cdot N(As_4S_6) + 6 \cdot N(As_XSb_YS_6) \tag{29}$$

$$N_{Sb} = 2 \cdot N(Sb_2S_3) + Y \cdot N(As_XSb_YS_6) \quad (30)$$

$$N_{As} = 4 \cdot N(As_4S_6) + X \cdot N(As_XSb_YS_6) \quad (31)$$

where:

$N(As_4S_6)$ = moles of As_4S_6 at equilibrium; and
$N(As_XSb_YS_6)$ = moles of $As_XSb_YS_6$ at equilibrium.

Finally, the total moles of equilibrated species can be related to the sum of the individual mole numbers:

$$N_T = N(H_2S) + N(H_2) + N(S_2) + N(Sb_2S_3) + N(As_XSb_YS_6) \quad (32)$$

Equations (26) to (32) represent a set of seven non-linear equations in seven unknowns. These equations were also solved by the method of Baker [20] to provide the mole numbers for the equilibrated gaseous species. The equilibrium partial pressures were calculated as noted above.

Results

Sb$_2$S$_3$(s,l) Vapor Pressure Measurements. The vapor pressure results for Sb_2S_3(s,l) are shown in Figures 3 and 4. A comparison of the calculated slopes with the theoretical slopes given in Table I shows that below 0.6 on the LOG (P_{H2S}/P_{H2}) scale the vapor transport reaction is:

$$Sb_2S_3(s,l) = Sb_2S_2(g) + \frac{1}{2} S_2(g) \quad (33)$$

and that above 0.8 on the LOG(P_{H2S}/P_{H2}) scale the vapor transport reaction is:

$$Sb_2S_3(s,l) = Sb_2S_3(g) \quad (7)$$

At values of LOG(P_{H2S}/P_{H2}) between 0.6 and 0.8, both Sb_2S_2(g) and Sb_2S_3(g) are predominant species, which leads to the bubble effect at the intersection of the two lines. Confirmation of equilibrium saturation for this series of experiments is shown in Figure 5.

The Sb_2S_3(s,l) vapor pressure results are summarized in Figures 6 and 7. The logarithm of the equilibrium constant for the indicated reaction is shown to be independent of the logarithm of the H_2S/H_2 partial pressure ratio at a constant temperature. This set of plots is useful to confirm that the correct reaction stoichiometry has been used and to assess the degree of scatter in the data. It should be noted that measurements above 823K are for Sb_2S_3(l) and below 823K are for Sb_2S_3(s).

A second law analysis of the data is given in Figure 8. Sigma function values were not calculated because of the absence of reliable heat capacity equations for Sb_2S_3(g) and Sb_2S_2(g). A least squares linear regression analysis was used to fit the lines through the data points for Sb_2S_3(g), with the result that the intersection of the lines for solid and liquid antimony trisulfide corresponds to the established melting point [6]. For the Sb_2S_2(g) results, a least squares linear regression analysis was used to define the line through the data

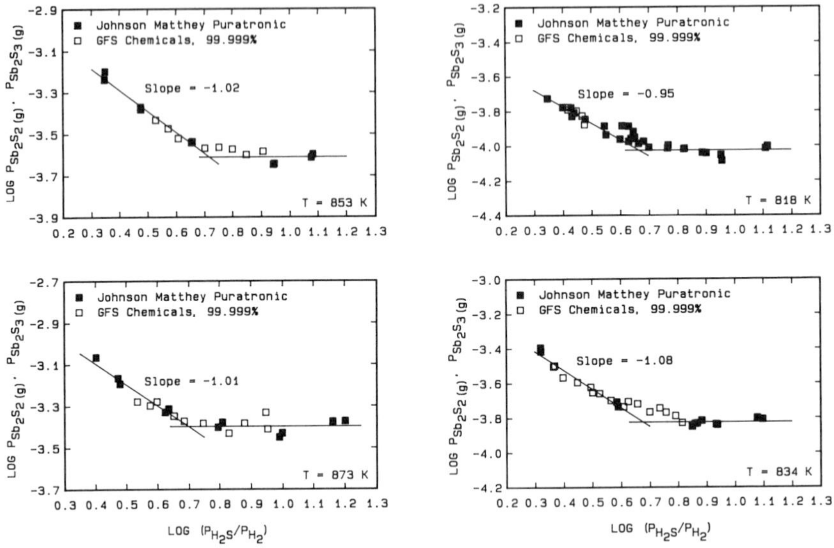

Figure 3 - Vapor pressure results for Sb_2S_3(s,l) at 873, 853, 834, and 818K.

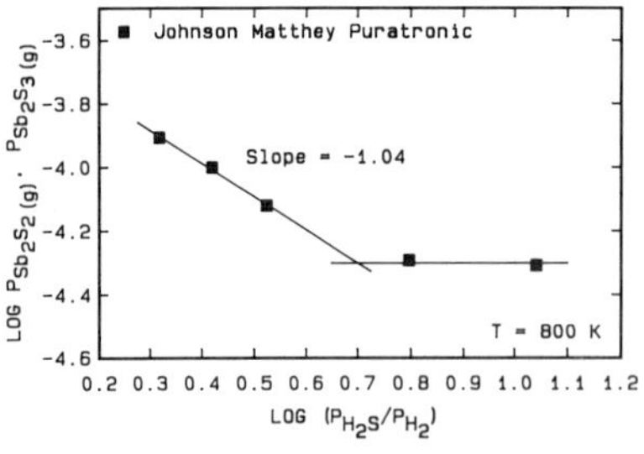

Figure 4 - Vapor pressure results for Sb_2S_3(s) at 800K.

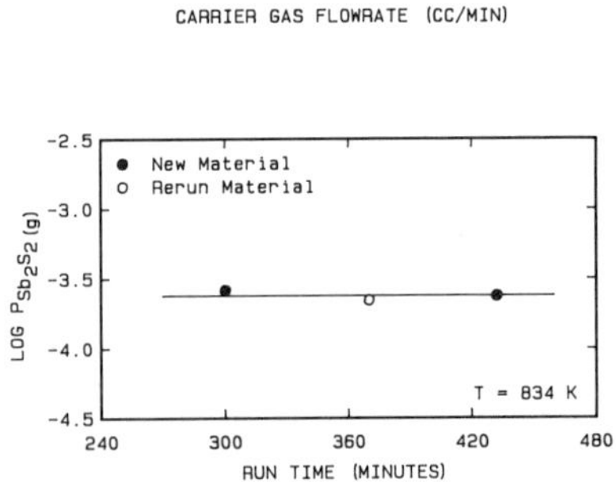

Figure 5 - Vapor pressure of Sb_2S_3(l) as a function of flow rate and run time.

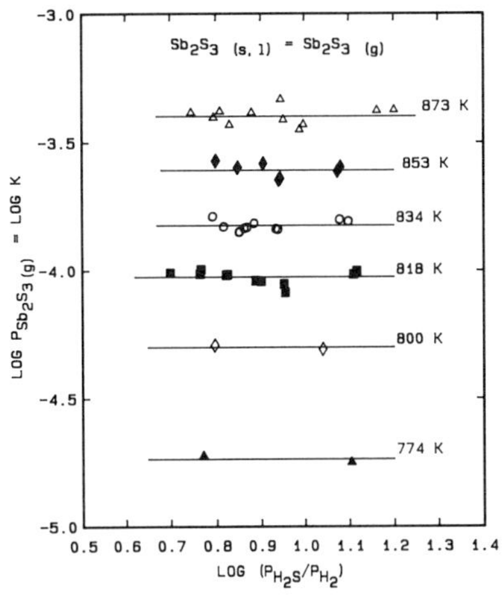

Figure 6 - Summary of results for $Sb_2S_3(g)$.

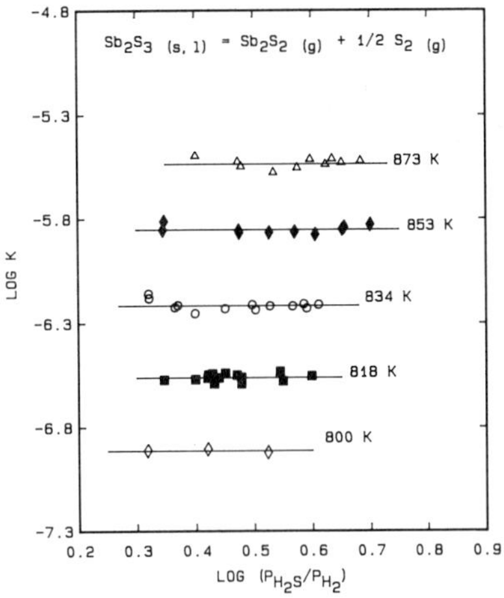

Figure 7 - Summary of results for $Sb_2S_2(g)$.

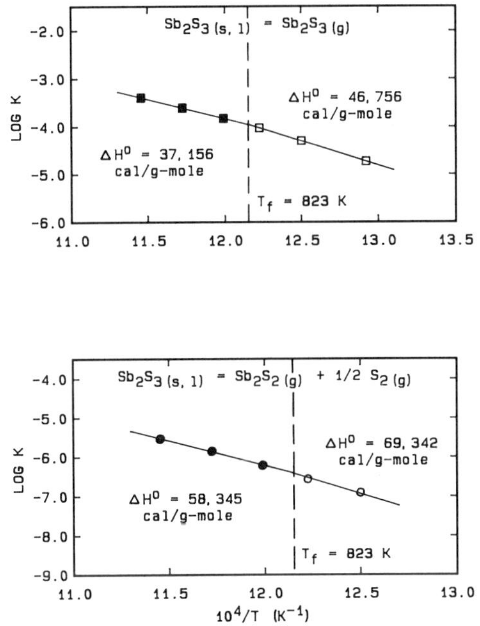

Figure 8 - Second-law plots for the formation of $Sb_2S_3(g)$ and $Sb_2S_2(g)$ from $Sb_2S_3(s,l)$.

points for $Sb_2S_3(l)$, whereas the line through the $Sb_2S_3(s)$ data points was constructed using the data points and the intersection of the derived line for the liquid trisulfide samples with the fusion temperature. The heat of fusion of antimony trisulfide can be calculated from the derived heats ofvaporization. The value based on $Sb_2S_3(g)$ should be considered to be the most reliable. The derived heats of fusion are compared in Table II.

The results shown in Figure 8 have been plotted on Figures 9 and 10 along with the results of previous investigators and the values published in USBM Bulletin 689 [3]. Since there was considerable uncertainty in the partial pressures of sulfur used by the previous investigators (the equilibrium pressure of S_2 was not known in most of the studies), the reported data were processed on the basis of each of the following reactions and the results plotted on Figures 9 and 10.

$$Sb_2S_3(s,l) = Sb_2S_3(g) \tag{7}$$

$$Sb_2S_3(s,l) = Sb_2S_2(g) + 0.5\ S_2(g) \tag{33}$$

TABLE II - Comparison of Derived and Literature Values for the Heat of Fusion of Antimony Trisulfide

Source	ΔH° (FUSION) cal/g-mole	J/g-mole
This Study (from Sb_2S_3(g))	9,600(1)	40,166
This Study (from Sb_2S_2(g))	10,997	46,011
USBM Bulletin 689 [6]	9,700	40,585
Meyers and Felty [21]	11,450	47,907
Glozov, et al. [22]	15,600	65,270

(1) - From this study:
$\Delta G°(FUSION) = 9{,}600 - 11.66T$ cal/g-mole
$40{,}166 - 48.78T$ J/g-mole

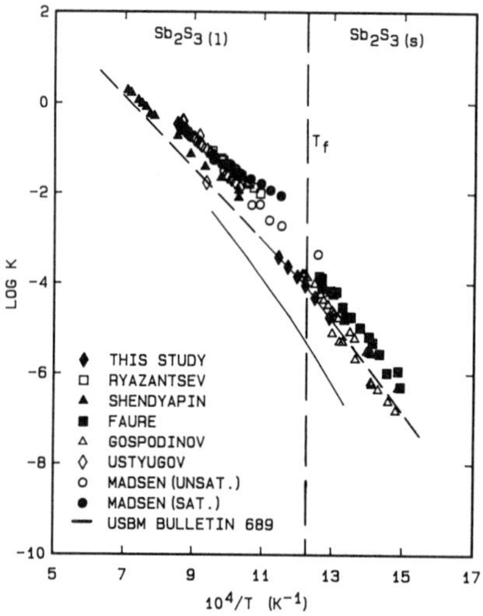

Figure 9 - Results for the reaction $Sb_2S_3(s,l) = Sb_2S_3(g)$.

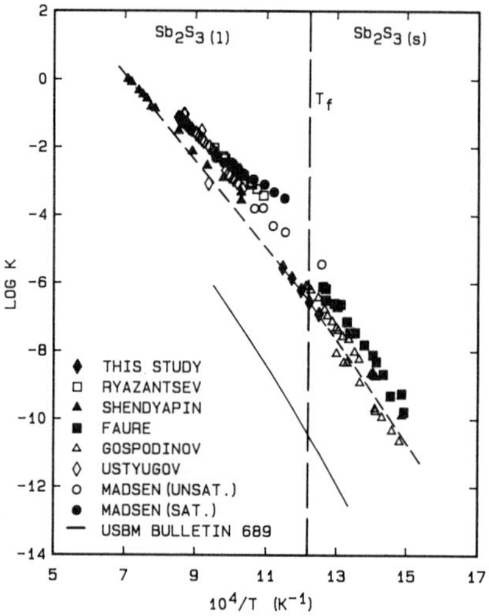

Figure 10 - Results for the reaction $Sb_2S_3(s,l) = Sb_2S_2(g) + 0.5\ S_2(g)$.

As can be seen, there is considerable scatter in the data from the previous investigators. However, several of the studies give results that are in general agreement with the results obtained by this study. The values taken from USBM 689 are, surprisingly, in rather poor agreement with the results of both this study and the results of previous investigators.

The thermodynamic results for the experiments on $Sb_2S_3(s,l)$ are given in equation form in Table III.

TABLE III - Thermodynamic Results for $Sb_2S_3(s,l)$

Reaction	ΔG° cal/g-mole	ΔG° J/g-mole	T; °K
$Sb_2S_3(s) = Sb_2S_3(g)$	46,756 - 38.72T	195,627 - 162.00T	784-818
$Sb_2S_3(l) = Sb_2S_3(g)$	37,156 - 27.05T	155,461 - 113.18T	834-873
$Sb_2S_3(s) = Sb_2S_2(g) + 0.5\ S_2(g)$	69,342 - 54.90T	290,127 - 229.70T	800-818
$Sb_2S_3(l) = Sb_2S_2(g) + 0.5\ S_2(g)$	58,345 - 41.54T	244,155 - 173.80T	834-873

Vapor Complex Measurements. The results for the vapor complex formation experiments are given in Figure 11. In all cases, the best fit of the data to the proposed complex formation reactions (reactions (4) to (6)) is obtained when the species $AsSb_3S_6(g)$ is chosen. This is consistent with the observation of Mauser [3] who found that the fraction of antimony in the complexes increased and the fraction of arsenic decreased as the temperature was increased from 573 to 773K. In fact, at 773K only $AsSb_3S_6(g)$ and $As_2Sb_2S_6(g)$ were reported, with the intensity ratio of $AsSb_3S_6$:$As_2Sb_2S_6$ being 3:1. Nevertheless, an attempt was made to include the contribution of $As_2Sb_2S_6$ to the calculation of the partial pressure of the complexes. However, no improvement was noted in the match of the observed slopes with the theoretical slopes, suggesting that in the present study $AsSb_3S_6(g)$ was indeed the principal vapor complex species.

The results of the complex formation experiments are summarized in Figure 12. This plot confirms the selected stoichiometry since, at constant temperature, the equilibrium constant is found to be independent of the partial pressure of $As_4S_6(g)$. Moreover, construction of other plots like Figure 13, where other complex species were chosen or a different arsenic sulfide species was used, showed the equilibrium constant to not be independent of the partial pressure of the arsenic sulfide. It should be noted that the selection of $As_4S_6(g)$ as the principal arsenic sulfide species is also consistent with the findings of Kaiser [4]. It is not surprising that in the presence of a large H_2S/H_2 partial pressure ratio that the species is $As_4S_6(g)$ as opposed to $As_4S_4(g)$, as reported by others [24]. For all vapor complex experiments, only $Sb_2S_3(s)$ was used to avoid contamination of the antimony sulfide by arsenic sulfide.

The results given in Figure 12 have been replotted in Figure 13, from which the following relationship has been derived:

$$As_4S_6(g) + 6\ Sb_2S_3(s) = 4\ AsSb_3S_6(g) \quad (34)$$

$$\Delta G^\circ = 108{,}792 - 75.94T \quad cal/g\text{-}mole \quad (35)$$

$$455{,}186 - 317.73T \quad J/g\text{-}mole$$

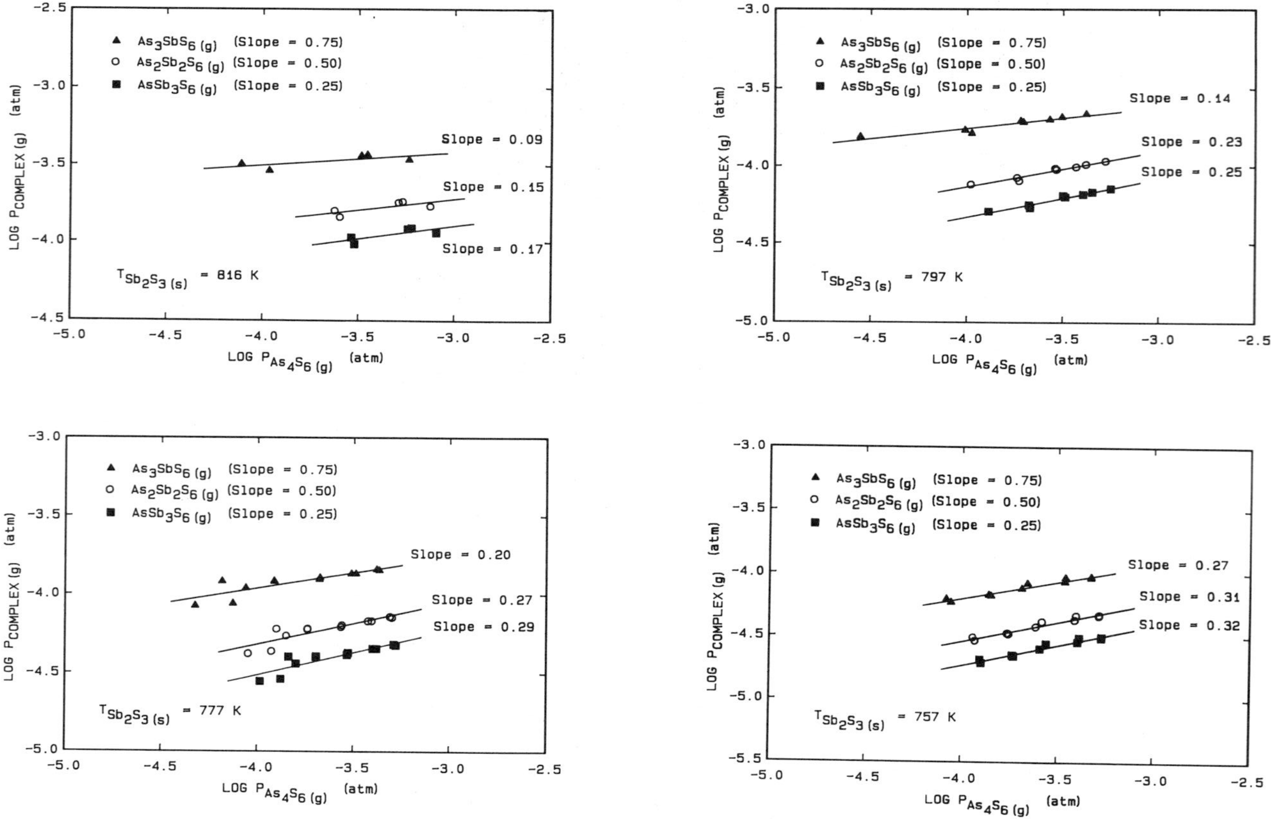

Figure 11 - Results of vapor complex formation experiments.

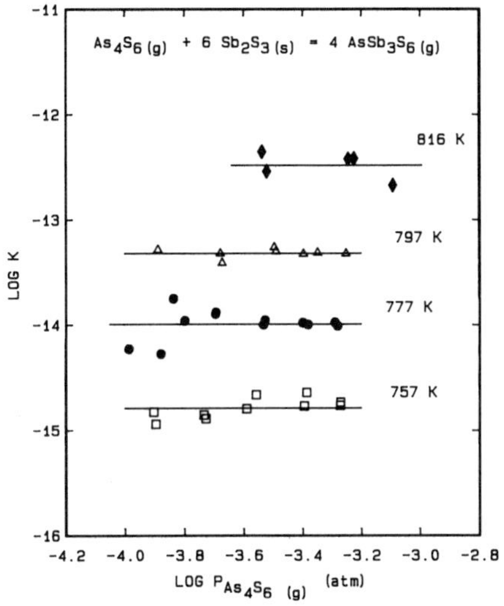

Figure 12 - Summary of results for the formation of $AsSb_3S_6(g)$.

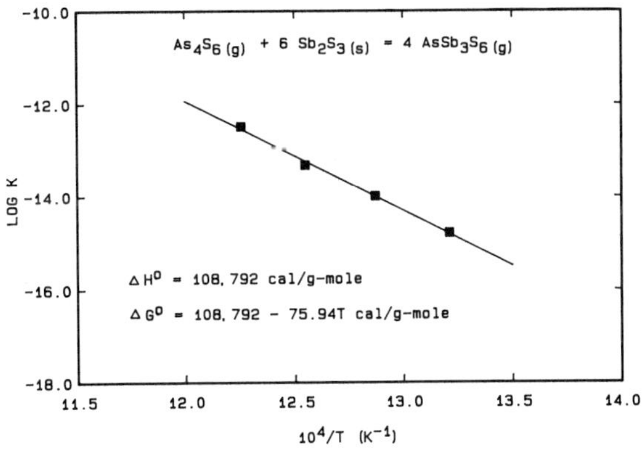

Figure 13 - Second-law plot for the formation of $AsSb_3S_6(g)$.

Using the equation derived for the free energy of fusion, the following result was derived for $Sb_2S_3(l)$:

$$As_4S_6(g) + 6\ Sb_2S_3(l) = 4\ AsSb_3S_6(g) \tag{36}$$

$$\Delta G^\circ = 51,192 - 5.98T \quad cal/g\text{-}mole \tag{37}$$

$$214,187 - 25.02T \quad J/g\text{-}mole$$

Discussion of Results and Conclusions

The principal difference between the $Sb_2S_3(s,l)$ vapor pressure results obtained by this study and the values reported in the literature appears to be a lack of knowledge of the equilibrium sulfur potential in the majority of the previous studies. For this reason, there is considerable uncertainty as to the vapor species that were present. The questionable data for $Sb_2S_3(g)$ and $Sb_2S_2(g)$ given in USBM Bulletin 689 [6] are based on the work of Prusaczyk [12], who estimated the molecular constants for these species. With these points in mind, the data obtained in this study were used to reconstruct the vapor pressure/stability diagram given in Figure 1. The revised diagram is given in Figure 14. The revised diagram differs substantially from the original diagram in that the line for $Sb_2S_3(g)$ has been shifted upward by a factor of 10 and the line for $Sb_2S_2(g)$ has been shifted upward by a factor of 5000. No other species are shown on Figure 14 since they would be present at partial pressures too small to be detected by the method used in this study. The boundary between Sb(s) and $Sb_2S_3(l)$ was calculated from the $Sb_2S_3(l)$ data given in [6]. The vapor pressure/stability diagram for the Sb-S system can be calculated at any desired temperature by using the thermodynamic results given in Table III and the Gibbs free energy of formation data for $Sb_2S_3(s,l)$ from [6].

Another application of the results of this study is to combine the data obtained for $Sb_2S_3(g)$ and $Sb_2S_2(g)$ with the data reported by Li and Hager [1] on $Sb_4O_6(g)$ to construct vapor pressure/stability diagrams for the Sb-S-O system. These diagrams would be useful in identifying processing conditions for wastes containing a mixture of antimony oxide and sulfide. These diagrams have been constructed at 700K and 900K and are given in Figures 15 and 16. At 900K, control of the SO_2 and O_2 pressures will provide for $Sb_2O_3(s)$ to be in equilibrium with $Sb_2S_3(l)$ to give a pressure of $Sb_4O_6(g)$ of 6.34 Torr and a pressure of $Sb_2S_3(g)$ of 0.59 Torr. The vapor transport rate of antimony would be a maximum along this boundary. The Gibbs free energy of formation data for $Sb_2O_3(s)$ and $Sb_2O_4(s)$ were taken from [23] and the data for $Sb_2S_3(s,l)$ were taken from [6].

The formation of the $AsSb_3S_6$ vapor complex may serve as the basis for a new process to extract antimony from sulfidic wastes. Since antimony-containing wastes frequently contain arsenic, the blending of feed-stocks could provide a material that would generate a sufficiently large pressure of $As_4S_6(g)$ that the enhanced volatilization of the antimony sulfide would be significant. The calculated enhancement in the vapor pressure of antimony trisulfide, based on the results of this study, is given in Table IV for different processing conditions. As can be seen, at a temperature of 850K and a pressure of $As_4S_6(g)$ of 100 Torr or greater, the apparent vapor pressure of $Sb_2S_3(g)$ will be in excess of 1 Torr. Of course, the partial pressure of $S_2(g)$ will need to be large enough to assure the stability of $Sb_2S_3(s,l)$: a pressure of 0.08 Torr at 850K.

A proposed process flowsheet based on the concepts presented above is given in Figure 17. The reactor off-gas will contain the combustion products from the fuel along with the S_2, As_4S_6, Sb_2S_3, and $AsSb_3S_6$. The gas will be sent to a gas treatment/separation system where air will be injected to decompose the vapor

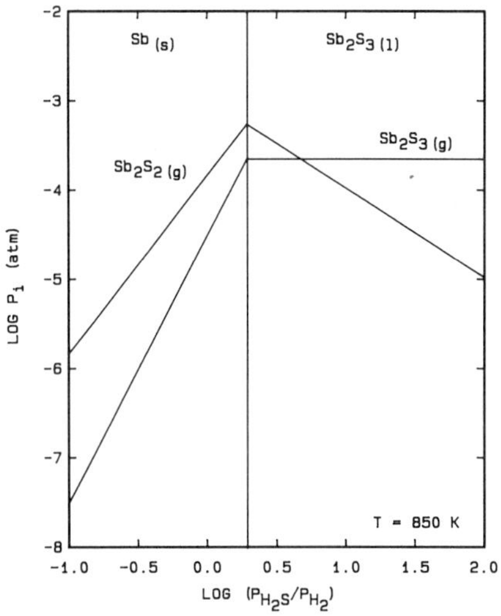

Figure 14 - Vapor pressure/stability diagram for the Sb-S system at 850K, based on the results of this study.

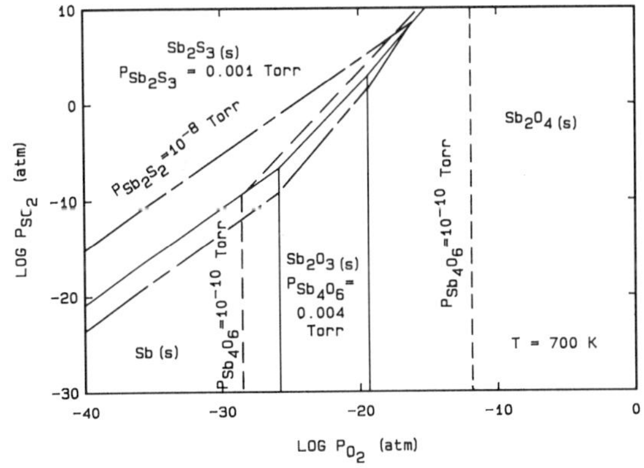

Figure 15 - Vapor pressure/stability diagram for the Sb-S-O system at 700K.

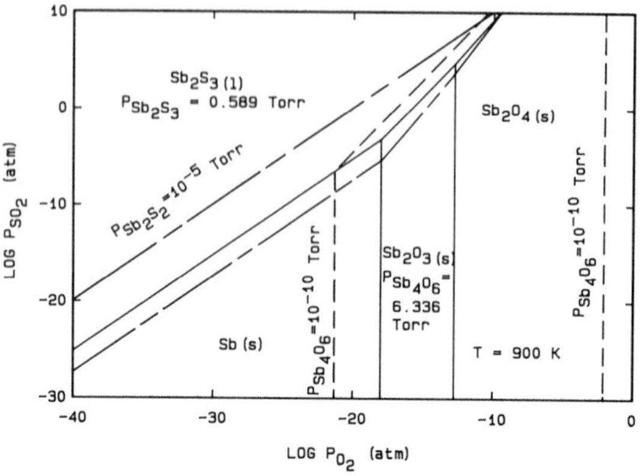

Figure 16 - Vapor pressure/stability diagram for the Sb-S-O system at 900K.

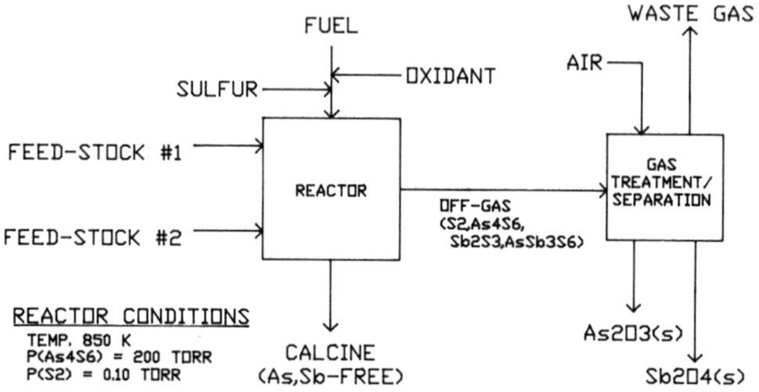

Figure 17 - Proposed process flowsheet.

TABLE IV - Calculated Enhancement in the Vapor Pressure of Antimony Trisulfide in the Presence of $As_4S_6(g)$[1]

		Pressure of $As_4S_6(g)$			
		10 Torr		100 Torr	
Temp. (K)	$P(Sb_2S_3)$ (Baseline)	$P(Sb_2S_3)$ (Apparent)	% Increase	$P(Sb_2S_3)$ (Apparent)	% Increase
700	0.001	0.019	1800	0.032	3100
800	0.037	0.240	547	0.397	973
900	0.588	1.227	109	1.724	193

1 - All pressures are in Torr (1 atm = 760 Torr); the baseline pressure of $Sb_2S_3(g)$ is the equilibrium pressure for $Sb_2S_3(s,l)$; the apparent pressure is the sum $P(Sb_2S_3)$ + 1.5 $P(AsSb_3S_6)$; the % increase is relative to the baseline pressure.

complex species and oxidize the gaseous sulfides. If air injection is as effective in the decomposition of the sulfide vapor complex as was observed by Li and Hager [1] in the decomposition of the oxide vapor complexes, chemical separation of the antimony and arsenic will be achieved. The two, separate condensation products will be $As_2O_3(s)$ and a mixture of $Sb_2O_3(s)$ and $Sb_2O_4(s)$. The proposed reactor conditions are given on Figure 17. The waste gas will contain all of the sulfur as SO_2 and will need to be treated for sulfur recovery.

Acknowledgments

The authors are pleased to acknowledge the support of this research by the U.S. Bureau of Mines Generic Mineral Technology Center for Mineral Industry Waste Treatment and Recovery, under grant G1175132. One of the authors (BRW) also received support from the Bureau of Mines through a MMRRI Fellowship.

References

1. T. Li and J. P. Hager, "New Chemistry for the Recovery and Separation of Arsenic and Antimony", in Proceedings of the Reinhardt Schuhmann International Symposium on Innovative Technology and Reactor Design in Extraction Metallurgy, edited by D. R. Gaskell, J. E. Hoffmann, J. P. Hager, and P. J. Mackey, TMS-AIME, (1986), 845-868.

2. T. P. Clement, T. Li, and J. P. Hager, "Process for the Recovery and Separation of Arsenic from Antimony", U.S. Patent 4,808,221, 28 February 1989.

3. J. E.Mauser, "Heteronuclear Compounds of Arsenic and Antimony", Met. Trans. B, 13B, (1982), 511-513.

4. D. L. Kaiser, "Construction of Vapor Pressure/Stability Diagrams for the As-S-O, Sb-S-O, and Bi-S-O Systems and Applications to Impurity Removal During Roasting", (M.S. Thesis, Colorado School of Mines, 1980), 131-292.

5. K. C. Mills, Thermodynamic Data for Inorganic Sulphides, Selenides, and Tellurides, Butterworths, London, (1974), 555-557.

6. L. B. Pankratz, A. D. Mah, and S. W. Watson, "Thermodynamic Properties of Sulfides", U.S. Bureau of Mines Bulletin 689, (1987), 326-343.

7. A. A. Ryazantsev, S. Pashinkin, and A. V. Novoselova, "Vapor Pressure Over Liquid Antimony Sulfide", Vest. Mosk. Univ. Khim., 23(5), (1968), 95-96.

8. G. P. Ustyugov, E. N. Vigdorovich, B. M. Kuadzhe, and I. A. Timoshin, "Saturated Vapor Pressure of the Antimony Chalcogenides", Russ. J. Inorg. Mat., 5(3), (1969), 589-590.

9. G. G. Gospondinov, A. S. Pashinkin, Z. Boncheva-Mladenova, and A. V. Novoselov, "Determination of the Saturated Vapor Pressure of Solid Antimony Sulfide", Russ. J. Inorg. Mat., 6(7), (1970), 1242-1247.

10. C. L. Sullivan, J. E. Prusaczyk, and K. D. Carlson, "Molecules in the Equilibrium Vaporization of Antimony Sulfide and Selenide", J. Chem. Phys., 53, (1970), 1289-1290.

11. V. S. Ban and B. E. Knox, "Mass-Spectrometric Study of the Laser-Induced Vaporization of Compounds of Arsenic and Antimony with the Elements of Group VIa", J. Chem. Phys., 52(1), (1970), 248-253.

12. J. E. Prusaczyk, "Dissociation Energies of New Chalocogenide Molecules", (Ph.D. Thesis, Case West. Reserve Univ., Cleveland, Ohio, 1971), Univ. Microfilms, Ann Arbor, Michigan.

13. F. M. Faure, M. J. Mitchell, and R. W. Bartlett, "Vapor Pressure Study of Stibnite (Sb_2S_3)", High Temp. Sci., 4, (1972), 181-191.

14. A. S. Shendyapin, V. N. Nesterov, and E. T. Ibragimov, "Vapor Pressure of Antimony Trisulfide", Inst. Met. Obog. Alma-Ata, U.S.S.R., (1975), 2-11.

15. A. S. Pashinkin, A. S. Malkova, and M. T. Pasternak, "Saturated Vapor Pressure of Liquid Antimony Sulfide and a Number of Alloys of the System Sb-S", Russ. J. Inorg. Mat., 13(6), (1977), 1094-1096.

16. V. N. Nesterov and R. A. Isakova, "The Vapor Pressure of Antimony Sulfide", Paper deposited at VINITI, No. 770-774.

17. L. Komorova, A. Holmstrom, and I. Imris, "Vaporization of Antimony from Synthetic Sulphosalts", Scand. J. Met., 14, (1983), 103-112.

18. B. W. Madsen, A. Adams, and P. A. Romans, "Determining Vapor Density and Gaseous Diffusion Coefficients of Sb_2S_3 Using a Modified Transpiration Apparatus", U.S. Bureau of Mines, R.I. 8812, U.S. GPO, (1983).

19. H. H. Kellogg, "Vaporization Chemistry in Extractive Metallurgy", Trans. of the Met. Soc. of AIME, 236, (1966), 602-615.

20. D. W. Baker, "Application of the Geometric Inequality to the Solution of Systems of Non-Linear Equations", (M.S. Thesis, Colorado School of Mines, 1980).

21. M. B. Meyer and E. J. Felty, J. Electrochem. Soc., 11(6), (1970), 818.

22. V. M. Glazov, R. A. Kuliev, and A. N. Krestovnikov, Russ. J. Phys. Chem., 45 (10), (1971), 1516.

23. L. B. Pankratz, "Thermodynamic Properties of Elements and Oxides", U.S. Bureau of Mines Bulletin 672, 672, (1982), 351-361.

24. A. D. Mah, "Thermodynamic Data for Arsenic Sulfide Reactions", U.S. Bureau of Mines Report of Investigations 8671, (1982), 50-52.

ALUMINUM - GENERAL

Session Chairman:
Ghyslain Dubé

ALUMINUM RECYCLING IN THE UNITED STATES - A HISTORICAL PERSPECTIVE

Patricia A. Plunkert

Bureau of Mines
Branch of Nonferrous Metals, MS#5208
2401 E Street, NW
Washington, D.C. 20241

Aluminum recovery from scrap has become an important component of the supply/demand relationship in the United States. Between 1950 and 1974, aluminum recovered from old scrap (post-consumer scrap) accounted for approximately 5% of the total domestic demand for aluminum. Increased costs for energy and growing concerns over waste management have provided the impetus for increased recycling rates. In 1988, recovery from old scrap supplied 20% of the total aluminum metal demand in the United States. This paper will provide an overview of the trends and developments in the domestic aluminum recycling industry.

Second International Symposium
Recycling of Metals and Engineered Materials
Edited by Jan H. L. van Linden,
Donald L. Stewart, Jr., and Yogeshwar Sahai
The Minerals, Metals & Materials Society, 1990

The secondary aluminum industry was born shortly before World War I at a time when the United States had only one primary aluminum producer. There was little need for a secondary industry prior to that time since the supply of aluminum scrap was limited. Although the secondary industry did grow during World War I and after, that growth was modest until World War II. World War II created many changes in the aluminum industry. The primary industry had changed from a monopoly to a triopoly, while the secondary industry had progressed from a remelter to a smelter. Secondary smelters thus emerged from the war years with the technology needed to process the huge quantities of aircraft scrap and other partially manufactured aluminum products. It was at this point in history that the secondary aluminum industry started its phenomenal growth.

Scrap is never purposely created. It is the unavoidable byproduct of fabricating operations or a product of obsolescence. Scrap resulting from semifabricating and fabricating operations is called "new scrap" while scrap resulting from obsolescence is called "old scrap".

New scrap generated by fabrication of aluminum products may be either home scrap (sometimes called runaround scrap) or prompt industrial scrap. Home scrap is recycled within the company generating the scrap and consequently seldom enters the commercial secondary market. Prompt industrial scrap, however, is new scrap from a fabricator who does not choose to, or is not equipped to, recycle the scrap. This scrap then enters the secondary market.

Old scrap is a product of obsolescence and becomes available to the secondary industry when consumer products (durable and nondurable) have reached the end of their economic life and have been discarded.

Figure 1 is a generalized view of the flow of aluminum scrap.

Aluminum recovery from scrap has become an important component of the supply/demand relationship in the United States. Between 1950 and 1974, aluminum recovered from old scrap (post-consumer scrap) accounted for approximately 5% of the total domestic demand for aluminum. Increased costs for energy and growing concerns over waste management have provided the impetus for increased recycling rates. In 1988, recovery from old scrap supplied 20% of the total aluminum metal demand in the United States.

Aluminum recovered from scrap has shown a tenfold increase since 1950. The recovery of aluminum from old scrap has shown an even more rapid expansion over the same period of time. In addition to improvements in recycling technologies, some of the increase in aluminum scrap recovery can be attributed to a changing and growing end use consumption pattern. Aluminum products, developed for the construction, transportation, and electrical industries, tend to have a fairly long life-cycle and are slow to enter the scrap supply stream. The emergence of the aluminum beverage can in the mid-1970's with a life cycle of approximately 1 year added dramatically to the potential aluminum scrap supply. In 1988, the Bureau of Mines estimated that over 2.1 million metric tons of metal was recovered from purchased aluminum scrap. Of this total, approximately one-half was recovered from old scrap.

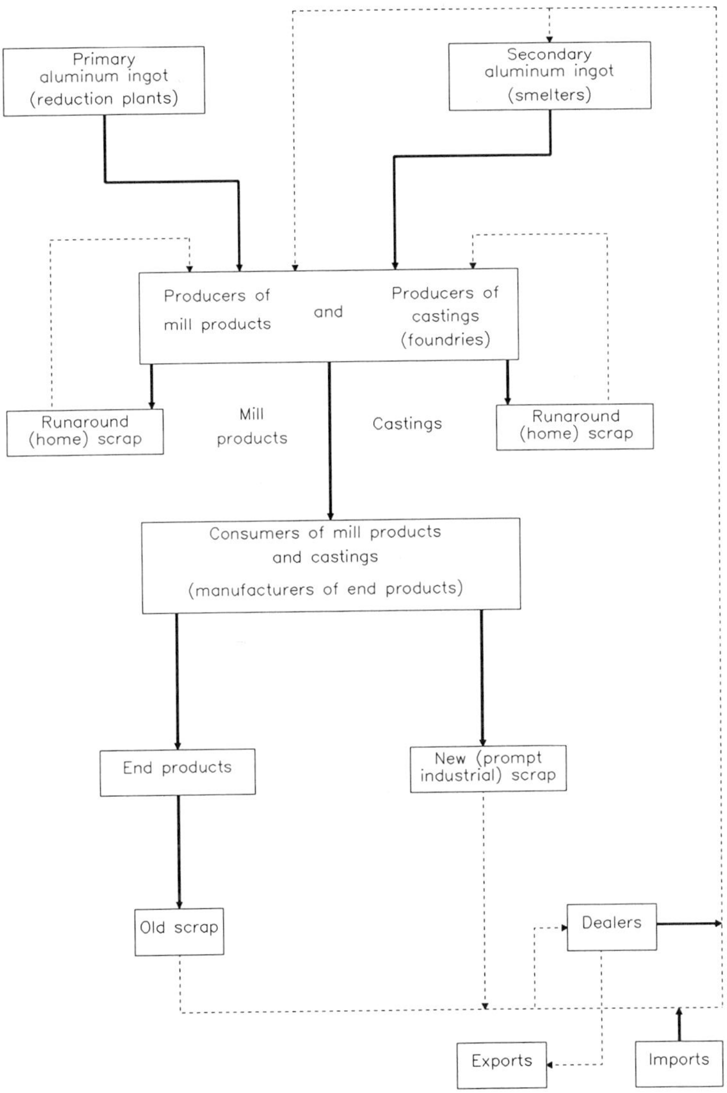

Figure 1 - Flow of aluminum scrap

TABLE I. - ALUMINUM SUPPLY-DEMAND RELATIONSHIPS, 1950-88
(Thousand Metric Tons)

	1950	1955	1960	1965	1970	1975	1980	1985	1988
COMPONENTS AND DISTRIBUTION OF U.S. SUPPLY									
Primary metal production	652	1,421	1,827	2,498	3,607	3,519	4,654	3,500	3,944
Secondary production (old scrap)	69	91	86	186	179	305	617	850	1,045
Shipments of Government stockpile excesses	---	---	---	34	22	2	---	---	---
Imports	232	217	178	562	425	499	647	1,420	1,620
Industry stocks, Jan. 1	327	508	689	1,270	1,716	2,339	2,325	2,651	1,894
Total U.S. supply	1,280	2,236	2,781	4,550	5,948	6,664	8,243	8,423	8,503
Distribution of U.S. supply:									
Industry stocks, Dec. 31	363	544	780	1,415	1,989	2,721	2,302	2,343	1,904
Exports	21	31	348	286	555	399	1,346	908	1,247
Government accessions	25	199	61	---	---	---	---	---	---
Industrial demand	871	1,462	1,591	2,849	3,404	3,544	4,594	5,174	5,352

Table II Aluminum Recovery From Scrap
(Thousand Metric Tons)

Year	Old Scrap	New Scrap	Total Metal Recovered
1950	69	152	221
1955	91	304	395
1960	86	368	454
1965	186	559	745
1970	179	778	957
1975	305	816	1,121
1980	617	960	1,577
1985	850	912	1,762
1988	1,045	1,077	2,122

Aluminum scrap, in one form or the other, is recovered by almost every segment of the domestic aluminum industry. Primary producers, independent secondary smelters, fabricators, foundries, and chemical producers can recover aluminum from scrap. Primary producers and independent secondary smelters are the major consumers of aluminum scrap.

The independent secondary aluminum smelter is the core of the commercial secondary aluminum industry. Unlike scrap dealers who buy and sell many different metals and die casters who cast other metals, the secondary aluminum smelter is totally dependent upon one metal - aluminum. The sole purpose of the secondary smelter is to transform aluminum scrap into a marketable product. Scrap is purchased by smelters wherever it can be purchased economically. Purchasing habits can also depend on how badly a smelter needs scrap. During periods of excess scrap supply, smelters tend to be very selective in their scrap purchases. When the supply of aluminum scrap becomes short, however, it is frequently necessary to buy scrap that would not ordinarily get melted down.

Independent secondary aluminum smelters, by definition, consume scrap and produce alloys for the die casting industry. A cursory look at the distribution of these secondary aluminum smelters in the United States consequently reveals a heavy concentration of smelters in the automotive and appliance manufacturing areas of the country.

The other major consumers of aluminum scrap are the primary aluminum producers. The types of scrap processed by the primary producers tend to be more segregated than those processed by the secondary smelters. Primary producers participate in either the collection or utilization of new aluminum scrap. They frequently purchase scrap from their industrial customers or on a contract conversion basis. Major primary aluminum producers also operate can recycling programs. These producers have set up literally thousands of collection centers around the country for used beverage containers. The large scale aluminum beverage can reclamation programs of these aluminum producers have added substantially to the rate of aluminum recovery from old scrap. The used beverage can component of old scrap consumption has doubled since 1975.

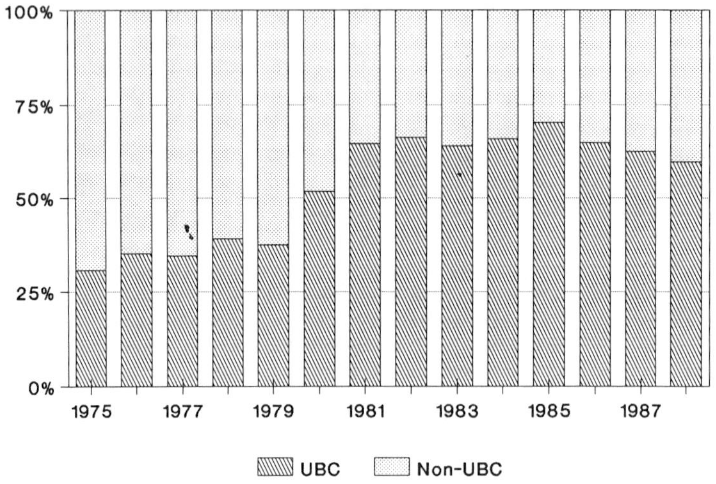

Figure 2 - UBC component of old scrap consumption

The pricing mechanism of the aluminum scrap industry is almost entirely dependent on the laws of supply and demand. During the past three years, the price of primary aluminum ingot went through a dramatic price swing. In January of 1987, the U.S. market price for primary aluminum metal, as published by Metals Week, was about 39 cents per pound. The price reached a peak of $1.26 per pound in June of 1988. By December of 1989, the price had fallen to 74 cents per pound. Figure 3 illustrates that the average price for aluminum scrap (mixed low copper clips), as published in the American Metal Market, followed the same general trend during this time period. Used beverage can scrap prices were also very volatile, and reached a high of 76 cents per pound in January 1989.

During 1987 and 1988, the aluminum industry as a whole was experiencing a period of high demand and tight supply. Aluminum scrap supplies were also tight, but the rise in price attracted more scrap into the marketplace. It was during this period that the press began reporting incidents of thefts of aluminum products that were apparently being sold for their scrap value. Aluminum guard rails, street signs, and aluminum siding became targets for these thefts.

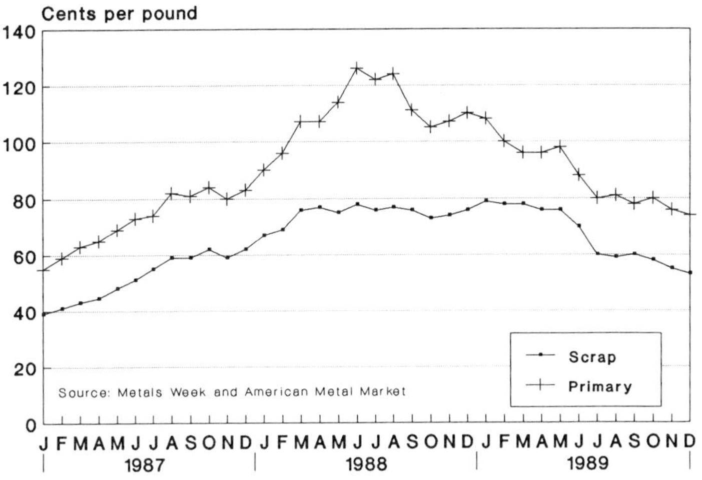

Figure 3 - Primary and scrap aluminum prices

Aluminum scrap is traded in the international marketplace. Price and shipping costs are usually the determining factors in choosing whether to sell scrap in the domestic or international markets. U.S. trade in aluminum scrap has grown dramatically over the last 30 years. Most of the scrap shipped into the United States comes from Canada. In 1988, the United States received about two-thirds of its scrap imports from Canada. Since the mid-1970's, Japan has been the major recipient of U.S. scrap exports. In 1988, over one-half of the U.S. exports of aluminum scrap were shipped to Japan.

Table III U.S. Aluminum Scrap Trade
(Thousand Metric Tons)

Year	Imports	Exports
1950	62	<1
1955	37	16
1960	5	73
1965	24	35
1970	34	52
1975	50	61
1980	54	404
1985	128	375
1988	201	487

The secondary aluminum industry has developed into a major market force in the domestic aluminum industry. The recycling of scrap provides a source of aluminum that not only helps the aluminum industry to maintain its growth but also helps to conserve energy and to slow the depletion of bauxite resources. Growing concerns over waste disposal and energy conservation have led many people to question the way in which we manage our mineral resources. Recycling is one way to alleviate many of these concerns.

References

1. Bureau of Mines Minerals Yearbook - Aluminum Chapter, various years.

2. Donald L. Siebert, "Impact of Technology on the Commercial Secondary Aluminum Industry" (Bureau of Mines, IC-8445, 1970).

3. "Aluminum Recycling Casebook", Aluminum Association Inc., December 1985.

EFFECT OF SALT FLUX ADDITIVES

ON ALUMINUM DROPLET COALESCENCE

Ray D. Peterson

Manufacturing Technology Laboratory
Reynolds Metals Company
P.O. Box 1200, Sheffield, Alabama, USA

Abstract

A laboratory system was set up to evaluate the coalescence ability of various additives to a standard NaCl-KCl molten salt flux for aluminum reclamation. The time and degree to which aluminum droplets were consolidated into a single pool under quiescent conditions were measured. A standard 50% NaCl-50% KCl (by weight) flux produced no coalescence of aluminum droplets. A similar result was demonstrated for the addition of other chloride constituents. Salt compounds containing fluoride ions encouraged more effective merging of the aluminum globules. Effects of other salt additives to the flux are discussed. As known and practiced by industry, cryolite (Na_3AlF_6) was observed to be one of the most effective aluminum coalescence agents. Various percentages of cryolite additions were scrutinized by examining coalescence and metal recovery of delacquered Used Beverage Containers (UBC) in laboratory scale furnaces. A critical cryolite addition to the salt flux must be exceeded to maximize coalescence to the molten aluminum pool and, thereby, to maximize recovery.

Second International Symposium -
Recycling of Metals and Engineered Materials
Edited by Jan H. L. van Linden,
Donald L. Stewart, Jr., and Yogeshwar Sahai
The Minerals, Metals & Materials Society, 1990

Introduction

Salt fluxes play an important role in the remelting of secondary aluminum whether it is in the form of finely divided fragments, shredded Used Beverage Containers (UBC), or dross. A review of the literature (in particular, see Rao and Dawson[1]) shows a bewildering array of fluxes being proposed for the melting and recovery of aluminum. Most fluxes contain a mixture of sodium chloride and potassium chloride (NaCl and KCl) to obtain low temperature melting near the eutectic point. Another salt may be added to the mixture to modify the flux. Over 120 non-patented fluxes are listed while over 30 patented fluxes are presented. Why should there be so many fluxes proposed and do they have unique properties?

Let's first review the requirements of a flux for aluminum melting. A flux has three main requirements:

1. A flux should cover the metal (either solid or molten) and prevent oxidation,

2. A flux should dissolve or suspend dirt, oxide, and other extraneous material adhering to the metal, and

3. A flux should promote coalescence of the molten metal by stripping away the adhering layer of aluminum oxide or magnesium oxide on the metal surface.

Other important physiochemical factors are listed in Table I. After a flux meets the three primary requirements, it should also meet the majority of the secondary requirements.

Table I Secondary Requirements of a Salt Flux

Melting point less than or equal to 660°C (1220°F)
Density less than 2.3 gm/cm^3
Should not react with or contaminate the metal
Should not attack the refractory walls
Non-poisonous
Low vapor pressure
Not hygroscopic
Inexpensive or easily recycled at low cost
Low viscosity
Should separate cleanly from the metal

Until recently, a significant portion of the fluxes chosen for a given operation were based on art or past experience rather than on a thorough understanding of the science of salt fluxes. This practice and the diversity of melting operations and types of furnaces used lead to many different flux requirements. The flux characteristics needed for a reverbatory furnace with an external charge well would not be expected to be the same as those needed for a rotary barrel furnace. Therefore, tailoring a salt for the latter application would require consideration of different factors than in the former case.

Most of the physical requirements for a salt flux, as listed in Table I, can be readily determined through an evaluation of tabulated data[2,3] or by a simple laboratory determination. However, the most important characteristic, that of a salt's ability to promote coalescence, is the least understood and may be the hardest to quantify. For coalescence of the metal droplets to occur, the oxide film of the droplets must be disrupted such that the internal liquid metal of the droplets can touch and merge into a single pool. The exact mechanism by which this rupture of the oxide film occurs is not fully understood at this time.

The goal of the work reported in this paper was to both qualitatively and quantitatively evaluate various additives to the standard sodium chloride-potassium chloride mixture commonly used as the base for various salt fluxes throughout the aluminum industry. Commonly-used additives include fluorite (CaF_2) and cryolite (Na_3AlF_6), but less commonly-used additives were to be screened as well. Additionally, the effect of concentration of an effective additive would be varied to assess its effect on metal recovery of a common scrap such as UBC. Since cryolite is one of the most common flux additives, it would be used for this purpose.

Experimental Procedure

The direct measurement of a salt's ability to promote coalescence is not easily obtained. Measurement of the wetting ability of a salt on aluminum or alumina by contact angle measurements might be one technique used to infer a salt's coalescence capability, while measurement of the interfacial tension[4,5] might be another. Neither of these is a direct measure of coalescence.

An alternative technique would be to measure the degree and amount of time required to attain coalescence of aluminum particles in various salt flux mixtures. This technique was chosen for this study to compare 5 weight % salt additions to a standardized flux mixture of 50 weight % NaCl-50 weight % KCl.

The experimental equipment used is shown in Figure 1. A reflective furnace (TransTemp Co., Cambridge, MA) was used to heat a fused quartz sample tube. The sample tube contained 50.0 grams of the premixed fluxing salt. The reflective furnace

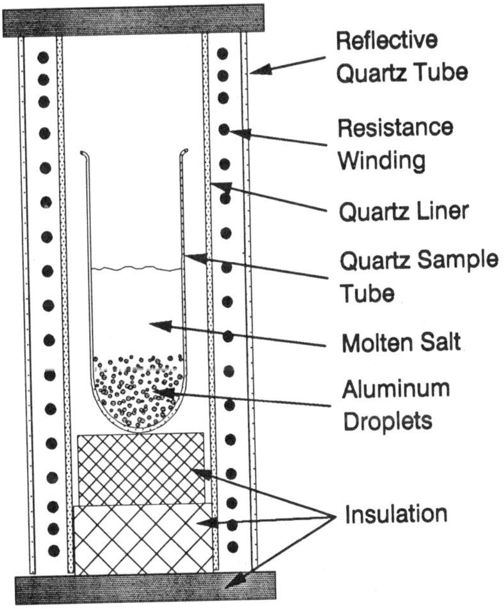

Figure 1 - Experimental set up for coalescence tests.

allowed the investigator to view the coalescence process while it was occurring. The reflective furnace contains an internal fused quartz tube which is wound by a resistance element. These components are totally surrounded by another tube with a thin metallic gold deposit which reflects back over 99% of the heat generated in the infra-red spectrum while still being partially transparent in the visible spectrum.

Varying Salt Additives

A basic test consisted of loading the sample tube with a pre-weighed and pre-mixed salt flux sample into the furnace. All salt flux samples were made of reagent-grade chemicals except where noted. The furnace was heated to 750°C (1382°F). The salt flux was melted and allowed to thermally stabilize. Then 5.000 grams of aluminum shot, produced by Reynolds Metals Co., of approximately 1-2 mm in size was added to the molten flux and timing was initiated. The aluminum pellets used for the test are shown in Figure 2. Observations were noted as changes occurred to the aluminum pellets or the flux. When it was possible to observe the many droplets, the time required for coalescence into a single pool was noted. No stirring of the fluxing system was made during the test. After 10 minutes the furnace was shut off and the salt-metal system was allowed to solidify. Upon cooling, the salt flux was dissolved in water for recovery of the aluminum metal. It was then possible to easily determine if, and to what degree, the metal had coalesced. All aluminum was collected and weighed at the end of each test by washing the salt away. Recovery calculations for the metal could then be performed based on the starting and ending weights. Each salt additive was compared under this same set of standard conditions.

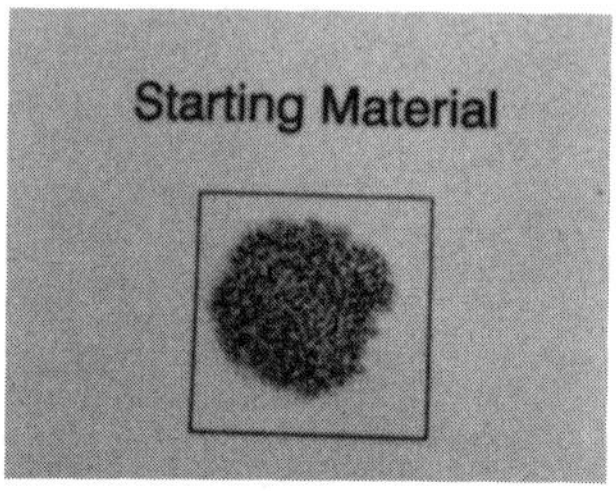

Figure 2 - Aluminum droplets used for coalescence.

The choice of possible flux additives came from the category of inorganic salts. As mentioned in the Introduction, a salt flux must not react with or contaminate the aluminum with which it comes in contact. This is true for the basic NaCl-KCl salt flux. An additive to the flux must meet the above-mentioned requirement, and also be compatible with the molten chloride media being used. These two factors limit the investigation to primarily halide compounds; particularly, fluorides and chlorides of the alkali and alkaline earth metals. Most other metal chlorides, such as copper chloride, or zinc chloride, would be reduced by the metallic aluminum. Other sodium salts with non-halide anions must be stable in the chloride melt. This eliminates salts such as sodium nitrate.

Both the cation and anion of the salts were varied in the test matrix. First, various chloride salt additions were examined (varying the cations). Also, various fluoride salt additions (varying the cations) were scrutinized including various AlF_3 compounds. These two classes make up the majority of salt flux additives. Additionally, a number of sodium salts with varying anions were examined for their effect on coalescence. The test matrix of salt additives examined is shown in Table II. Most of these additives meet the criteria for salt fluxes outlined earlier.

Table II Test Matrix of Salts Examined

		Cations (Increasing Atomic Number ->)								
		Li	Na	Mg	Al	Si	K	Ca	Sr	Ba
Anions	F	LiF	NaF	MgF_2	AlF_3	Na_2SiF_6	KF	CaF_2	SrF_2	BaF_2
	Cl	LiCl	NaCl*	$MgCl_2$	$AlCl_3$		KCl*	$CaCl_2$		$BaCl_2$
	I		NaI							
	Br		NaBr							
	CO_3		Na_2CO_3							
	SO_4		Na_2SO_4							
	AlF_3		$NaAlF_4$ Na_3AlF_6				$KAlF_4$			

* The actual test consisted of no addition to the NaCl-KCl mixture.

Varying Cryolite Percentage

For the tests examining the effect of cryolite addition on metal recovery from UBC, an excess salt flux technique was used. Once again, a 50 weight % NaCl-50 weight % KCl mixture was used. To this salt, additions of Greenland cryolite of 0, 1, 2, 3, 4, 6, 8, 10 weight % were made. Two hundred grams of premixed NaCl/KCl/Cryolite were melted in a graphite crucible at 825°C (1517°F). Two hundred grams of decorated UBC was then added to the molten salt. This action took several minutes to complete and required that the shreds be pushed under the flux with a graphite rod. The mixture was then stirred several times to assist coalescence of the aluminum droplets. Twenty minutes after the test was begun, the molten mixture was poured out into an Inconel chill pan. Pieces of aluminum larger than 0.5 cm (0.25 inches) were hand picked and washed to remove any adhering salt. This aluminum was designated "easily recoverable." The spent salt flux was washed with water and then filtered to recover any residue not soluble in water. The residue was dried, weighed, and screened at 4 Mesh and 8 Mesh. All of the 8 mesh and larger material was assumed to be aluminum metal. The undersize material was checked for metallic aluminum content by hydrogen evolution.

Results and Discussion

As outlined in Table II, 22 salt additives (excluding the base condition) were tested plus several replicates to verify the results. The results of the coalescence tests are presented in Table III. The table lists the extent of droplet coalescence, the metal recovery, and the coalescence time. Also shown are the color of the molten salt flux and the observation of surface flares after the addition of the metal to the salt. Photographs of the recovered aluminum after the coalescing stage are presented in Figures 3 through 6. A short description of the various additive classes and the results is made below.

Table III Qualitative Assessment of Halide Flux Additions

Test Number	Additive (%)	Color of Flux	Extent of Coalescence	Coalescence Time	Aluminum Recovery	Gas Formation
1	None	Clear	None	-------	99.7	None
2	5% Gr. Cryolite	Slightly Cloudy	Complete	30 sec.	101.8	None
3	5% CaF_2	Undissolved Material	Moderate	222 sec.	100.1	Small w/flares
4	5% AlF_3	Black & Cloudy	Complete	> 270 sec.	100.6	None
5	5% AlF_3 (87% Purity)	Clear Brown w/Al_2O_3	Moderate	> 60 sec.	100.6	Violent
6	5% $MgCl_2$	Slightly Cloudy	None	-------	100.0	None
7	5% $CaCl_2$	Clear	None	-------	100.0	None
8	5% MgF_2	Clear w/Red Ppt.	Moderate	> 300 sec.	99.7	None
9	5% LiF	Clear w/White Ppt.	Complete	?	100.6	None
10	5% $NaAlF_4$	Light Yellow	Near Complete	24 sec.	100.5	Small
11	5% LiF	Clear	Complete	114 sec.	100.6	Small
12	5% AlF_3	Light Tan	Near Complete	36 sec.	100.7	Small
13	5% LiCl	Clear	Poor	> 600 sec.	100.1	Violent
14	5% $AlCl_3$	Cloudy Tan	Poor	> 600 sec.	100.2	Small
15	5% CaF_2	White Residue at Bottom	Moderate	216 sec.	99.8	Small w/flares
16	5% KF	Clear	Complete	41 sec.	100.2	Small w/flares
17	5% SrF_2	Clear	Near Complete	52 sec.	100.2	Small w/flares
18	5% BaF_2	Clear	Near Complete	37 sec.	100.9	Small
19	5% Na_2SiF_6	Decomposes	Near Complete	?	100.0	Moderate
20	5% $KAlF_4$	Slightly Cloudy	Minor	-------	101.0	Flaring
21	5% NaI	Clear Tan	None	-------	100.0	Minor
22	5% NaBr	Clear	Poor	-------	100.0	Small
23	5% $BaCl_2$	Clear	None	-------	100.0	Flaring
24	5% Na_2CO_3	Cloudy w/bubbling	None	-------	91.1	Bubbling
25	5% Na_2SO_4	Clear	None	--------	99.6	None

Blank Test

The actual test consisted of no addition being made to the NaCl-KCl salt flux. The aluminum droplets melted in the clear flux, but they did not coalesce (as can be clearly seen in Figure 3) despite the fact that they were in intimate contact for over 10 minutes. Obviously, this flux has no ability to strip away the protective oxide film. The apparent reason that this flux does consolidate aluminum in an industrial recovery operation is that the oxide film is disrupted during the stirring and mixing of the salt flux, and this allows the metal droplets to coalesce. These results are consistent with observations which will be presented later which show only 50% metal recovery for UBCs under laboratory conditions when no additions are made to the NaCl-KCl flux.

Chloride Salts

In general, the chloride salts do very little to increase the ability of the salt flux to strip away the oxide film and allow coalescence of the aluminum droplets. This can be plainly seen in Figure 3. Only LiCl and $AlCl_3$ promoted any coalescence at all. Additions of the chloride salts generally lower the freezing point of the salt flux, which can be a useful property in some circumstances. Unfortunately, some of the chloride salts are hygroscopic such as LiCl and $AlCl_3$. This property can cause problems when the salts are added to an already hot flux or if the added salts convey dissolved water into the molten salt where it can react with the aluminum. More strongly-hydrated salts such as $MgCl_2$ and LiCl may retain their associated water in the melt by simple dissolution or by hydrolyzing through such a reaction as:

$$LiCl + H_2O \rightarrow LiOH + HCl \quad (1)$$

The hydroxide compound may react further with other salt species or the aluminum metal.

As stated above, only LiCl and $AlCl_3$ promoted coalescence, but both caused violent gas formation when the aluminum pellets were added. The mixing created by the gas release may have been responsible for the partial coalescence. Addition of $AlCl_3$ is often quoted in the literature as being a "good coalescing agent," especially with regard to drosses. It seemed to have only a minor effect under the laboratory conditions. In circumstances where the violent bubbling action of $AlCl_3$ (due to its high vapor pressure) could be utilized, the coalescing effect might be greater.

Fluoride Salts

In general, all of the fluoride salts seem to aid coalescence to some extent. The degree to which they promote coalescence can be seen in Figure 4. Apparently, the presence of the fluoride ion in the molten salt helps to strip away the oxide film on the aluminum droplets. Not all of the fluoride salts are readily soluble in the NaCl-KCl system so the extent of their effect can be limited; examples of this would be CaF_2 and MgF_2.

The degree to which the fluoride salts promote coalescence, as well as the speed at which they complete the coalescence process, can be subdivided into three rather subjective categories: Superior, Good, and Moderate. The "Superior" additives--AlF_3, BaF_2, KF, and SrF_2--all had complete or near complete coalescence in under 60 seconds. The next lower level of effectiveness, "Good," consisted of LiF and CaF_2. They promoted moderate to complete coalescence, but required a longer time interval of up to 4 minutes. The "Moderate" group of salts included NaF, MgF_2, and Na_2SiF_6. They were placed in this category due to the length of time required to achieve coalescence, which was greater than 4 minutes. It is possible that Na_2SiF_6 does not belong in this group. Due to cloudiness in

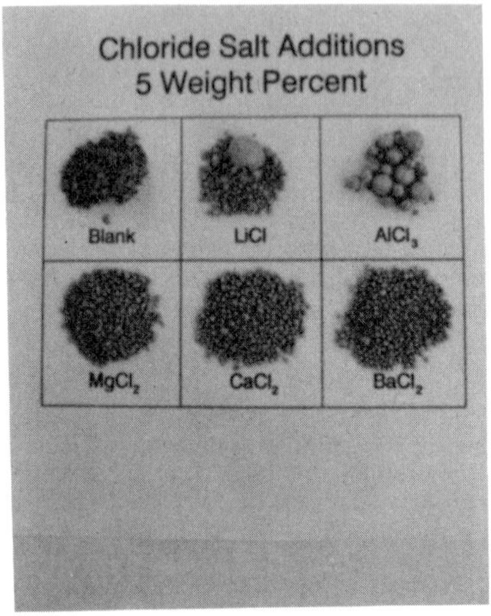

Figure 3 - Coalescence due to chloride salt additions.

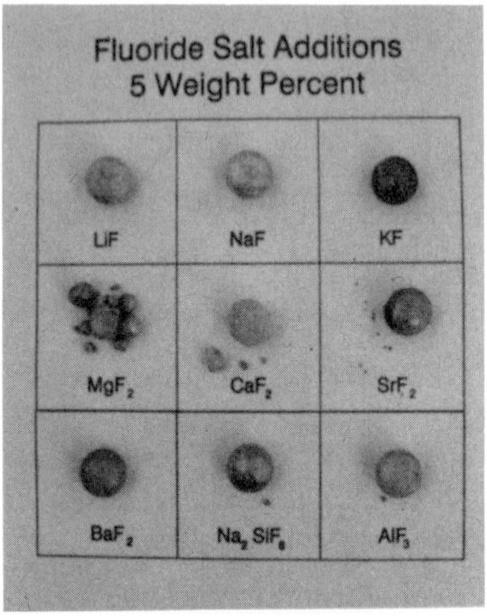

Figure 4 - Coalescence due to fluoride salt additions.

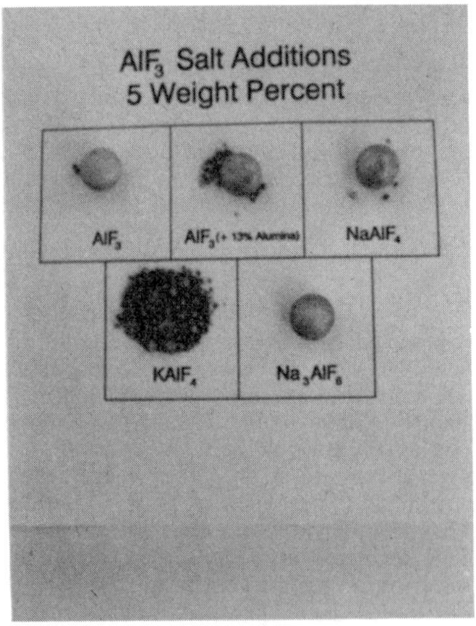

Figure 5 - Coalescence due to AlF_3 salt additions.

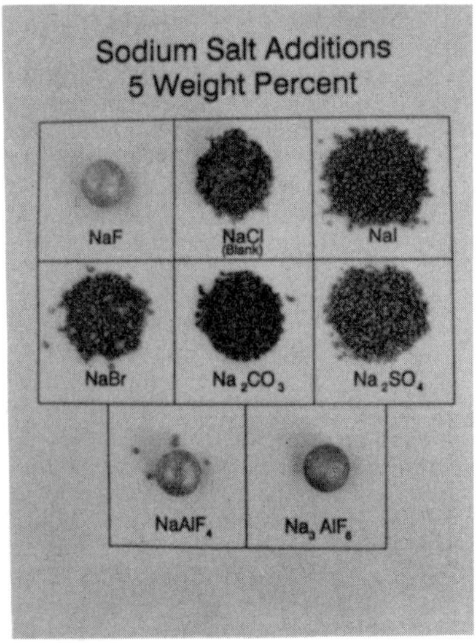

Figure 6 - Coalescence due to sodium salt additions.

the salt after the addition of the aluminum pellets, the exact time of coalescence could not be determined. Additionally, this salt decomposed during the heating process. Once the salt melted, the Na_2SiF_6 began to decompose by the reaction:

$$Na_2SiF_6 \rightarrow 2\,NaF + SiF_4\uparrow \qquad (2)$$

This reaction leaves the sodium fluoride behind in the melt, but the silicon fluoride is lost to the atmosphere since its boiling point is -86°C.

It is interesting to note that the three alkali metal fluoride salts all achieved complete coalescence of the aluminum droplets though in differing lengths of time. Their coalescence ability was superior to MgF_2 and CaF_2. This observation is in agreement with the work of van Linden and Stewart[6] using these same additives in fluxing Al-Mg alloys (AA5182) at 760°C. The only other additive to achieve this level of success was the Greenland cryolite which will be addressed in the next section.

Two tests were conducted with aluminum fluoride of varying purity. In the first case, an industrial AlF_3 with 13% alumina as an impurity was used while in the second case a high purity AlF_3 was used. In the first test, the salt had a cloudy appearance due to the presence of the alumina. The salt experienced very violent bubbling when the metal was added to the molten salt. This test gave moderate coalescence which took over 60 seconds to occur. In the second test, the molten salt was light tan and clear in appearance. The aluminum coalescence was nearly complete and required only 36 seconds. Very little bubbling was observed.

The differences noted in the two AlF_3 tests indicate the important role that an inert additive to the system, like alumina, can play. The excess alumina can prevent intimate contact between aluminum droplets and prevent coalescence even in quantities as low as 0.65% in the salt flux! The alumina may also tie up the active aluminum fluoride thereby preventing it from removing the oxide film. Other investigators[6,7] as well as field practitioners in the secondary industry, have noted that aluminum oxide in the flux can lessen the fluidity of the flux, increase the entrapment of the aluminum droplets, and decrease the recovery. It has also been noted that stirring can overcome this effect to some degree since the flux mixture is thixotropic.[8] The alumina impurity in the first test was probably in a hydrated form which introduced moisture into the salt system and caused the later violent reactions.

The fluoride salts are not as useful as chloride salts in lowering the melting point of the salt fluxes. However, the fluoride compounds generally are not hydrated nor hygroscopic so that water entering the salt flux is not a problem.

Aluminum Fluoride Containing Salts

Including the previously-discussed AlF_3, a total of four aluminum fluoride-containing salts were tested: AlF_3, $NaAlF_4$, Na_3AlF_6, and $KAlF_4$. The degree of coalescence achieved with each of these additives is presented in Figure 5. With the exception of the $KAlF_4$, these salts were the best coalescing agents of all the salts tested. The coalescence times were rapid and the degree of completeness was high. Only the $KAlF_4$ did not produce a high degree of coalescence. The anomaly of the $KAlF_4$ cannot be fully explained at this time. Since this salt should form a reciprocal salt flux to the NaCl-KCl-$NaAlF_4$, its abilities should be very similar. The material used in the test was an industrial by-product rather than a reagent-grade chemical and consequently contained a minor amount of other oxides (approximately 4%). These oxides may have interfered with coalescence as they did in the earlier reported AlF_3 test work.

Sodium Salt Additives

The sodium salt additives were used to examine the effect that varying the anion has on the salt flux. The coalescence abilities of sodium fluoride and sodium chloride were just presented, but will be repeated with the other sodium anion salts in this section for completeness. The results of the test work are shown in Figure 6. Of the newly tested salts, only NaBr promoted coalescence to any degree. Sodium iodide (NaI) had no effect like NaCl or Na_2SO_4. The addition of sodium carbonate, Na_2CO_3, had a detrimental effect on metal recovery. With the addition of the aluminum to the cloudy melt, the carbonate ions reacted with and oxidized the aluminum causing a loss in metal recovery. Only 91.1% of the original input metal was recovered.

Effect of Cryolite on Metal Recovery of UBC

The earlier results with clean aluminum droplets show the necessity of having a coalescing agent, such as cryolite, in the flux to produce a single pool of metal. From the extremes of 0% to 5% cryolite addition, a shift from negligible to full coalescence was observed in a quiescent pool. Obviously, these are not the conditions in an industrial recovery operation. As outlined in the Experimental Section, the recovery of decorated UBC in an agitated molten salt was used to evaluate the effect of increasing cryolite concentrations on recovery. Previous work[9] has shown the need for cryolite when fluxing the more reactive UBC alloy system as opposed to fluxing pure aluminum. The tendency to form $MgAl_2O_4$ spinels necessitates the use of an effective coalescing agent.

The results of the test work with varying cryolite additions are shown in Table IV. The amount of easily-recovered aluminum is strongly influenced by the addition of cryolite.

Table IV Decorated UBC Recovery for Increasing Cryolite Additions

% Cryolite in Flux	% Easily Recovered Aluminum	Metal Recovered in Finer Sizes +4 Mesh	Metal Recovered in Finer Sizes +8 Mesh	+8 Mesh Aluminum Recovered	Total Lost in -8 Mesh Solids	%Al Grams Oxide Produced
0	55.27	4.67	7.38	67.31	24.10	21.32
0	49.90	8.88	9.98	68.76	27.47	23.33
0	35.365	15.48	10.50	61.35	28.61	26.79
1	53.97	17.21	5.6	76.78	14.87	22.65
2	82.31	4.76	1.36	88.44	6.71	9.05
3	89.80	1.53	1.01	92.34	2.27	10.98
4	84.50	7.68	0.57	92.74	2.51	11.19
4	92.65	2.51	0.16	95.31	0.40	8.76
6	89.91	0.75	0.34	90.99	0.43	7.94
8	94.95	0.00	0.06	95.01	0.35	10.50
10	94.30	0.58	0.12	95.00	0.66	9.17

This is presented visually in Figure 7. With no cryolite addition, the recovery is quite variable due to the lack of coalescence and is generally low: 35 to 55%. As the cryolite content increases, the recovery increases. The recovery begins to level off at 95% around the 5% cryolite addition point. This 95% recovery level represents the metal content of the UBC shred; the 5% loss is due to the lacquer decoration on the can as well as moisture and dirt adhering to the can scrap.

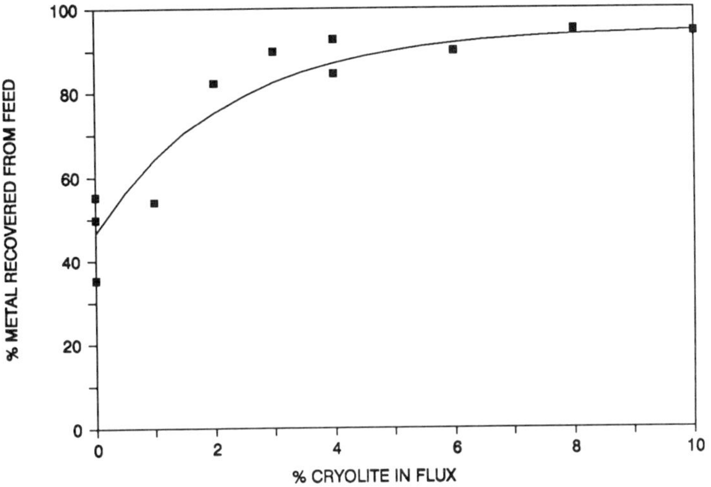

Figure 7 - Easily recovered metal from UBC.

Figure 8 presents the total metal recovery for the +8 Mesh aluminum as a function of the cryolite content. Once again, the recovery increases significantly with increasing cryolite. The dramatic effect of cryolite on aluminum loss is shown in Figure 9 where the lost aluminum is defined as the -8 Mesh aluminum entrapped in the salt flux. As the cryolite content is increased, this metal is coalesced and recovered in the main aluminum pool instead of being lost to the salt flux.

The last column in Table IV lists the grams of metal oxide recovered from the washed and filtered spent salt flux. With no cryolite addition, the amount of oxide produced averages 24 grams. With cryolite additions of 2 % or more, this value drops by about 50%. It is known that cryolite will strip the protective oxide layer from the aluminum droplets and could possibly enhance further oxidation of the metal. Balancing this effect is the reduction of the total surface as the many droplets are now able to coalesce and provide less surface area for oxidation. Under these fairly-quiescent test conditions, increased cryolite reduces the total amount of oxide formed.

While these metal recovery values are not directly applicable to a particular furnace operation, the trend of increasing recovery followed by a leveling off should be true of any recovery furnace. The cryolite level at which maximum recovery occurs will vary from operation to operation, depending on many differing conditions unique to each furnace. Nevertheless, increasing cryolite in the salt flux will increase the aluminum recovery up to some point and then additional cryolite will be of no further benefit. Similar recovery curves were observed for fluoride additions (NaF, KF, CaF_2, and MgF_2) in laboratory tests by van Linden and Stewart[6]. They found a leveling off in recovery at the 10% addition level. Racanas[8] determined similar behavior for cryolite additions in rotary barrel furnaces processing furnace skim drosses.

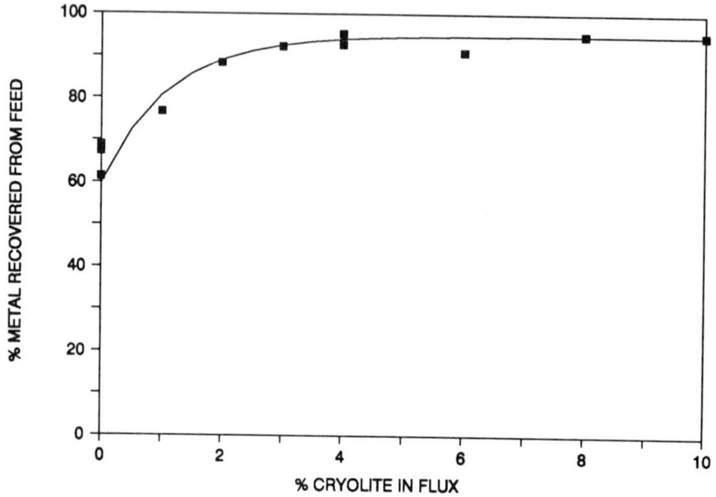

Figure 8 - Total metal recovery of +8 mesh aluminum.

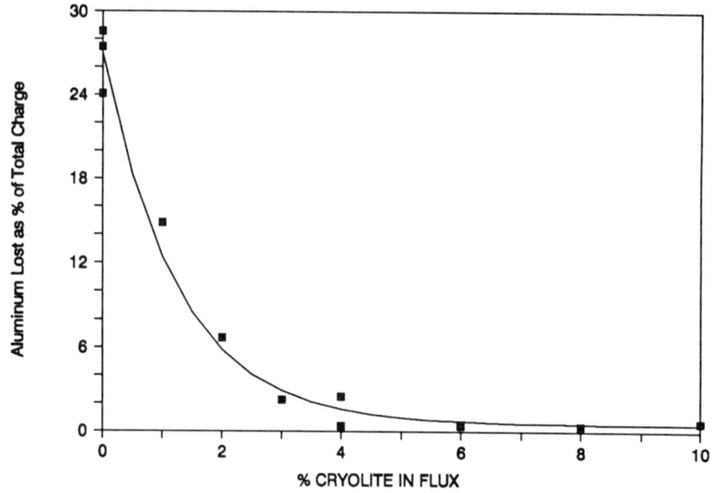

Figure 9 - Metal lost in the flux due to lack of coalescence.

Discussion

An attempt to better understand the coalescence mechanism for aluminum droplets will be made in this section. The relationships will be based on the observations of the coalescing ability of the various salt additives and the consideration of various salt properties

Table V lists the free energies of formation, ΔG_f, at 1000 K for some of the salt additives tested. All free energy values are based on the amount of halide ion present (kcal/mole halide ion). For the chloride salts, LiCl has the most negative free energy while $AlCl_3$ has the least negative free energy, yet they both coalesce the aluminum to a small extent while the other chloride compounds do nothing. Obviously, thermodynamics alone does not help in the evaluation of the coalescing ability of chloride compounds.

Table V Free Energies of Formation for Selected Salts[10,11]

Compound	ΔG_f Free Energy of Formation (kcal/mole of Halide Ion)
$AlCl_3$	-37.1
$MgCl_2$	-57.6
NaCl	-76.2
$CaCl_2$	-76.8
LiCl	-78.1
KCL	-81.6
$BaCl_2$	-83.4
AlF_3	-99.6
$NaAlF_4$(g)	-102.8
Na_3AlF_6	-109.9
KF	-111.6
NaF	-112.6
MgF_2	-113.6
SrF_2	-124.0
LiF	-124.3
BaF_2	-125.7
CaF_2	-126.3

The free energies of formation for fluoride compounds are also shown in Table V. Many investigators in the field of aluminum reduction technology have postulated on what chemical species are formed when alumina dissolves in cryolite[9]. At this time there is no consensus as to the exact aluminum fluoroxy compound formed, but for the sake of the following discussion, the species AlOF will be used. Similar analyses could be made for any species. For compounds without aluminum fluoride present, the reaction to form the aluminum fluoroxy compound would be:

$$Al_2O_3 + 2/x\, MeF_x \rightarrow 2AlOF + Me_{\frac{2}{x}}O \qquad (3)$$

Similarly for aluminum fluoride, the reaction would be:

$$Al_2O_3 + AlF_3 \rightarrow 3AlOF \quad (4)$$

Since the free energy of formation for the aluminum fluoroxy compound would be constant, only the free energy of the metal fluorides and oxides would affect the reaction. When the thermodynamic values are accounted for, it can be seen that the equilibrium for the reactions is shifted to the left side of the equation, or alternatively, favors the stability of the aluminum oxide. Based on thermodynamics, LiF would react the least followed by CaF_2 and MgF_2. This was not observed experimentally. Once again, thermodynamics alone cannot be used to predict a salt's coalescing power.

Interfacial tensions between NaCl-KCl melts and liquid aluminum at 740°C have been measured by Martin-Garin, Dinet, and Hictor;[4] as well as Ho and Sahai.[5] In an equimolar melt, the measured interfacial tension was found to be 710 dynes/cm. Additions of NaF and KF up to 15 weight % caused a significant decrease in interfacial tension to the range of 400 dynes/cm. The first researchers found only half as much decrease in interfacial tension when LiF and Na_3AlF_6 were added as when NaF and KF were added. In orally-presented results accompanying the paper, Ho and Sahai[5] showed that MgF_2 and CaF_2 had a slightly less effect on decreasing the interfacial tension than either NaF or KF, while AlF_3 had the least effect of the additives tested. These interfacial tension results do not coincide with the coalescence measurements made in this study. While NaF and KF were effective at promoting coalescence, MgF_2 was one of the poorer agents. CaF_2 with nearly identical changes to the interfacial tension as MgF_2 was more effective in droplet coalescence. The most effective coalescing agents, AlF_3 and Na_3AlF_6, made the least change in interfacial tension. The factor of interfacial tension, alone, cannot be used to explain a salt's coalescing power.

Another relationship that might have a bearing on a salt's coalescing ability is the solubility of alumina in the molten salt. Unfortunately, very little data on this aspect has been recorded in the literature. Cryolite, Na_3AlF_6, can dissolve up to 12 weight % alumina at 955°C. Pure sodium fluoride can dissolve about 5% alumina at 985°C; while CaF_2 can take as much as 25% alumina into solution, but only at temperatures of 1275°C. The decrease in alumina solubility of these fluoride salts matches their coalescing ability. It is also known[12] that the addition of NaCl to cryolite greatly reduces the solubility limit of alumina in cryolite. In fact, alumina is virtually insoluble in NaCl. Picard and co-workers[13] showed that in a LiCl-KCl eutectic salt mixture at 470°C, the solubility product for α - alumina is 10^{-44}. Additions of NaF increased the alumina solubility. It would appear that the solubility of alumina in the salt flux may be the determining factor in coalescence of aluminum droplets.

The hypothesis of alumina solubility seems valid for the cryolite additions tests as well. Cryolite is one of the best coalescing agents, especially for UBC, due to its ability to dissolve a portion of the oxide. Increasing amounts of cryolite in the chloride salt flux will increase the amount of aluminum oxide which can be dissolved. Once enough cryolite is present to strip away all of the oxide films, further additions produce no benefit.

Since there is not a complete oxide solubility data set available for verifying the hypothesis, the observations from this study should be useful for choosing fluxing salt additions.

Conclusions

While a profusion of salt flux additives have been proposed for use in the aluminum industry, only a few actually promote rapid coalescence of aluminum droplets. Chloride salts seem to have no coalescing ability under quiescent conditions. Chloride salts which contain moisture, may encourage slight coalescence due to the mixing effect caused by moisture reacting with metallic aluminum and the liberation of hydrogen gas. Fluoride salts are much more effective in promoting coalescence, especially those salts containing the AlF_3 compound, such as cryolite. Other sodium salts, such as the rest of the halides and the sulfates, do not enhance coalescence. Even small amounts of oxide particles can have a detrimental effect on coalescence under quiescent conditions, even when effective coalescing agents are present.

It may be that the coalescing effect of the fluoride salts can be explained by the salt flux's ability to dissolve the aluminum oxide film from the droplet surface rather than by other salt properties. Increasing the amount of a useful coalescing salt, such as cryolite or NaF, will initially increase metal recovery by suppressing metal losses in the dross due to incomplete coalescence. Above a critical addition value, more coalescing salt additions will not be beneficial.

References:

1. S. Rao, and P. R. Dawson, "A State of the Art Report on Secondary Aluminum Production with Particular Emphasis on Fluxes and Emission Control" (Report LR 359 (ME), Warren Spring Laboratory, 1980).

2. George J. Janz, Molten Salts Handbook (New York, NY: Academic Press, 1967).

3. G. J. Janz et al., "Physical Properties Data Compilations Relevant to Energy Storage. II Molten Salts: Data on Single and Multi-Component Salt Systems" (Report NSRDS-NBS 61, Part II, U.S. Department of Commerce, 1979).

4. L. Martin-Garin, A. Dinet, and J. M. Hicter, Journal of Material Science, 14 (1979), 2366-2372.

5. F. K. Ho, and Y. Sahai, "Interfacial Tension in Molten Aluminum and Salt Systems," Light Metals, 1990, ed. C. M. Bickert (Warrendale, PA: The Metallurgical Society, 1990) 717-720.

6. J. H. van Linden and D. L. Stewart, "Molten Salt Flux Composition Effects in Aluminum Scrap Remelting," Light Metals, 1988, ed. L. G. Boxall (Warrendale, PA: The Metallurgical Society, 1988) 391-398.

7. A. H. Sully, K. K. Hardy, and T. J. Heal, "An Investigation of Thickening and Metal Entrapment in Light Alloy Melting Flux," Journal of the Institute of Metals, 82 (1952), 49-58.

8. B. J. Racunas, "Recovery of Aluminum from Secondary Sources by Means of Agitated Molten Salt," (Paper presented at the Molten Salt Symposium, Sponsored by the Electrochemical Society, May 1976).

9. T. J. Johnston and R. D. Peterson, "The Role of Magnesium in Fluxing UBC," Recycle and Secondary Recovery of Metals, ed. P. R. Taylor, H. Y. Sohn, and N. Jarrett (Warrendale, PA: The Metallurgical Society, 1985) 417-428.

10. "JANAF Thermochemical Tables," (Report NSRDS-NBS 37, U.S. Department of Commerce, 1971).

11. H. Alan Fine and Gordon H. Geiger, Handbook on Material and Energy Balance Calculation in Metallurgical Processes, (Warrendale, PA: The Metallurgical Society, 1979).

12. K. Grjotheim, et al., Aluminium Electrolysis, (Düsseldorf, FDR, Aluminium-Verlag, 1982) 123-124.

13. G. F. Picard, F. Seon, B. Tremillon, and Y. Bertrand, "Effect of the Addition of Fluoride on Conditional Solubility of Alumina in LiCl-KCl Eutectic Melt," Electrochemica Acta, (25) 1980, 1453-1462.

INTERFACIAL PHENOMENA IN MOLTEN ALUMINUM AND SALT SYSTEMS

Francis K. Ho and Yogesh Sahai

Department of Materials Science and Engineering
The Ohio State University
Columbus, Ohio

Abstract

The remelting of aluminum scrap such as Used Aluminum Beverage Containers (UBC) is normally done under a cover of molten salt flux usually based on sodium and potassium chlorides. During the melting, the salt flux prevents oxidation of metal and may assist in the coalescence of molten aluminum alloys. The wetting characteristics and the interfacial phenomena between the melt and salt strongly influence the behavior of the dispersed metal in the salt matrix. In this paper, an apparatus for the measurement of interfacial tension of molten aluminum alloys salt systems by drop weight method is described. The interfacial tension results of molten aluminum and its alloys in salts of varying compositions are presented. The potential of salt flux composition on remelting of aluminum scrap is discussed.

Second International Symposium -
Recycling of Metals and Engineered Materials
Edited by Jan H. L. van Linden,
Donald L. Stewart, Jr., and Yogeshwar Sahai
The Minerals, Metals & Materials Society, 1990

Introduction

Recycling of Aluminum and UBC

Recycling of aluminum is economically sound and ecologically advantageous. Less than 5% of the energy required to produce aluminum ingot from its ore is needed to produce an ingot from recovered aluminum. In addition to this important energy saving, recycling aluminum has lower capital requirements for a given output of ingots (1). Over the past twenty five years, consumer aluminum recycling programs provided one of the most dramatic basic materials recycling success stories. Billions of UBC entered the recycling plants every year, recycled into new cans and other aluminum products. Recycling of this hard-to-melt aluminum scrap has gained more attention and the percentage of UBC recycled is progressively increasing (2). The continuing growth of UBC recycling is mainly the result of high value of aluminum, and the increasing concern of environment and energy conservation. These advantages are further enhanced by the fact that aluminum can be recycled and consumed again and again.

UBC Remelting

Usually the UBC first were crushed and shredded then processed at temperature within the melting range of the alloy for drying and delacquering. After this they were still in solid form but coated with different oxides. After screening and passing through magnetic separator, the UBC fragments were charged into the rotary salt furnace kept at much higher temperature (3). This resulted in suspended molten metal droplets in flux with different shapes and size distribution. The droplets were coated with high melting point metallic oxides, e.g. alumina, magnesium oxide and spinel ($MgO{\cdot}Al_2O_3$). During the remelting the metal drops coalesced with each other and grew bigger and finally were collected at the bottom of the furnace.

It had been noticed by Sully *et al* (4) long ago that during the remelting of aluminum alloy a thin and clear flux became progressively viscous and opaque, and eventually it must be renewed. A large amount of aluminum alloy globules were found entrapped in the so called the "spent" flux. West (5) suggested that the alkali halide mixture was used to attack the base metal at the metal/oxide interface. The oxide film was detached mechanically and removed as a suspension, rather than being dissolved into the flux.

The molten UBC alloy primarily has two major alloying elements, about 1.91% magnesium and 0.79% manganese (table 1). It is a mixture of UBC's end (AA5182) and its body (AA3004). The latter weighs about 75% of the whole can (3). In remelting, a salt flux based on an equimolar KCl-NaCl melt close to its eutectic composition was commonly used with cryolite or other halides as a cover of the molten metal. But considerable amount of fine droplets of metal was found entrapped in the "spent" UBC flux during the practice.

The recovery efficiency or metal yield of UBC recycling process depends upon the amount of the metal (in a fine separated form coated by different oxides) entrapped in the "spent" salt. As a matter of fact how many times the salt flux can be used for such purpose is also an important index for the cost and efficiency of the remelting.

Johnston and Peterson (6) carried out laboratory experiments and studied the metal yield in systems of equimolar KCl-NaCl plus 5% cryolite with both pot metal and UBC shreds. They found the high magnesium content in the UBC metal yield a low metal recovery. By X-ray analysis, double oxide spinel ($MgO{\cdot}Al_2O_3$) and α alumina (Al_2O_3) were found in the flux. In their scanning electron microscopy (SEM) and backscatter electron imaging (BEI) studies, it was shown that the entrapped aluminum dispersed in irregular shapes and wetted to the oxides in flux. This stoichiometric buildup of spinel in flux was substantiated by the depletion of magnesium in the metal charge.

It is understood that interfacial properties, such as interfacial tension and wetting characteristics between molten metal and flux, play an important role in the remelting process (4-6). The salt flux is applied to prevent the molten aluminum from reoxidation, but more importantly to "strip" or disrupt the oxide film on the metal drop surface into the salt and assist the coalescence of the metal drops.

Interfacial Variables of UBC Remelting

The interfacial properties such as interfacial tension between molten metal and the salt melt and contact angle between metal and oxide when immersed in salt depend upon the following process variables:

(1) Aluminum alloy (UBC) composition.

(2) Process temperature.

(3) Salt melt composition.

The aluminum and UBC compositions are shown in table 1. The interfacial tension is a temperature dependent property. For most systems surface tension and interfacial tension have negative temperature coefficients, i.e. they decrease with the increase of the temperature. This

Table 1: Composition and Melting Data of Aluminum and UBC Alloys

Metal	%Mg	%Mn	%Si	%Fe	Melting Range Approx.,°C
Al (99.999% pure)	-	-	-	-	659(mp)
AA5182	4.50	0.35	0.20	0.35	581-637
AA3004	1.05	1.25	-	-	629-654
UBC (Calculated)	1.91	1.03	0.28	0.61	-
UBC Melt (Alcoa Practice)	1.30	1.03	0.33	0.53	-
Alcoa UBC (This study)	1.91	0.79	0.03	0.04	-

is another way of showing that the surface entropy is positive (7). The key variable affecting the interfacial tension is the salt melt composition. In fact it depends on the choice of alkaline or alkaline earth halides. The composition effect on the interfacial properties is the direct result of interaction between all metal cations and halide anions in the salt melt taking place on the metal surface. As it has been stated before, the common practice is to use an equimolar or equal weight percent KCl-NaCl melt with the addition of different alkaline or alkaline earth halides. The contact angle values of the systems of interest are limited by the experimental difficulty. In addition, mechanical agitation can also influence the kinetic factors of oxide removal and drop coalescence.

Interfacial Tension Measurement

Measurement of molten metal surface tension at elevated temperature is difficult to carry out. White (8) did an exhaustive review on high temperature measurement of liquid metal surface tension and interfacial tension. Molten metal/flux interfacial tension measurements further limit the availability of the methods and increase the difficulty of measurement. Some· commonly used methods are (9):

(1) Drop shape method (10, 11): X-ray is applied to produce an image of a molten metal drop immersed in a second liquid phase. The interfacial tension is calculated from the image of the sessile or pendent drops.

(2) Maximum drop pressure method (9): The liquid metal is forced through a capillary with increased pressure until a drop is formed and detaches itself from the capillary tip immersed in the second liquid phase. The applied pressure is related to the interfacial tension.

(3) Capillary depression method (12, 13): The interface in a capillary tube is examined and related to the interfacial tension.

(4) Meniscus elongation method (14): A flat blade is used to depress the metal/flux interface and the force balance is related to the interfacial tension.

(5) Drop weight method (15, 16): The weight of a drop during its formation at an orifice and after its detachment in another liquid is measured and used to calculate the interfacial tension. This method is used in the present study.

Special remarks should be given to both the drop shape method and drop weight method. Only these two methods can examine the dynamic interfacial phenomena. *In situ* dynamic interfacial tension can be attained by drop weight method. A plot of interfacial tension versus time or sequential drop number can be used to analyze the nonequilibrium phenomena in such systems.

Theoretical Analysis

For removal of oxide film from the metal drop surface into the salt to occur, Sully *et al* (4) suggested that the following interfacial tension criterion must be satisfied:

$$\gamma_{metal/flux} + \gamma_{oxide/flux} < \gamma_{metal/oxide} \cdots\cdots(1)$$

Here $\gamma_{metal/flux}$, $\gamma_{oxide/flux}$, and $\gamma_{metal/oxide}$ stand for interfacial tension between molten metal and flux, interfacial tension between oxide and flux, and interfacial tension between molten metal and oxide, respectively. Different steps involved in oxide film removal and metal drop coalescence in salt flux are schematically shown in figure 1. According to Kozakevitch and Olette (17), the change in the Gibbs free energy of the system at constant temperature and pressure can be related to the change in the interfacial energies. Free energy changes of the system (oxide covered or naked metal drops in salt matrix), ΔG, associated with situations shown in figure 1 were described and calculated as follows:

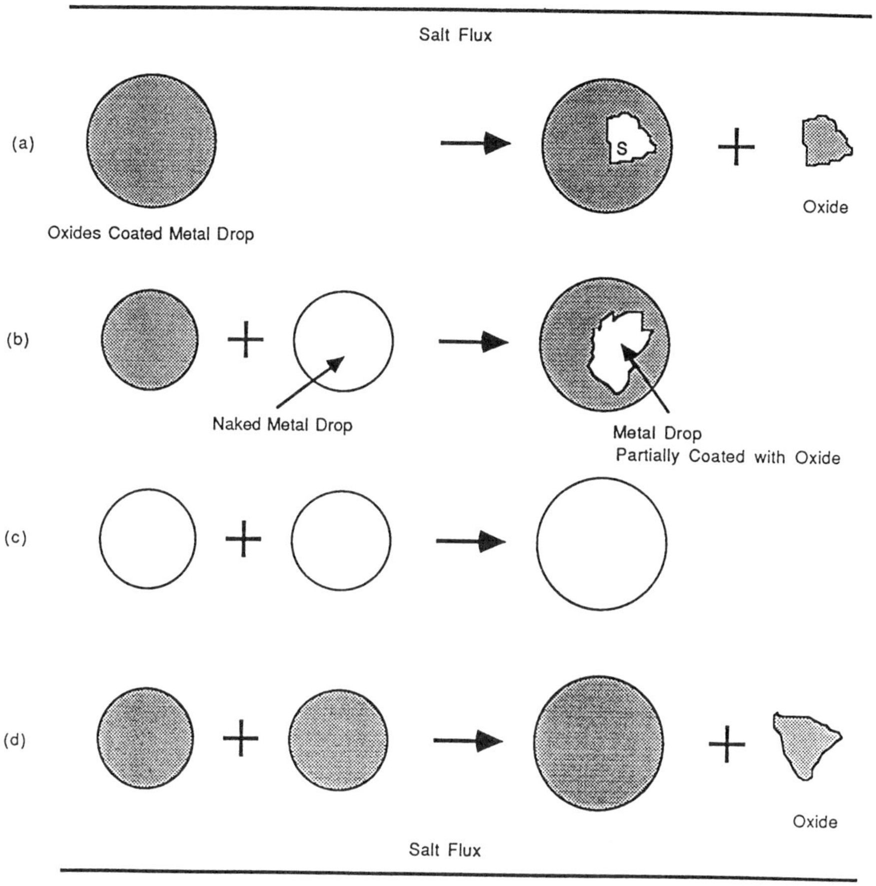

Figure 1: Schematic of oxide film removal and metal drop coalescence

Case (a):

A thin oxide layer of area S was "stripped" away from a metal drop coated with thin oxide film, leaving the drop to have area S in direct contact with salt.

$$\Delta G_a = S\,(\gamma_{metal/flux} + \gamma_{oxide/flux} - \gamma_{metal/oxide})\text{.....(2)}$$

When step (a) is spontaneous, it gives the above Sully's wetting criterion. Applying Young's relation into equation (2):

$$\Delta G_a = S\,\gamma_{metal/flux}\,(1 + \cos\theta)\text{.....(3)}$$

θ in equation (3) and hereafter is the contact angle between metal and oxide when immersed in salt flux. From equation (3), it can be seen that the removal of oxide film is not spontaneous. But larger contact angle, lower interfacial tension value between metal and flux, and smaller area of oxide film being removed will promote this oxide film removal.

Case (b) and (c):
In case (b), a fully thin oxide layer covered metal drop coalesces with a naked metal drop forming a larger drop partially covered with oxide. In case (c), two naked metal drops coalesce together forming a bigger drop in the salt flux, here r is the radius of the spherical metal drops before coalescence.

$$\Delta G_b \text{ or } \Delta G_c = -\ 1.65\ \pi\ r^2\ \gamma_{metal/flux} .....(4)$$

The coalescence is thermodynamically spontaneous. Higher interfacial tension values between metal and flux, and larger diameter of the drops facilitate these coalescence.

Case (d):
Two metal drops completely covered with thin oxide layer coalesce together forming a larger drop fully covered with oxide and leaving a piece of oxide suspended in salt flux.

$$\Delta G_d = 1.65\ \pi\ r^2\ \gamma_{metal/flux} \cos\theta .....(5)$$

In this case when $\theta \geq 90°$, i.e. metal does not wet oxide when immersed in salt flux, higher interfacial tension values between metal and flux, and larger diameter of the drops facilitate the coalescence. When wetting occurs between metal and oxide while immersed in salt flux, lower interfacial tension values between metal and flux, and smaller diameter of the drops should facilitate this coalescence.

From above it can be concluded that the interfacial tension between molten metal and salt melt and the contact angle in salt between the molten metal and oxide govern the thermodynamic spontaneity of oxide film removal and metal drops coalescence.

(1) The oxide film removal, is not spontaneous and is difficult. Larger contact angle between metal and oxide film and lower interfacial tension values between metal and flux are preferred for the oxide film removal. Whether molten metal wets the oxide ($\theta<90°$) or not (i.e. $\theta \geq 90°$), lower interfacial tension values between metal and flux will always be needed to promote the oxide film removal.

(2) Higher values of interfacial tension between metal and oxide facilitate the coalescence. Larger size drops coalesce more easily. One exception should be noted when molten metal wets the oxide ($\theta<90°$), for metal drops completely covered with oxide to coalesce (case d), the low values of interfacial tension between metal and oxide facilitate the coalescence and smaller size drops coalesce more easily. This indicates the oxide film removal is a necessity for coalescence. Once the oxide film is disrupted, higher interfacial tension would assist to coalesce the drops.

(3) External mechanical agitation should be applied to promote the difficult oxide film removal and also to help more oxide film disrupted drops coalesce each other.

Experimental

The drop weight method was first introduced and applied for surface tension measurements in organic liquids (18). It was later used by El Gammel *et al* (15) to measure interfacial

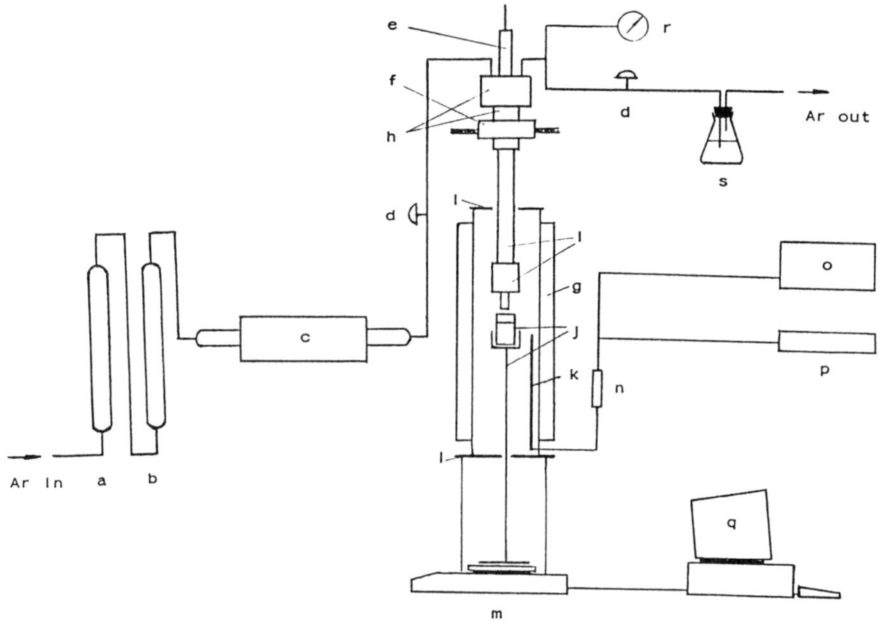

a Indicating-drierite trap

b Molecular-sieve trap

c Copper furnace at 580°C

d Needle valve/flow meter

e Piston moving unit

f Position adjusting unit

g Furnace

h Stainless steel cap/tube

i BN tube/tip

j Quartz crucuble/holder

k Thermocouple

l Baffle

m Electronic balance

n Electronic ice point

o Temperature controller

p Millivoltage meter

q PC

r Pressure gage

s Mineral oil tank

Figure 2: Schematic of drop weight method apparatus

tension in molten steel/slag systems. Lilius and co-workers (16) used it to investigate lead-lead silicate systems. The schematic of the apparatus is shown in figure 2. Metal sample was contained in a boron nitride tip with a small orifice (about 1.3 mm diameter) at the lower end. The boron nitride tip was immersed in the salt melt contained in a quartz crucible. The crucible was held in a gold plated, resistance heated quartz tube furnace. The crucible was kept on top of a quartz holder which rested on an electronic balance. The balance was connected to an IBM-PC for continuous weight change recording. A photograph of the main part of the apparatus is shown in figure 3.

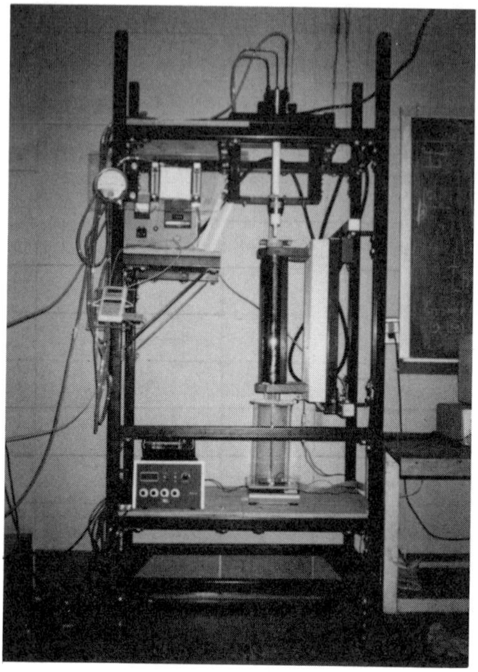

Figure 3: Photograph of the apparatus

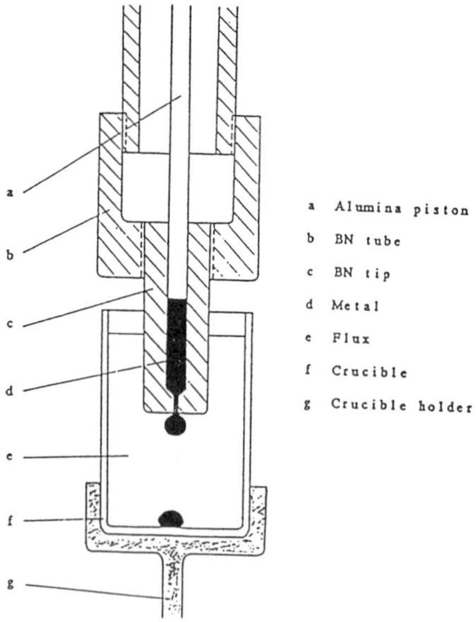

Figure 4: Schematic of metal drop formation in salt melt

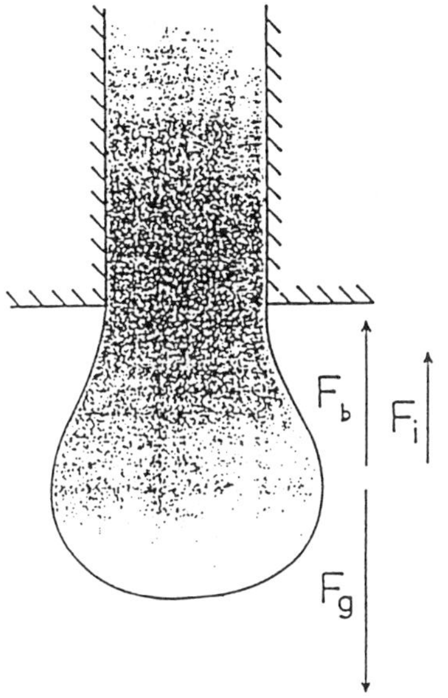

Figure 5: The forces acting on a detaching drop

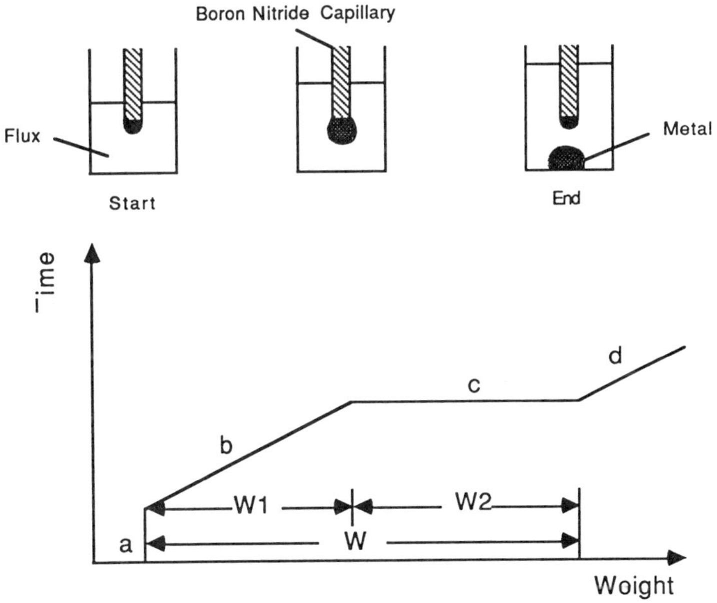

Figure 6: Typical weight change vs. time during formation and detachment of a drop

Figure 4 shows, schematically, a metal drop forming at the boron nitride capillary immersed in a quartz crucible. Figure 5 shows an enlarged capillary with drop and the forces acting on the drop. The force balance of the drop is affected by three forces: the interfacial tension force (F_i), the gravitational force (F_g), and the buoyancy force (F_b). Figure 6 shows a typical weight change versus time curve during the formation and detachment of a metal drop in molten salt. In figure 6, section b corresponds to the drop growth and section c shows the weight change when the drop is falling. The details of this method are described by several authors (15,16,18,19). The weight change during drop detachment is related to interfacial tension by Tate's law, and the interfacial tension can be expressed as:

$$\gamma_{metal/flux} = \frac{W_2\, g}{2\,\pi\, r_o\, \varphi} \quad .....(6)$$

Here W_2 is the weight change during the detachment of the drop and r_o is the radius of the orifice. In the calculation, a correction factor, φ, is needed since every time there is a small portion of the drop left at the tip of the capillary. This correction factor is a function of the detaching drop volume, radius of the tip orifice, and density of the molten metal. Its values can be obtained from tabulated or plotted values in the literature (20-22). In the present experiments, Wilkinson's extended values of correction factors were used ranging from 0.897 to 0.916.

In a typical experiment the molten aluminum was displaced by an alumina piston to form a drop at the boron nitride tip. The metal drop formation and detachment in the molten salt could be observed through the semitransparent furnace tube, while heated during an experiment. An inert atmosphere was maintained above the molten metal in the boron nitride cylinder by flowing purified argon gas. The experimental temperature was maintained at 740°C. Before every experiment, the salts and the boron nitride tip were dried at about 450°C for long period of time to eliminate any moisture.

Results and Discussion

The results of this work are presented in tables 2 and 3. Interfacial tension results with pure aluminum are given in table 2 where as table 3 gives interfacial tension results for the UBC alloy. These results are also plotted in figures 7 to 14, showing variation of interfacial tension with different amount of halide additions.

Figure 7 shows the effect of KF in lowering the interfacial tension between pure aluminum and salt flux. Similarly figures 8, 9 and 10 show the effects of NaF, LiF, and cryolite, respectively. In figures 7-12, it can be seen that addition of fluorides to the equimolar salt melt decrease the interfacial tension. Figures 7-10 also compare the present work with those of Martin-Garin *et al* (14) and Kurdyumov *et al* (24). The drop weight method provided the results in good agreement with those in literature.

Figure 11 shows the effect of AlF_3, cryolite and NaF in lowering the metal/salt interfacial tension. The results of CaF_2, MgF_2, LiF, NaF and KF additions to the salt flux are presented in figure 12. Finally, figure 13 shows results of $MgCl_2$, $CaCl_2$ and LiCl additions to the salt. From figures 11-13 it can be seen that all fluorides and chlorides addition decreased the interfacial tension between pure aluminum and the equimolar KCl-NaCl salt, except the addition of 1% $MgCl_2$.

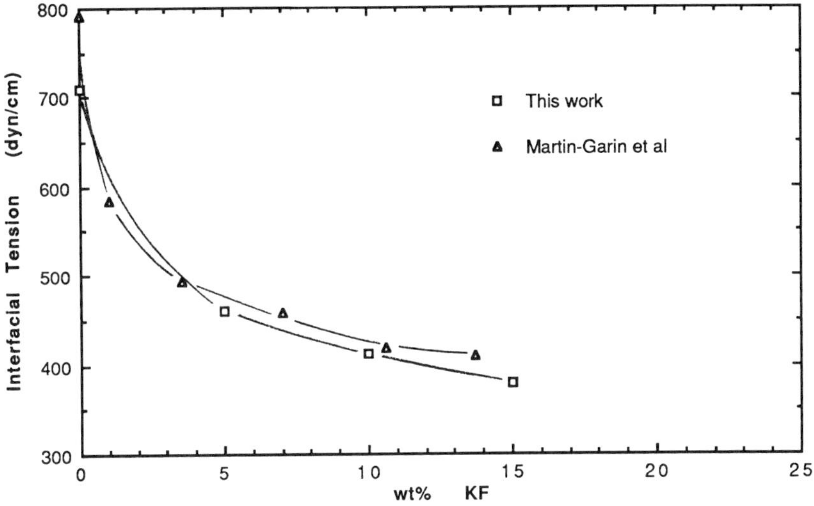

Figure 7: Variation of aluminum/equimolar KCl-NaCl salt interfacial tension as a function of potassium fluoride additions at 740°C

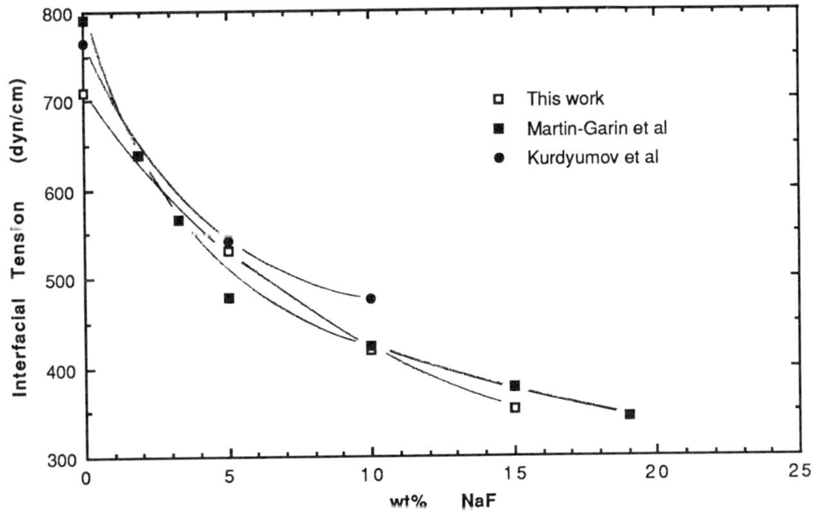

Figure 8: Variation of aluminum/equimolar KCl-NaCl salt interfacial tension as a function of sodium fluoride additions at 740°C

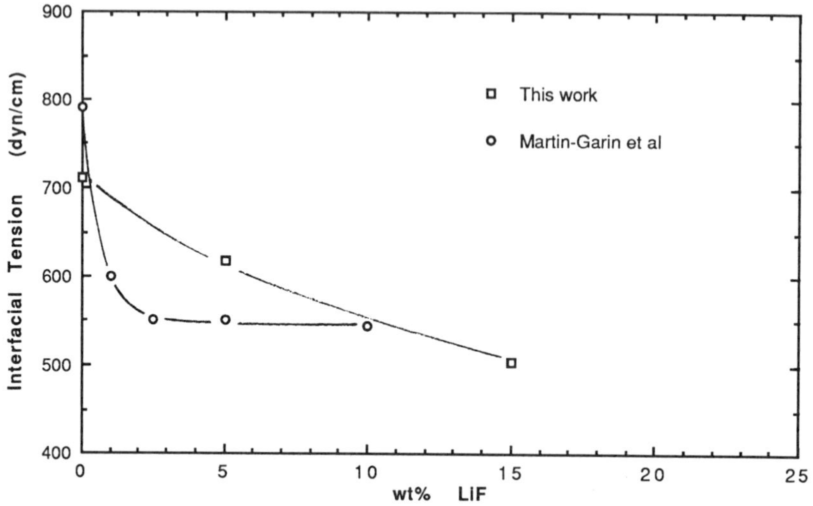

Figure 9: Variation of aluminum/equimolar KCl-NaCl salt interfacial tension as a function of lithium fluoride additions at 740°C

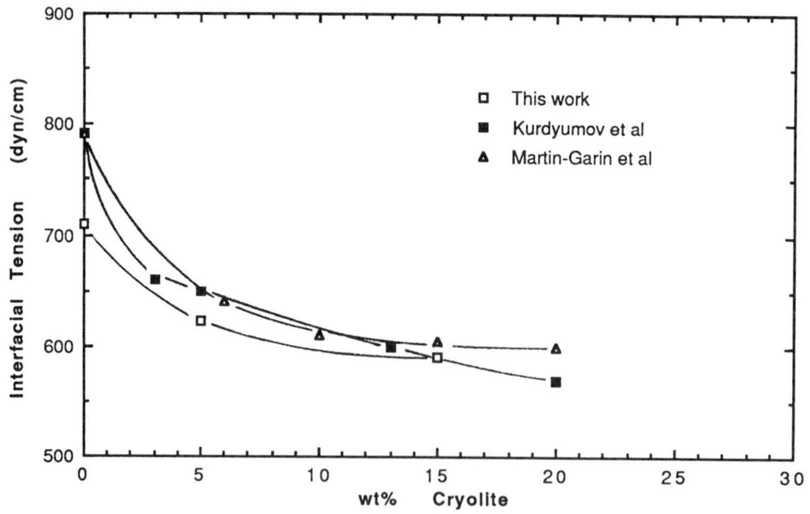

Figure 10: Variation of aluminum/equimolar KCl-NaCl salt interfacial tension as a function of cryolite additions at 740°C

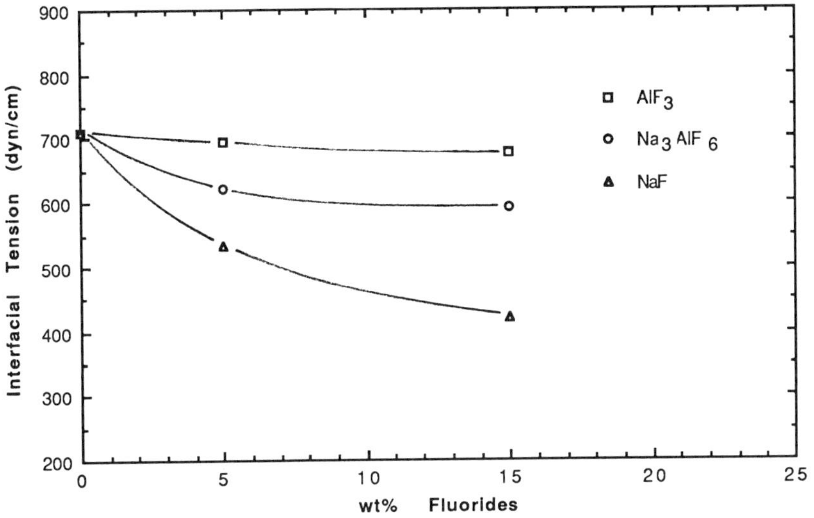

Figure 11: Variation of aluminum/equimolar KCl-NaCl salt interfacial tension as a function of NaF, AlF_3 or cryolite (Na_3AlF_6) additions at 740°C

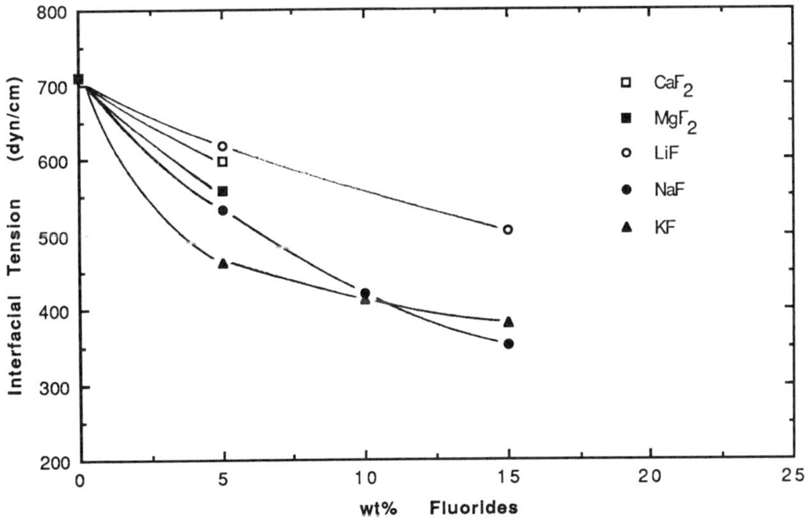

Figure 12: Variation of aluminum/equimolar KCl-NaCl salt interfacial tension as a function of KF, NaF, LiF, CaF_2, or MgF_2 additions at 740°C

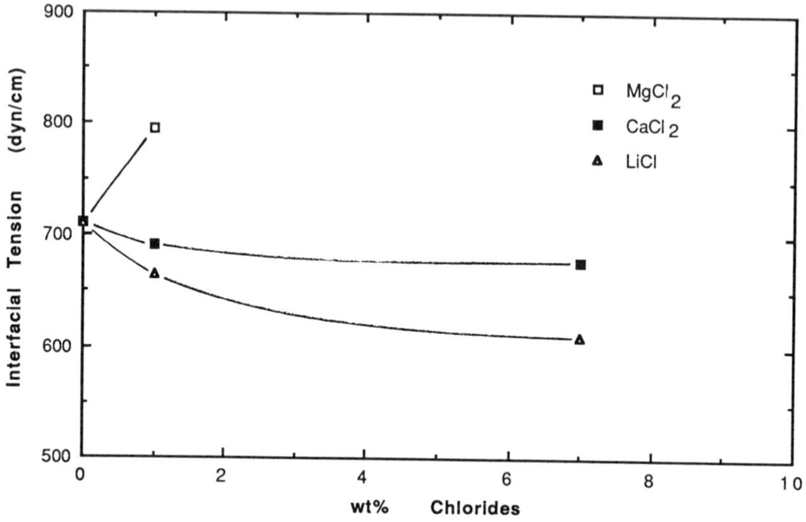

Figure 13: Variation of aluminum/equimolar KCl-NaCl salt interfacial tension as a function of (Mg, Ca or Li) chlorides additions at 740°C

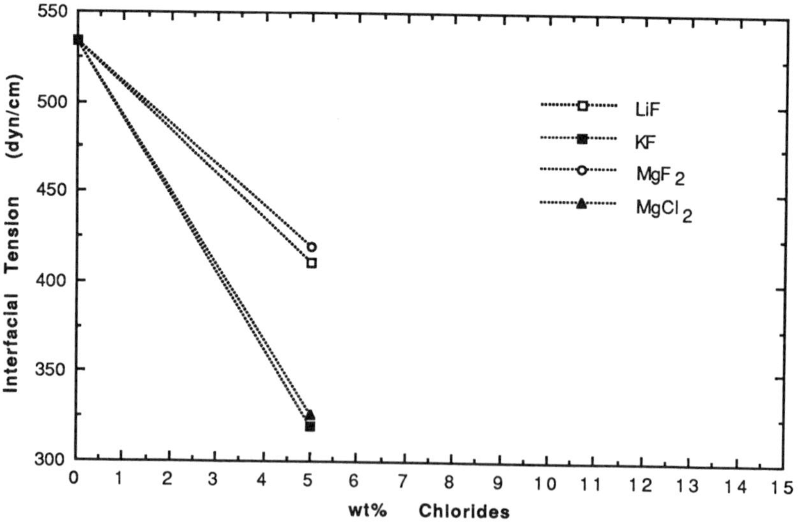

Figure 14: Variation of UBC alloy/equimolar KCl-NaCl salt interfacial tension as a function of halides additions at 740°C

Unlike pure aluminum which has an interfacial tension of 710 dyn/cm, UBC alloy has a lower interfacial tension (534 dyn/cm) with the equimolar KCl-NaCl salt (Figure 14). Addition of KF and LiF further lower this interfacial tension value.

Table 2: Interfacial Tension between Aluminum and Equimolar KCl-NaCl Salts with Different Halides Additives at 740°C by Drop Weight Method

Salt Melt Composition KCl-NaCl-X X	wt%	Interfacial Tension dyn/cm
None	0	710
NaF	5	532
NaF	10	420
NaF	15	352
KF	5	461
KF	10	412
KF	15	380
Na_3AlF_6	5	622
Na_3AlF_6	15	590
LiF	5	618
LiF	15	505
CaF_2	5	597
AlF_3	5	695
AlF_3	15	676
MgF_2	5	560
LiCl	1	665
LiCl	7	610
$CaCl_2$	1	692
$CaCl_2$	7	677
$MgCl_2$	1	796

Table 3: Interfacial Tension between UBC alloy and Equimolar KCl-NaCl Salts with Different Chlorides Additives at 740°C by Drop Weight Method

Salt Flux Composition KCl-NaCl-X X	wt%	Interfacial Tension dyn/cm
None	0	534
KF	5	320
LiF	5	411
$MgCl_2$	5	326*
MgF_2	5	420*

* Dynamic interfacial phenomena was observed and initial values are presented if equilibrium values are not available.

The following factors could change the interfacial tension:

(1) Metal atom adsorption on the metal/salt interface.

(2) Surface tension of the salt melt.

(3) Mass transport taking place on the metal surface.

Table 4: Surface Tension of Metals at Their Melting Temperatures (23)

Element	mp (°C)	Surface Tension (dyn/cm)	$-d\gamma/dT$ (dyn cm^{-1} deg^{-1})
Al	659	914	0.35
Ca	833	838	0.10
K	63	115	0.08
Li	453	398	0.14
Mg	649	559	0.35
Mn	1244	1090	0.20
Na	98	191	0.10
Si	1412	865	-
Zn	420	782	0.17

Dewing and Desclaux (12), Utigard and Toguri (10), proposed that the surface active elements like sodium are adsorbed on the molten aluminum surface in cryolite melts due to an increase in the sodium activity. The two factors which govern whether the addition of an alkaline halide will reduce or increase the interfacial tension are:

(1) The adsorption of the metal (Me) atoms on the aluminum surface which is related to the activity of the metal (Me) in equilibrium with the aluminum and the salt melt. The activity of Me is governed by the following reaction:

$$Al + 3/n\ MeX_n = 3/n\ Me + AlX_3 \quad .....(7)$$

where X is fluoride or chloride anion and n is an integer.

(2) The surface tension of the pure metal (Me) adsorbed on the surface (table 4).

The addition of a metal halide should reduce the interfacial tension when the activity of metal (Me) is increased and the pure metal, Me, has a low value of surface tension compared to pure aluminum. The additions of KF, NaF and LiF in equimolar KCl-NaCl melt, which decrease the interfacial tension, may have the following reactions resulting in adsorption of Na, K or Li atoms.

$$Al + 3\ NaF = 3\ Na + AlF_3 \ldots\ldots(8)$$

$$Al + 3\ KF = 3\ K + AlF_3 \ldots\ldots(9)$$

$$Al + 3\ LiF = 3\ Li + AlF_3 \ldots\ldots(10)$$

K, Na, and Li have lower surface tension than that of aluminum. K has lowest surface tension of the three (table 4). Only a very small amount of surface active atoms is needed to form a monolayer to cover the aluminum surface (10,12) and to reduce the interfacial tension. This probably can explain why the addition of KF decreases the interfacial tension most effectively. It can be seen in figure 12 that up to 10 wt%, KF is the most effective in reducing the interfacial tension which is followed by the NaF and LiF. These results are in accord with the surface tension values given in table 4. In figure 11, results of AlF_3, Na_3AlF_6 and NaF are presented. The addition of Na_3AlF_6 may undergo the following reaction:

$$Al + Na_3AlF_6 = 3\ Na + 2\ AlF_3 \ldots(11)$$

It can be seen that Na_3AlF_6 is more effective than AlF_3 but less effective in comparison to NaF in reducing the interfacial tension. The AlF_3 has almost no effect since no surface active metal atoms can be formed on the aluminum surface. The same mechanism can also explain the effects of CaF_2 and MgF_2 additions.

The interfacial tension value between the UBC alloy and the equimolar KCl-NaCl salt (figure 14) was found be much lower than that of pure metal. In this case, the magnesium in the alloy could act as a surface active element and lower the interfacial tension. The above mentioned mechanism still can be applied to explain the addition of KF and LiF in reducing the interfacial tension between UBC metal and the salt solution.

Table 5: Surface Tension of Molten Chlorides at Their Melting Temperatures (23)

Chlorides	mp(K)	Surface Tension (dyn/cm)
$MgCl_2$	987	67
$CaCl_2$	1045	148
LiCl	883	126

The surface tension of the chloride salts may play an important role in influencing the interfacial tension. It can be seen in table 5 that $MgCl_2$ has the lowest surface tension value. This may explain why 1 wt% addition of $MgCl_2$ (figure 13) increases the interfacial tension between pure aluminum and equimolar KCl-NaCl salt.

Conclusions

The drop weight method has been successfully used to measure the molten aluminum alloy (UBC) interfacial tension. Interfacial tension between pure aluminum and equimolar KCl-NaCl salt at 740°C was found to be about 710 dyn/cm. Interfacial tension between UBC alloy and equimolar KCl-NaCl salt was found to be about 534 dyn/cm.

Partial removal of oxide films from the completely oxide covered metal drops is an important prerequisite for the coalescence. The higher value of the contact angle between the metal and oxide immersed in salt and lower value of interfacial tension between metal and salt are preferred for the oxide film removal.

The effect of metal halide additions on lowering the interfacial tension can be used to evaluate a salt's power in removing the oxide films from the metal drops. Addition of halides, with the exception of $MgCl_2$, decreased the interfacial tension in pure aluminum cases. This decrease of interfacial tension could be caused by adsorption of low surface tension atoms on the aluminum surface. For interfacial tension between pure aluminum and salt, this effect decreases in the following order: $KF > NaF > MgF_2 > CaF_2 > LiF > Na_3AlF_6 > AlF_3$.

The interfacial tension is used to analyze the metal drop coalescence under different conditions. Higher interfacial tension value is required to promote the coalescence of oxide film disrupted drops.

Acknowledgements

This work was funded by the Aluminum Company of America. The authors would like to thank Dr.J. H. L. van Linden and Mr.Donald L. Stewart, Jr. of Alcoa Technical Center for the discussion and support of this program. One of the authors (F.K.H.) would like to acknowledge the assistance of Mr.Thomas Kelch on materials preparation using the facility in Physics Department at OSU.

References

1. The Aluminum Association, Aluminum Recycling Casebook (Washington, D.C.: The Aluminum Association, Inc., 1979), No.95, 5-12.

2. J. H. L. van Linden, "Aluminum Recycling from Junkyard to Boardroom," Recycle and Secondary Recovery of Metals, ed. P. R. Taylor, H. Y. Sohn and N. Jarrett (Warrendale, PA: The Metallurgical Society, 1985), 35-45.

3. K. A. Bowman, "Alcoa's Used Beverage Can (UBC) Alloy Separation Process," *ibid.*, 429-43.

4. A. H. Sully, H. K. Hardy and T. J. Heal, "An Investigation of Thickening and Metal Entrapment in a Light Alloy Melting Flux," J. Inst. Met., **82**(1953-54), 49-58.

5. E. G. West, Trans. Inst. Weld., **3**(1940), 93.

6. T. J. Johnston and R. D. Peterson, "The Role of Magnesium in Fluxing UBC," Recycle and Secondary Recovery of Metals, ed. P. R. Taylor, H. Y. Sohn and N. Jarrett (Warrendale, PA: The Metallurgical Society, 1985), 417-428.

7. M. Shimoji, An Introduction to the Physics and Chemistry of Metals in the Liquid State (London: Academic Press, 1977), 347.

8. D. W. G. White, "The Surface Tension of Liquid Metals and Alloys," Met. Rev., **13**(124), 1968, 73-96.

9. F. K. Ho and Y. Sahai, "Interfacial Tension Measurement of Liquid Metal/Flux Systems" (Report to Alcoa Technical Center, July, 1988).

10. T. Utigard and J. M. Toguri, "Interfacial Tension of Aluminum in Cryolite Melts," Metall. Trans., **16B**(1985), 333-38.

11. C. S. Kanetkar, A.S. Kacar, and D. M. Stefanescu, "The Wetting Characteristics and Surface Tension of Some Ni-Based Alloys on Yttria, Hafnia, Alumina, and Zirconia Substrates," *ibid.*, **19A**(1988), 1833-39.

12. E. W. Dewing and P. Desclaux, "The Interfacial Tension Between Aluminum and Cryolite Melts Saturated with Alumina," *ibid.*, **8B**(1977), 555-61.

13. L. H. van Vlack, Trans. AIME, **191**(1951), 251.

14. L. Martin-Garin, A. Dinet and J. M. Hicter, "Liquid-Liquid Interfacial Tension Measurements Applied to Molten Al-Halide Systems," J. Matls. Sci., **14**(1979), 2366-72.

15. T. El Gammal and R-D. Mullenberg, "Simultane Ermittlung der Dichte und Grenzflächenspannung von flüssigen Schlacken und Metallen," Arch Eisenhuttenwes, **51**(1980), 221-226.

16. K. Keskinen, A. Taskinen, and K. Lilius, "Interfacial Tension between Lead and Lead Silicates," Scand. J. Metall., **13G**(1984), 11-14.

17. P. Kozakevitch and M. Olette, "Rôle des phénomènes superficiels dans le mécanisme d'élimination des inclusions solides," Rev. Metallurg., **68**(1971), 635-46.

18. W. D. Harkins and F. E. Brown, "The Determination of Surface Tension (Free Surface Energy), and the Weight of Falling Drops: the Surface Tension of Water and Benzene by the Capillary Height Method," J. Amer. Chem., **41**(1919), 499-524.

19. F. K. Ho and Y. Sahai, "Interfacial Tension in Molten Aluminum Salt Systems," Light Metals, ed. C. M. Bickert (Warrendale, PA: The Metallurgical Society, 1990), 717-20.

20. J. L. Lando and H. T. Oakley, "Tabulated Correction Factors for the Drop-Weight-Volume Determination of Surface and Interfacial Tensions," J. Colloid Interface Sci., **25**(1967), 526-30.

21. M. C. Wilkinson and R. L. Kidwell, "A Mathematical Description of the Harkins and Brown Correction Curve for the Determination of Surface and Interfacial Tensions," *ibid.*, **35**(1971), 114-19.

22. M. C. Wilkinson, "Extended Use of, and Comments on, the Drop-Weight (Drop-Volume) Technique for the Determination of Surface and Interfacial Tensions," *ibid.*, **40**(1972), 14-26.

23. E. T. Turkdogan, Physical Chemistry of High Temperature Technology (London: Academic Press, 1980), 60-125, 179-199.

24. A. V. Kurdyumov, G. A. Grigoriev, S. V. Inkin, E. V. Vygoskii and B. I. Stepanov, Sov. Non-Ferrous Met. Res., **4**(1973), 211.

RECENT DEVELOPMENTS IN MOLTEN METAL PUMPING APPLICATIONS

IN ALUMINUM RECYCLING

David V. Neff
Metaullics Systems Division
The Carborundum Company
31935 Aurora Road
Solon, Ohio 44139

While molten metal pumps are well estblished in secondary aluminum smelting practice, recent expansion of UBC recycling and continued evolution of secondary processing has put greater emphasis on pumping productivity, reliability and quality than ever before.

Molten metal pumps are now used integrally with a number of scrap submergence systems with excellent results. Advances in drive systems, shaft configuration, and materials of construction have improved pump performace and reliability resulting in greater melting rates and production capabilities. The metals refining capability of gas injection pumping systems in lowering hydrogen content, reducing alkali metal impurity, and removing inclusions by flotation is well recognized.

This paper reviews various current molten metal pumping applications in secondary aluminum and UBC recycling operations.

Second International Symposium -
Recycling of Metals and Engineered Materials
Edited by Jan H. L. van Linden,
Donald L. Stewart, Jr., and Yogeshwar Sahai
The Minerals, Metals & Materials Society, 1990

INTRODUCTION

The recyclability of aluminum has contributed significantly to the industrial prowess of aluminum as a widely successful commercial material. Of course, it is well known that remelting a pound of aluminum scrap takes only 5% of the energy in which that pound was first produced through electrolysis. This factor certainly has created economic impetus for the recycling of aluminum. However, a variety of contributing technologies - collection, sorting, scrap preparation, and furnacing, have certainly also contributed greatly to making the recycling of aluminum the extensive business it is today.

Secondary recycling - the remelting of scrap from public consumption has been practiced for quite some time, and the use of centrifugal pumping systems in secondary recycling is common industrial practice in the secondary smelting industry. More recently, with expanded mill recycling - remelting of producer scrap - and most certainly the explosion in UBC recycling, centrifugal pumping systems have become firmly established in these remelting operations, also. Centrifugal pumping systems improve energy efficiency, productivity and quality in aluminum remelting.[1]

Remelting/recycling with molten metal pumping systems is principally enhanced through increased forced convectional heat transfer. This paper presents a current review on the use of such pumping systems in remelting/recycling operations.

Heat Transfer In Aluminum Remelting

Heat transfer occurs in aluminum melting conventionally by radiation, conduction and convection. Initial heat up of a cold charge is usually by radiation alone, i.e. a "reverberatory furnace."

Most reverberatory furnaces employed in the aluminum remelting industry are either rectangular or round. The latter are top charged and are primarily used by the primary producers and mill recyclers where scrap coils and sheet are charged. Secondary remelters use rectangular furnaces. The high radiation heat transfer and potentially high thermal heads (roof and sidewall temperatures) offer ample opportunity for a bath circulation system to improve the radiation heat flux into the bath by providing forced convection within the melt overcoming rather small natural convection forces. High velocity burners do aid somewhat in forced convection, but only as a surface phenomena at the solid charge and just a few millimeters into a molten bath.

In the melting of solid aluminum, approximately 500 BTU are required per pound to bring the solid to the melting point and to overcome a rather large latent heat of fusion. A forced convectional system greatly assists the heat transfer in two fundamental ways. Firstly, within the molten bath itself the greater forced convectional heat transfer permits a greater heat flux into the bath, through the surface, from the radiant heat source, namely the burners and reflectors (hence reverberatory)-roof and sidewalls. Without the ability of the pumping system to aid in this heat flux, the bath surface temperature and the thermal head temperatures would rise with continuing heat input. Hot spots within the melt itself, which retard further burner efficiency, can also be alleviated by a forced convectional system. When melting scrap this is especially important in minimizing oxidation and dross losses. Secondly, a forced convectional system within the bath allows solid charge to be melted much more quickly, both by heat transfer and by relative momentum or motion. This becomes especially important in the so called flat-bath melting stage, wherein "sunken solids" charge metal coexists with surrounding liquid metal at very nearly the same temperature (Fig. 1). There is, in this case, little thermal gradient to continue the melt down process at a high rate, hence further melting proceeds somewhat slowly with great inertia. A liquid metal pumping system assists in such "ablative" melting by virtue of increasing the kinetic energy at the solid-liquid interface, improving the rate of heat transfer and hence the melting rate.[2,3]

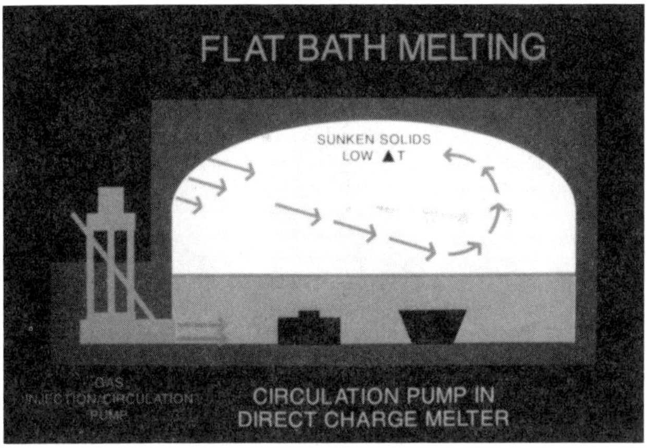

Figure 1, Schematic, Sunken Solids at Flat Bath Melting Condition

The heat transfer coefficient applied within a molten aluminum bath can be determined by conventional convective heat transfer theory which makes use of various dimensionless number relationships. The end result is that the heat transfer coefficient, h, is proportioned to a fractional power $^{(k)}$ of the induced fluid velocity:

$$h = f(v)^k \qquad (1)$$

For liquid metals, k is usually near 0.8.

There are several means by which forced convection can be applied to an aluminum melting furnace - stirring (either manually or electromagnetically), or pumping (electromagnetically or mechanically). Other devices such as fluxing tubes and porous plugs provide only minimal means of true convectional heat transfer. Indeed, such degassing devices can actually lower bath temperatures.

The main reason that mechanically induced forced convectional systems, as typified by the centrifugal pump (Fig. 2) have proven to be so effective in scrap melting is a result of the induced velocity. An electrically driven centrifugal pump has been measured to have a discharge velocity near 1000 ft./min. and a flow rate of 15,000 lb./min. By contrast, other devices, such as electromagnetic pumps and stirrers are often an order of magnitude less in equivalent discharge velocity, 100-200 ft./min.[4] Naturally then, the heat transfer coefficient, equation 1, is considerably greater for devices with higher velocities, such as the centrifugal pump.

Figure 2, Centrifugal M-30 Circulation Pump

Consequently a centrifugal mechanical pumping system is to be preferred due to its greater impact on forced convectional heat transfer as influenced by the higher heat transfer coefficient.

Metals Refining

There are other functions that molten metal pumping systems perform in recycling/remelting other than providing forced convectional heat transfer. Metals refining - demagging, alkali removal, degassing - are significant melt treatment practices often necessary in remelting and which are uniquely provided by centrifugal pumping systems versus other technologies.

Demagging

Demagging is commonly practiced in secondary smelter recycling of consumer scrap. Often the available scrap charge mix includes a predominance of wrought scrap products whose overall magnesium content exceeds that required for producing specification die cast or foundry alloy ingot. Hence, the melt often must be "demagged". The use of chlorine injection has been found most beneficial; chlorinating a molten aluminum alloy containing magnesium produces a reaction which ultimately results in the formation of a liquid magnesium chloride which usually separates to the surface or dross phase given enough residence time.

Historically, this reaction has been accomplished by flux tube injection, the use of salt fluxes (Derham process), or the scrubber bell process. These techniques were generally found wanting, either possessing somewhat low reaction efficiency, environmental or disposal problems. The favored technology for demagging within the secondary industry employed today is the gas injection pump (Fig.3).[5] This device is commonly used in reverberatory furnace remelting of consumer scrap. Key benefits include operating at near theoretical efficiency of chlorine usage with virtually no effluent (free aluminum chloride and its derivatives) released to the atmosphere when properly operated. The high demagging efficiency and non-polluting nature of this technology is a result of high operating speeds of the pump (600-1100 rpm) wherein the high velocity discharge from the impeller shears the incipient gas injection just ahead of it, thereby creating an extremely fine bubble size with high bubble surface area-to-volume ratio, for best gas-liquid reaction kinetics. Hence demagging can proceed very efficiently. Figure 4 depicts a typical demagging curve for a 150,000 lb. melt (75,000 lb. heel) with continuous charging of high magnesium bearing scrap.

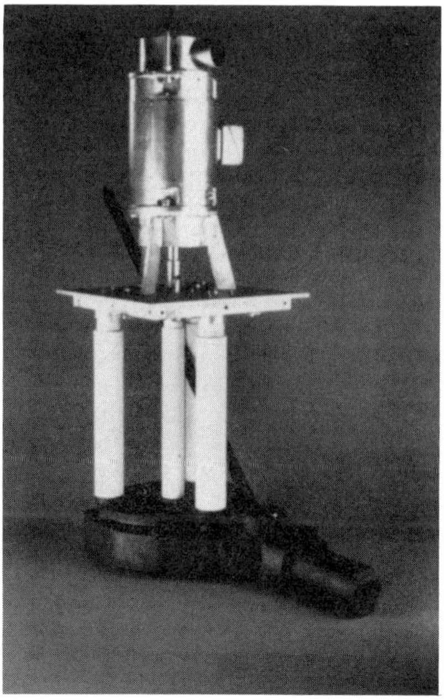

Figure 3, Electric Drive M-30 Gas Injection Pump

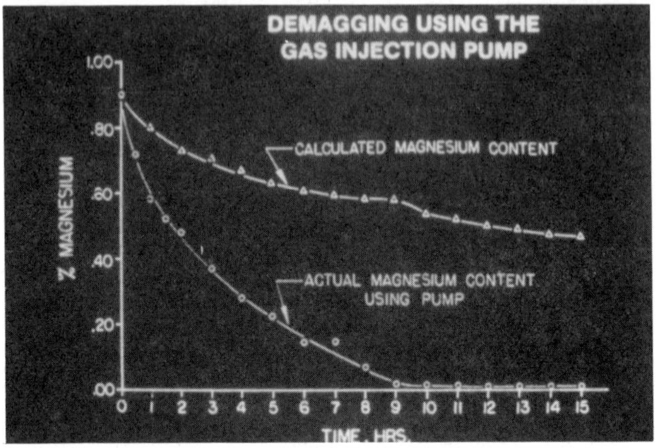

Figure 4, Demagging a 150,000 Lb. Secondary Alunminum Melt

Most mill or producer scrap is usually remelted to the same chemical specification and hence demagging is not generally employed here. Likewise, UBC recycling and the growth of dedicated facilities committed to regeneration of UBC can body and lid stock alloys usually do not require demagging. Indeed, economics virtually demand that beverage can alloys be reproduced from the same high magnesium alloys, with only alloy "sweetener" added in small amounts to maintain preferred composition.

Alkali Removal

Certain impurities other than magnesium often exist in mill producer scrap, and in consumer scrap earmarked for remelting. The alkali metals lithium, sodium, potassium, and calcium may enter the scrap stream as a result of original primary production, or later casthouse processing. Primary producers often do deliberately refine lithium, sodium and calcium from most melts, using chlorine injection treatment of potline vessels and/or in-line degassers, but it must be remembered that this is relatively recent emphasis. Varieties of consumer scrap entering the remelt arena after 20-40 years of service were not originally refined in the same manner and hence such contamination is possible.

The gas injection pump is suitable for continuously refining alkali metal impurities from scrap metal during the remelting process. Figure 5 represents typical refining of lithium from a mixture of sow, mill scrap, and customer scrap in a 100,000 lb. reverberatory furnace charge.

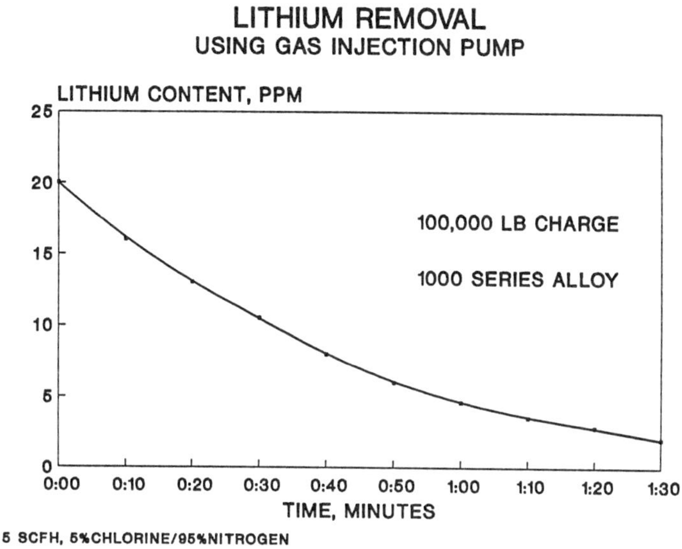

Figure 5, Typical Refining of Lithium

Degassing

Secondary smelting of consumer scrap, remelting of good quality producer scrap, and melt recycling of UBC usually require extra measures of melt cleanliness to be taken. Non-metallic inclusions or hydrogen gas pose probable problems in the end product. Hence degassing and filtration are usually employed, often in various in-line systems positioned between furnace and casting pit. Other degassing measures are often used in the furnace as well. Hydrogen may be removed from the melt in the furnace using hexachloroethane tablets, flux tubes, or with the gas injection pump. The latter provides degassing in-situ during melting and avoids the need for opening doors or a separate degassing period at the end of the heat. The bubble-shearing action of the rotating impeller permits effective degassing throughout the heat with typical 5% chlorine, 95% nitrogen mixtures. The chlorine addition helps wet non-metallic inclusions and deliver them to the surface, so providing additional melt cleaning. A period of bath quiescence or holding time is desirable to allow a majority of inclusions in the bath to separate by sedimentation (heavier particles) or flotation (lighter particles), two "natural" filtering phenomena.

In-furnace degassing with the gas injection pump can result in final gas levels near 0.15ml H_2/100g, thereby facilitating or even eliminating the use of a separate in-line degasser. Performance of in-line degassing and filtration, if practiced, are usually enhanced by the preliminary cleaning treatment in the furnace.

Benefits of Mechanical Pumping Systems in Remelting

The benefits of centrifugal pumps in secondary aluminum recycling have been well described recently.[1,2,5] Improvements in energy efficiency, productivity, and melt rate are well documented, briefly summarized in Table I. Energy efficiencies depend of course on the starting point. Several years ago it was not uncommon for energy differences of 3500 BTU/lb. (without a pump) and 2000 BTU/lb. (with a pump) to be experienced.

TABLE I

General Benefits of Adding Centrifugal Pumping Systems to Scrap Remelting Processes*

Charge Rate	Increased up to 100%
Melt Rate	Increased 25-50% (to 16,000-18,000 lbs./hr.)**
Energy Efficiency	To 1600-1800 BTU/lb. (20-50% improvement)
Productivity	As much as 100% improvement
Temperature Stratification	Reduced to 10° F. or less throughout
Thermal Heads	Roof temperature reduction - 2300-1900° F. Sidewall temperature reduction - 1900-1650° F. Bath surface temperature reduction - 1600-1400° F.

*Based on 100,000 lb. or greater furnace size.
**Also dependent on burner capacity.

Today even a good energy efficient secondary operation starting at 2200 BTU/lb. can generate a reduction to 1600-1800 BTU/lb. (50% heel, cold charge) by adding a centrifugal pumping system. Even such modest reduction of 10-15% energy savings make pumping system installation cost-justifiable.

Melting rates are highly dependent on thermal input capacity to the furnace and on specific operating circumstance. It is not uncommon to achieve 25-50% increase in melt rate. Most secondary aluminum 100,000 lb. well-charge furnaces experience 12,000 to 18,000 lb./hr. melting rates with a variety of scrap metal feeds.

Most of these foregoing results apply to secondary smelting of consumer scrap, with perhaps some produce scrap mixed in. The usual heterogeneous nature of secondary scrap - clips, chips, crushed castings, architectural scrap, fabrications, etc., plus unknowns, places a burden on pumping operations to handle such a diverse scrap mix, especially when the scrap is not consolidated by briquetting or baling. Nevertheless, pumping systems perform admirably well and are very cost-effective in secondary scrap smelting operations.

The mill products recycling facilities have a higher class of controlled scrap, (i.e. coils, extrusions, plate, sow, etc.) and where these facilities are located within primary plants, there is a source of hot (potline) metal as well to make up the furnace charge. Melt rate, energy efficiencies, etc. are usually a notch above those for the secondary operations, principally because of the better controlled, narrower scrap feed mix. Likewise, pump performance and longevity are also improved.

Table II describes pumping performance in one such dedicated UBC recycling facility, IMSAMET, Hauser, Idaho. Of special note is that in this type of dedicated remelting:

- melting rate is considerably higher (22,000 lbs./hr.) than for conventional secondary melting.
- pump performance life is very good (35 days average) without parts replacement
- overall pumping costs are very modest

TABLE II

Centrifugal Molten Metal Pumps in UBC Recycling at IMSAMET, Hauser, Idaho

Pump Type	4-M-42 (Metaullics Systems)
Number in Continuous Use	2
Melting Rate	22,000 lb./hr. maximum
Average Metal Temperature, Charge Well	1375° F. (18,000 lb./hr.)
Average Metal Temperature, Furnace Interior	1400° F.
Furnace Firing Capacity	28,000,000 BTU
Energy Efficiency (Hot UBC)	1150 BTU/lb.
Energy Efficiency (Cold UBC)	1650 BTU/lb.
Pump Life (on-line time)	35 days average
Average Monthly Pump Maintenance Cost	$4,520 total (4 pumps)

UBC recycling, or dedicated, homogeneous scrap remelting, thus represents the ideal basis for pumping systems to provide enhancement upon, especially from the standpoint of pump service life. There is considerable economic benefit in melting performance with very modest capital and operating cost, all of which provides continued impetus for the successful commercial recycling of UBC materials.

Recently there have been a number of molten metal pump engineering design improvements being explored which promise even greater performance capabilities and reliability. These include a torque coupling system which limits torque "spikes" and lessens premature shaft failure; new shaft adaptor coupling and alignment techniques, and special shaft/-impeller designs. In addition, ways to increase the amount of gas injection possible are also under development.

Conclusions

Mechanical centrifugal pumping systems are well suited to enhance forced convectional heat transfer in remelting/recycling of scrap aluminum. The improved heat transfer greatly assists the smelting of consumer scrap by the secondary aluminum industry, remelt of producer scrap by the mill products producers, and the recycling of UBC in dedicated facilities.

Centrifugal pumping systems increase the melt rate, reduce energy consumption, improve productivity and reduce thermal heads in furnace refractory, aiding in longer furnace life. Most importantly, centrifugal pumps enhance melt metallurgical quality by homogenizing bath temperatures, overcoming temperature stratification, and creating greater uniformity in alloy composition.

In the remelting of a homogeneous scrap source, such as with processed UBC, pumping performance can assist in achieving sustained melt rates greater than 20,000 lbs./hr. Pump longevity is extended significantly with homogeneous UBC feed metal and with proper on-line maintenance, such that continuous usage averaging 35 days or more is possible without necessitating pump replacement or repairs.

References

1. D. V. Neff, "Molten Metal Pumping Systems - Current Applications and Benefits", Light Metals, AIME, 1987, P. 805

2. D. V. Neff, "Molten Metal Pumping and Pipeline Transport in Aluminum Casthouses" Proceedings, Energy Conservation Workshop X, Aluminum Association, 1987, P. 145

3. T. Cunard, "Application of Graphite Pump to Direct Charged Aluminum Melting Furnace", Proceedings, Energy Conservation Workshop X, Aluminum Association, 1987, P. 159

4. P. J. Bamji, F. W. Pierson, "Electromagnetic Circulation of Molten Aluminum", Journal of Metals, AIME, 1985, P. 44

5. D. V. Neff, "The Use Of Gas Injection Pumps in Secondary Aluminum Metal Refining" Conference Proceedings, Recycle and Secondary Recovery of Metals, AIME, 1985, P. 73

FLUID FLOW PROBLEMS IN ALUMINUM REMELTING FURNACES

J. Szekely and O.J. Ilegbusi

Massachusetts Institute Of Technology
Department of Materials Science and Engineering
Cambridge, Massachusetts

A major portion of the aluminum that is being continuously cast is melted in gas-fired reverbatory type furnaces. The efficient utilization of fuel in these systems depends critically on the fluid flow and mixing in the molten regions of these furnaces. The furnace geometry (shallow, but wide pool) is far from ideal for providing conditions conducive to good mixing or vigorous recirculation. An analysis is presented of this problem, exploring possible alternatives to improve mixing in these systems.

Second International Symposium -
Recycling of Metals and Engineered Materials
Edited by Jan H. L. van Linden,
Donald L. Stewart, Jr., and Yogeshwar Sahai
The Minerals, Metals & Materials Society, 1990

COPPER/SUPERALLOY SCRAP

Session Chairman:
R. Bruce Tippin

A COBALT COMMODITY RECYCLING FLOW MODEL

By Robert C. Gabler, Jr.

Albany Research Center
U.S. Bureau of Mines
1450 SW Queen Avenue
Albany, Oregon 97321

Abstract

World resources of cobalt appear ample for the foreseeable future, however, political and economic events have caused concern over the availability and reliability of an uninterrupted supply. Because of the concern, Congress requested that the Bureau of Mines initiate a study to model the flow of cobalt in the economy that would delineate and quantify the production areas where metal values are lost. A major attribute of the model is that it be amenable to updating as supply, demand, or production data change. This report follows the flow model of cobalt through its metallurgical and chemical applications and high-lights areas where significant losses occur due to downgrading, export, or disposal. The study indicates that about 1,967 Mg of cobalt-bearing materials are lost each year, representing 24 pct of U.S. domestic demand. (Mg is an SI unit that equals 10^6 g or the commonly used term "metric tons.")

**Second International Symposium -
Recycling of Metals and Engineered Materials
Edited by Jan H. L. van Linden,
Donald L. Stewart, Jr., and Yogeshwar Sahai
The Minerals, Metals & Materials Society, 1990**

Introduction

Cobalt is an indispensable element in a variety of critical and strategic applications such as superalloys, permanent magnets, cemented carbides, alloy and tool steels, catalysts, and chemicals. Cobalt has few viable substitutes in most of its major end uses. The United States currently is entirely dependent on foreign sources for its primary cobalt demands. Although world resources of cobalt are ample for the foreseeable future, political and economic events have raised questions about the uninterrupted availability and reliability of supplies of this commodity. The largest suppliers of cobalt are the central African countries of Zaire and, to a lesser extent, Zambia. In a 1983 publication, the National Materials Advisory Board (NMAB) stated that, future chances of disruption of supply from these sources are very high" (1). There have been no disruptions of supply prior to and since the NMAB publication, and current opinion of the Bureau of Mines' commodity specialists is that there is little if any probability of disruptions in the future (2).

Various literature studies have reported the strategic importance of a number of commodities, their availability, and their vulnerability to interrupted supplies (1,3-9). Most of these studies are either out of date or do not supply information on losses of the strategic and critical commodities due to waste disposal, downgrading, and/or exportation.

Industry officials in the production, manufacturing, and recycling of strategic and critical commodities are reluctant to discuss the quantities of scrap that go to disposal, downgrading, and export. This information is often considered to be proprietary in nature.

Even when data are available, there is confusion between disposition of materials and recycling. For example, in a survey of 7 scrap recyclers in 1987 (10) the following results were obtained for the superalloy industry:

1. Estimates ranged from 60 to 100 pct for domestic recycling of superalloy scrap;

2. Estimates ranged from 0 to 40 pct for the amounts of superalloy scrap exported;

3. Estimates ranged from 75 to 100 pct for the amount of recycled superalloy scrap that is remelted domestically;

4. Estimates ranged from 38 to 90 pct for the recycle of superalloy scrap to the same superalloy;

5. Estimates ranged from 5 to 38 pct for the recycle to other high-value superalloys;

6. Estimates ranged from 10 to 25 pct for downgrading; and,

7. Estimates ranged from 3 to 5 pct for the discard of waste.

These estimates are typical of the data on processing, manufacturing, disposal, downgrading, and recycling of strategic and critical materials. Therein lies the difficulty in determining the amounts of critical and strategic commodities that are discarded, downgraded, or exported and the reasons for conducting this study.

In 1980 the Bureau of Mines reported the results of contracted studies by Inco Research and Development Center, Suffern, NY, and Arthur D. Little, Inc., Cambridge, MA, to assess the domestic availability of chromium in scrap and the amount that was recycled (11, 12). These reports contain data up to 1978; no updates of these studies were conducted following their publication. In 1983, the National Materials Advisory Board (NMAB) published a study of the flow of cobalt in all end uses and discussed the amounts of downgraded scrap and waste produced by the cobalt industry (1). The report also discussed the need for improved recycling of cobalt-containing scrap.

In May, 1985, the U.S. Congress, Office of Technology Assessment (OTA) produced a study entitled, "Strategic Minerals: Technologies to Reduce U. S. Import Vulnerability" (13). Included in the study was set of recommendations to the Bureau of Mines to conduct a survey of recycling-related activities. OTA recommended a scrap recycling study to update the chromium scrap metal information contained in 2 Bureau of Mines publications (11, 12), as well as to expand the scope of the previous studies to include information on scrap generated by cobalt, manganese, and the platinum-group metals industries as well as the wastes generated by all the various industrial and Department of Defense (DOD) users of these four commodities.

In 1987, the Bureau of Mines initiated this study of the four commodities. The objective of the study is to produce a commodity-orientated structural model tracing flow, recycling, and final disposition of the four identified commodities, which can be updated in subsequent years. A second objective of the study is to provide, in an understandable manner, an overview of the commodity flow that can be used by Congress and industry associations as a tool to help in the study of that commodity's vulnerability to political and availability factors outside the control of the United States. Another objective of the study is to be able to highlight significant commodity loss areas where further research is required. This will enable the development of project proposals to address specific recycling areas.

To meet the objectives of the study a hierarchical model was developed for cobalt and the data for cobalt are presented here. This report consists of a general description of the model and its components. This particular model is a generic model that can be applied to other strategic and critical commodities.

General Description

The cobalt commodity flow model is composed of a hierarchical structure and a materials balance for each component of the hierarchy. An overview of the commodity flow model is shown in figure 1. The model structure provides a means of organizing the overall industry data and of estimating material flow and waste generation by subcomponents of the industry. The entire commodity industry is represented in the first level of the structure with subsequent consumers in the lower levels of the structure. This hierarchical structure lends itself to a coding system for identifying the industries and their relationships. The cobalt industry, for example, splits into eight industries at the highest level: superalloys; magnetic alloys; cemented carbides;[1] stainless steels, heat-resistant steels and other alloys; tool steels; miscellaneous and unspecified industries; cobalt catalysts and other chemicals.

The first level of industry components would thus be coded: Co.1 for superalloys, Co.2 for magnetic alloys, Co.3 for cemented carbides, etc. Under cobalt superalloy processing would be cobalt parts manufacturing - Co.1.1. A new digit is added for each descending level in the hierarchy. For example, the end use industries for cobalt superalloys would be Co.1.1.1 - jet engine turbine blades, Co.1.1.2 - compressor rings, etc.

[1]Cemented carbides are used for cutting tools and wear resistant materials such as dies.

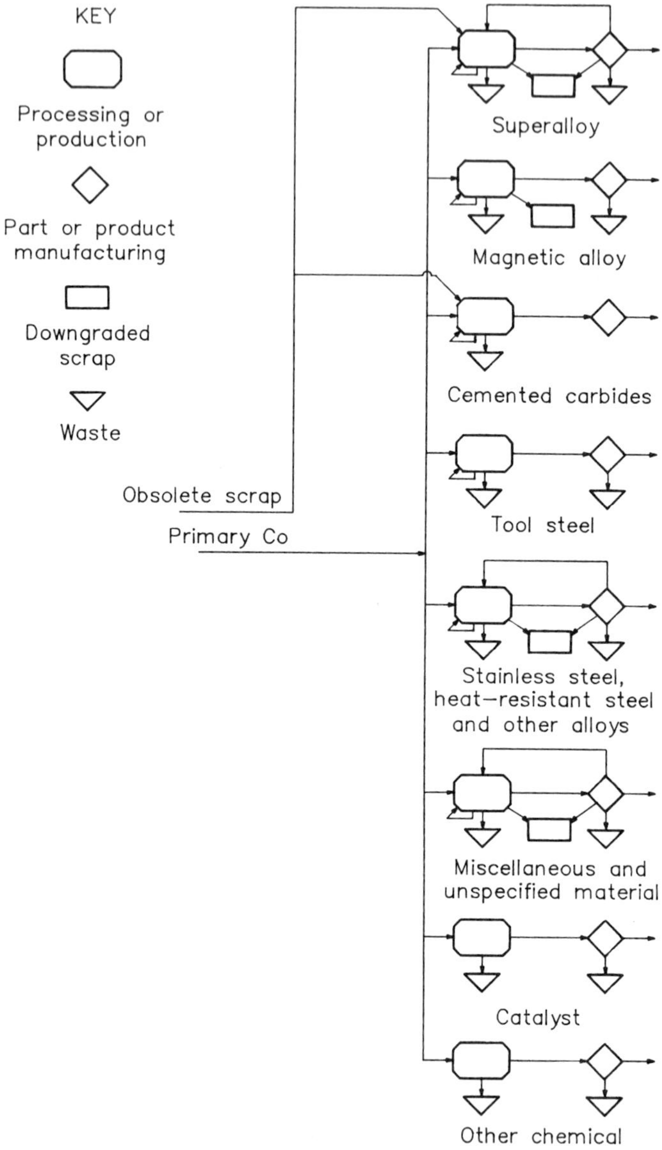

FIGURE 1 - The cobalt commodity flow model.

With adequate input data, this system permits one to simply, quickly, and unambiguously identify the subcomponents of any component of the cobalt industry. The system permits the identification of cobalt consumption or waste by industry component simply by adding all waste or consumption for the subcomponents. For example, to list all the waste by the cobalt catalyst industry, one simply adds the waste by all components whose label starts with "Co.7." If one wants to know the production of scrap and waste in the cobalt superalloy parts manufacturing industry one adds the scrap from all components whose label starts with "Co.1.1."

The data organized by the hierarchy structure include material balances by industry. This system permits "upward" and "downward" calculation of consumption and waste quantities. That is, using this structure one can distribute domestic consumption of cobalt over the numerous industries that consume cobalt. This is "downward" calculation. One could also start by calculating consumption for the lowest components and add to get total domestic consumption. This is "upward" calculation. Using both upward and downward calculation provides a method for checking self consistency. Using upward and downward calculation also permits the estimation of limits on data that is not otherwise attainable. For example, if 85 pct of the companies in an industry report data and the total consumption in this industry is known, then the estimates of the contributions of the other 15 pct of the companies can be calculated by difference.

Minimally, for the purpose of identifying scrap and waste generation sources, recycling routes, and recycling research project potential, each component should have a quantified cobalt input and output, in contained weight and in gross weight by material. The difference between input and output in contained weight is identified waste. One can balance input and output in contained cobalt units to determine whether identified waste streams are significant.

Each industry material balance is the result of material balances for individual processes. For example, in the production of superalloys, a detailed materials balance should be produced for each superalloy production operation. It is likely that each company has a different furnace, uses a different melting procedure, produces to different chemical specifications, recycles differently, and has different economic restrictions. All of these factors affect the fraction of input materials that reports to output product, recyclable material, or waste. It is at this level that information on recycling potential is most valuable. It is at this level that individual waste streams and scrap that leave the production process are identified and characterized (quantity, concentration, contamination, etc.), and that identified waste is quantified.

With the hierarchy structure of the industry, the input-output materials balance for each component, and the detailed materials balance for each process that makes up an industry component, a flow model for commodity data collection and organization has been developed. The base of this system is the process material balance. The process material balances are then aggregated to form the industry material balance. The industry components are organized by the hierarchy structure which identifies material flows among industry components.

The model developed here for the study of scrap and waste generated by cobalt processing and consuming industries is a "blueprint" or guideline for organizing a logical study of cobalt flow. The waste and scrap generated through the production and use of cobalt materials is identified through application of the material balance concept. This model divides the cobalt industry into smaller parts which should make data collection and analysis as well as data revision more manageable. Probably there are problems that will arise in application of this model, for example, some companies may be reluctant to provide data. But the model is general and flexible enough to be adapted to these problems.

Industry Flow Models

Neither the reported consumption nor the apparent consumption of cobalt in any given year necessarily equals the actual consumption of cobalt in that year. "Reported consumption" is that data reported to the Bureau of Mines in response to the Bureau's industry consumption survey questionnaires. The "apparent consumption" equals primary production plus production from old scrap plus (imports - exports) plus (beginning inventories - ending inventories). Apparent consumption is closer to actual consumption, but reported consumption provides a breakdown by end use.

In developing the overall cobalt flow models, the following assumptions were made:

1. The difference between reported consumption and apparent consumption was prorated among the various level 2 industry components, because there are insufficient data to determine where the difference belongs;

2. The amount of cobalt in each of the level 2 industry flows that reported to loss, downgrade, prompt scrap, flow to manufacturing, and flow to final product was proportional to the 1980 data reported by the NMAB (1) and that this proportionment is still valid at this time;

3. The portions of loss and downgraded material for most of the metals that appear as solid scrap, mixed scrap, turnings, grindings, and waste were assumed to be proportioned as reported by Curwick, Petersen, and Makar (11).

The cobalt industry consumption data for 1987 are shown in table I.

TABLE I. - Reported and apparent consumption of cobalt in the United States (Mg contained cobalt)

Industry	Reported[4]	Apparent
Metallurgical:		
Superalloy	2,845	3,358
Magnetic alloy	780	920
Cutting and wear resistant	533	629
Tool steel	174	205
Stainless steel, heat-resistant steel, and other alloys (excluding alloy steels and superalloys)	68	81
Miscellaneous and unspecified materials[1]	401	473
Chemical:		
Catalyst	497	587
Other chemical uses[2]	1,552	1,833
Total	[3]6,849	[3,4] 8,085

[1]Full alloy steel; high-strength, low-alloy steel; nonferrous alloys; welding materials (structural and hard facing); and metal powder products were combined to avoid disclosing company proprietary information.

[2]Driers, feed or nutritive additives, glass decolorizers, ground coat frit, and pigments.

[3]Data do not add to total shown because of independent rounding.

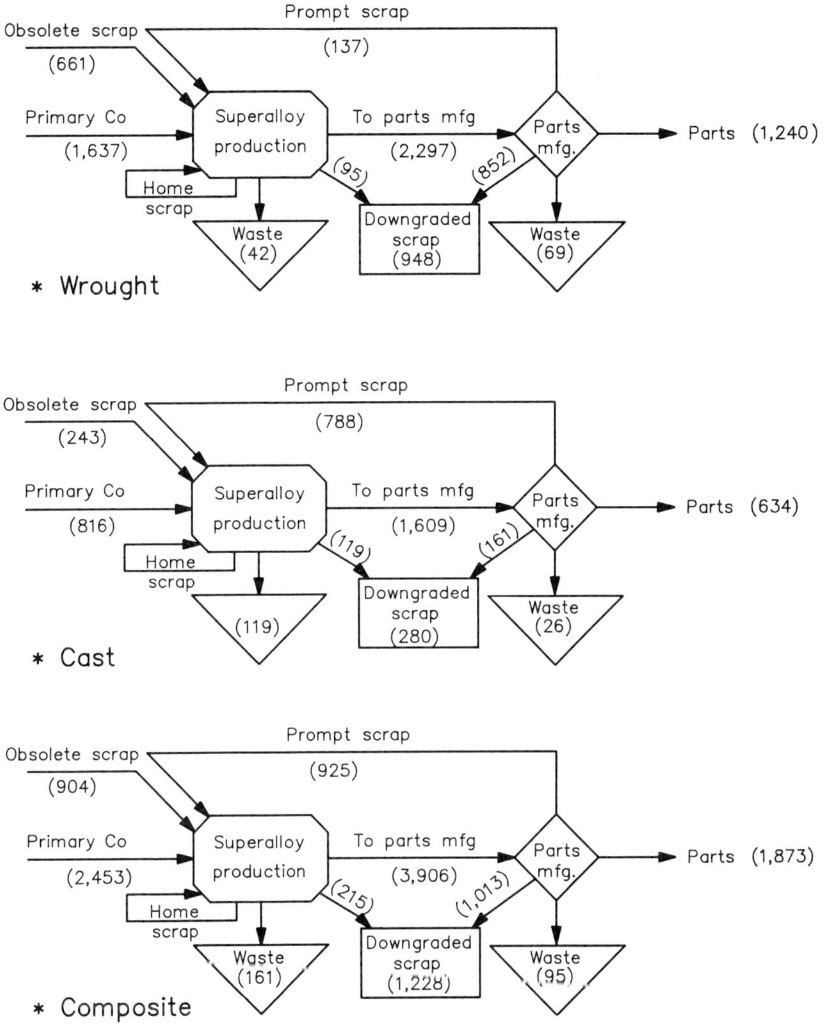

FIGURE 2 - Superalloys; wrought (top), cast (middle), and composite (bottom) (quantities in Mg contained cobalt). Data may not balance because of Mg independent rounding.

<u>Superalloys</u>

The flow of cobalt in the superalloy industry is shown in figure 2. The following assumptions were made in preparing these data:

1. The reported amount of cobalt contained in recycled obsolete scrap (14) was proportioned only to superalloy and cemented carbides processing.

2. The input of cobalt in obsolete scrap to superalloy production is the difference between contained cobalt in total consumed obsolete scrap and contained cobalt in obsolete scrap consumed in production of cemented carbides.

3. The difference between total apparent consumption of cobalt in superalloy production and that part from obsolete scrap was primary cobalt.

4. The ratio of primary cobalt consumed in wrought products to that consumed in cast products is 2:1. (1)

The composite flow of superalloys in the manufacture of parts is the sum of two independent flows, which are the production of wrought alloy parts and cast parts. (Fig. 2, top and middle). The composite flows for the superalloy industry are also shown in figure 2. Quantities in the figures were generated by the model.

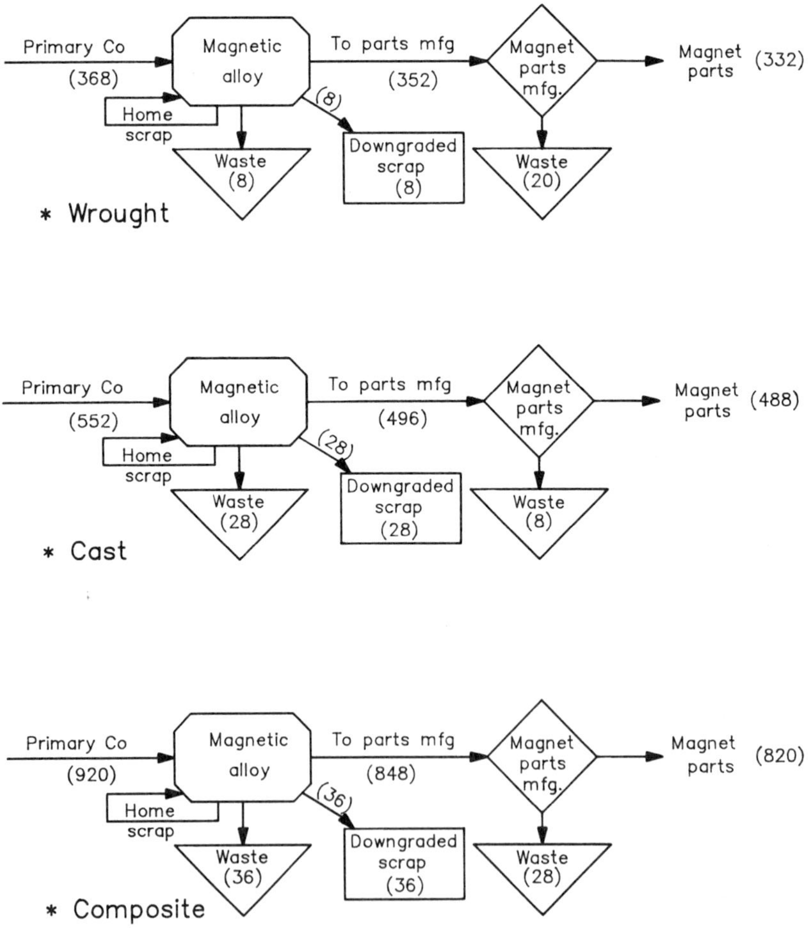

FIGURE 3 - Magnet alloys; wrought (top), cast (middle), and composite (bottom) (quantities in Mg contained cobalt).

In the production of wrought alloy parts, metal is processed to an alloy and this material, in the form of a billet or ingot, then goes to producing a manufactured part. In the cast parts section of the industry, the alloy is prepared by the melting of pure metals by the parts manufacturer immediately prior to casting the parts. To simplify the model, these two different operations are summed up as a composite flow sequence of superalloy processing and parts manufacturing.

From the model the following figures were generated. In 1987, the superalloy industry consumed materials estimated to contain 3,358 Mg cobalt, generated waste material estimated to contain 256 Mg of cobalt, and downgraded material estimated to contain 1,228 Mg of cobalt. The distribution of this material is shown in table II. In many cases the loss to downgrade is due to the fact that specifications for certain end-use products require the use of virgin metal to produce the alloys. These requirements cause large amounts of scrap material to be used as a cobalt or nickel source for alloys with lower concentrations of cobalt and lower economic values (downgraded.)

TABLE II. - Estimated distribution of waste and downgraded material in the superalloy industry, 1987

Material	Mg contained cobalt
Loss material: Waste....	256
Downgraded:	
Solid scrap........	168
Mixed scrap........	98
Turnings...........	494
Grindings..........	467
Total....	[1] 1,484

[1]Data do not add to total shown because of independent rounding.

Magnetic Alloys

The manufacture of magnets involves two major processes, wrought alloy processing and casting. The flows of cobalt in the manufacture of magnets with wrought alloys are shown in figure 3 (top), the flows of cobalt in the production of magnets from cast alloys (middle), and the composite flow of cobalt in all magnet manufacturing in (bottom). The flows were estimated based on the data of the NMAB (1) and Curwick, Petersen, and Makar (11). The data in figure 3 show that an estimated 920 Mg of cobalt were used in magnet manufacture in 1987. Of this amount, an estimated 64 Mg of cobalt reported to waste, 36 Mg of cobalt to downgrade, and 820 Mg of cobalt to magnets. The distribution of waste and downgraded materials is shown in table III. The NMAB states that essentially no obsolete cobalt magnet scrap was recycled to the production of cobalt-containing products (1).

TABLE III. - Estimated distribution of waste and downgraded scrap in the magnetic alloy industry, 1987

Material	Mg contained cobalt
Loss material: waste.....	64
Downgraded scrap: mixed..	36
Total....	100

Cemented Carbides

This category consists of cutting tools and wear-resistant materials such as dies. The flow of cobalt in this industry is based on the NMAB data (1) and is shown in figure 4. The material first goes through powder processing where the only reported loss occurs. Then, the material goes to parts amnufacturing. About 3-25 pct cobalt is used as a matrix/cement for carbide cutting tools and about 6-25 pct cobalt in carbide drawing dies (1). In the production of cemented carbide inserts and parts, obsolete scrap containing an estimated 234 Mg of cobalt and 395 Mg of primary cobalt were used, producing waste containing an estimated 10 Mg of cobalt and parts and inserts containing an estimated 618 Mg of cobalt.

Tool Steels

The primary use for this product is in high-speed tools. Cobalt is added to tungsten-chromium hot-work tool steels to increase the toughness and strength of the alloy at high temperatures (15). The flows of cobalt in this industry are based on NMAB data (1) and are shown in figure 5. In this industry an estimated 205 Mg of cobalt is used to produce high-speed tools containing an estimated 168 Mg of cobalt. The production and manufacturing losses of waste are estimated to contain 37 Mg of cobalt.

Stainless Steels, Heat-Resistant Steels, and Other Alloys (Excludes Alloy Steels and Superalloys)

The primary end use of products of this industry is in applications where heat and corrosion resistance are required (16). The flows of cobalt in this industry are shown in figure 6. In this industry, a total of 95 Mg of contained cobalt was estimated to go into production of the final products. Of this, 81 Mg was primary cobalt and the remainder was prompt scrap containing 14 Mg of cobalt. During production operations, on estimated 5 Mg of contained cobalt reported as waste and 18 Mg of contained cobalt reported as downgraded scrap. Using the data of Curwick, Petersen, and Makar (11), the downgraded material is estimated to break down into solid scrap (13 Mg cobalt), mixed scrap (4 Mg cobalt), and grindings (1 Mg cobalt) (table IV). The final products were estimated to contain 58 Mg of cobalt. The NMAB reports that essentially no obsolete cobalt-containing scrap is used by this industry.

TABLE IV. - Estimated distribution of waste and downgraded scrap in the stainless steel, heat resistant steel, and other metals industries, 1987

Material	Mg contained cobalt
Loss material:	
Grindings....	0.5
Waste........	5
Downgraded:	
Solid scrap..	13
Mixed scrap..	4
Grindings....	1
Total...	[1] 23.0

[1]Data do not add to total shown because of independent rounding.

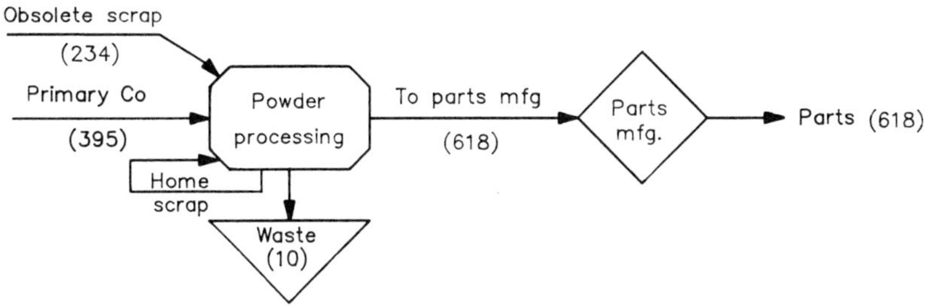

FIGURE 4 - Cemented carbides (quantities in Mg contained cobalt). Data may not balance because of independent rounding.

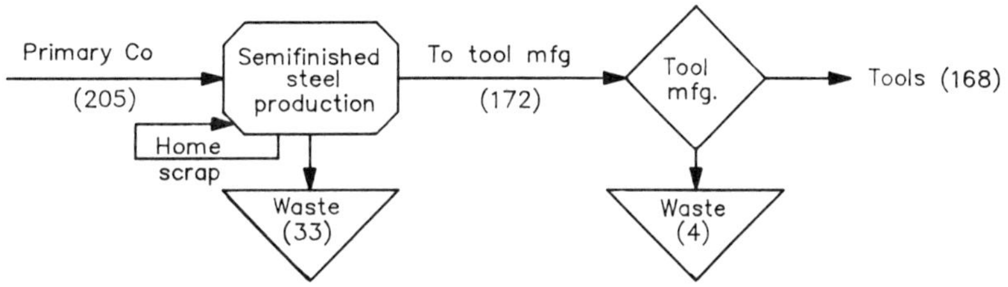

FIGURE 5 - Tool steels (quantities in Mg contained cobalt).

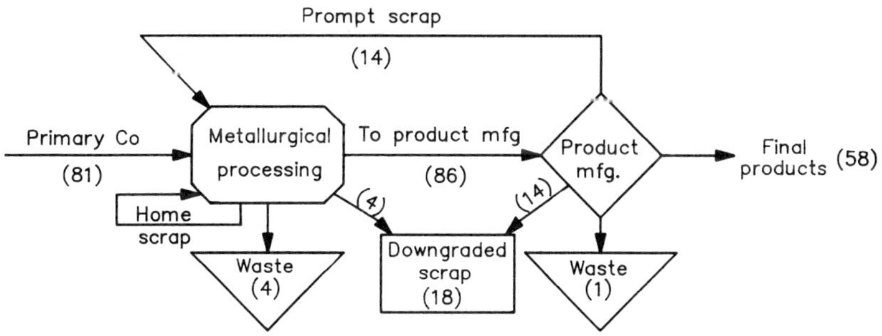

FIGURE 6 - Stainless steels, heat-resistant steels and other alloys (excludes alloy steels and superalloys) (quantities in Mg contained cobalt). Data may not balance because of independent rounding.

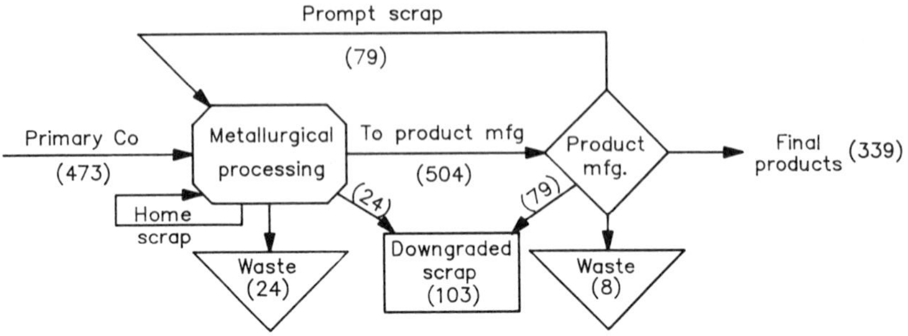

FIGURE 7 - Miscellaneous and unspecified products (quantities in Mg contained cobalt). Data may not balance because of independent rounding.

Miscellaneous and Unspecified Uses

This series of end use industries produces the following products: welding materials for structural and hardfacing applications, high-strength low-alloy steels, full alloy steels, nonferrous alloys, mill products made from metal powder, and other unspecified uses. These data are combined by the Bureau of Mines to avoid disclosing company proprietary data. Information in the 1979 to 1984 Minerals Yearbooks (17) indicates that an average of 82 pct of the currently designated miscellaneous and unspecified materials were known to be metal alloys. Thus, a metallurgical flow model used by NMAB (1) was used as a "blueprint" for this portion of the cobalt industry. The data are shown in figure 7. The NMAB states that essentially no obsolete cobalt-containing scrap is used to produce these products (1). An estimated 473 Mg of primary cobalt was used by these industries in 1987. From this input, the processing and manufacturing sectors of the industries were estimated to produce final products containing 339 Mg of cobalt. During the processing and manufacturing operations waste material containing 32 Mg of cobalt and downgraded scrap containing 103 Mg of cobalt were generated. Estimates, based on the data of Curwick, Petersen, and Makar (11), show that the downgraded scrap can be broken down into solids containing 73 Mg cobalt, mixed scrap containing 20 Mg cobalt, and grindings containing 9 Mg of cobalt (table V).

TABLE V. - Estimated distribution of waste and downgraded scrap in miscellaneous and unspecified end uses, 1987

Material	Mg contained cobalt
Loss material:	
Grindings....	3
Waste........	29
Downgraded:	
Solid scrap..	73
Mixed scrap..	20
Grindings....	9
Total..	[1] 135

[1]Data do not add to total shown because independent rounding.

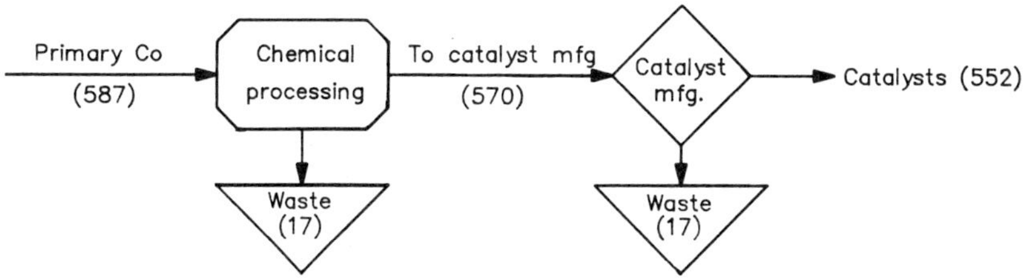

FIGURE 8 - Catalysts (quantities in Mg contained cobalt). Data may not balance because of independent rounding.

Catalysts

A major use of cobalt in chemical products is as catalysts. The catalysts are used for hydrogenation, hydration, desulfurization, oxidation, reduction, and synthesis of hydrocarbons (15-17). The flows of cobalt in the catalyst industry are shown in figure 8. The amounts in the flow chart are estimated using data from the NMAB (1). An input of materials containing an estimated 587 Mg of cobalt in 1987 produced catalysts containing an estimated 552 Mg of cobalt. Process and manufacturing waste materials contained an estimated 34 Mg of cobalt. Some of these catalysts can be regenerated several times, but ultimately the fate of most is disposal or metal recovery. The Bureau of Mines developed technology to recover cobalt from the spent catalysts (18). There is currently some reported use of recycling technology by industry. Deering reports that between 227 and 454 Mg of cobalt is recovered from spent catalysts annually (19). The difference between the estimated cobalt content of catalysts produced and the content of cobalt from this recycling activity leaves an estimated disposal of spent catalysts containing in the range of 98 to 325 Mg of cobalt per year.

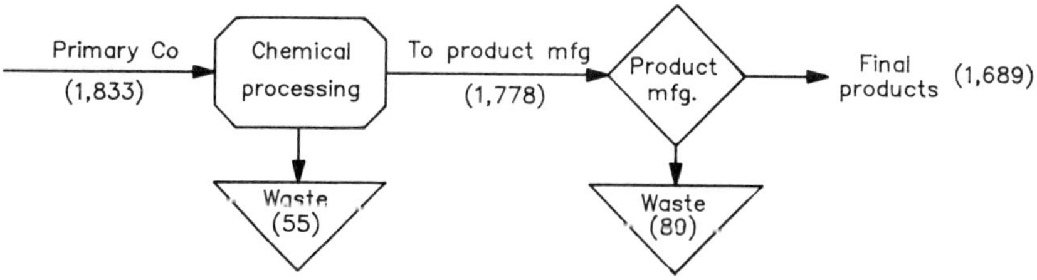

FIGURE 9 - Other chemicals (quantities in Mg contained cobalt).

Other Chemicals

This industry produces chemicals in the form of paint driers for lacquers, enamels and varnishes, feed additives, decolorizers for soda-lime-silicate glasses, ground glass frits to improve the adherence of porcelain enamels to sheet metal, pigments in colors of violet, blue, green, and pink, and in salts used in other industries (15). The flows of cobalt in this industry are shownin figure 9. Chemical processing efficiency is about 97 pct and chemical manufacturing efficiency is somewhat lower, about 95 pct (1). These values were used in defining flows within this industry. The model estimates that the industry used about 1,833 Mg of primary cobalt in 1987. This reports as waste material containing 144 Mg cobalt, and final products containing 1,689 Mg cobalt. The model's estimated distribution of cobalt among these products based on reported consumption is shown in table VI. About 2/3 of the products are paint driers.

TABLE VI. - Estimated amounts of cobalt in non-catalyst chemical products, 1987

Product	Mg contained cobalt
Paint driers.......	971
Ground coat frit...	391
Pigments...........	281
Feed additives.....	26
Glass decolorizers.	19
Total..	[1]1,689

[1]Data do not add to total shown because of independent rounding.

TABLE VII. - Estimated losses of cobalt by industry, 1987

Industry	Operation	Loss: Mg contained cobalt	Loss: pct of U.S. consumption
Superalloy..........	Manufacturing downgrade....	1,013	12.5
Do................	Processing downgrade.......	215	2.7
Do................	Processing waste...........	161	2.0
Do................	Manufacturing waste........	95	1.2
Other chemical.......	Do.................	89	1.1
Miscellaneous and unspecified........	Manufacturing downgrade....	79	1.0
Other chemical.......	Processing waste...........	55	.7
Magnetic alloy.......	Processing downgrade.......	36	.4
Do................	Processing waste...........	36	.4
Tool steel..........	Do.................	33	.4
Magnetic alloy.......	Manufacturing waste........	28	.3
Miscellaneous and unspecified........	Processing downgrade.......	24	.3
Do................	Processing waste...........	24	.3
Catalyst............	Manufacturing waste........	17	.2
Do................	Processing waste...........	17	.2
Stainless steel, heat-resistant, and other alloy.....	Manufacturing downgrade....	14	.2
Cemented carbide.....	Processing waste...........	10	.1
Miscellaneous and unspecified.........	Manufacturing waste........	8	.1
Stainless steel, heat-resistant, and other alloy.....	Processing downgrade.......	4	.05
Do................	Processing waste...........	4	.05
Tool steel..........	Manufacturing waste........	4	.05
Stainless steel, heat-resistant, and other alloy.....	Manufacturing waste........	1	.02
Total..................................		1967	24.27

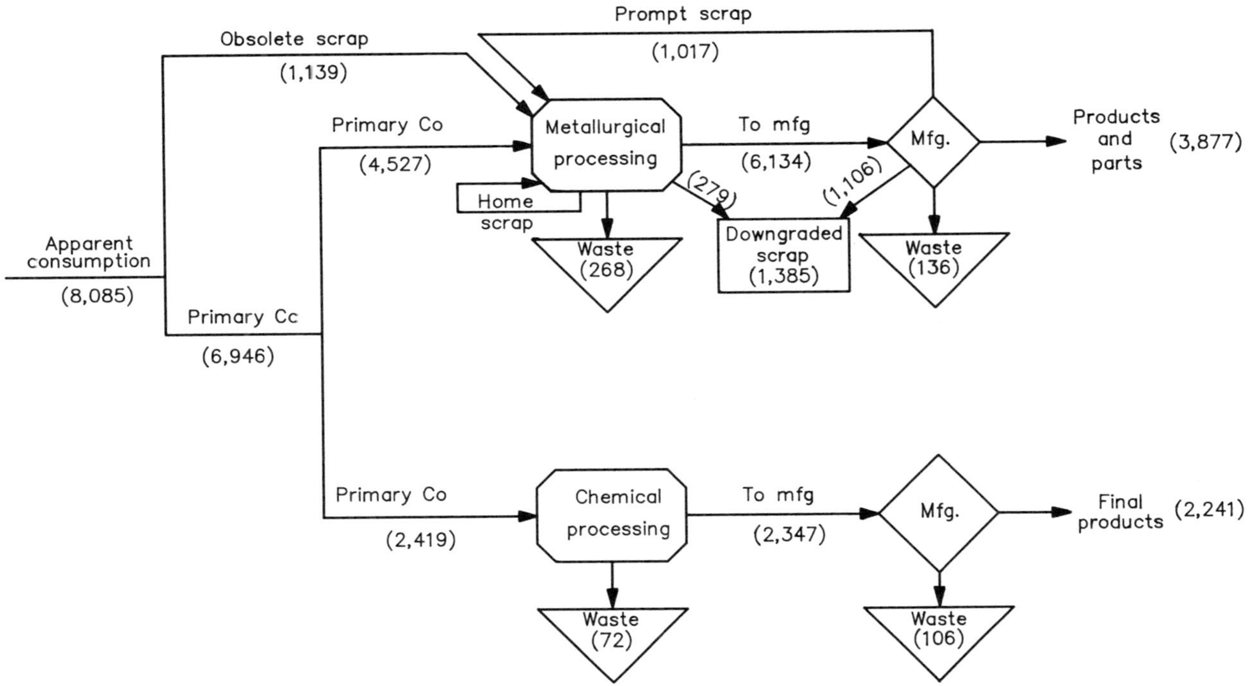

FIGURE 10 - Composite cobalt commodity flow sheet (quantities in Mg contained cobalt). Data may not balance because of independent rounding.

Cobalt Material Flows

The overall flows of cobalt in the industries using cobalt are shown in figure 10. In the chemical portion of the flow chart, in 1987 the industries consumed an estimated 2,419 Mg of cobalt. During processing and manufacturing operations, materials containing an estimated 178 Mg of cobalt are lost as waste. The final products contain an estimated 2,241 Mg of cobalt. In the metallurgical portion of the flow chart, in 1987, the industries consumed obsolete scrap containing an estimated 1,139 Mg of cobalt, an estimated 4,527 Mg of primary cobalt, and prompt scrap containing an estimated 1,017 Mg of cobalt. During processing and manufacturing, materials containing an estimated 404 Mg of cobalt were lost as waste and scrap materials containing an estimated 1,385 Mg of cobalt were downgraded with significant loss of cobalt values. The final products contain an estimated 3,877 Mg of cobalt. The wastes and downgraded materials from both the chemical and metallurgical industries caused resource loss equivalent of about 24 pct of the 1987 apparent consumption of cobalt. Downgraded scrap is either melted to other alloys where the cobalt may be lost or greatly diluted, or exported for refining to recover nickel and cobalt. The Bureau of Mines reported that material containing about 366 Mg of cobalt was exported in 1987 (14). This includes unwrought cobalt materials, waste and scrap.

Conclusions

A cobalt commodity flow model has been developed. The model is set up in a hierarchical manner with levels for the various industries and subindustries. The model shows the distribution of apparent 1987 consumption of 8,085 Mg of cobalt into 8 subindustries. Estimates from the data in table VII show total production of downgraded scrap containing 1,385 Mg cobalt and loss/waste material containing 582 Mg cobalt. The largest producer of scrap and waste is the superalloy industry where scrap and waste produced during alloy processing and parts manufacturing contain 2,409 Mg of cobalt. (Table VII and figure 2). Of this, 925 Mg of contained cobalt was in prompt scrap which was reused by the industry. The remaining material containing 1,484 Mg cobalt is divided into downgraded scrap with 1,228 Mg of cobalt and waste with 256 Mg of cobalt. These two loss areas are places where prime concern should be addressed in studies of recycling. The principal cause of the materials downgrading is that certain end uses require the use of virgin metal, and thus, any scrap is downgraded to production of lower cobalt value alloys or alloys where the cobalt is lost in processing.

Based on the estimated life times of various metal products (1), and the estimated amounts of products produced in 1977 (10 years), 1982 (5 years), and 1985 (2 years), in 1987 there should have been potentially recyclable metal products containing 3,886 Mg of cobalt (17). Table VIII shows these products' potential life span, the amount of cobalt content potentially recyclable, and the amount estimated to have been recycled to the same alloy without downgrading. The data in Table VIII indicates that about 54 pct of obsolete superalloy scrap and 51 pct of the obsolete cemented carbide scrap are estimated to be recycled.

This leads to an overall estimated 29 pct of the total obsolete scrap that was efficiently recycled. The estimate superalloy data do not appear to be in agreement with survey estimates obtained by Papp (10).

The development of this flow model has highlighted areas where data are needed to make the model more effective. There are few or no current data available on the amounts of material that report as solid scrap, mixed scrap, turnings, grindings, and waste. The amounts calculated in this model are based on data from 1980 and 1976. New, up-to-date data are needed to keep the model current. There are some Environmental Protection Agency data that are useful in determining the amounts of waste and downgraded material, but even these data have substantial gaps.

TABLE VIII. - Estimated recyclable and recycled materials by industry, 1987, Mg contained cobalt

Product	Lifetime, yrs[1]	Potentially recyclable[2]	Estimated recycle
Superalloys..........	10	1,684	904
Magnetic alloys........	10	1,578	0
Cemented carbides......	2	461	234
Tool steels............	5	73	0
Stainless steel, heat-resistant, and other alloys (excluding alloys steels and superalloys).........	5	23	0
Miscellaneous and unspecified..........	5	67	0
Total.........	-	3,886	[3]1,139

[1]Reference (1), page 24.
[2]Reference (17).
[3]Data do not add to total shown because of independent rounding.

References

1. National Materials Advisory Board -- National Research Council. Cobalt Conservation Through Technological Alternatives. (U.S. Dept. of the Interior, Bureau of Mines contract J 0113103). National Academy Press Pub. NMAB-406, 1983, 204.

2. K. B. Shedd and W. S. Kirk. Private communication with author, U.S. Bureau of Mines, 1 February 1990.

3. G. R. Petersen, D. I. Bleiwas, and P. R. Thomas. Cobalt Availability - Domestic: A Minerals Availability System Appraisal. U.S. Bureau of Mines IC 8848, 1981, 31.

4. C. P. Mishra, et al. Cobalt Availability - Market Economy Countries: A Minerals Availability Program Appraisal. U.S. Bureau of Mines IC 9012, 1985, 33.

5. U. S. Dept. of Commerce. Critical Materials Requirements of the U.S. Steel Industry. U.S. Government Printing Office, 1983, 259.

6. National Materials Advisory Board -- National Research Council. Basic and Strategic Metals Industries: Threats and Opportunities. National Academy Press Pub. NMAB-425, 1985, 151.

7. C. C. Kilgore and P. R. Thomas. Manganese Availability - Domestic: A Minerals Availability System Appraisal. U.S. Bureau of Mines, IC 8889, 1982, 14

8. T. F. Anstett, D. I. Bleiwas, and C. Sheng-Fogg. Platinum Availability - Market Economy Countries: A Minerals Availability System Appraisal. U.S. Bureau of Mines, IC 8897, 1982, 16.

9. U.S. Bureau of Mines. An Appraisal of Minerals Availability for 34 Commodities. U. S. Bureau of Mines, B 692, 1987, 300.

10. J. F. Papp. Letter report (USAF Military Interdepartmental Purchase Request Number FY615-87-05292), 29 July 1987, 19.

11. C. L. Kusik, H. V. Makar, and M. R. Mounier. Availability of Critical Scrap Metals Containing Chromium in the United States: Wrought Stainless Steels and Heat-Resisting Alloys. U.S. Bureau of Mines, IC 8822, 1980, 51.

12. L. R. Curwick, W. A. Petersen, and H. V. Makar. Availability of Critical Scrap Metals Containing Chromium in the United States: Superalloys and Cast Heat- and Corrosion-Resistant Alloys. U.S. Bureau of Mines, IC 8821, 1980, 51

13. Office of Technology Assessment. Strategic Materials: Technologies to Reduce U.S. Import Vulnerability. (OTA-ITE-248, U.S. Congress, 1985).

14. W. S. Kirk. Cobalt. U.S. Bureau of Mines Mineral Yearbook 1987, v. 1, 8. (Preprint).

15. F. Planinsek and J. B. Newkirk. Cobalt and Cobalt Alloys. Kirk-Othmer Encyclopedia of Chemical Technology, (6) 1979, 481-494.

16. W. S. Kirk. Cobalt. Mineral Facts and Problems, (1985 Edition. U. S. Bureau of Mines, B 675), 1985, 171-183.

17. U.S. Bureau of Mines. Minerals Yearbooks 1977-1986. Chapter on Cobalt.

18. R. E. Siemens, B. W. Jong, and J. H. Russell. "Potential of Spent Catalysts as a Source of Critical Metals." Conservation and Recycling, 9, (2) (1986) 189-96.

19. W. Deering. "No Title" (Presentation on catalyst recycling.) (Paper Presented at the Metals Week Ni-Mo-Co Conference, Tucson, Arizona, 26-28 October 1988, 8.

METAL RECOVERY FROM COPPER PROCESSING WASTES

K.S. Gritton, D.K. Steele, J.E. Gebhardt

Bureau of Mines

729 Arapeen Drive

Salt Lake City, UT 84108

The U.S. Bureau of Mines has investigated the recovery of valuable metals from copper-processing flue dusts and refinery-bleed solutions. Flue dusts were leached with sulfuric acid and refinery-bleed solution to solubilize arsenic and copper. Arsenic was recovered from the leach liquors as As_2O_3 by reduction and precipitation using SO_2. Copper was recovered as a sulfate by crystallization and as a precipitated sulfide. Leach residues contained significant amounts of lead, and small quantities of bismuth and antimony. The residues were deemed suitable for recycle to a lead smelter. Other metals, such as zinc and molybdenum, were concentrated in byproduct streams for subsequent purification and recovery.

Second International Symposium -
Recycling of Metals and Engineered Materials
Edited by Jan H. L. van Linden,
Donald L. Stewart, Jr., and Yogeshwar Sahai
The Minerals, Metals & Materials Society, 1990

BLISTER COPPER PRODUCTION FROM SECONDARY MATERIALS

Milton I. Schwab — Franklin Smelting & Ref.Co., Castor Ave. E.of Richmond St., Philadelphia, PA 19134

Albert W. Spitz — Consulting Eng., 614 N. Easton Rd., Glenside, PA 19038

Robert A. Spitz — Consulting Eng., 28 Weymouth St.#14, Holbrook, MA 02343

Abstract

Our raw materials are copper bearing scrap, slags, skimmings, electronic equipment,etc. which contain almost every common and precious metal. They are blended to give a uniform charge and run in a blast furnace using oxygen enrichment. The metal produced, approximately 70% copper, is transferred to a rotary converter where the copper is concentrated to 98% with oxygen. This blister copper, containing the precious metals, is shipped to an electrolytic refinery. The gases from the blast furnace are incinerated and filtered. The gases from the rotary furnace are also filtered and then scrubbed to convert any sulfur compounds to a sulfur-nitrogen solution that is used in fertilizer formulations. The collected particulate from both furnaces is sold for the contained metals. The blast furnace slag is used for roofing granules and blasting grit.

Second International Symposium -
Recycling of Metals and Engineered Materials
Edited by Jan H. L. van Linden,
Donald L. Stewart, Jr., and Yogeshwar Sahai
The Minerals, Metals & Materials Society, 1990

Introduction

A substantial percentage of the copper used in the United States is produced from scrap. This material comes from a wide variety of sources including old automobiles, obsolete electronic, communications and process equipment, demolition sites and various manufacturing operations. If the scrap is clean, practically all metallic and is all of the same or similar analysis, it can usually be remelted and blended to produce a brass or bronze alloy. This is the function of ingot makers. If it contains a wide range of metals such as iron, aluminum, zinc, lead, nickel,antimony and precious metals, along with oxides, slags, organic compounds and dirt, it must go to a refiner.

Economics of Raw Materials

Copper scrap is classified and priced in accordance with the amount of contained copper and the other metals and the ease of recovering the desired alloy or pure copper. Number one classification is the top grade. It consists of clean copper wire, bus bar, etc. which can readily be reused. It therefore usually commands the highest price of the copper scraps. An exception could be electronic scrap with a large amount of gold or platinum.Intermediate grades can be melted, alloyed and cast into ingots or billets. Our plant is designed to process the lowest grades of copper bearing scrap which consist of mixed brass and bronze turnings, clips, electronic scrap, some containing precious metals, copper catalysts, copper clad steel wire, oxides, slags, grindings: in short every kind of copper bearing material that cannot simply be melted and alloyed.

The economic success of any plant using these materials depends not only on efficient operation but also on careful purchasing of the copper bearing scrap.When the copper price is high,scrap with lower copper content may be economically processed. When copper prices are low, many of the lower copper content materials have depressed values.

The plant has a more or less fixed capacity in terms of tonnage of scrap materials per day that can be processed. With better grades of scrap the copper production is greater and the fuel, labor and overhead costs per pound of copper output are reduced. Since the finished metal output varies directly with the percentage of copper in the scrap. The value of the scrap goes down as the cost of copper produced from it goes up.

Preparation for Processing

Once the raw materials are on hand,carefully segregated,the preparation process begins. Primary copper smelters, work with concentrates of constant composition,which are ideal for any furnace operation.We try to approximate, with varying success, this situation by preparing and blending the different materials we purchase before charging them into our blast furnace.

Large assemblies are sheared to suitable size, some materials are compressed into small bales, slags are broken and screened to minimize fines which would be blown out of the furnace, and then these and other scrap materials are blended to produce an approximately uniform charge. Purchased fines and those from our own screening operation are briquetted, to be added to the blending operation as required.

Copper Concentration

Having produced a blended material within the desired size range, our first step in recovering the copper content is accomplished in a blast furnace using oxygen enriched air in the tuyeres. The scrap charge is fed onto a belt conveyor which in turn discharges into one of two skip hoist buckets These, in turn are hoisted and alternately dumped into opposite sides of the furnace. Coke, fluxes, etc. are similarly fed

While blending the blast furnace charge alleviates some of the variation, the blast furnace operator must adjust the addition of coke and fluxes. His goal is not only to recover the maximum amount of copper but also to minimize the costs involved. The amount of coke required per ton of charge is related to both the heat required for melting and to the amount of oxide in the charge that must be reduced to metal. Sufficient flux must be added to provide a fluid slag.
The blowing rate and the oxygen enrichment level must match the charge rate and at the same time, maintain a proper atmosphere in the furnace.Improper control of these parameters can lead to difficulties in operating the blast furnace such as low thruput, viscous slags and excessive costs.

The blast furnace crucible is continuously tapped thru a liquid seal, to prevent the blast air from blowing out the bottom. A mixture of molten copper and slag flows down a launder into an oil fired rocking furnace that can rotate thru an approximately 120 degree arc.This furnace is large enough to give the slag sufficient time to separate from the copper. Rotating the furnace in one direction allows the liquid copper to fill a preheated ladle on a rail car below the rocking furnace. Rotation in the opposite direction allows the slag to pour into a granulating trough. Granulation is accomplished by hitting the liquid slag stream with a high pressure jet of water. The slag and water are collected in a pit which is large enough to remove the slag with a clamshell bucket on a crane. The liquid metal in the ladle, known as black copper, is approximately 70% copper.

Purification

The rail car with the ladle is pulled to the converter building, where the ladle is picked up by a traveling crane and then poured into one of two tilting rotary converters. These converters operate with pure oxygen which is produced in our own plant. Oxy-fuel burners are utilized if the metal is not sufficiently hot. When the metal has reached the desired temperature, the fuel supply is discontinued and the successive oxidation of the various metals such as zinc, lead, tin, iron, etc. is accomplished with oxygen .

Some of the metallic oxides leave the furnace in the off gases, most remain in the slag. The furnace slag is tapped several times during the cycle and accumulated to rerun in the blast furnace. When the copper has reached the desired purity, it is poled with green logs to removed dissolved gases, then poured into molds and sent to an electrolytic refinery for further purification and precious metal recovery

The operation of the converters is dependent on the composition of the black copper produced in the blast furnace. While this composition tends to be relatively stable with regard to copper, the levels of the other metals can vary. This, in turn, may require changes in the blowing time and slagging practice. High levels of nickel, for instance, require longer blowing times. Overblowing the furnace leads to high losses of copper in the final slag. We have begun to develop a strategy for programming the converter operation based on the analysis of the molten black copper feed.

Environmental Controls

Blast Furnace

The blast furnace offgas contains unburned organics and particulate consisting of coke, fines from charged materials, metallic fumes and oxides.It is first passed thru a cyclonic classifier, which is designed to collect particulate larger than approximately 150 micrometers;the metal fumes, organics and oxides pass thru the classifier. The collected fines are combined with others from the plant which are briquetted and returned to the blast furnace. The effluent from the classifier is oxidized in an afterburner using a combustion burner if necessary. Frequently the CO and the organic materials in the gas stream provide sufficient fuel to maintain a temperature of at least 1800 degrees F. The gases from the afterburner are cooled in a spray cooling chamber with water and then dilution cooled to a temperature of 450 degrees F. The gas is then filtered in a reverse air cleaned baghouse using treated glass bags.The filtered gases are discharged to the atmosphere. The collected dust is shipped to a zinc smelter to recover the metal values. Gases from charging and tapping hoods are collected; they are used as dilution air to reduce gas temperature prior to filtering.

Rotary Converters

The volume of gases leaving the rotary converters is relatively low, since approximately 99% oxygen is used for combustion and oxidizing the metallic impurities. The gases contain metal and sulfur oxides.Hoods are used to collect fugitive emissions at the charging and tapping locations.

The converter effluent is cooled by dilution with air which is pulled into the ductwork around the furnace exit and the air from charging and tapping hoods.It is then filtered in a baghouse using glass bags and reverse air cleaning. The collected dust shipped to zinc smelters or tin and lead refiners for metal recovery.

Sulfur Oxide Removal from Converter Gases

The gases leaving the baghouse are scrubbed with a solution of ammonium sulfite, ammonium bisulfite and ammonium sulfate, which removes better than 90% of the sulfur oxides.The gases from the scrubber are discharged to the atmosphere.

This process, which we developed and patented, consists essentially of recirculating the ammonia salt solution until the maximum desired salt concentration, approximately 45%, is attained. This concentration can be safely shipped without crystallizing if the temperature drops. While it is being recirculated, precise ph control between 6 and 7 is maintained by adding anhydrous ammonia. The concentrated solution has a freezing point below 0 degrees F, a nitrogen content of approximately 9% and a sulfur content around 12% to 13%.

When the solution reaches the desired salt concentration, approximately one third is removed and replaced with water. The removed solution is filtered, accumulated, and then shipped to fertilizer manufacturers to be made into liquid fertilizer by the addition of additional nitrogen and other elements,such as phosphorous and potassium. It is of interest to note that the constituent most desired in the scrubber liquid is the sulfur. The most unwanted ingredient,lead, fortunately, is practically insoluble in the solution and is filtered out to levels less than 5 parts per million.

Thus ammonia, which is extensively used for fertilizer, is more or less borrowed to collect the sulfur which would, with more conventional processes, be wasted. The ammonia, obviously is also utilized in the final fertilizer.

Prior to offering this material to fertilizer manufacturers, it was carefully analyzed and studied by various state and university agricultural agencies.

Blast Furnace Slag

The granulated blast furnace slag is dried,crushed and screened to produce a variety of products based on particle size. The slag is composed mainly of calcium, iron, and silicon,

The slag granules are hard and dense, which makes them ideal abrasives, jet black in color and extremely inert.They are used as a substitute for sand in sand blasting, a non-skid product for use on icy roads, a filler for asphalt shingles, black roofing granules and fillers for various coatings.

Future Projects

While it may appear that we know most of the answers, in truth we do not even know all the questions. We are constantly striving for better yields, more efficient energy utilization, better pollution control and more precise operation with computerization.Projects in all these directions are well under way. We hope to present some answers at the next symposium.

HIGH-PURITY COBALT AND NICKEL PRODUCTION FROM SUPERALLOY SCRAP

By L. D. Redden and D. K. Steele

Salt Lake City Research Center
Bureau of Mines, U.S. Department of the Interior
729 Arapeen Drive
Salt Lake City, Utah
USA

Abstract

As part of the Bureau of Mines program to reduce U.S. reliance on foreign supplies of critical and essential metals, Bureau researchers have conducted studies to improve technology for recovering critical and essential metals from superalloy scrap (SAS). This report describes a combined pyro- hydrometallurgical process for recovering high-purity cobalt and nickel from SAS. Pyrometallurgical pretreatment simplifies subsequent hydrometallurgical procedures by altering the scrap matrix through carburization of the more refractory SAS constituents. The hydrometallurgical treatment involves electrodissolution of carburized SAS anodes in a novel Bureau-developed double membrane electrolytic cell (DMEC). The resulting anolyte is treated by a series of hydrometallurgical procedures to produce a purified cobalt chloride solution. This solution is utilized as a catholyte in a DMEC to win high-purity cobalt. Laboratory tests have produced cobalt cathodes meeting National Stockpile Specification P 13-R6, for Grade A cobalt. Additional test work has produced high-purity nickel in a DMEC.

Introduction

A goal of the Bureau of Mines is to reduce the United States' reliance on foreign supplies of strategic and critical minerals and metals (1). This goal can be achieved in part by recovering cobalt and nickel from superalloy scrap (SAS), provided the metals are recovered in forms that can be utilized in strategic and critical applications. The objective of this investigation was to develop technology to recover high-purity cobalt and nickel meeting Department of Defense Stockpile Specifications from underutilized SAS.

A significant portion of processed SAS is underutilized in the sense that it is downgraded or used as feed stock for alloy classes having less stringent compositional requirements than the original superalloy (2,3). Most downgrading operations either deliberately or inadvertently oxidize chromium and several minor alloying elements such as aluminum, molybdenum, niobium, titanium, and tungsten, and the resulting oxides are lost to the slag. Although the nickel and unoxidized chromium are utilized in downgrading operations, contained cobalt is often lost through dilution in steels where it is a relatively ineffective alloying element (4). From a practical

Second International Symposium - Recycling of Metals and Engineered Materials
Edited by Jan H. L. van Linden,
Donald L. Stewart, Jr., and Yogeshwar Sahai
The Minerals, Metals & Materials Society, 1990

standpoint, downgrading makes it possible to recycle some scrap that could not otherwise be economically processed; however, from a strategic perspective, the cobalt in downgraded SAS is underutilized or wasted.

A considerable amount of research has been conducted and numerous processes proposed to recover critical metals from complex, contaminated SAS. The bulk of the recycling research reported in the literature has been conducted in three areas: pyrometallurgical processing (5), hydrometallurgical processing (6-10), and processes combining both pyrometallurgical and hydrometallurgical treatments (4,11-15).

This investigation utilized the combined pyrometallurgical-hydrometallurgical approach and introduces a novel double-membrane electrolytic cell (DMEC) which is a key component of the process under investigation. This report describes the operation of the DMEC and the integration of the DMEC into a proposed process to recover high-purity cobalt and nickel from SAS.

Process Description

The proposed process utilizes pyrometallurgical techniques to melt, carburize, and cast SAS into anodes. A chloride system was selected for the subsequent hydrometallurgical treatment which involves SAS anode electrodissolution, electrolyte purification, and the electrodeposition of cobalt and nickel. High-purity cobalt and nickel are recovered in separate DMEC's; however, all DMEC's are utilized to dissolve SAS. A conceptual flowsheet of the process is given in figure 1. Actual process solutions were used in the cobalt recovery investigation and simulated process solutions were used in the nickel recovery investigation.

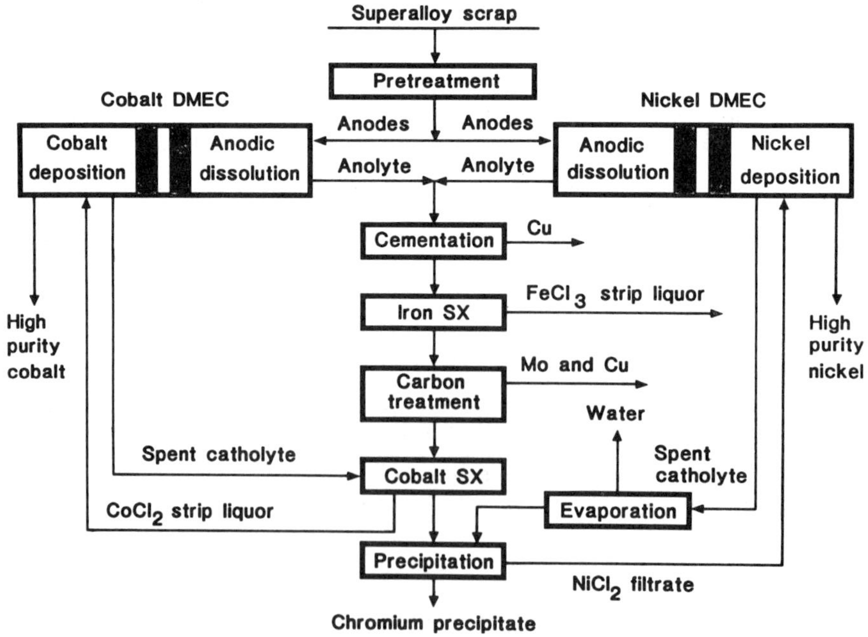

Figure 1 - Generalized process flowsheet.

DMEC Configuration and Operation

The DMEC is composed of an anode compartment and a cathode compartment separated by two commercially available anionic membranes. The separation between the two membranes forms a third compartment which is referred to as a membrane compartment. An impure SAS anode is electrolytically dissolved in the anode compartment, and the resulting anolyte is treated by hydrometallurgical techniques to produce purified chloride electrolytes. The purified electrolytes are circulated to the cathode compartment of the DMEC where high-purity metal is electrolytically deposited at the cathode. DMEC operation is shown schematically in figure 2.

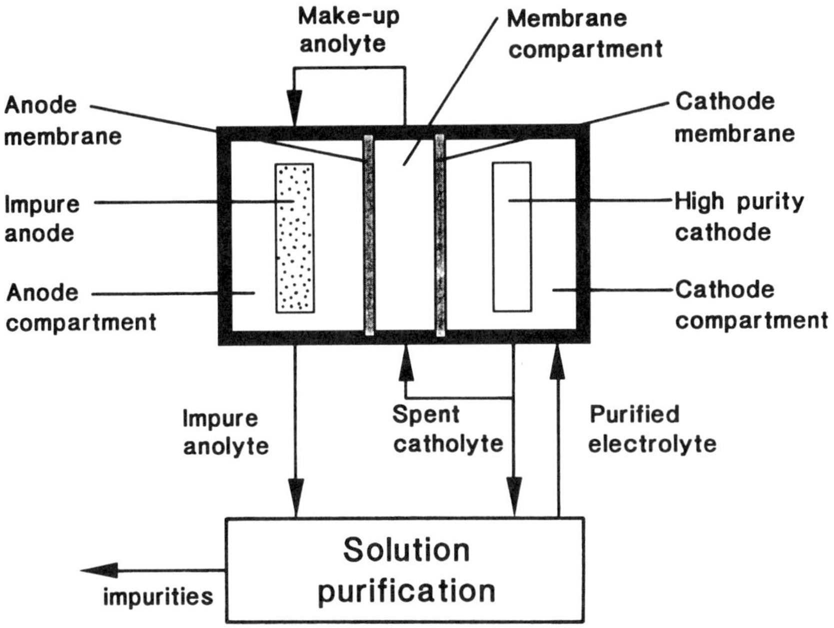

Figure 2 - Schematic of DMEC operation.

The function of the DMEC membranes is to separate the impure anolyte from the purified catholyte solutions while allowing chloride ions to pass from the catholyte to the anolyte. A stream of diluted, spent catholyte solution flows through the membrane-compartment, and compartment effluent is recycled to the anode compartment. This configuration nearly eliminates the transfer of anolyte impurities to the catholyte and makes it possible to produce very high grade cathode products.

Membrane impurity transfer is related to the impurity concentration gradient across a single membrane. Under normal operating conditions, the impurity concentration gradient across the anolyte membrane is considerable. Consequently, impurity transfer across the anode membrane into the membrane compartment is relatively high. However, the majority of the impurities that migrate through the anolyte membrane are carried away by the spent catholyte solution before they reach the cathode membrane. The concentration gradient across this membrane is low resulting in low impurity transfer to the catholyte.

The DMEC, like conventional electrorefining cells, uses the same electrical energy to deposit metal on the cathode that is utilized to dissolve the anode. However, the DMEC can produce higher purity products than electrorefining operations because anode impurities do not contaminate the purified catholytes. In addition, the anionic membranes allow anions, in this case chloride, to migrate from the catholyte to the anolyte. This migration maintains the electrolyte charge balance and thus reduces reagent costs by recycling anions within the cell.

Anode Preparation

Two types of material were utilized as anodes in this investigation, simulated composite SAS and Haynes 263 alloy. The simulated composite SAS anodes were prepared by melting elemental materials in a vacuum induction furnace. The composition of these anodes was selected to simulate a carburized SAS melt prepared from a composite blend of investment cast nickel-and cobalt-base alloys, nickel- and cobalt-base hardfacing alloys and wrought nickel-iron base alloys. The melt was cast into anodes measuring approximately 16 by 7 by 3 cm.

The second type of anode was Haynes 263 alloy, 0.5-cm thick plate, cut into 6.5 by 13-cm anodes. The compositions of the two anode-types are given in table I.

Table I. - Anode composition, pct

Constituent	Simulated SAS	Haynes 263 Alloy
Al	0.96	1.39
C	1.59	.06
Co	10.73	19.8
Cr	14.32	19.7
Cu	1.11	.04
Fe	5.07	.36
Mn	.02	.42
Mo	4.48	5.82
Nb	1.04	ND
Ni	54.9	49.96
Ta	1.4	ND
Ti	1.12	2.23
W	2.12	ND

ND = Not determined.

Experimental Equipment and Procedures

Electrolytic operations were conducted in a laboratory-scale DMEC measuring 13-cm high by 15-cm long and 11-cm deep. Membrane separation was 2.2 cm, and each membrane had a 5.0 by 7.0 cm cross-sectional area. Anode to cathode spacing was 8.0 cm, and anodes were placed in porous nylon bags to contain the anode sludge. Agitation was provided in the cathode compartment by air sparging, and 18-gauge, 316 stainless steel was utilized as cathodes starting sheets. Hot water was circulated through glass coils immersed in the anode and cathode compartments to maintain the cell operating temperature at 55°C.

Iron powder was utilized to cement the copper from the anolyte at room temperature in an open reactor. Agitation was provided with a turbine-type impeller, and the cementation product was separated from the anolyte by filtration.

Solvent extraction (SX) operations were carried out in continuous countercurrent circuits. The circuits were designed based on equilibrium extraction and stripping isotherms and McCabe-Thiele Analyses. Mixer-settler-type contactors were used in the circuits. Individual contactors had a 130-mL mixing volume and a 370-mL settling volume.

Carbon adsorption studies were conducted with Calgon CPG (12 by 40 mesh) activated carbon. The carbon was treated prior to use by boiling in a simulated copper-molybdenum-free anolyte to remove entrapped and dissolved air. The treated carbon was packed into two glass columns connected in series. SAS anolyte was pumped through the columns at a rate of 1 bed volume (BV) per hour, and column effluents were periodically sampled to determine carbon adsorption characteristics.

Chromium precipitation studies were conducted to selectively precipitate chromium as a basic chromium sulfate. Sodium sulfate was added to the cobalt circuit raffinate at a dosage of 0.18 mol/mol Cr. The raffinate pH was raised to 2.0 with 170 g/L Na_2CO_3 solution. The resulting solution was cooled to below 6°C, and the pH was slowly adjusted to 4.7 with additional sodium carbonate solution while stirring vigorously. After the desired pH was achieved, slurry temperature was increased to 85°C and maintained for 1 hr; the slurry was then filtered. The chilling step was incorporated into the procedure to improve chromium precipitation and reduce the co-precipitation of nickel.

Membrane Selection

Several manufacturers were contacted with respect to selecting anionic membranes for the DMEC; however, only two membranes were recommended for use in acidic chloride solutions with high-ionic strengths, Types 204-SXZL-386 and 204-UZL-386, manufactured by Ionics Incorporated of Watertown, MA. Both membranes are anion selective comprising cross-linked copolymers of vinyl monomers containing quaternary ammonium anion exchange groups.

The membranes were evaluated at process conditions to determine long-term chemical and physical stability. Because the manufacturer indicated that both membrane electrical resistance and cation flux increase as the membranes degrade, these parameters were measured every 48 to 72 hrs. After 600 hrs of testing, results indicated that neither membrane had degraded significantly; however, the impurity flux through the SXZL membrane, although low, was four times greater than that of UZL membrane. Because of the superior performance of the UZL membrane, testing of the SXZL membrane was discontinued. The UZL membrane was evaluated for a total of 1,216 hrs and showed no significant degradation. Subsequent test work was performed with UZL membranes.

Integrated Circuit Operation

A DMEC was integrated with copper cementation, iron SX, carbon adsorption, and cobalt SX operations to determine the grade of cobalt that could be produced from SAS using DMEC technology. Unless otherwise specified, actual SAS process solutions were utilized in the operation of the integrated circuit.

Anolyte Production

A DMEC was utilized to produce the anolyte required to operate the integrated circuit. Although cobalt was simultaneously won as anodes were dissolved, the objective was to produce anolyte, and cobalt electrodeposition was not studied during this phase of the investigation. Unfortunately, the availability of SAS anodes was not sufficient to produce the quantities of anolyte required to operate the integrated circuit, consequently, Haynes 263 alloy anodes were used in place of the SAS anodes in the initial production of anolyte. The composition of the resulting anolyte was then adjusted to match the composition of anolytes produced from SAS anodes.

A two anode- one cathode-DMEC was operated for 350 hrs at a cathode current density of 430 amps/m^2. The anode compartments were initially filled with anolyte produced in previous tests. The cathode and membrane compartments were filled with cobalt chloride solutions prepared from reagent grade cobalt chloride chemicals, containing 30 and 20 g/L Co, respectively. Deionized water was metered into the anode compartments, and a $CoCl_2$ solution (40 g/L Co) was metered into the cathode compartment. The anodes and cathodes were changed as required, and the $CoCl_2$ solution in the membrane compartment was static and was changed every 24 hrs of operation.

The anodic dissolution operation produced approximately 12 L of anolyte. To match the anolyte composition to that of anolytes produced from SAS anodes, the Haynes 263 anolyte composition was altered by water dilution and reagent grade chemical addition. The composition of a SAS anolyte and the altered Haynes anolyte are given in table II.

Table II. - Anolyte compositions, g/L

	Haynes	SAS		Haynes	SAS
Al	0.1	1.0	Nb	<.5	<.8
Co	12	14	Ni	75	80
Cr	9.5	8.7	Ta	0.02	0.07
Cu	1.3	1.1	Ti	<.2	<.4
Fe	2.7	3.4	W	.05	.2
Mn	.1	0.07	Zn	.02	.01
Mo	.5	.04	Cl	130	145

Solution Purification Operations

A series of solution purification operations were utilized to produce from the Haynes anolyte a cobalt chloride catholyte from which high-grade cobalt could be won. Copper, molybdenum, and iron were removed from the anolyte prior to cobalt SX. Otherwise these elements would coextract with the cobalt and contaminate the cobalt chloride catholyte.

Copper Cementation. A two-stage cementation procedure was developed to remove copper from the anolyte while minimizing iron consumption. The first stage of the procedure contacted the anolyte with 1.5 times the theoretical iron requirement for 10 min. The resulting anolyte contained 0.06 g/L Cu. The second stage utilized 0.5 times the theoretical iron requirement for the first stage feed, and the contact time was 10 min. The second cementation stage further reduced the anolyte copper concentration to 0.006 g/L. The cement copper produced in the first stage contained nearly 95% copper, whereas that produced in the second stage contained only 12% copper. The first stage cementation effluent contained 4.5 g/L Fe and the second stage effluent contained 4.8 g/L Fe.

Anolyte Oxidation. The cementation effluent was sparged with chlorine gas to oxidize iron to the ferric state. Complete oxidation was ensured by sparging the anolyte until the oxidation potential was above 1000 mV as measured using platinum and calomel electrodes.

Iron Solvent Extraction. Iron was removed from the oxidized anolyte by SX. The organic phase consisted of Amberlite LA-2, a secondary amine, diluted to a concentration of 0.5M with Shell Cyclo-Sol 63 Diluent. Water was used to strip the loaded solvent. A total of eight mixer-settler type contactor units were used in the continuous circuit, four in each of the loading and stripping sections.

The circuit was operated at loading and stripping A:O ratios of 1.5 and 0.65, respectively. After reaching steady state conditions, the circuit extracted over 99% of the anolyte iron and produced a raffinate that contained 0.001 g/L Fe. Some cobalt, molybdenum, and tungsten coextracted with the iron; however, cobalt losses due to coextraction were less than 0.3%. Strip liquors typically contained, in g/L, 11 Fe, 0.02 Co, 0.1 Mo, and 23 Cl.

Carbon Adsorption. The iron-circuit raffinate was treated by carbon adsorption to remove molybdenum and copper which were present at concentrations of 0.08 and 0.006 g/L, respectively. Greater than 77% of the molybdenum and 83% of the copper were removed by carbon adsorption. The composite effluent from the second column contained, in g/L, 0.09 Al, 12 Co, 9.5 Cr, 0.001 Cu, 0.001 Fe, 0.1 Mn, 0.018 Mo, 75 Ni, and 0.02 Ta.

Cobalt Solvent Extraction. The carbon column effluents were treated by SX to concentrate and purify the cobalt. Cobalt was extracted with 0.6M Alamine 336 in Shell Cyclo-Sol 63 Diluent. A total of 12 mixer-settler type contactors were used in the continuous circuit. Six contactors were utilized in the loading section, one in the scrubbing section and five in the stripping section. The single scrubbing stage was incorporated into the circuit to remove entrained aqueous phase from the loaded solvent which would contaminate the strip liquor with nickel and chromium. A cobalt chloride (40 g/L Co) - 1.5M HCl solution was used to scrub the loaded solvent. The scrub solution composition was selected to be in equilibrium with the loaded solvent which contained 6 g/L Co, thus preventing any net cobalt transfer between the aqueous and organic phases.

The stripping section of the circuit was operated in two stages because spent cobalt chloride catholyte from the DMEC did not completely strip the loaded solvent. The spent catholyte, 30 g/L, was utilized in the first three stripping stages to remove 85% of the cobalt from the scrubbed solvent. Water was used to strip the remaining cobalt in the final two stripping stages. After pH adjustment, the primary strip liquor was utilized as feed to the cathode compartment of the DMEC. Cobalt in the secondary strip liquor was recycled after concentration by evaporation to the iron solvent extraction circuit. The circuit was operated at loading, scrubbing, primary stripping, and secondary stripping A:O ratios of 0.5, 0.25, 0.25, and 0.2, respectively.

Essentially all of the cobalt was recovered, and the circuit typically produced raffinates containing, in g/L, 0.09 Al, 0.004 Co, 9.5 Cr, <0.001 Fe, 0.1 Mn, 0.01 Mo, 75 Ni, 0.02 Ta, and 116 Cl. The primary and secondary strip liquors typically contained 57 and 7 g/L Co, respectively. The primary strip liquor had a pH of 0.87 which was raised to a pH of 1.0 with a sodium hydroxide solution. The resulting liquor contained 56 g/L Co and 65 g/L Cl. The impurity profile of the liquor is presented in Table III.

Table III. - Primary strip liquor impurity concentrations, mg/L

Ag	<0.1	Cu	<0.1	Pb	0.015
Al	<1	Fe	<.1	P	<.15
As	<.5	Hg	<.002	Se	.2
Bi	<.2	Mg	.31	Si	.52
Ca	2.0	Mn	13.4	Sn	<.1
Cd	<.01	Na	224	Tl	<.2
Cr	2.0	Ni	47	Zn	.79

DMEC Cobalt Electrodeposition and Operation

A DMEC was utilized to recover high-purity cobalt metal from the treated primary strip liquor. The cell was operated with one anode and one cathode compartment. A SAS anode was utilized in this phase of the investigation for production of actual SAS anolyte, and the back side of the single stainless steel cathode was masked allowing cobalt to plate only on the side facing the anode. The anode compartment was initially filled with anolyte produced in a previous test containing, in g/L, 20 Co, 9.1 Cr, 4.9 Fe, 1.8 Mo, 68 Ni, and 141 Cl. The membrane compartment was filled with a cobalt chloride solution (20 g/L Co) prepared from reagent grade chemicals, and the catholyte compartment was filled with primary strip liquor diluted with deionized water to a concentration of 30 g/L Co.

During cell operation, primary strip liquor (56 g/L Co) was metered to the cathode compartment. A small portion of the spent catholyte (33 g/L Co) was diluted with water to a concentration of 15 g/L Co and pumped to the membrane compartment. Effluent from the membrane compartment was utilized as anolyte make-up solution. This scheme provided a bleed for catholyte impurities.

The cell was operated for 42 hrs at a cathode current density of 430 amps/m^2 and produced a smooth adherent cobalt deposit. The cell initially operated at 5.5 volts; however, within 2 hrs the cell voltage had decreased to 4.6 volts as the membranes reached equilibrium metal ion loadings. A cathode current efficiency of 94% was achieved with an energy consumption of 2.0 kW·h/lb of cobalt metal produced. This level of energy consumption ranges from comparable to significantly better than that reported for some commercial cobalt electrowinning plants operating at similar current densities (16,17).

The cobalt deposit was analyzed by Analytical Associates, Inc., of Detroit, MI. Results indicate that the cobalt met Department of Defense Stockpile Purchase Specification P-13-R6 dated 12-18-85 for grade R6-A cobalt. Grade R6-A cobalt is specified for the most critical applications such as cobalt base- and nickel base-vacuum processed superalloys for jet aircraft engines (18). Table IV gives the analytical results with respect to the Stockpile Specifications.

Table IV. - Comparison of DMEC cobalt against National Defense Stockpile Purchase Specification

	P 13-R6 Specification	DMEC Cathode		P 13-R6 Specification	DMEC Cathode
	Percent			Percent	
Co + Ni	99.95	>99.96	N	0.0014	0.00015
Co	99.85	>99.95	Sb	.0005	<.00003
Ni	.065	.014	P	.0034	<.0010
Fe	.01	.003	Sn	.0014	.0003
Mn	.001	.00085	Se	.0005	.00005
Cu	.0044	.0028	Ag	.0005	<.00002
Pb	.0005	.00025	Zn	.0034	.0010
Si	.0054	.0004	Cd	.0010	.00035
Al	.0054	.0007	Te	.0005	<.00002
S	.0020	.0015	Tl	.0005	<.0001
C	.0140	.0060	Hg	.0005	<.0001
H	.0014	.00068	Cl	.0400	<.0050
O	.0200	.0080	Na	.0200	.0002
Bi	.0001	<.00002	Mg	.0100	.00005
As	.0010	<.00005	Ca	.0200	.0002

The cell dissolved high percentages of anode cobalt and nickel and produced a grey, black anode sludge. Table V presents the electrodissolution results in terms of analyte and anode sludge composition plus the percentage of anode constituents that were dissolved and reported to the anolyte.

Table V. - SAS electrodissolution results

	Composition		
	Anolyte, g/L	Anode sludge, (%)	Dissolution, (%)
Al	1.1	0.06	98.8
C	ND	5.9	ND
Co	29	.21	99.6
Cr	9.5	30	59.5
Cu	1.2	.02	99.7
Fe	5.6	.96	96.3
Mo	1.9	22.5	2.9
Nb	<.8	4.9	28.0
Ni	70	3	98.9
Ta	<.7	7	3.2
Ti	<.4	6.1	.05
W	.2	11.4	.05

ND Not determined.

Total ion migrations and average ion flux across the anode membrane to the membrane compartment and across the cathode membrane to the cathode compartment are shown in table VI for the 42 hrs of cell operation. More than 3 g of nickel migrated across the anode membrane, but due to the decreased concentration gradient across the cathode membrane only 0.12 g of nickel migrated into the catholyte. Only traces of other impurities migrated into the catholyte compartment. Material balance calculations indicate that chloride ions and water migrated across the cathode membrane, through the membrane compartment, and across the anode membrane into the anolyte. No significant migration of cobalt was observed.

Table VI. - Ion and water migrations across membranes

	Ni	Cr	Fe	Cu	Mo	Cl*	H_2O*
Anode Membrane							
Total grams	3.4	0.19	0.19	0.011	0.0019	-102	-81
Flux, g/m^2/day	555	31	31	1.80	.310	-16,650	-13,290
Cathode Membrane							
Total grams	.12	.0009	.0086	.0018	.0018	-108	-40
Flux, g/m^2day	19.6	.15	1.40	.29	.29	-17,630	-6,560

*Chloride and water migrated in the opposite direction of the other ions.

Nickel Recovery

Research on nickel recovery from cobalt solvent extraction circuit raffinate was conducted concurrently with the previously described work; consequently, actual process solutions were not available. Simulated solutions prepared from reagent grade chemicals were used in this phase of the investigation.

Nickel - Chromium Separation. Several methods of separating the nickel and chromium were investigated. The first area investigated was chromium hydroxide precipitation which was unsuccessful because the gelatinous chromium-hydroxide precipitates were unfilterable. Solvent extraction with aliphatic oximes (19) was next investigated; however, the high nickel-to-chromium ratio in the cobalt SX circuit raffinate made this approach impractical.

The most successful procedure tested was an approach previously developed by the Bureau (7) to selectively precipitate a basic chromium sulfate. Tests were conducted with a simulated cobalt-circuit raffinate which contained, in g/L, 9.5 Cr, 80 Ni, and 120 Cl. Test results indicated that the procedure precipitated over 99% of the chromium; however, 27% of the nickel coprecipitated with the chromium. A filtrate containing <0.001 g/L Cr and 50 g/L Ni was produced. Subsequent test work was unsuccessful at significantly improving the chromium - nickel separation. Additional research is needed in this area.

DMEC Nickel Electrodeposition. A DMEC was utilized to recover high-purity nickel metal from simulated nickel chloride filtrates. The cell was operated with one anode compartment, a SAS anode, one cathode compartment, and a stainless steel cathode masked on one side. The anode compartment was initially filled with a simulated anolyte containing, in g/L, 15 Co, 9.5 Cr, 5 Fe, 80 Ni, and 150 Cl. The membrane and cathode compartments were initially filled with nickel-chloride solutions containing 15 and 50 g/L Ni, respectively.

Three separate tests were conducted at current densities of 107.5, 215, and 430 amps/ft^2 for 48, 24 and 12 hrs, respectively. Water was utilized as anolyte makeup solution. The pH of the cathode compartment feed was 5.5 and the pH of the cathode compartment discharge was 2.5 to 3.0.

Visual inspection of the nickel deposits indicated that they were highly stressed. The deposit produced at a current density of 430 amp/ft^2 partially peeled off the stainless-steel cathode and grew through the cathode membrane. Although the other two deposits did not grow through the cathode membrane, they did partially separate from the cathode. Apparently a change is needed in the catholyte composition, such as the addition of a leveling agent, to reduce the stresses within the nickel deposits.

The deposits produced at 107.5 and 215 amps/ft^2 were analyzed by analytical Associates, Inc., of Detroit, MI. Results indicate that both nickel deposits were very similar in composition and with the exception of hydrogen and oxygen, met DOD Stockpile Purchase Specification, P-36-R4 dated 26 Jan., 1983. Table VII gives the analytical results for the anode produced at 215 amps/ft^2 with respect to the DOD specification. Steady-state cell voltages, current efficiencies and energy consumptions are presented in table VIII for the tests conducted at cathode current densities of 107.5 and 215 amps/ft^2. DMEC cell voltages, current efficiencies, and energy consumptions are comparable to those reported by Mantell for the electrolytic refining of nickel (20).

Table VII. - Comparison of DMEC nickel cathode against DOD Stockpile Purchase Specification, percent

Constituent	DOD Specification	DMEC Nickel Cathode
Ni	≥99.95	>99.97
Ag	≤.0005	.00008
As	≤.0002	.00008
Bi	≤.00003	.00002
C	≤.015	.0025
Co	≤.005	<.001
Cu	≤.001	<.001
Fe	≤.02	.005
H	≤.00007	.0033*
Mn	≤.0001	.00003
N	≤.0005	.00034
O	≤.008	.017*
P	≤.0005	<.0005
Pb	≤.0001	<.0001
S	≤.002	.0003
Sb	≤.0001	.0001
Se	≤.00005	<.00001
Si	≤.001	<.0002
Sn	≤.0005	.0001
Te	≤.00005	<.00001
Tl	≤.00005	<.00002
Zn	≤.001	.00095

*Exceeds specification.

Table VIII. - Nickel DMEC operating results

	Cathode Current Density, Amps/ft^2	
	107.5	215
Steady-state DMEC Voltage, V	1.6	2.9
Current Efficiency, Percent	98.3	97.1
Energy Consumption, kW·h/lb Nickel	.67	1.2

Conclusions

This work has demonstrated that a DMEC, coupled with appropriate solution purification technology, can be utilized to produce high purity metal from impure anodes. High-purity cobalt meeting DOD specifications and high-purity nickel were produced in a DMEC while SAS anodes were simultaneously dissolved.

- In spite of the voltage drop across the DMEC membranes, laboratory tests have indicated that energy consumptions in the DMEC are comparable or better than those reported in current electrowinning practice. In addition, the DMEC does not produce the troublesome acid mists or the corrosive gases that are often associated with electrowinning operations.

- The DMEC membranes showed long-term stability under process conditions, prevented anode impurities from contaminating the cathode products, and allowed chloride ions to migrate from the cathode to the anode, thus reducing reagent costs by recycling chloride within the process.

- Copper cementation, anolyte oxidation, iron solvent extraction, carbon adsorption and cobalt solvent extraction produced from a complex SAS anolyte a high-purity cobalt chloride solution from which high-purity cobalt was won.

- Improvements in the chromium-nickel separation are needed to effectively integrate nickel recovery with the proposed process for treating SAS. Additional research is needed in this area.

REFERENCES

1. R. S. Kaplan. An Overview of the Bureau of Mines Recycling Research. Paper in Recycle and Secondary Recovery of Metals., ed. by P. R. Taylor, H. Y. Sohn and N. Jarret (Proc. Int. Symp. Recycle and Send. Recovery Met. and Fall Extr. and Process Metall. Meet., Fort Lauderdale, FL, Dec. 1-4). Metall. Soc. of AIME, 1985, pp. 3-11.

2. J. F. Papp. Superalloy Recycling 1976-1986. Paper in Superalloys 1988. ed. by S. Reichman, D. N. Duhl, G. Maurer, S. Antolovich and C. Lund (Proc. 6th Int. Symp. on Superalloys, Champion, PA, Sept. 18-22, 1988). Metall. Soc. of AIME, 1988, pp. 367-376.

3. L. R. Curwick, W. A. Peterson, and H. V. Makar. Availability of critical Scrap Metals Containing Chromium in the United States. Superalloys and Cast Heat- and Corrosion-Resistant Alloys. BuMines IC 8821, 1980, 51 pp.

4. J. J. deBarbadillo, J. K. Pargeter and H. V. Makar. Process for Recovering Chromium and Other Metals from Superalloy Scrap. BuMines RI 8570, 1981, 73 pp.

5. C. L. Kusik, K. Parameswaran, D. J. Kinneberg, and H. V. Makar. Pyrometallurgical Recovery of Chromium from Scrap Metals: Laboratory Studies. BuMines RI 8571, 1981, 41 pp.

6. J. D. Hall. Method of Preparation of Metal Salts. U.S. Pat. 2,716,588, Aug. 30, 1955.

7. P. T. Brooks, G. M. Potter, and D. A. Martin. Chemical Reclaiming of Superalloy Scrap. BuMines RI 7316, 1969, 28 pp.

8. E. R. Baggott, A. W. Fletcher, and T. A. W. Kirkwood. Recovery of Valuable Metals from Nickel-Cobalt Alloy Scrap. Paper in Proceeding of Ninth Commonwealth Min. and Met. Congress. Institute of Mining and Metallurgy, London, 1970, pp. 803-821.

9. D. Pearson. Recovery of Cobalt from Metallurgical Wastes. Paper in Proceedings of International Conference on Cobalt: Metallurgy and Uses. Benelux Metallurgie, Brussels, Belgium, 1981, pp. 139-148.

10. L. D. Redden, R. D. Groves, and D. C. Seidel. Hyrometallurgical Recovery of Critical Metals From Hardface Alloy Grinding Waste: A Laboratory Study. BuMines RI 9210, 1988, 31 pp.

11. J. A. Thomas, D. L. Jones, and V. A. Ettel. Chromium Recovery from Superalloy Scrap by Selective Chlorine Leaching. U.S. Pat 4,377,410, Mar. 22, 1983.

12. B. H. Rosof. Process for Recovering Valuable Metals from Superalloy Scrap. U.S. Pat 4,138,249, Feb. 6, 1979.

13. A. Aue, L. Skjutare, G. Bjorling, H. Reinhardt, and J. Rydberg. Separation of Iron, Cobalt and Nickel from Scrap Alloy by a Solvent Extraction Process. Paper in Proceedings of the International Solvent Extraction Conference. Society of Chemical Industry, London, 1971, v. 1, pp. 447-450.

14. M. Hayashi, D. R. Peterson, and D. W. Bridges. Electrochemical Recovery of Cobalt-Nickel Alloy from Superalloy Scrap. BuMines RI 6645, 1964, 35 pp.

15. J. L. Holman, S. A. Molloy, and L. A. Neumeier. Recovery of Nickel-Cobalt Alloy From Superalloy Scrap. Paper in Process Mineralogy VII, ed. by A. H. Vassiliou, D. M. Hausen, D. J. T. Carson (Proceedings of Process Mieralogy Symposium., Denver, CO, Feb. 23-27, 1987). Metallurgical Soc. of AIME, 1987, pp. 347-360.

16. J. Aird, R. S. Celmer, and V. A. May. "New Cobalt Production From R.C.M.'s Chambishi Roast-Leach-Electrowin Process," Mining Magazine, 143, Oct. 1980, pp. 320-336.

17. R. Agnew, P. Garritsen, and D. MacVicar. Production of Electrolytic Cobalt Rounds at INCO's Cobalt Refinery. Paper in Extractive Metallurgy of Nickel and Cobalt, ed. by G. P. Tyroler and C. A. Landolt (Proceedings of Symposium, Phoenix, AZ, Jan. 25-28, 1988). Metallurgical Soc. of AIME, 1988, pp. 555-561.

18. American Society for Metals Panel. Quality Assessment of National Defense Stockpile Cobalt Inventory (Federal Emergency Management Agency contract EMW-C-1023). Am. Soc. Met., Metals Park, OH, 1983, 40 pp.

19. R. D. Groves and L. D. Redden. Nickel Extraction From Acidic Chloride Solutions With Aliphatic Oximes, Hydrometallurgy, 24 (1990), in press.

20. C. L. Mantell, Electrochemical Engineering (New York, NY: McGraw-Hill Book Company, Inc., 1960), pp. 142-197.

PROCESS TO RECYCLE CONTAMINATED SUPERALLOY SCRAP

L. J. Lutz[1], S. A. Parker[2], J. L. Holman[3], J. B. Stephenson[4]

Rolla Research Center - U.S. Bureau of Mines
1300 North Bishop
Rolla, Missouri 65401

Abstract

The Bureau of Mines has developed a process to recover Ni, Co, Cr, and other metals from mixed and contaminated superalloy scrap. The superalloy scrap is melt-carburized and cast into anodes for electrodeposition. The carburizing/electrorefining process recovers approximately 90 pct of the Ni-Co available from the scrap as a 99+ pct pure Ni-Co alloy. Carburization results in Cr and other refractory metals reporting to the anode alloy primarily as carbide phases. These carbides form an insoluble anode sludge during electrolysis. Electrorefining data and operating parameters are given for a 10-L seven-electrode compartmented cell. Techniques for the removal of Fe, Cr, and trace impurities from the recycled electrolyte are outlined.

[1]Chemical engineer
[2]Physical science technician
[3]Research physicist (retired)
[4]Supervisory Physical Scientist

Second International Symposium -
Recycling of Metals and Engineered Materials
Edited by Jan H. L. van Linden,
Donald L. Stewart, Jr., and Yogeshwar Sahai
The Minerals, Metals & Materials Society, 1990

Introduction

The single largest use for Co is the production of superalloys (SA). Because of stringent specifications for SA, relatively little scrap is charged to SA melts (1). Mixed or contaminated, SA-type scrap is commonly reprocessed by secondary metal refiners and downgraded for use as Ni-enriched feedstock for stainless steel (SS), cast Fe, or alloy steel.

A number of processes for the recycling of Ni- and Co-based alloys or separation and recovery of specific constituents from these alloys have been investigated, but only a limited number have been used commercially. These include Kenworthy's research that demonstrated Cr removal from SA scrap by melt oxidation (2) and Would's oxidation removal of Al, Ti, Zr, Hf, and other reactive elements from B1900+Hf scrap with subsequent vacuum melt refining and alloy reconstitution (3). Kusik (4) investigated a scheme for melt oxidation whereby the Cr and Fe from SS scrap were oxidized to the slag for subsequent reduction to ferrochromium. Bureau-sponsored research by deBarbadillo led to a procedure (5) whereby SA scrap was melt oxidized and sulfidized under controlled conditions. Beneficiated chromium sulfide was roasted and reduced; whereas, cobalt and nickel sulfides were leached, solvent-extracted, and electrowon as pure metals.

Bureau research on SA reclamation in the 1960's used a variety of constant-current electrodeposition and hydrometallurgy techniques (6-7). Various mixtures of Ni-Co-Cr-bearing SA scrap were examined; these were comprised of turnings, grinding sludge, and bulk scrap. The scrap was melted into soluble anodes for electrolysis. Magnetic or other separations techniques were examined to remove portions of the grinding media from grinding sludges. Electrolysis tests were also conducted in which scrap was contained in Ti anode baskets and directly electrowon. This approach encountered numerous problems with passivity of the scrap surface. Also, no satisfactory procedure was devised to recover the Cr in this scrap, and the byproduct recovery of Mo and tungsten was not addressed. Additionally, recovery of minor elements, such as Cb and Ta, was not specifically addressed nor were overall material balances for continuous operations.

Brooks and others at the Bureau (7) developed and demonstrated, on a laboratory scale, a process for treating particulate SA scrap to recover Ni, Co, Mo, and Cr. The principal features of the process were dissolution of the scrap in acid with solvent extraction separation of Mo, Fe, Co, and Mn, then precipitation of chromium sulfate, and finally precipitation of nickel carbonate. The Mo, Co, and Ni were calcined to metallurgical grade oxides with the final Cr product being a crude chromium basic sulfate. Recoveries on the order of 90 pct were reported for Ni, Co, Mo, and Cr.

The current investigation includes melting scrap charges, refining the melt, and casting soluble anodes for electrorefining to produce a recyclable Ni-Co alloy. This combined pyrometallurgical and electrometallurgical approach utilizes controlled-potential electrolysis (CPE) to selectively deposit Ni-Co alloy and reclaim Cr, W, Mo, and other metals from the process as byproducts (see figure 1). Various aspects of this research were previously reported (8-10). The approach presented here was selected based on its application to various forms of scrap and an economic advantage by producing a Ni-Co alloy over the production of pure Ni and Co.

CPE was utilized to enhance the selectivity of elemental deposition. In the past, CPE has been used primarily as an analytical tool and has not been developed for commercial selective electrodeposition of metals and alloys. CPE is based on the fact that different metals ideally exhibit different reduction potentials (11). Standard reduction potentials for selected metallic species of interest are shown in table I.

Table I. - Selected standard reduction potentials (12)

Cathode reaction	E_o[1], V
Cu^{2+} + 2e = Cu..	+0.340
Mo^{3+} + 3e = Mo..	-0.20
Ni^{2+} + 2e = Ni..	-0.23
Co^{2+} + 2e = Co..	-0.28
Fe^{2+} + 2e = Fe..	-0.409
Cr^{3+} + 3e = Cr..	-0.74

[1] Reference normal hydrogen electrode = 0.0

The standard reduction potentials for individual metals are derived for highly idealized and controlled conditions not realized in actual practice. Combinations of metals, variations of electrolyte ions, pH, and factors such as overvoltage, passivity, electrolyte composition, and temperature all lead to shifts in potential and to anomalous behavior for a particular metal (14). However, CPE provides a more controlled and reproducible deposition and can restrict codeposition of some elements that are more electronegative than the alloy being deposited. A compartmentalized cell was needed to help control the transfer of impurities that tend to codeposit with the Ni-Co alloy.

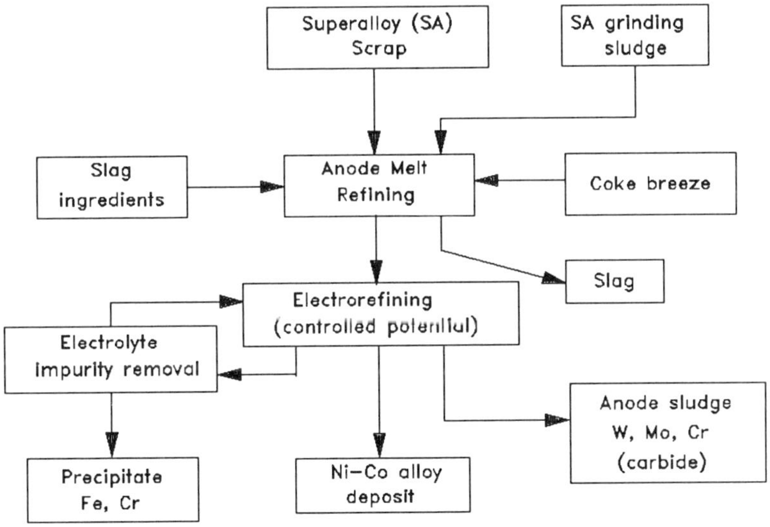

Figure 1 - Flowchart for recycling superalloy scrap

Materials and Methods

Melt-Refining

The mixed and contaminated SA turnings and the contaminated SA grinding sludge used in this work were obtained from an SA scrap processor. The batch analyses of turnings and deoiled sludge are shown in table II.

Table II. - Typical compositions of deoiled superalloy turnings scrap and grinding sludge, weight percent

Elements	Turnings	Deoiled grinding sludge
Ni......	59	26
Co......	14	12
Cr......	19	8
Fe......	1.7	2
W.......	0.3	6
Mo......	3.8	4
Si......	0.0	2
$Al2O_3$...	2.2	40

The direct carburized melt-refining experiments were conducted in a 4.5 kg capacity, 17kV•A indirect-arc electric rocking furnace lined with an alumina refractory. A typical furnace charge was made up of 1,200-g SA turnings scrap and 667-g deoiled SA grinding sludge, along with additions of (by wt pct) 0.3 Ta, 0.4 Cb, and 0.7 V. These Ta, Cb, and V additions were made to the melt in order to follow these elements through the process. Also added were slag-forming ingredients (excluding the alumina contained in the grinding sludge): 110 g CaO, 125 g SiO_2, 40 g CaF_2. Fifty grams of coke breeze were added with the furnace charge, an additional 140 g added after the charge had been melted. After stirring the coke breeze into the melt, it was held at ~1,550°C for 15 min before pouring into a conical carbon mold. The slag rose to the surface during solidification. The alloy product was remelted, under a partial He vacuum (-254 mm Hg), and cast into 13-cm by 18-cm by 1-cm anodes for use in the electrorefining step. The remelting was performed for experimental ease; in commercial practice, the melt refining and anode casting would be done in a single operation.

Electrorefining

The 2-L cell design consisted of two anodes and one cathode separated by a membrane. The membrane, a commercial synthetic cloth, permitted a "head" to be maintained on the catholyte so that net electrolyte flow was from the cathode to the anode chamber. The cell was operated at 55° C at a cathode current density (CD) of ~30 A/ft^2.

The 10-L cell used to electrorefine the cast anodes was constructed of polypropylene; the cell contained four anodes and three cathodes. The cathode compartments utilized acrylic diaphragm cloth (permeability rating of 7.5 to 23.0 cm^3/min per cm^2 of cloth) on the sides adjacent to the anodes. Purified electrolyte containing Ni and Co (the catholyte reservoir initially contained ~200-L) was pumped into the three catholyte compartments at a rate of 0.67 L/h per compartment. A slight pressure "head" of 1.5 cm was maintained in the catholyte compartments as the electrolyte flowed through the diaphragms into the anolyte compartment. Anolyte was maintained at a fixed overflow level. The flow of electrolyte away from the cathode directed impurities dissolved at the anode away from contact with the cathode.

The starting electrolyte contained: 45 g/L Ni, 4.5 g/L Co, 90 g/L Cl, 18 g/L SO_4, and 5 g/L B (added as boric acid). Starter cathodes of 316 SS or Ni 200 were used, and the cell temperature was maintained at 55°±5° C by surrounding the cell with a constant temperature coil. The pH was monitored and controlled in each catholyte compartment to a pH of 1.00 ± 0.5. The cell voltage was periodically adjusted to maintain the cathode potential of the center cathode in the range of -0.53 to -0.63 V versus standard hydrogen electrode. When used in this manner, the power supply behaved as a potentiostat.

Anolyte Purification

The anolyte overflow reservoir was purified at 40-h intervals by adding H_2O_2 to oxidize the Fe^{2+} to Fe^{3+}, then raising the pH to 4.0 by an addition of 6M NaOH. After stirring for 1 h at this pH, the Fe-Cr precipitate was removed by filtering. The filtrate was reheated to 55° C, reacidified with a mixture of HCl and H_2SO_4, and then pumped through a wound polypropylene filter and through an activated carbon filter back into the catholyte reservoir for recycling to the cell. The filtering was done to remove minor impurities and particulates

Results and Discussion

Melt-Refining

The direct melt-carburization of the scrap provided relatively high recovery of all of the metals present in the scrap. The Cr, W, Mo, Cb, Ta, and V were retained in the anode alloy as carbidic compounds that remained insoluble during electrolysis. This provided an opportunity to recover these highly oxidizable metals which are difficult to reduce once oxidized. Previous reports (9,10) addressed the other melt-refining processes. Both the metal and slag were separated from the mold, weighed, and chemically analyzed. The average elemental distribution between the metal phase and the slag phase in the SA scrap melts is shown in table III. Table IV shows the average composition of the cast SA scrap anodes.

Table III. - Average metal distribution in SA scrap melts, weight percent

	Ni	Co	Fe	Cr	W	Mo	Ta	Cb	V
Slag...	3	3	8	4	5	4	37	12	27
Metal..	97	97	92	96	95	96	63	88	73

Table IV. - Composition of cast scrap superalloy anodes

Element	Wt pct	Element	Wt pct
Ni.....	52.0	Ta.....	0.19
Co.....	13.0	Cb.....	0.35
Cr.....	15.4	Zr.....	0.44
Fe.....	2.2	V......	0.51
Mo.....	3.4	Si.....	4.7
W......	1.0	C......	3.4

Electrorefining and Anolyte Purification

Anolyte Purification: The anolyte purification procedure described previously and the use of CPE enabled the deposit composition to be consistently maintained at <0.2 pct Fe and <0.2 pct Cr, with the balance essentially a Ni-Co alloy. The typical compositions for the electrolyte before and after purification is given in table V.

Table V. - Anolyte purification concentration

Element	Anolyte composition (g/L)	
	Initial	Final
Co.....	11.1	10.3
Ni.....	37.1	37.9
Fe.....	0.05	0.002
Cr.....	1.78	0.028
Mo.....	0.01	0.004
W......	0.12	0.017

The precipitate from the anolyte purification procedure contained large amounts of Cr and Fe along with smaller amounts of Mo, W, and other metals. Minor amounts of Ni and Co were also present in the precipitate. The anolyte purification procedure was very successful in keeping the solution impurity level low during electrorefining.

2-L Cell Experiments: Initial electrochemical experiments conducted in a 2-L compartmented cell used two scrap anodes (~250 g each) and one 316 SS cathode. Cell temperature was maintained at 55° C with an anolyte volume of 850 mL and catholyte volume of 375 mL. The pH was controlled at 1.0 or 0.5 and the CPE setting was -0.85 V versus SCE. The data given in table VI illustrate the results for 75- to 100-h tests using direct-carburized (DC) anodes. Tests I and II differed in the controlled-electrolyte pH, being pH 1.0 for test I and pH 0.5 for test II. Test III, which used high-Fe (17 pct) anode material, was also conducted at pH 0.5.

Table VI. - Data for 2-liter electrolytic cell

	Experiments		
	I	II	III
Duration.................h..	100	100	75
Anode type..................	DC	DC	DC[1]
Electrolyte.............pH..	1.0	.5	.5
Anode dissolved..........g..	136	209	[2]192
Cathode deposit..........g..	160	172	[2]153
Sludge formed...........g..	61	90	[2]51
Anode CE...............pct..	97	83	89
Cathode CE.............pct..	89	59	66
Cathode CD...........A/ft^2..	20	31	27
Average cell voltage.....V..	1.11	1.50	1.04
Anode dissolution..g/(A•h)..	.82	.79	.90
Ni-Co deposit......g/(A•h)..	.97	.65	.72
Energy/kg Ni-Co.......kW•h..	1.14	2.32	1.44

[1]Iron added to furnace charge.
[2]Calculated on 100-h basis.

It is evident from the data in table VI that, at a pH of 1.0, the cathode current efficiencies (CE) is about 90 pct. As the electrolyte pH is decreased to 0.5, a decrease occurs in the cathode CE and the cell current and voltage increase, which is reflected in higher cathode CD values. The drop in CE is attributed to increased H_2 evolution at the cathode. The calculated cell energy needed to produce 1 kg of Ni-Co deposit is about 0.9 to 1.1 kW•h if a pH of 1.0 is maintained. The required energy rises for lower electrolyte pH as more H_2 is given off at the cathode reducing the current efficiency. Thicker anode sludge buildup causes increased cell voltage, which also results in higher energy consumption.

Deposit and anode sludge compositions for experiments I through III are given in table VII. Nickel in the deposits ranged from ~77 to 89 pct and the Co ranged from ~10 to 23 pct. The amounts reflect the anode compositions in only a general sense as the electrolyte used for the tests had been used previously and had varying starting concentrations of Ni and Co. The electrolyte Ni:Co ratio has a strong influence on the deposit Ni:Co ratio. The deposits for experiments I through III each contained <0.1 pct Fe. Chromium in the deposits ranged from 0.15 pct for test I to <0.01 pct for test III. The Mo and tungsten were 0.01 pct or less in each of the deposits and the carbon was 0.02 pct or less.

Table VII. - Deposit and Anode Sludge Compositions from 2-Liter Cell Experiments, weight percent

Test No.	Ni	Co	Cr	Fe	Mo	W	C
I-D....	76.8	23.0	0.15	0.05	<0.01	<0.01	.02
I-S....	5.3	5.1	35.2	1.0	6.5	2.8	8.3
II-D...	80.6	18.9	.13	.03	<.01	<.01	.01
II-S...	4.5	5.1	38.6	1.1	7.1	4.2	7.6
III-D..	86.6	12.4	<.01	.05	<.01	<.01	.01
III-S..	2.4	1.8	40.6	5.8	11.2	5.4	6.7

D Deposit. S Anode sludge.

Elemental distributions during electrolysis are given in table VIII. The data in table VIII show that >98 pct of the total Cr+Fe+Mo+W is in the anode sludge and purification precipitate. Comparison of tests I and II shows that the lower pH caused a decrease in the amount of Ni+Co deposited relative to the amounts dissolved. This was implied in the data in table VI, where more metal dissolved than was deposited for test II, whereas the reverse was true for test I. The relative distributions for test III differ from those for test II, also run at pH 0.5 mainly in the deportment of the high Fe dissolved, which essentially reported to the electrolyte precipitate.

Table VIII indicates negative percentages of Ni+Co in the final electrolytes. This reflects the fact that Ni+Co was deposited faster than it was dissolved from the anodes. At the cathode, while essentially only Ni+Co was being deposited, energy was being consumed at the anode to dissolve material containing 80 pct or less Ni+Co, plus Fe and other soluble constituents. There is, therefore, a mass transfer imbalance of Ni+Co for the conditions described. The anode would have to dissolve at a considerably faster rate or deposit Ni-Co at a slower rate in order to maintain a constant Ni+Co concentration in the electrolyte. When the electrolyte pH was lowered from 1.0 to 0.5, a better Ni+Co balance was achieved at the expense of reduced cathode efficiency.

Table VIII. - Percentage distribution of metal for electrolysis of direct-carburized anodes

	Test I		Test II		Test III	
	Ni+Co	Cr+Fe+Mo+W	Ni+Co	Cr+Fe+Mo+W	Ni+Co	Cr+Fe+Mo+
Dissolved........	93	32	94	29	98	65
Precipitate[1]....	9	31	11	28	10	64
Deposit[1]........	175	<1	115	.5	114	<.1
Electrolyte[1-2] ..	-91	<1	-32	<1	-26	~1
Anode sludge.....	7	68	6	71	2	35

[1]Distribution of metals dissolved from anodes.
[2]After deposition and purification.

<u>10-L Cell Experiments:</u> The 200-h test described in the following is a representative run of the 10-L cell. Four anodes of composition shown in table IV, anode dimensions 13 cm by 18 cm by 1 cm, were attached to SS hangers and hung on Cu bus bars with three SS starter cathodes (cathode dimensions 16 cm by 19 cm by 0.075 cm). The anodes were sandblasted and the cathodes were cleaned with No. 240 grit paper and holes punched around the bottom and sides at 4.5 cm intervals to avoid premature peeling of the cathode deposit. The edges of the cathodes were masked with tape to avoid high current density points where dendrites might form. The following parameters were recorded every 1/2 h: temperature in each cathode compartment, cell voltage, cell current, center cathode potential, pH in each cathode compartment, and accumulated A•h. A simplified flow diagram of the process is shown in figure 2, and results from the 200-h test are shown in table IX.

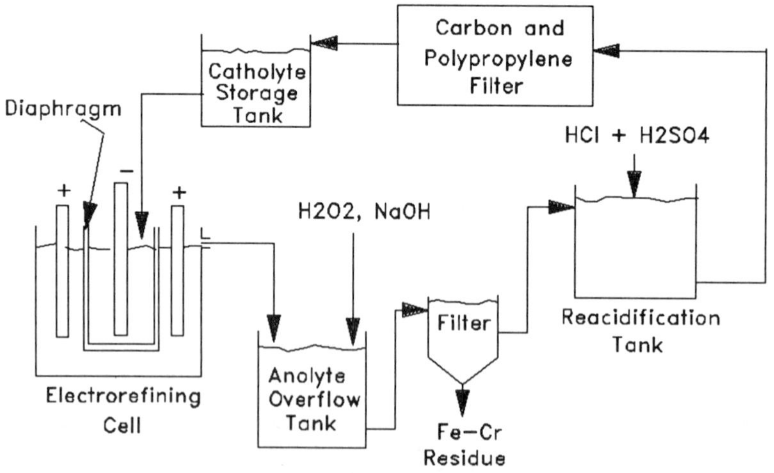

Figure 2 - Schematic of Electrolytic Flow Circuit

Table IX. - Data for 200-hour electrorefining experiment, 20-liter cell

Electrolyte..........................pH..	1.0
Anode dissolved........................g..	4,445
Cathode deposit........................g..	4,864
Anode sludge...........................g..	2,133
Electrolyte purification precipitate...g..	1,464
Anode CE.............................pct..	88
Cathode CE...........................pct..	81
Cathode CD.........................A/ft^2..	18
Average cell voltage...................V..	1.47
Anode dissolution...............g/(A•h)..	.81
Ni-Co deposition................g/(A•h)..	.88
Energy used per kilogram of Ni-Co deposited.............kW•h/kg..	1.67

After 48 h into the 200-h test, the anodes in the end compartments were exchanged with those in the middle of the cell so that the sides not facing cathodes could be dissolved more uniformly. At the end of 100 h, the first set of four anodes were removed from the cell and replaced with a new set, since a build-up of anode sludge on the anode surface reduced cell efficiency if the sludge was left in place. Three and one-half liters distilled water were added each day to the catholyte reservoir to make up for evaporation loss.

At the conclusion of the test, the catholytes and anolyte were sampled for chemical analysis and then transferred to the final anolyte overflow to be purified along with the last batch of unpurified electrolyte. Anode sludge was scraped or peeled from the anodes. The sludge was then ball milled and washed in distilled water to remove all of the water-leachable Ni+Co. The leach water was added to the final batch of electrolyte for purification. The small amount of anode sludge that could not be removed was sandblasted from the anodes with the anode weight change noted and added to the total weight of anode sludge. Drillings were taken from the deposits for chemical analysis.

Experimental Summary: During anode dissolution and cell operation, various degrees of anode passive-active behavior were observed. When dissolution represented active behavior, the cations dissolved as W^{6+}, Mo^{6+}, Cr^{3+}, Fe^{2+}, Co^{2+}, and Ni^{2+}. When the dissolution reflected passive behavior, Fe^{3+} and Cr^{6+} were formed instead of Fe^{2+} and Cr^{3+}. The single most important factor affecting the passive-active nature of the anodes was the carbon content. The presence of 0.9 pct carbon or more resulted in essentially active dissolution, decreasing carbon resulted in increasing passive dissolution. Mixtures of Fe^{2+} with Fe^{3+} and Cr^{3+} with Cr^{6+} resulted in mixed active-passive dissolution. Because the anodes contained constituents other than Ni+Co and anode and cathode (CE) varied according to conditions, nearly twice as much Ni+Co was deposited in initial experiments as was dissolved.

During dissolution and deposition of Ni-Co alloy, some 13 to 18 pct of the anode remained as insoluble anode sludge, which was composed of carbides, nitrides, intermetallics, and slag inclusions. The sludge contained essentially all of the Cb, half or less of the tungsten, and a fourth or less of the Cr and Mo. Unless considerable tungsten was present, the sludge mostly dropped from the anodes. The cell voltage increased with time because of the sludge formation.

In the sulfate-chloride electrolytes, Co was found to deposit more readily than Ni. The pct Co in the deposit also increased with increasing temperature, agitation, and CD. The Ni:Co ratio of deposits increased with increasing deposition period. The target for the experiments was nominally 80-20 Ni-Co. The actual composition of the final deposit is shown in table X. Electrolyte

Table X. - Deposit and anode sludge composition, weight percent

Element	Deposit	Anode sludge
Co.....	18.5	1.2
Ni.....	81.3	1.1
Fe.....	0.2	0.52
Cr.....	<0.01	43.4
Mo.....	0.01	9.8
W......	0.06	2.6
Si.....	0.52	8.8
C......	0.03	10.7
Cb.....	0.01	1.2
Ta.....	0.02	0.6
V......	<0.00	1.9
Zr.....	0.04	0.3

temperature ranged from 55° to 60° C. The Mo and Fe (and Cu and Zn when present) readily codeposited with the Ni-Co; Mn did not codeposit. If the catholyte pH was <1, Cr up to 2.4 g/L did not codeposit. At a pH of 2, however, the presence of Cr impaired Ni-Co deposition so severely that no more than 0.05 g/L Cr could be tolerated.

Electrolytes contained substantial concentrations of both SO_4^{2-} and Cl^-, but normally more SO_4^{2-}. When Cl^- was increased relative to SO_4^{2-}, deposits became more stressed. When Cl^- was decreased, deposits increasingly contained "tree" dendrites at cathode edges. Some 50 to 55 g/L Ni+Co was considered near optional. Higher or lower concentrations tended to produce inferior deposits. Stainless steel cathodes were strippable when electrolyte pH was >1.6. For tests with more acidic electrolyte, Ni-Co starting cathodes were used. Cathode CE was near 100 pct at pH 2 and decreased as electrolyte acidity was increased.

Conclusions

This work has demonstrated that it is possible to recover 90+ pct of the Ni-Co in SA scrap as a 99+ pct pure Ni-Co deposit.

The melt refining step of this process partitions approximately 92 pct of the metallic constituents in the scrap to a cast anode where these metallic elements are found as either metallic carbides or as elemental constituents. The electrowinning of a Ni-Co product was shown to be possible on a continuous basis using the purification and electrorefining processes described.

Future Research

Since melt-refining accounts for most of the estimated capital and operating costs associated with this process, the future direction of this work will be to eliminate or improve the melt-refining step of this process, thus improving the economics of the process.

References

1. K. B. Shedd, COBALT (BuMines Minerals Yearbook, 1988), 9.

2. H. Kenworthy, V. A. Nieberlein, and A. G. Starliper, "Experimental Extraction of Strategic Components From S-816 Alloy Scrap" (BuMines RI 5786, 1961), 27-28.

3. M. J. Woulds, "Recycling of Engine Serviced Superalloys" (Presentation at Fourth International Symposium on Superalloys, Sept. 21-25, 1980, Champion, PA), 11.

4. C. L. Kusik, K. Parameswaran, D. J. Kinneberg, and H. V. Makar, "Pyrometallurgical Recovery of Chromium From Scrap Metals: Laboratory Studies" (BuMines RI 8571, 1981), 41.

5. J. J. deBarbadillo, J. K. Pargeter, and H. V. Makar, "Process for Recovering Chromium and Other Metals From Superalloy Scrap" (BuMines 8570, 1981), 73.

6. M. Hayaski, D. R. Peterson, and D. W. Bridges, "Electrochemical Recovery of Cobalt-Nickel Alloy From Superalloy Scrap" (BuMines RI 6445, 1964), 35.

7. P. T. Brooks, G. M. Potter, and D. A. Martin, "Chemical Reclaiming of Superalloy Scrap" (BuMines RI 7316, 1969), 28.

8. J. L. Holman and L. A. Neumeier, "Deposition of Nickel-Cobalt Alloys by Controlled-Potential Electrolysis" (BuMines RI 8931, 1985), 17.

9. J. L. Holman and L. A. Neumeier, "Experimental Nickel-Cobalt Recovery From Melt-Refined Superalloy Scrap Anodes" (BuMines RI 9034, 1986), 17.

10. J. L. Holman, S. A. Molloy, and L. A. Neumeier, "Recovery of Nickel-Cobalt Alloy From Superalloy Scrap" (Process Mineralogy VII, Proceedings of Process Mineralogy Sessions at the AIME Annual Meeting, Denver, CO, 23-27 February 1987), 347-360.

11. J. J. Ligane, Electroanalytical Chemistry, 2nd Edition (NY,NY: Interscience Publishers, Inc. 1958), 669.

12. R. C. Weast, ed., Handbook of Chemistry and Physics (Cleveland, OH: The Chemical Rubber Company, 1972-1973), D-111.

13. R. H. Perry, ed., "Cost and Profitability Estimation," Perry's Chemical Engineers Handbook, 5th ed. (NY: McGraw-Hill 1973), 47.

14. A. Brenner, Electrodeposition of Alloys - Principles and Practice, vol. 1 and 2 (NY and London: Academic Press, 1963), 1,370.

CHEMICAL PROCESSING OF HARDFACE ALLOY GRINDING WASTE

L. D. Redden

Salt Lake City Research Center
Bureau of Mines, U.S. Department of the Interior
729 Arapeen Drive
Salt Lake City, Utah 84108

PROCESS ECONOMICS

R. Swarup

Office of Process Evaluation
Bureau of Mines, U.S. Department ofthe Interior
Washington, DC 20241

Abstract

The U.S. Bureau of Mines has investigated technology to recover critical metals from hardface alloy grinding waste, a form of superalloy scrap. This report describes a hydrometallurgical process for the recovery of values such as cobalt, chromium, manganese, nickel, and tungsten. The process involved deoiling the grinding waste followed by chloride-based dissolution of the deoiled material. A high-grade calcium tungstate product was recovered from the resultant leach residue. Iron, cobalt, manganese, and nickel chlorides were sequentially recovered from the leach liquor by solvent extraction. Market-grade cobalt chloride, manganese chloride, and nickel chloride products can be produced directly from the strip liquors, and the final raffinate, a relatively pure chromic chloride solution, compared favorably with commercial solutions. An economic evaluation for a plant processing 455 kg/day of deoiled grinding waste is presented.

Introduction

A major goal of the Bureau of Mines is to reduce the United States' reliance on imported critical and strategic minerals and metals (1). This goal can be achieved in part by developing technology to recover critical metals from domestic superalloy scrap. The Bureau has investigated hydrometallurgical procedures to recover cobalt, nickel, chromium, tungsten, and manganese from hardface alloy grinding waste, a form of superalloy scrap. This report describes a recovery process that was developed as a result of the investigation.

Each year significant quantities of hardface alloy grinding waste are produced by various domestic industries. Approximately 3.5 million lb of hardface alloys are annually converted through grinding operations to a finely divided waste material (2). In addition to containing significant quantities of nickel and chromium, these hardface alloy grindings contain

Second International Symposium -
Recycling of Metals and Engineered Materials
Edited by Jan H. L. van Linden,
Donald L. Stewart, Jr., and Yogeshwar Sahai
The Minerals, Metals & Materials Society, 1990

over 1 million lb of cobalt which is approximately 7% of the 1987 U.S. cobalt consumption (3).

A considerable amount of research has been conducted and numerous processes proposed to recover critical metals from superalloy and related types of scrap. The bulk of the recycling research reported in the literature has been conducted in three areas: pyrometallurgical processing (4), hydrometallurgical processing (5-7), and processes combining both pyrometallurgical and hydrometallurgical treatments (8-12).

To avoid pretreatment difficulties and to utilize the fine grindings high surface area, a chloride-based hydrometallurgical approach was selected for this investigation. The research objective was to develop technology to produce purified solutions from which marketable chemicals or high-grade metal products could be recovered. In most cases, research did not go beyond the production of purified solutions because technology is available for subsequent recovery operations.

Sample Description and Pretreatment

A sample considered to be typical of hardface alloy grinding waste was obtained from a midwestern manufacturer. The grindings were contaminated with moisture, cutting oil, iron-based alloy grindings, and grinding media. Drying and solvent deoiling test results showed that the sample contained 2.4% water and 23.7% oil.

The grinding waste was deoiled to provide a clean feed material for laboratory testing. The deoiling procedure consisted of slurrying grinding waste with perchloroethylene at 20% solids for 15 min at room temperature. The slurry was then filtered and the cake reslurried with clean perchloroethylene. Five cycles of slurrying and filtering were carried out with each sample. The deoiling procedure removed 98.3% of the contaminating oil.

Deoiled grindings were screened on a 35-mesh sieve to break up clumps and remove large pieces of foreign material. A screen analysis of the minus 35-mesh fraction showed that 95.7% of the material was finer than 100 mesh. Chemical analysis of the deoiled material is presented in table I.

TABLE I Analyses of cleaned grinding waste

	Analyses, %		Analyses, %
Element:		Element:	
Co	15.1	Ni	11.5
Cr	19.8	W	5.1
Fe	31	Acid insoluble[1]	13.1
Mn	3.6	Residual oil	.54
Mo	.20		

[1]Principally alumina, silica, and silicon carbide.

Process Description

The process for treating the deoiled grinding waste included the following operations. A generalized process flowsheet is shown in figure 1. Cleaned grindings were dissolved with chloride lixiviants producing a residue and a chloride leach liquor. Tungsten was recovered from the leach residue as calcium tungstate, $CaWO_4$. Iron, cobalt, manganese, and nickel chlorides were sequentially recovered from the leach liquor by solvent extraction (SX). The SX operations produced strip liquors from which salable

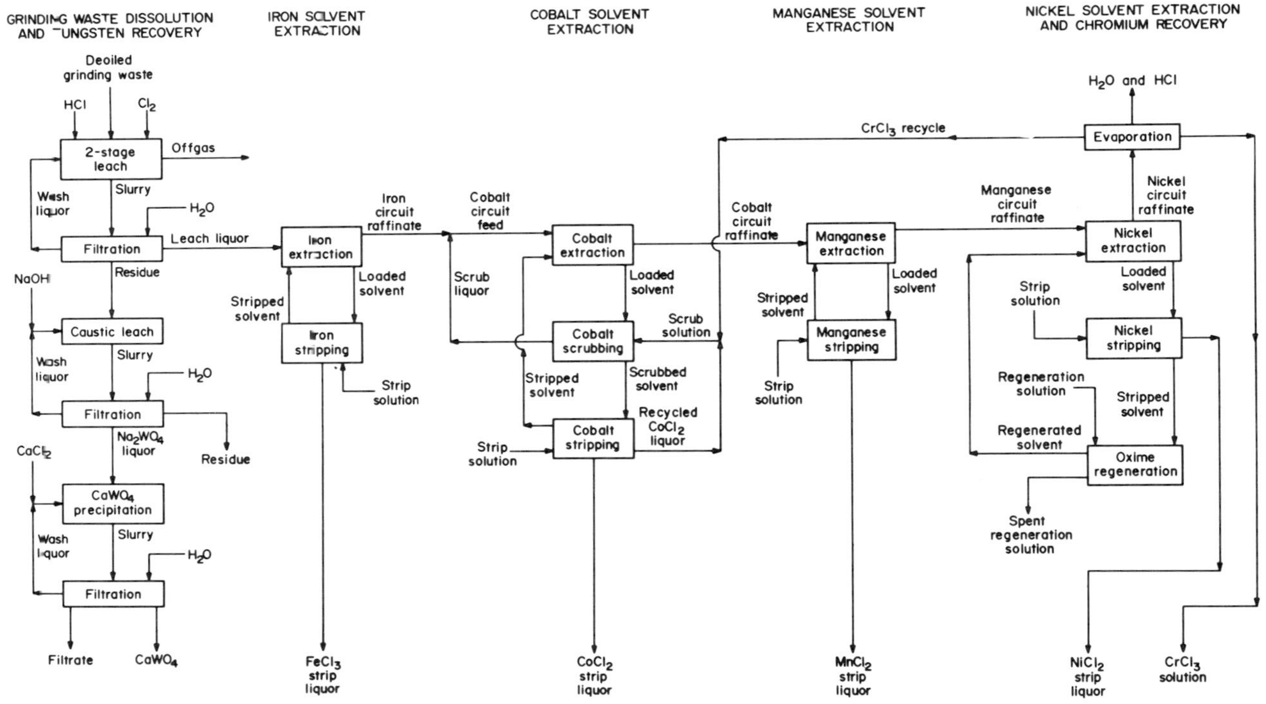

FIGURE 1. - Generalized process flowsheet.

chemicals or high-grade metal products could be recovered. The final raffinate was a purified $CrCl_3$ solution.

The individual SX operations within the process were not fully integrated; however, grinding waste leach liquor and raffinates were sequentially treated in the SX circuits. Details of the process operations are available (13), and only a summary is presented here.

Grinding Waste Dissolution

Deoiled grinding waste was leached using a two-stage $HCl-Cl_2$ dissolution procedure (2), which consisted of first leaching the grindings with hydrochloric acid followed by a period in which leach pulp was sparged with chloride gas. Chlorine was sparged into the pulp to oxidize ferrous iron, improve dissolution, and precipitate tungsten. The leach solution was composed of approximately 2.1 L of wash water recycled from the previous leach and 3.55 L of hydrochloric acid solution, 37.25% HCl. After 1,200 g of grinding waste had been added to the room temperature leach solution, external heat was applied to the reactor, and the grindings were agitation leached for 6 hr at boiling temperatures of 100° to 104°C. Chlorine gas was sparged into the leach pulp at a rate of 1.8 g/min during the last 2.3 hr of leaching. After 6 hr of leaching, the hot-leach pulp was filtered and the filter cake was washed with approximately 2 L of deionized water. The resulting wash water was diluted to 2.1 L, if necessary, and recycled to the subsequent leaching operation.

Average test results indicated that grinding waste constituent dissolutions were, in percent, 98.0 Co, 79.8 Cr, 99.2 Fe, 98.9 Mn, 75.0 Mo, 99.3 Ni, and 0.7 WO_3. Tungsten dissolutions were low because dissolved tungsten precipitated as tungstic acid, H_2WO_4, during leaching and reported with the leach residue. Chemical analyses of the oxidized two-stage leach liquor is given in table II, and analyses of the residue is given in table III.

TABLE II Analyses of grinding waste leach liquor

Constituent:	Analyses, g/L	Constituent:	Analyses, g/L
Cl................	348.0	H^+................	1.3
Co................	35.3	Mn................	8.5
Cr................	37.8	Mo................	0.33
Fe[1]................	73.4	Ni................	27.3
Fe^{2+}..............	<.03	WO_3................	.048

[1]Total Fe^{2+} and Fe^{3+}.

TABLE III Analyses of composite grinding waste leach residue

Constituent:	Analyses, %	Constituent:	Analyses, %
Al_2O_3.............	37.8	Mn................	.16
C[1]................	3.4	Ni................	.31
Co................	1.2	SiC...............	2.5
Cr................	15.6	SiO_2..............	10.9
Fe................	1	WO_3...............	25
Mo................	.20		

[1]Organic carbon.

Tungsten Recovery From Acid Leach Residue

The tungsten-bearing residue was leached with 0.84M NaOH solution at 24.3% solids at 25°C. The slurry was agitated for 15 min and filtered. Approximately 25% of the liquor was retained in the residue. The retained liquor was displaced by washing the filter cake with deionized water. The resulting wash liquor was diluted to 1.0 L and recycled to the subsequent caustic leach. The procedure was repeated for five leaching cycles. At steady-state conditions, 90.0% of the tungsten reported to the caustic leach liquor. A chemical analysis of the caustic leach liquor is presented in table IV.

TABLE IV Analyses of caustic leach liquor

Constituent:	Analyses, g/L	Constituent:	Analyses, g/L
C[1]	0.99	Mo	.57
Cl	3.1	Na	22.5
Co	.002	Ni	.003
Cr	.002	Si	.95
Fe	.001	WO_3	100.1

[1]Organic carbon.

Calcium Tungstate Precipitation. The sodium tungstate leach liquor was transferred to a distillation flask and heated to 80°C. Calcium chloride was added to the hot liquor as a 100-g/L solution at a dosage of 0.5 g $CaCl_2$/g WO_3. After $CaCl_2$ addition, the slurry was mixed for 1 hr at 80°C. The slurry was then filtered, and the filter cake was water washed.

The procedure precipitated essentially all of the tungsten as a filterable $CaWO_4$ product. The filtrates contained only 0.002 g/L WO_3. Nearly 90% of the grinding waste tungsten was recovered as $CaWO_4$. Table V gives the typical composition of the $CaWO_4$ produced. The table also gives the commercial specification for tungsten ore concentrates (14). A comparison of the two indicates that the $CaWO_4$ product is high grade and should be readily marketable.

TABLE V Calcium tungstate comparison with commercial specification

Constituent	Analyses, g/L	
	Calcium tungstate	Commercial specification
As	0.06	<0.2
C	.3	NS
Cu	<.01	<.4
Mo	.14	<.4
P_2O_5	.01	<.08
S	.1	<.75
Sn	.01	<1
WO_3	76	>65

NS No specification.

Solvent Extraction Operations

Values in the acid leach liquor were separated and recovered in a series of continuous countercurrent SX circuits. The circuits were designed based on equilibrium extraction and stripping isotherms and McCabe-Thiele analyses. Unless otherwise specified, actual grinding waste solutions were used in the investigations. Due to the sequential operation of the SX circuits, a commercially-marketed chromium chloride solution was used to make up the cobalt circuit scrub solution rather than recycling nickel circuit raffinate as indicated in figure 1. The composition of the commercial chromium chloride solution is given in table VI in a subsequent section of this paper.

Iron Solvent Extraction. Iron was removed from the leach liquor by SX. The solvent used to extract the iron was composed of Amberlite LA-2, a secondary amine, diluted to a concentration of 0.5M with Shell Cyclosol 63. A total of seven mixer-settler-type contactors were used in the continuous circuit, four in the loading section and three in the stripping section. Water was used to strip the loaded solvent.

The iron circuit was operated intermittently for over 200 hr with grinding waste leach liquor. Average test results for 114.2 hr of steady-state circuit operation indicated that the circuit extracted over 99% of the leach liquor iron and produced a raffinate containing, in grams per liter, 0.005 Fe, 38.6 Co, 30.1 Ni, 41.6 Cr, 9.4 Mn, 0.1 Mo, and 225 Cl. Average cobalt loss due to coextraction with iron was slightly less than 1.0%. Strip liquors typically contained, in grams per liter, 47 Fe, 0.21 Co, 0.015 Cr, 0.17 Mo, 0.015 Ni, 0.01 WO_3 and 92 Cl.

Cobalt Solvent Extraction

After iron was removed from the grinding waste leach liquor, cobalt was recovered from the resulting raffinate. The solvent selected to recover the cobalt was Alamine 308, a tertiary amine, diluted to 0.6M with Shell Cyclosol 63, an aromatic naphtha. Tributylphosphate, 2.5 vol pct, was added as a phase modifier.

The SX circuit was composed of 12 mixer-settler-type contactors, 5 in the loading section, 2 in the scrubbing section, and 5 in the stripping section. A coalescing separator was utilized between the scrubbing and stripping sections to remove entrained scrub liquor from the solvent. The circuit was operated intermittently for over 220 hr. Aqueous feed was a blend of 41 vol pct iron circuit raffinate and 59 vol pct scrub liquor. The blended feed contained, in grams per liter, 24.8 Co, 4.1 Mn, 12.2 Ni, 83.5 Cr, and 248 Cl. Prior to stripping, the loaded solvent was scrubbed to crowd coextracted manganese from the solvent. The scrub solution was made up from reagent-grade chemicals and contained, in grams per liter, 18.3 Co, 112 Cr, 0.002 Mn, and 265 Cl. Water was used to strip the loaded solvent.

Average test results for the last 136 hr of circuit operation indicated that the circuit extracted essentially all of the cobalt and produced a raffinate that contained, in grams per liter, 0.003 Fe, 0.006 Co, 12.2 Ni, 83.5 Cr, 3.8 Mn, 0.001 Mo, and 217 Cl. A strip liquor containing, in grams per liter, 82.3 Co, 0.009 Mn, 0.026 Mo, and 99.1 Cl was produced. Ninety-seven percent of the grinding waste cobalt was recovered in the cobalt chloride strip liquor.

Manganese Solvent Extraction. Manganese, like cobalt and iron, forms an anionic chloride complex (15) and can therefore be extracted with amine extractants. In the absence of iron and cobalt, amines have a strong affinity for manganese; consequently, the same solvent that was used to extract cobalt was selected to extract manganese from the cobalt circuit raffinate.

The manganese SX circuit was composed of six mixer-settler-type contactors, three in a loading section, and three in a stripping section. A coalescing separator was utilized between the loading and stripping sections to remove entrained liquor from the loaded solvent. The circuit was operated intermittently for 97.6 hr. Aqueous feed was cobalt circuit raffinate, and water was used to strip the loaded solvent.

Average test results for the last 45 hr of circuit operation indicated that the circuit extracted essentially all of the manganese in the aqueous feed producing a raffinate containing, in grams per liter, 0.002 Co, 0.003 Mn, 12.2 Ni, 83.5 Cr, 0.003 Fe, and 200 Cl. Nearly 99% of the grinding waste manganese was recovered in the manganese chloride strip liquors which contained, in grams per liter, 10.6 Mn, 0.012 Co, 0.09 Cr, and 0.03 Ni.

Nickel Solvent Extraction. Nickel was recovered from manganese circuit raffinate with decanal oxime (DOX), an experimental extractant. Preston (16-18) has used similar oximes in mixed extractant systems; however, DOX has not been previously used to extract nickel from acidic chloride solutions. The DOX used in this investigation was prepared in the laboratory from decyl aldehyde, $CH_3(CH_2)_8CHO$ and hydroxyl amine hydrochloride, $NH_2OH{\cdot}HCl$ (19). The investigation was conducted with 1.04$\underset{\sim}{M}$ DOX in Shell Cyclosol 63 diluent.

The nickel SX circuit was composed of ten mixer-settler-type contactors, three in the loading section, and six in the stripping section and one in a solvent regeneration section. A solvent regeneration stage was incorporated into the circuit because previous work indicated that the DOX partially degraded in the acidic process solutions. The DOX degradation products were identified as $CH_3(CH_2)_8CHO$, and $NH_2OH{\cdot}HCl$, the reagents from which the DOX was synthesized. This work also indicated that partially degraded DOX can be regenerated by contacting the solvent with a neutralized $NH_2OH{\cdot}HCl$ solution.

The ten-stage circuit was operated for 72 hrs utilizing manganese circuit raffinate, a 7.5-g/L HCl strip solution, and a regeneration solution containing 14.0 g/L $NH_2OH{\cdot}HCl$ and 10.7 g/L Na_2CO_3. During the final 36 hr of steady state operation, product analyses were relatively consistent. Average analyses were, in grams per liter of nickel, feed 12.1, raffinate 0.002, loaded solvent 10.3, and stripped solvent 0.018. The average composition of the strip liquor produced was, in grams per liter, <0.001 Co, 0.03 Cr, 0.002 Fe, <0.001 Mn, and 27.2 Ni. Over 99% of the grinding waste nickel was recovered in the nickel chloride strip liquor.

Stripped solvent and regenerated solvent were analyzed regularly for DOX. The DOX concentration was determined using a procedure in which the maximum nickel loading of an unknown was compared to that of several standards. The average DOX concentration before regeneration was 0.92$\underset{\sim}{M}$, and after regeneration was 1.04$\underset{\sim}{M}$. Therefore, 12% of the extractant was degraded and regenerated each cycle.

Subsequent batch testing has indicated that DOX can be stabilized if appropriate concentrations of $CH_3(CH_2)_8CHO$ are added to the solvent and if a dilute solution of $NH_2OH{\cdot}HCl$ is used for stripping (20). Additional research is needed to test this concept in a continuous countercurrent SX circuit. Although the long-term stability of the DOX extractant modified with degradation products has not been determined in a continuous countercurrent circuit, the modified extractant shows promise for separating and recovering nickel from manganese circuit raffinate and other acidic chloride solutions.

Chromium Recovery

The nickel circuit raffinate was essentially a purified chromium chloride solution which contained, in grams per liter, 83.5 Cr, 0.002 Co, 0.003 Fe, 0.003 Mn. 0.002 Ni, and 185 Cl. After evaporation to 160 g/L Cr, the raffinate compared favorably to a commercially marketed chromium chloride solution as indicated in table VI.

TABLE VI Comparison of evaporated nickel circuit raffinate with commercial chromium chloride solution

Constituent	Analyses, g/L	
	Evaporated raffinate	Commercially marketed solution
Cl[1].........	340.0	260.0
Co..........	.004	.003
Cr..........	160	156
Fe..........	.006	.005
Mn..........	.006	.001
Ni..........	.004	.006

[1]The difference in chloride concentration is due to a higher acidity in the raffinate.

Process Economics

A preliminary economic evaluation of the process to recover critical metals from hardface alloy grinding waste was prepared by the Bureau's Process Evaluation Group. The purpose of this cost estimate was to evaluate the Bureau's superalloy scrap program and to identify high-cost processing steps where economic improvements may be likely.

The estimated capital cost for a small plant processing about 200,000 kg of grinding waste per year (1,000 lb deoiled grinding waste per day) was approximately 4.0 million dollars based on second quarter 1989 costs. This plant would produce annually 23,500 kg chromium as chromic chloride, 21,900 kg cobalt as cobaltous chloride, 5,300 kg manganese as manganous chloride, 17,000 kg nickel as nickelous chloride, and 6,800 kg tungsten as calcium tungstate. The plant operating cost is $11.70/kg ($5.30/lb) of deoiled grinding waste processed; however, the cost of crystallizing and dehydrating cobalt, manganese, and nickel chloride chemicals from the purified solutions has not been estimated and must be added to the operating cost.

Based on August 1989 commodity prices, earnings of about $8.80/kg ($4.00/lb) of deoiled grinding waste processed are expected if the products are sold as calcium tungstate and chloride chemicals. At current commodity prices, the process does not offer an economic benefit; however, under conditions of a national emergency, the process offers a viable source of several commodities. If market conditions change and commodity prices significantly improve, the process could have economic potential. Realization of a 15% interest rate of return after taxes would require an increase in earnings to $17.55/kg ($7.96/lb) of deoiled grinding waste processed.

A summary of the economic evaluation is presented in the Appendix. A brief description of a 455 kg per day plant is given, followed by a discussion of the capital and operating costs.

Summary and Conclusions

- Laboratory-scale research has demonstrated that hardface alloy grinding waste can be treated by a hydrometallurgical process to separate and recover cobalt, chromium, manganese, nickel, and tungsten.

- The solvent deoiling procedure effectively removed 98.3% of the grinding waste oil and produced a clean feed material that was suitable for hydrometallurgical processing.

- The two-stage dissolution procedure successfully dissolved the grinding waste while providing an efficient way to separate and subsequently recover nearly 90% of the tungsten.

- Essentially all the iron was selectively removed from the grinding waste leach liquor by SX with a secondary amine.

- Ninety-seven percent of the cobalt was successfully recovered from the grinding waste as a relatively pure cobalt chloride solution by SX with a tertiary amine.

- Ninety-nine percent of the manganese was successfully recovered from the grinding waste as a relatively pure manganese chloride solution by SX with a tertiary amine.

- Ninety-nine percent of the nickel was successfully recovered from the grinding waste as a high-grade nickel chloride solution with DOX, an experimental nickel extractant.

- The modified DOX extractant shows promise for recovering nickel from acidic chloride solutions; however, additional testing is needed to determine the stability of the modified extractant in a continuous countercurrent SX circuit.

- Seventy-nine percent of the chromium was successfully recovered from the grinding waste as a commercial grade chromic chloride solution.

- Although the process is not currently economic, under conditions of a national emergency, it offers a viable source of several commodities.

References

1. R. S. Kaplan, "An Overview of the Bureau of Mines Recycling Research," Paper in Recycle and Secondary Recovery of Metals, ed. by P. R. Taylor, H. Y. Sohn, and N. Jarret (Proc. Int. Symp. Recycle and Send. Recovery Met. and Fall Extr. and Process Metall. Meet., Fort Lauderdale, FL, Dec. 1-4). Metall. Soc. of AIME, 1985, pp. 3-11.

2. L. D. Redden, et al., Chloride-Based Dissolution of Hardface Alloy Grinding Waste (The Metallurgical Society, AIME Preprint A85-24, 1985), 15 pp.

3. W. S. Kirk, Cobalt, (Ch. in BuMines Minerals Yearbook 1987, vol. 1), 271-278.

4. C. L. Kusik, et al., Pyrometallurgical Recovery of Chromium From Scrap Metals: Laboratory Studies (BuMines RI 8571, 1981) 41 pp.

5. E. R. Baggott, A. W. Fletcher, and T. A. W. Kirkwood, Recovery of Valuable Metals from Nickel-Cobalt Alloy Scrap (paper in Proceedings of Ninth Commonwealth Mining and Metallurgy Congress, Institute of Mining and Metallurgy, London, England, 1970), 803-821.

6. P. T. Brooks, G. M. Potter, and D. A. Martin, Chemical Reclaiming of Superalloy Scrap (BuMines RI 7316, 1969), 28 pp.

7. D. Pearson, Recovery of Cobalt from Metallurgical Wastes, (paper in Proceedings of International Conference on Cobalt: Metallurgy and Uses. Benelux Metallurgie, Brussels, Belgium, 1981), 139-148.

8. A. Aue, et al., Separation of Iron, Cobalt and Nickel from Scrap Alloy by a Solvent Extraction Process (paper in Proceedings of the International Solvent Extraction Conference, Society of Chemical Industry, London, England, vol 1, 1971), 447-450.

9. J. J. deBarbadillo, J. K. Pargeter, and H. V. Makar, Process for Recovering Chromium and Other Metals from Superalloy Scrap, (BuMines RI 8570, 1981), 73 pp.

10. J. L. Holman, and L. A. Neumeier, Deposition of Nickel-Cobalt Alloys by Controlled-Potential Electrolysis (BuMines RI 8931, 1985), 17 pp.

11. B. H. Rosof, Process for Recovering Valuable Metals from Superalloy Scrap (U.S. Pat. 4,138,249, Feb. 6, 1979).

12. J. A. Thomas, D. L. Jones, and V. A. Ettel, Chromium Recovery from Superalloy Scrap by Selective Chlorine Leaching (U.S. Pat. 4,377,410, Mar. 22, 1983).

13. L. D. Redden, R. D. Groves, and D. C. Seidel, Hydrometallurgical Recovery of Critical Metals From Hardface Alloy Grinding Waste: A Laboratory Study (BuMines RI 9210, 1988,) 31 pp.

14. "Tungsten: Need for Alternate Ore Pricing Basis to Result in MW US Quote on Jan. 1, 1977," Metals Week, (Nov. 22, 1976), 9 pp.

15. D. M. Muir, "The Application of Thermodynamics to Extractive Metallurgy With Chloride Solutions--A Review," Warren Spring Lab. Stevenage Hertsfordshire, U.K., 1984, pp. 18-36.

16. J. S. Preston, "Solvent Extraction of Cobalt and Nickel by Organophosphorus Acids, I. Comparison of Phosphoric, Phosphonic and Phosphinic Acid Systems," Hydrometallurgy, 9, (1982), 115-133.

17. J. S. Preston, "Solvent Extraction of Base Metals by Mixtures of Organophosphoric Acids and Non-chelating Oximes," Hydrometallurgy, 10 (1983), 187-204.

18. J. S. Preston, Non-chelating Oximes in the Solvent Extraction of Base Metals, (paper in Proceedings of the International Solvent Extraction Conference, AIME, Denver, CO, 1983), 357-358.

19. A. H. Blatt, Organic Syntheses, Collective Volume II (John Wiley and Sons, Inc., New York, NY, 1943), 313-315.

20. R. D. Groves and L. D. Redden, "Nickel Extraction From Acidic Chloride Solutions With Aliphatic Oximes," Hydrometallurgy, 24 (1990), in press.

21. J. B. Weaver and H. C. Bauman. Cost and Profitability Estimation. Sec. 25 Perry's Chemical Engineer's Handbook, ed. by R. H. Perry and C. H. Chilton. McGraw-Hill, 5th ed., 1973, p. 47.

Appendix--Process Economics

By R. Swarup

Plant Description

To determine capital and operating costs, a plant processing 615 kg of grinding waste, 455 kg of deoiled grinding waste, per day was designed. A process material balance (13) was utilized to size equipment and estimate utility and reagent requirements. The plant was designed to operate, 1 shift per day, 7 days per week, 330 days per year. The remaining 35 days per year were intended for scheduled and unscheduled downtime.

The plant design is based on the flowsheet presented in figure 1; however, a distillation column not considered in the process flowsheet is incorporated into the plant design to recover oil-free perchloroethylene for reuse in the grinding waste deoiling process. The plant design also includes the necessary environmental control equipment wherever needed to treat spills and off-gases prior to environmental discharge.

The primary solid waste produced in the process, grinding waste leach residue, is relatively inert and showed no signs of leaching when contacted with a dilute acidic acid solution; however, the waste will likely be considered hazardous due to the contained chromium and other heavy metals. In addition to the solid leach residue, three liquid wastes are produced in the process: 1) a filtrate from the tungsten recovery operation, 2) an iron SX circuit strip liquor, and 3) a spent nickel SX circuit regeneration solution.

The tungsten recovery filtrate is basically a sodium chloride solution contaminated with traces of cobalt, chromium, iron molybdenum, nickel, and tungsten. The iron SX circuit strip liquor is a ferric chloride solution contaminated with traces of cobalt, chromium, molybdenum, nickel, and tungsten. The spent regeneration solution contains unutilized hydroxyl amine hydrochloride, sodium chloride, and traces of nickel and chromium chloride. Prior to disposal, excess water is evaporated from these solutions in steam jacketed tanks provided in the design. The residues are placed in drums and due to the heavy-metal content will most likely be considered hazardous wastes. All wastes are disposed of by an outside contractor; disposal costs are included in the economic analysis.

Capital Costs

Capital costs for a plant based on the preceding description are presented. The capital cost estimate is of the general type called a study estimate by Weaver and Bauman (21). This type of estimate is typically expected to be within 30% of the cost to build the plant.

The estimated capital cost on a second quarter 1989 basis (Marshall and Swift index of 894.7) for a plant processing 615 kg per day of grinding waste is about $4.0 million, as shown in table A-I. The table arbitrarily divides the plant into six sections: grinding waste deoiling, dissolution and tungsten recovery, iron extraction, cobalt extraction, manganese extraction, and nickel and chromium extraction. The total cost for each section includes equipment, foundation, structure, insulation, instrumentation, electrical, piping, painting, miscellaneous, and state sales tax costs. Equipment costs used in these estimates are based on informal cost quotations from equipment manufacturers and on capacity cost data. Cost for the other items were obtained by multiplying the total uninstalled equipment cost by an appropriate factor.

Land costs are excluded from this estimate, and the financing costs on the capital borrowed for construction is included as interest during construction. Working capital is defined as funds in addition to fixed capital and startup costs that must be provided to operate the plant. Working capital, as shown in table A-I, is estimated from the following items: 1) Raw materials and supplies (cost of raw materials and operating supplies for 30 days); 2) product and in-process inventory (total operating cost for 30 days); 3) accounts receivable (total operating cost for 30 days); and 4) available cash (direct expenses for 30 days).

Startup costs are estimated as 10% of the fixed capital costs, of which 1% is shown in table A-I as capitalized startup costs. The remaining 9% are assumed to be first year operating costs. This operating cost is not included in the operating cost table, which represents the expenses for an average year. However, they are included in the discounted cash flow calculations used to determine rates of return on investment.

Operating Costs

The estimated annual operating cost of about $1.8 million, as shown in table A-II, is based on 1 shifts per day, 7 days per week, 330 days of operation per year over the life of the plant. The operating costs are divided into direct, indirect, and fixed costs.

Direct costs include raw materials, utilities, direct labor, plant maintenance, payroll overhead, waste disposal, demurrage, and operating supplies. Electricity, water, and natural gas are purchased utilities, and payroll overhead costs include: vacation, sick leave, social security, and fringe benefits.

The indirect costs include the expenses of the control laboratories, accounting, plant protection and safety, plant administration, marketing, and company overhead. Fixed costs include taxes, insurance, and depreciation. The annual cost of taxes and insurance are each estimated to be 1% of the plant construction cost. Depreciation is based on a straight-line, 10-yr period. A $0.07/kg credit is given in the operating cost for grinding waste, since a major producer is presently paying $50 to $55/st to dispose of the material.

Profitability

The estimated operating cost of this process is $11.70/kg ($5.30/lb) of deoiled grinding waste processed; however, the cost of crystallizing and dehydrating cobalt, manganese, and nickel chloride chemicals from the purified solutions has not been estimated and must be added to the operating cost.

Based on August 1989 commodity prices, earnings of about $8.80/kg ($4.00/lb) of deoiled grinding waste processed are expected if the products are sold as calcium tungstate and chloride chemicals. At current commodity prices, the process does not offer an economic benefit; however, under conditions of a national emergency, the process offers a viable source of several commodities. Since most corporations require at least a 15% interest rate of return after taxes, earnings would need to be at least

$17.55/kg or $7.96/lb of deoiled grinding waste processed in order for this process to be considered profitable.

The economics would be greatly improved by running two or three shifts instead of one, providing sufficient grinding waste was available. This would reduce the operating costs per unit of grinding waste processed, since the fixed costs would be spread over a larger production rate. In addition, upscaling is possible if scrap is collected from several producers; however, this may introduce additional alloying elements which have not been considered by this study. To realize scale-up benefits, it would be necessary to design a flexible processing scheme capable of handling varying feed compositions, which would increase the capital and operating costs to some extent.

TABLE A-I. - Estimated capital cost

Fixed capital:	
Grinding waste deoiling section	$656,700
Dissolution and tungsten recovery section	426,490
Iron extraction section	560,260
Cobalt extraction section	356,490
Manganese extraction section	124,500
Nickel and chromium extraction section	627,950
Subtotal	2,752,390
Plant facilities, 10% of above subtotal	275,240
Plant utilities, 12% of above subtotal	330,290
Total plant cost	3,357,920
Interest during construction period	139,990
Total fixed capital cost	3,497,910
Working capital:	
Raw material and supplies	46,150
Product and in-process inventory	144,730
Accounts receivable	144,730
Available cash	95,780
Working capital cost	431,390
Capitalized startup costs	34,980
Subtotal	466,370
Total capital cost	3,964,280

TABLE A-II. - Estimated annual operating cost

	Annual cost	Cost per kg deoiled grinding waste
Direct cost:		
Raw materials	480,580	3.21
Utilities	29,610	.20
Materials and utilities total	510,190	3.40
Direct labor:		
Labor at $16 per hour	232,960	1.56
Supervision, 15% of labor	34,940	.23
Direct labor total	267,900	1.79
Plant maintenance:		
Labor	81,150	.54
Supervision, 20% of maintenance labor	16,230	.11
Materials	81,150	.54
Plant maintenance total	178,530	1.19
Payroll overhead, 35% of above payroll	127,850	.85
Operating supplies, 20% of plant maintenance	35,710	.24
Disposal Costs	45,160	.30
Total direct cost	1,165,340	7.79
Indirect cost, 20% of direct labor and maintenance	178,570	1.19
Fixed cost:		
Taxes, 1% of total plant cost	33,580	.22
Insurance, 1% of total plant cost	33,580	.22
Depreciation, 10-yr life	349,790	2.34
Total operating cost	1,760,860	11.76
Credit:		
Grinding waste at $0.07 per lb	9,410	.06
Net operating cost	1,751,450	11.70

USED BEVERAGE CONTAINERS

Session Chairman:
Donald L. Stewart, Jr.

RECYCLING OF LIGHTWEIGHT ALUMINUM CONTAINERS:

PRESENT AND FUTURE PERSPECTIVES

R. E. Sanders, Jr. and A. B. Trageser

Alcoa Laboratories
Alcoa Center, PA 15069

C. S. Rollings
Tennessee Operations
Aluminum Company of America
Alcoa, TN 37701

Abstract

The importance of recycling to the aluminum beverage can industry and the environment is paramount. This paper will consider some of the factors which influence the technology and economics of recycling aluminum beverage cans. Models are used to describe the utilization of metal units in the can manufacturing, recycling and alloying processes. Future trends in lightweight cans and aluminum food cans are examined with respect to their impact on the recycling process. It is concluded that advances in recycling technology will keep pace with projected trends in lightweight containers to insure a continued source of high quality metal for the aluminum industry.

Second International Symposium -
Recycling of Metals and Engineered Materials
Edited by Jan H. L. van Linden,
Donald L. Stewart, Jr., and Yogeshwar Sahai
The Minerals, Metals & Materials Society, 1990

Introduction

The aluminum beverage can has been overwhelmingly successful in competition with other materials due to consistently high material quality and the recognition that recycling is an important component of an effective package. As aluminum's role in the can industry has grown, it has been possible to "lightweight" (reduce the metal content of) the can through innovative designs and material selection. The process of lightweighting has been driven by the canmaker because the metal value makes up nearly 70% of his unit cost.

The impetus for recycling comes from several economic and environmental sources. Available metal units and reduced energy consumption (versus smelter metal) are inherent advantages for the aluminum sheet producer while consumers benefit from the economics of turning used beverage cans (UBCs) in for cash ($900,000,000 in 1989 in USA). In 1989, Alcoa alone recycled 611 MM pounds of UBCs, an amount equivalent to the output of two typical aluminum smelting plants. Figure 1 summarizes the growth in can production and recycling since 1970. Data from the U.S. Bureau of Mines (Reference 1) shows that the aluminum recovered from UBCs makes up more than 10% of total U. S. aluminum consumption, Table 1.

Thus, it is evident that recycling is not a luxury or promotional gimmick for the aluminum industry but a vital necessity to insure a steady supply of metal. From the environmental standpoint, state and local legislation have begun to make recycling a high national priority as concern over solid waste and limited landfill capacity grows.

With the knowledge that aluminum cans are a growing market and can-makers will continue to pursue opportunities to reduce the metal content of their product, we need to consider as an industry how future trends in lightweighting and recycling will be interrelated. How will the reduced net metal content of future lightweight cans impact recycling? And the converse question: How will increases in recycling rate impact the ability of the aluminum industry to supply high quality sheet to facilitate further lightweighting? We will consider the current metal balance situation and what possible future changes in either recycling rate or lightweighting due to can design or alloy will have on the necessary balance between the aluminum can and recycling. Models developed for looking at the metal unit balance within the aluminum beverage can system will be used to make these determinations. Where possible, actual data will be used for the period 1983-1989 to check the validity of the model predictions.

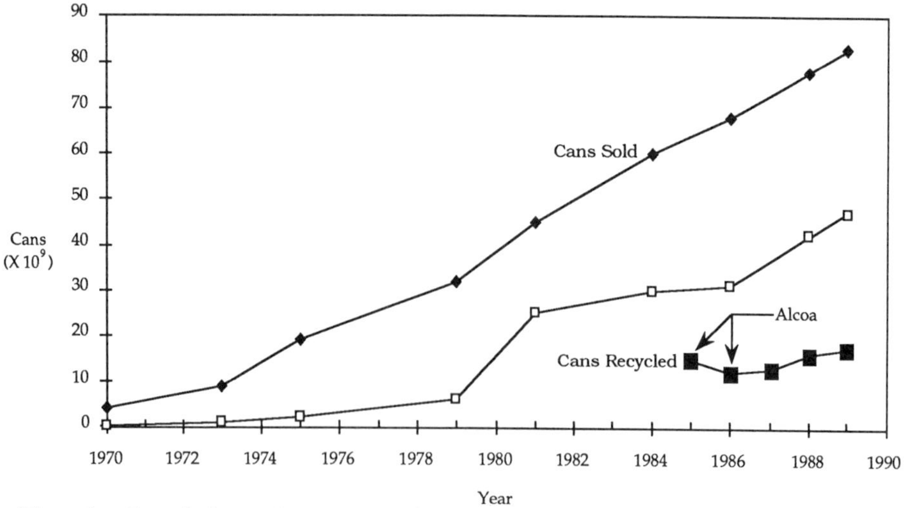

Figure 1 - Growth in production and recycling of aluminum beer and beverage cans, 1970-1989.

TABLE 1

United States Aluminum Production, 1978-1989

(Thousand Metric Tons)

	1978	1980	1982	1984	1986	1988	1989
Primary Production	4,358	4,654	3,274	4,099	3,037	3,944	4,030
Secondary Recovery	1,518	1,577	1,666	1,760	1,773	2,122	1,470
Old Scrap	522	617	782	825	784	1,045	1,100*
UBCs (part of old scrap)	164	303	445	473	429	551	572
Apparent Consumption				5,279	5,143	5,352	5,087
World Production	14,131	15,383	13,433	15,705	15,354	17,304	18,000

Source: P. A. Plunkert, U.S. Department of the Interior, Bureau of Mines.

*Estimate based on preliminary 1989 data compiled at the Bureau of Mines.

NOTES: UBC figures are included in Old Scrap. Other components are not listed here.

Apparent consumption is calculated by:

Primary + Import + Old Scrap - Export +/- Inventory Changes = Apparent Consumption

Background

Lightweighting

Many mass-produced consumer products--paper clips, record albums, eyeglasses--have undergone considerable but often imperceptible cost-cutting design changes. Sparked by geometric changes or improvements in materials, these changes often involve the reduction in weight of the product without changing its form or function.

Twelve-ounce beverage containers were once three-piece seamed wraparound cans of tin-plated steel. The weight per can was 30-50 g. Early two-piece seamless cans made of aluminum weighed 15-20 g each, or 35-40 pounds per thousand cans. Today's aluminum can weighs only 25 pounds per thousand, with a starting thickness less than 0.012 inch. Figure 2 shows the trend in downgauging of cans over the past ten years.

The structural requirements on beverage cans have changed somewhat during this lightweighting. Pasteurization in beer, or high transportation temperatures in soft drinks, results in an internal pressure requirement of 90 psi. Can bodies must also withstand 250 pounds of axial column load (formerly 300-350 pounds) for filling and seaming. Additional tests for dropping the can without rupture or buckling of the bottom dome have been developed only since lightweighting has occurred. Computer simulations of these tests are performed to assist designers with package weight reduction (Reference 2).

The design changes which allowed this lightweighting to occur stem largely from the improvements in the necking process which reduces the diameter of the open end of the can. Increased necking allows the lid of the can to be smaller in diameter, reducing the lid's cost. (The alloy used in the lid is more expensive than that of the can body.) Early lids were "211" diameter (2-11/16 inches) when no necks were used, then were reduced to 209 diameter for single neck, and are 206 diameter today. The reduction in lid diameter has an additional lightweighting effect in that smaller diameters can withstand more internal pressure without buckling, so lid thickness can also be reduced.

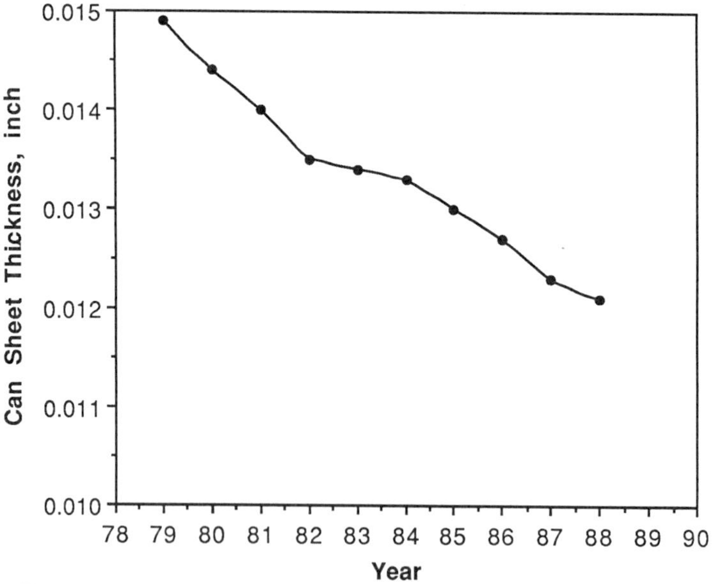

Figure 2 - Trend showing progressive reduction in thickness of aluminum body stock in domestic U.S. market.

As lid diameter was decreased, the base diameter was also decreased in order to maintain stackability between cans (Figure 3). Like lid diameter, the smaller base diameter can withstand more pressure without buckling, so the bottom thickness can be reduced. With a roughly constant 67% total wall thickness reduction in a bodymaker, the thinner bottom leads to a thinner wall. Thin can sidewalls mean lower necking loads, so column load requirements can be reduced. This lightweighting cycle is depicted in Figure 4.

The net effect of this continuous improvement in reducing the weight of the beverage can has been a greater reduction in the weight of the lid than in the weight of the body of the can.

To date, lightweighting of aluminum cans, lids and tabs has been achieved primarily through geometry and process improvements, with the alloys for each component remaining largely unchanged. Alloy development for rigid container sheet (RCS) has been limited to a large extent by the forming requirements of the individual forming processes. It is possible that as process improvement continues and design changes are made for lightweighting, the necessity for higher strength alloys will arise.

<u>The Aluminum Can Alloy System</u>

Because of the unique forming and strength requirements for beverage can bodies and lids, they are typically produced from two different work-hardened aluminum alloys, 3004 and 5182. The can is approximately 75% by weight 3004 body and 22% 5182 lid. The tabs make up less than 3% of the can weight and are produced from medium-strength 5XXX alloys such as 5042, 5082 or 5017. All alloys are part of the Al-Mg-Mn system with the major impurity elements being Fe and Si. A schematic diagram of the aluminum alloys used in domestic RCS production is shown in Figure 5. Nominal chemical compositions for these alloys are listed in Table 2. A more detailed description of RCS alloys and processing is provided in Reference 3.

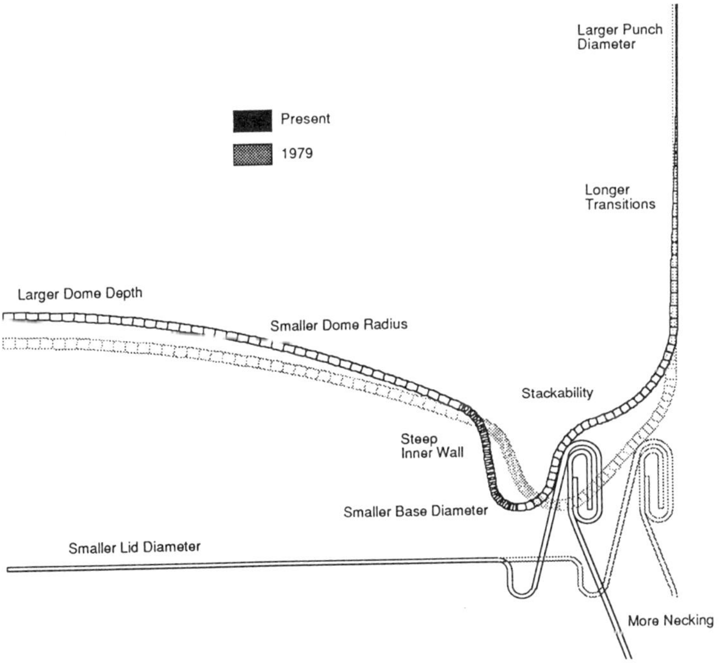

Figure 3 - Trends in design of aluminum beer and beverage cans, 1979-1990.

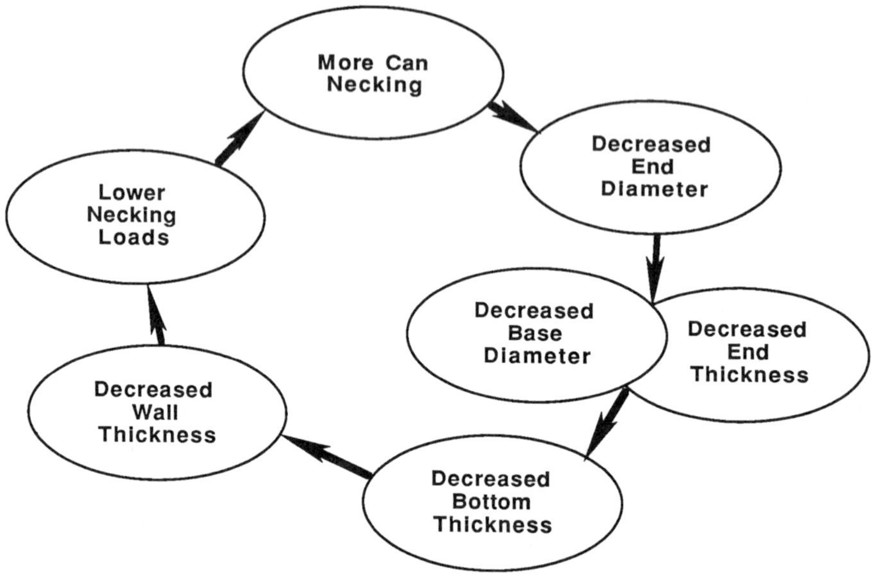

Figure 4 - Beverage can lightweighting cycle.

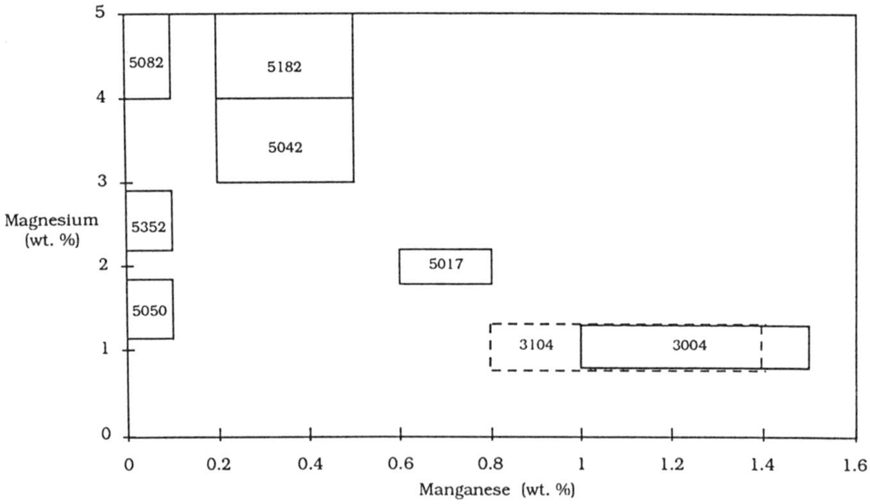

Figure 5 - Typical alloy composition ranges for domestic U.S. rigid container sheet.

Compatibility in composition among the various alloys is important in that recycling necessarily brings them together; i.e., recycled cans will eventually be used to remake one or more of the components of a new can. The most important effect of the 5XXX lid in the recycling process is to "sweeten" the UBC composition by diluting the Fe and Si impurities. Even with this sweetening, it is apparent from Table 2 that alloys with low levels of Mn, Fe, or Si will absorb very few UBCs. Consequently, most UBCs are recycled into body stock alloys 3004 or 3104. [Alloy considerations for UBC utilization will be discussed in a later section of this paper.]

TABLE 2

Nominal Chemical Compositions for Domestic RCS alloys

	Si	Fe	Cu	Mn	Mg
3004	0.18	0.45	0.13	1.1	1.1
3104	0.18	0.45	0.15	0.9	1.25
5182	0.10	0.24	0.03	0.35	4.5
5042	0.10	0.24	0.03	0.30	3.5
5082	0.10	0.17	0.03	0.05	4.5
5017	0.30	0.35	0.25	0.75	1.85
UBC (theoret.)	0.16	0.38	0.11	0.88	1.94
UBC (actual)	0.23	0.46	0.15	0.89	1.41

Some work (Reference 4) has focused on maximizing the absorption of recycled cans by using a single-alloy system (e.g., 5017) for body, lid and tab. While cans can be successfully formed from alloy 5017, complete commercial success of this approach has not been demonstrated as yet. Considerable production experience will be required before can line productivity, can performance and surface quality are demonstrated to be equivalent to the current 3004/5182 package.

Current Metal Unit Balance and Recycling

The current metal balance for aluminum RCS depends upon the complex interrelationships of a number of factors. Figure 6 attempts to describe this system graphically. It is apparent that new D&I alloy body stock can be cast from various combinations of: used beverage cans, mill scrap, skeleton scrap and spoilage from the can-makers, metal units from secondary reprocessing of skim and pure metal from the smelter. The diagram shows that metal units are lost from cans not returned, skim reprocessing, UBC processing and ingot casting. Pure metal from the smelter makes up these losses each time around the loop. With such a nearly closed loop, the economics of recycling depend crucially on the costs of the different types of metal units and the efficiency with which UBCs can be converted to new 3004 alloy.

The efficiency and economics of UBC conversion are impacted negatively by factors which contribute to weight loss during the process (Reference 5). A large portion of the total system loss during the process stems from the presence of "dirt," moisture, organic lacquers and coatings. The total weight of organic material may vary from 300-800 mg per can depending upon the type of can, decoration, etc. The melt loss due to aluminum oxidation during the process also results in lost pounds of recycled aluminum. Obviously the presence of steel, cardboard and other non-aluminum materials which must be separated or burned off will also result in total system losses.

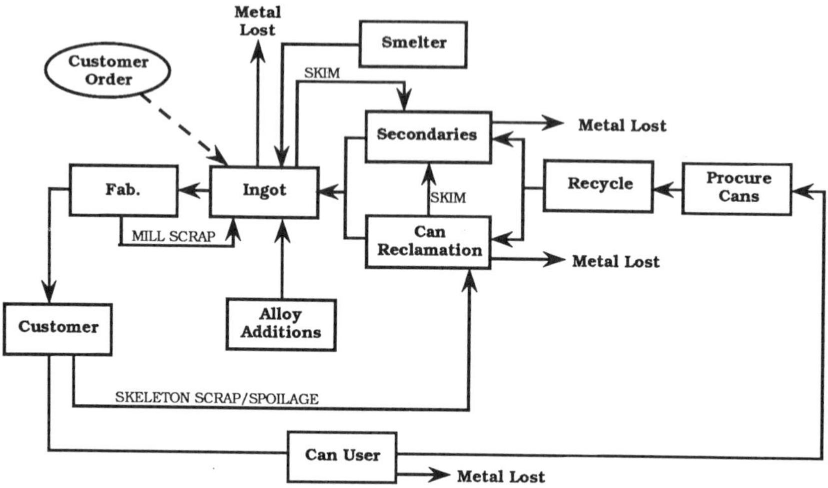

Figure 6 - Schematic of metal unit system for production of aluminum rigid container sheet.

When higher-priced smelter metal must be added to bring the composition on target, UBC contamination is a definite economic concern, but there is also an important quality consideration. Factors which degrade the consistency of UBC composition (and thus limit the ability of UBCs to be absorbed into new body stock ingot) have a deleterious effect on the recycling process. Since the alloy composition controls the strength and workability of the final RCS product to a large extent, low variation around that target is vital to the can-maker. Broad alloy composition limits are set by the Aluminum Association for such alloys as 3004 or 3104, but each RCS producer must control tighter than those limits to make a competitive product. Thus, any degradation in the consistency of UBC composition due to the factors listed below makes the manufacture of a highly consistent body stock more difficult and/or costly:

- dirt (contributes to Si pickup)
- steel, iron and other non-Al metals
- non-RCS aluminum alloys containing such elements as Zn, Cr and high levels of Cu, Fe, or Si.

Analytical Techniques

Can Weight Analysis

A spreadsheet was written which calculates the can body, lid and tab weight based on the following manufacturing parameters:

Gauge*	the starting thickness of the sheet
Number Out	number of units made per width of the sheet
Blank Diameter*	the circle diameter used to make each unit
Web Width	the distance between blanks
Edge Allowance	the distance between the edge of the sheet and the first blank
Trim Diameter*	typically the diameter of the can body to be trimmed
Trim Length*	the length of the trimmed portion of the can body
Trim Thickness*	typically the "thickwall thickness"
→ Coil Width	calculated from blank diameter, number out and offal dimensions
Spoilage	the percentage of cans lost in the can plant

To calculate the weight of the body, the weight of the annular trim is subtracted from the blank weight, so only the parameters flagged with a * are required. For the can lid there is no trim, so only blank diameter and sheet thickness are needed to calculate weight. Tab design analysis is not done here; the tab weight is estimated to be a constant 0.97 pounds per thousand.

The other manufacturing parameters (e.g., coil width, spoilage) are used to calculate the weight of the offal per 1000 cans produced. This can be used to evaluate the total amount of scrap aluminum entering the remelt rather than only the UBCs.

Using the nominal compositions from Table 2, the alloy composition of the UBCs is calculated by analyzing the weight percent of each of the components in the melt (used cans, used lids, body scrap, etc.)

Further analysis is possible by putting in other components such as dirt in the form of added silicon, skeleton scrap, food cans, etc.

Metal Unit Model

The schematic diagram in Figure 6 is the basis for the quantitative model used to calculate the input metal units necessary to produce a new ingot of alloy 3004 composition. Linear programming software was used to develop a model which utilizes data and functions from a standard Lotus spreadsheet. An objective statement is optimized by the model according to the limits set for the available inputs. Inputs for this model include smelter metal, UBCs, mill scrap, used food cans (UFCs), etc. The objective statement for the model is made up of a combination of alloys to be cast, the pounds to be produced and the elemental analysis associated with each alloy. The software analyzes the elemental analysis for each alloy in the input product mix and then will attempt to make the required poundage of each alloy to be cast. A cost factor can be introduced to account for each input source in order to calculate the total cost of the desired alloy. Typically it is desirable to maximize the usage of UBCs or mill scrap at the expense of higher cost smelter metal.

The application of this or any similar model for a given UBC/RCS production facility will be unique. Facilities will differ in their metal sources, capacities and exact processing sequences and flexibilities. To arrive at generalized conclusions for this paper, the model was simplified to calculate maximum UBC absorption potential. Mill scrap and clean can plant scrap were ignored and only UBC, UFC and smelter metal sources were considered.

Trends in Lightweighting, UBC Composition and Utilization

To investigate lightweighting effects on recycling, trends in lightweighting were estimated for four points in time: 1983, 1989, 1993 and 1997. Table 3 contains "typical" manufacturing information for each year.

The 1983 can design is a 211 diameter can with a 209 diameter lid (single neck). There are two 1989 can designs: a "standard" 211 diameter can with a 206 lid and a "tall 12" 209 diameter can with a 206 lid. The 1993 can design is predicted to be 211 diameter with considerable downgauging in bottom and lid (from the introduction of reforming techniques); by 1997, it is assumed that the next generation in necking (to 204 diameter) will allow further lightweighting.

When these data are put through the can weight and recycling models, the results in Tables 4 and 5 are obtained. Table 4 can be used to predict the impact of current and future can designs on the component weights, theoretical UBC compositions, system losses and Si pickup. The recycling model was used to formulate Table 5 from the predicted compositions in Table 4. The various alloy and can design combinations were used to investigate their ability to be recycled (theoretical absorption) back into 3004 body stock. The absorption analyses in Table 5 were run under two different sets of assumptions. The first case simply assumed the theoretical compositions (plus the predicted Si pickup values) in Table 4. A second case thought to be somewhat more realistic also included the Si pickup but allowed for a 25% reduction in Mg due to oxidation losses during recycling and molten metal processing.

Table 3
Trends in Can and Lid Manufacturing Information
(all dimensions in inches)

	Year	1983	1989 Standard	1989 Tall 12	1993	1997
BODY	Gauge	0.0133	0.012	0.0113	0.0112	0.01
	Number out	6	12	5	14	14
	Blank diameter	5.46	5.50	5.70	5.50	5.60
	Web width	0.05	0.02	0.05	0.02	0.02
	Edge allowance	0.07	0.045	0.07	0.045	0.045
	Punch diameter	2.595	2.6	2.47	2.6	2.6
	Trim length	0.50	0.30	0.30	0.25	0.25
	Thickwall thickness	0.0072	0.0067	0.0075	0.006	0.006
LID	Gauge	0.0135	0.0113	0.0107	0.0096	0.0092
	Number out	6	22	24	24	24
	Blank diameter	3.26	2.984	2.895	3.047	2.954
	Web width	0.03	0.02	0.02	0.02	0.02
	Edge allowance	0.05	0.05	0.04	0.05	0.04
Body or Lid	Spoilage	5%	5%	5%	5%	5%

With the current 3004/5182 alloy combination, the differences in can design between the standard and tall cans currently in use have a negligible impact on final UBC composition or absorption. However, the effect of using different alloys in the body or lid are more noticeable. When alloy 3104 is used for body stock, the composition shifts to higher Mg and lower Mn. The higher Mg of the 3104/5182 combination limits absorption back into 3004 for the theoretical case, but when Mg burnoff is accounted for, the absorption figures are much higher and more similar to the base 3004/5182 case. When 5042 is used to make the lid (for a lower strength can), Mg is reduced substantially compared to the standard 5182 lid. Lower Mg in the UBC increases absorption substantially. In fact, Si, rather than Mg, is predicted to be the element limiting absorption when Mg burnoff is taken into consideration.

When alloy 5017 is introduced, the recycling situation changes dramatically. Using 5017 as the lid alloy increases the Si content substantially and expectedly Si is the element which limits 5017 absorption, Table 5. Since Si cannot be effectively eliminated in the process, these high-Si UBCs must be diluted with pure metal to bring the melt to the required composition. When both body and end are produced from 5017, the UBC absorption into conventional 3004 alloy is quite limited, falling to nearly 40%.

TABLE 4

Results of Can Weight Analysis Model Calculations for Various Can Designs and Alloys

Can Design	1983		1989						1993		1997	
			Standard 12-oz		Tall 12-oz Can, 204 Lid							
Lid (#/1000, %)	10.80	(27.4)	7.57	(21.7)	6.75	(19.7)			6.71	(20.6)	6.04	(20.2)
Body (#/1000, %)	27.70	(70.2)	26.38	(75.5)	26.60	(77.5)			24.93	(76.5)	22.98	(76.6)
Tab (#/1000, %)	0.97	(2.4)	0.97	(2.8)	0.97	(2.8)			0.97	(2.9)	0.97	(3.2)
Can Total	39.47	(100)	34.92	(100)	34.32	(100)			32.61	(100)	29.99	(100)
Cans/#			28.64		29.14				30.67		33.34	
Grams/can (calc.)			15.85		15.58				14.80		13.62	
(meas.)			16.10		15.85							
Alloy (Body/Lid)	(3004/5182)		(3004/ 5182)	(3104/ 5182)	(3004/ 5182)	(3004/ 5042)	(3004/ 5017)	(5017/ 5017)	(3004/5182)		(3004/5182)	
Theoretical UBC Composition (%)												
Si	0.16		0.16	0.16	0.16	0.16	0.20	0.30	0.16		0.16	
Fe	0.39		0.40	0.40	0.40	0.40	0.42	0.35	0.40		0.40	
Cu	0.10		0.11	0.12	0.10	0.10	0.15	0.25	0.11		0.11	
Mn	0.88		0.92	0.76	0.91	0.90	1.00	0.75	0.92		0.92	
Mg	2.09		1.90	2.02	1.93	1.70	1.34	1.85	1.87		1.86	
Coating Loss (%)	2.18		2.46*		2.50				2.63		2.86	
Melt Loss (%)	7.08		8.0*		8.14				8.6		9.3	
Si Pickup (wt %)	0.044		0.050*		0.051				0.054		0.058	

* Assumed as Base Case

TABLE 5

Theoretical Absorption Values for Can Designs/Alloys
Using the Recycling Metal Unit Model

Can Design Alloy (Body/Lid)	1989 Standard (3004/5182)	1989 Standard (3104/5182)	1989 Tall (3004/5182)	1989 Tall (3004/5042)	1989 Tall (3004/5017)	1989 Tall (5017/5017)	1993 Future (3004/5182)	1997 Future (3004/5182)
% Absorption (theoretical)	56.9	54.4	56.9	64.6	62.0	40.4	58.8	59.1
Limiting Element	Mg	Mg	Mg	Mg	Si	Si	Mg	Mg
% Absorption (with Mg loss)	75.8	71.4	76.2	78.9	62.0	40.4	77.3	75.3
Limiting Element	Mg	Mg	Mg/Si	Si	Si	Si	Mg	Mg

Since the differences in can design shown in Table 4 have little impact on the weight per can of typical beverage cans in the 1989 market, the effects on coating and melt loss figures and Si pickup are also quite small. Variability in organic coating weights (beer vs. soft drink, base-coat vs. clear-coat, Reference 5) can play as large a role in total system loss as the changes in can design.

As lightweighting proceeds, a number of potential changes can be predicted. Melt loss will increase with the number of cans/pound given the assumption that can surface area remains constant. With an assumed 1989 melt loss of 8% (cans at 15.85 grams each), a projected increase to 9.3% would occur with a 1997 can at 13.62 grams. System losses due to the organic lacquers (inside and outside coating and decoration) will also increase somewhat. Using the figure of 0.39 grams of coating per can (Reference 6), the data in Table 4 suggests this component of system loss will rise from 2.46% presently to a predicted 2.86% in 1997.

The increase in impurity elements such as Si can be treated in a similar manner. If Si in UBCs currently exceeds that in the original can by 0.05%, then lightweighting of the can to 13.62 grams would increase Si proportionately to 0.058%.

As lightweighting of the body and lid proceeds at different paces, the alloy composition of the UBC melt will change to reflect the new proportions of lightweight body and lightweight lid as shown in Table 4. In 1983, the can and lid designs resulted in a UBC composition containing 0.88 Mn and 2.09 Mg. As the proportion of the lid is reduced with time from 27.4 to 20.2% of total can weight, the UBC is enriched in Mn and reduced in Mg. The absorption predictions in Table 5 suggest that utilization of UBCs to make 3004 body stock will not be greatly affected.

Future Recycling Scenarios

Future demands for production of aluminum can stock will depend on an increase in the recycling rate. Demand for aluminum is increasing while smelting capacity is limited by the immense capital costs associated with new facilities. In order to meet projected demands, the recycling rate will have to reach 75% by the year 2000. The effects of the increased rate on UBC quality and utilization are not clear. Recycling programs in communities and municipalities will lead to cleaner cans as these cans will be kept clean in homes, stored in clean containers and go directly to collection points run by the cities. In contrast, attempts to reclaim recyclable materials from disposed waste will result in dirtier cans such as the case of co-mingled material (UBCs that have been separated from garbage). Depending on the growth of these two sectors, returned cans may become cleaner or dirtier on the average. The effects of dirtier cans should, however, be counteracted as metal cleaning technology is expected to keep up with increasing contaminant levels.

There are other important factors to consider as we look to the future of recycling. As the use of aluminum food cans grows, these will be introduced into the recycling stream and will provide additional challenges to the recycling industry. System losses may increase somewhat due to the presence of food residues in the cans and external paper labels. Based upon measured weights of 18.45 grams for a 6-oz. pet food can and 11.00 g for a 3-oz. can, paper labels (weighing 0.85 g. and 0.60 g, respectively) make up nearly 5% of the total UFC weight. It appears likely that some means of cleaning the UFCs and removing the paper labels during the recycling process may be necessary.

The impact of a growing number of UFCs on the composite composition of recycled cans was examined using the can weight analysis model. The conditions assumed are shown in Table 6. The increased number of UFCs in the recycling stream has only minor impacts on the UBC composition, even with the generous assumptions above for food can return. [The 2% figure considered for the 1989 scenario is probably high as it would require nearly a billion aluminum food cans being recycled today!] With the assumed food can growth from 2 to 10% of total cans, the composite UBC composition will only change by 0.06% in Mn and 0.12% in Mg. Thus, we conclude that the introduction of food cans into the recycling stream, while it might produce some perturbations on a local level, will not appreciably affect the overall recycling picture from a composition standpoint.

TABLE 6

Composition of Recycled Metal Stream as Affected by Introduction of Used Food Cans (UFCs)

	1989	1993	1997
Composition, wt %			
Si	0.16	0.16	0.15
Fe	0.39	0.39	0.38
Cu	0.10	0.10	0.10
Mn	0.90	0.88	0.84
Mg	1.95	1.98	2.07
Assumptions			
UFC unit weight, g	22.0	21.0	20.0
UFC % of metal stream	2%	5%	10%
UBC unit weight, g	15.85	14.80	13.62

Food Can Alloy is 5042 (Body and Lid)
Beer/Beverage Can Alloys are 3004 and 5182

As recycling becomes a greater legislative issue, there is the possibility that other packages will become aluminum to take advantage of the recycling systems already in place. These packages may be permanently converted to aluminum or may switch back to other materials when their recycling becomes viable. This scenario would pose important challenges to the existing beverage can recycling system since a wider variety of alloys and unused contents of these containers would potentially be introduced into the recycling stream.

Summary and Conclusions

Simple mathematical models have been used to investigate whether trends in design, material selection and recycling of aluminum beverage containers are in conflict. The following conclusions are warranted:

- Present differences in can design and alloy have only a negligible impact on the overall viability of recycling.
- Introduction of significant amounts of high Si alloy 5017 could limit the absorption of UBCs into alloy 3004 body stock.
- Future projections for recycling of lightweight containers indicate that some increases in system losses (due to melt loss + lacquer) will take place.
- Differential lightweighting of body and lid will have some impact on the composition of recycled UBCs but should not substantially affect absorption back into body stock.
- As used food cans are introduced into the recycling stream, composition of the total metal stream will not be significantly affected. System losses due to paper labels and food residues will increase.

References

1. P. A. Plunkert, U.S. Department of the Interior, Bureau of Mines, 1989-1990.

2. A. B. Trageser and R. E. Dick, "Aluminum Can Design Using Finite Element Methods," Society of Manufacturing Engineering Can Manufacturing Technology Symposium, Schaumburg, IL, 1988 September 14-16.

3. R. E. Sanders, Jr., T. L. Hartman and D. J. Lege, "Aluminum Rigid Container Sheet for the Packaging Industry," *Aluminium*, Vol. 65 (1989), No. 9, pp. 941-947.

4. D. McAuliffe, "Production of Continuous Cast Can Body Stock," presented at TMS-AIME Meeting, Las Vegas, NV, 1989 February 27.

5. R. E. Sanders, Jr., J. K. McBride and M. W. Milner, "Recycling of Used Beverage Cans into 3004 Alloy Can Sheet," presented at TMS-AIME International Symposium for Recycle and Secondary Metals, Ft. Lauderdale, FL, 1985 December 01.

6. J. R. Greenwald, Alcoa Laboratories, Private Communication, 1989 September.

THE ALCAN DECOATER PROCESS FOR UBC DECOATING

B. McAvoy, J. McNeish
Alcan Recycling Ltd, Greensboro, Ga
W. Stevens
Alcan International Limited
Arvida Research and Development Centre
Jonquière, Québec, Canada, G7S 4K8

Abstract

Alcan has developed an economical decoater process to delaquer shredded Used Beverage Container material prior to melting. This paper describes the origin of the Alcan process, discusses why decoating is necessary, and outlines the operating results obtained with this technology.

Second International Symposium -
Recycling of Metals and Engineered Materials
Edited by Jan H. L. van Linden,
Donald L. Stewart, Jr., and Yogeshwar Sahai
The Minerals, Metals & Materials Society, 1990

Introduction

Alcan Aluminum has developed an economical means to recycle returned Used Beverage Containers (UBC). The heart of the recycling process is a decoating apparatus to remove the organic coatings from shredded UBC before the material is melted. The recycled containers have become a major source of raw material for the production of aluminum can body sheet ingots. This source of material is not only economically advantageous but an environmentally sound process since it reduces the quantity of consumer waste disposed of in landfill as well as reducing the energy requirement as compared with the production of primary aluminum from bauxite.

Before economical Decoating Processes were developed, painted aluminum scrap was normally remelted in open well furnaces. As the painted materials burned, large quantities of smoke, fumes and dross were generated, thereby reducing the aluminum recovery.

The Alcan Decoating Process is proprietary and protected with US patent 4,654,088 (1).

Why Decoat?

Before examining the Alcan decoater process it is important to review why painted and lacquered aluminum should be decoated before melting.

Clean metal does not oxidize when submerged into molten metal for melting. Coated metal, however, is evolving organic gases thereby making submergence difficult, leading to excessive oxide formation. As well, due to the oxide envelope formed, for thin gauge material it becomes very difficult to drain the molten metal from the inside of the oxide envelope.

Therefore, the principal purpose of decoating is to destroy the organic layer to permit rapid submergence and reduce the formation of oxides. This leads to the principal advantage of decoating of UBC material being the reduction in melt loss.

It is reported that direct melting of UBC results in 25-30% material loss, i.e. only 70 to 75% of the initial weight being recovered as useable metal.

Figure 1 (2) shows the results of laboratory scale melting experiments of painted and bare metal samples. Also shown is the increased melt loss resulting from the melting of a high magnesium alloy - a major component of the UBC material. Effectively melt losses are reduced from 13.5 to 3% for the AA 3003 alloy and from 17% to 4% for the AA 5182 alloy. Note that the magnesium alloy required the addition of a salt flux to reduce the melt loss to the level obtained with the bare non-magnesium alloy.

Another very important advantage of decoating is the reduction of a serious safety hazard - reducing the risk of molten metal-water contact. A Decoating Process, as it ensures that all the material is raised to well above the boiling point of water, eleminates this risk.

A third evident advantage of decoating is to reduce the environmental effect of the creation of smoke and fumes as occurs with direct addition of this material to a well furnace. Even when fumes are efficiently collected in a baghouse, volatile compounds are released to the atmosphere. The Alcan Decoating Process includes an efficient fume incinerator reducing the majority of the volatile carbon compounds to CO_2 and water vapour.

COATING EFFECT ON MELT LOSS

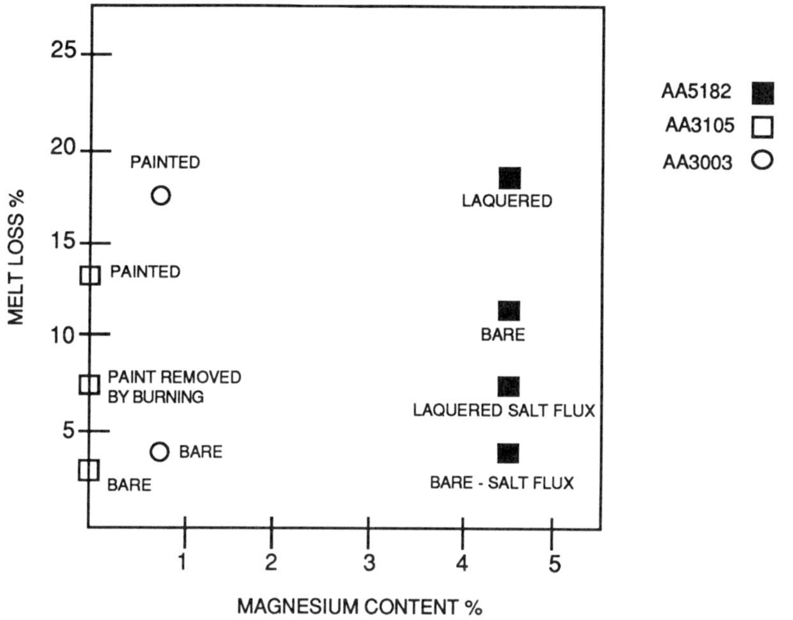

Figure 1

Decoating before addition of material to the furnace well reduces baghouse maintenance required by minimizing the loading on the baghouse. The safety of the baghouse operation is also improved as soot formed from painted UBC is eliminated, because soot can lead to baghouse fires.

If the transport of treated material to the furnaces is carried out in a manner to conserve the heat content of the material at the completion of the Decoating Process the process also serves to reduce the energy required in the subsequent melting process or to increase the throughput of a given size of furnace as compared to operating with cold feed material. Figure 2 shows the proportion of heat energy added by the Decoating Process as a proportion of the heat required for melting aluminum and heating to normal furnace process temperatures (730 °C). Thirty-five per cent of the total energy required for melting is available if the decoated material is delivered to the melting furnaces at 450 °C.

Development of the Decoating Process

The first step in the development of the decoating process was the development of a technique to determine the process requirements of a Decoating Process. Work was carried out to analyse the composition of the majority of the coatings in use as well as to determine the conditions required to remove the coatings. A laboratory scale fluidized bed using sand as the fluidizing agent was used for the initial test work. Samples of painted cans were submerged in the fluidized bed at various temperatures and for varying residence times to determine the conditions required for acceptable decoating. It was found that it is necessary to not only remove the volatile compounds but to ensure that the residual carbon had been pyrolised.

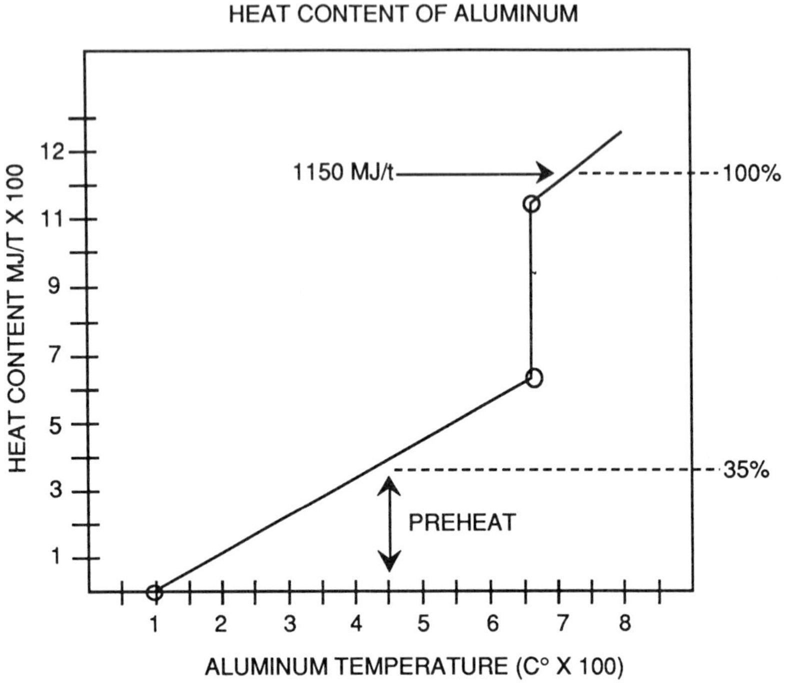

Figure 2

The criteria developed specified that, for decoating to be considered effective, the decoated sample must not produce a "bubble" on submergence into molten aluminum. The formation of a bubble is indicative of volatisation at the solid-liquid interface which would reduce the ease of melting of the sample.

As shown in figure 3, for the organic coatings tested, all the coatings were completely decoated in one minute with a temperature of 550 °C. The numbers refer to various organic coatings found on UBC and other painted scrap material. As the solidus temperature of the can end alloy (AA 5182) is only 586 °C, a Decoating Process must be highly controllable.

It was determined a plug flow process was required. Three plug flow processes were considered - the rotary kiln; the shaft furnace; and the shallow packed bed.

The thin or shallow bed process was chosen as the Decoating Process is exothermic and a thin bed is required to ensure that the temperature at the bottom of the bed does not exceed the solidus temperature. The experimental parameters of a thin bed process were developed with a stationary bench scale model. This work showed that specific relationships existed between the air inlet temperature, peak bed temperature and air flow rates. The experimentation also demonstrated that results were reproducible from test to test.

TIME - TEMPERATURE RELATIONSHIPS FOR GOOD DECOATING

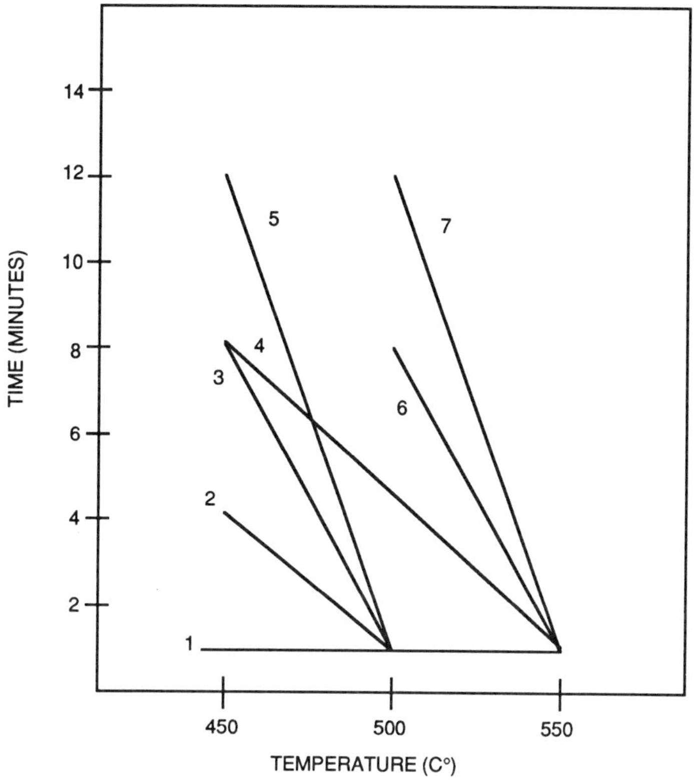

Figure 3

Figure 4 shows typical time temperature curves at the upper surface, middle, and bottom of a stationary test packed bed. This work was followed by trials using a moving bed cross flow apparatus.

A conventional pilot scale sinter strand machine located in Lurgi-Chimie's Frankfurt installation was used to test the process. The pilot machine had a bed surface area of 0.58 m^2 and was able to produce acceptably decoated material at a rate of up to 473 kg/h/m^2 of bed area. The material was decoated in a reproducible manner over a wide range of bed densities, air flow rates and residence times. The pilot equipment dealt easily with knife shredded or hammer milled scrap including UBC, siding, and multiple compacted layers of heavily lacquered foil.

These tests allowed the design parameters of the pilot plant to be located at Alcan Rolled Products Oswego works to be determined.

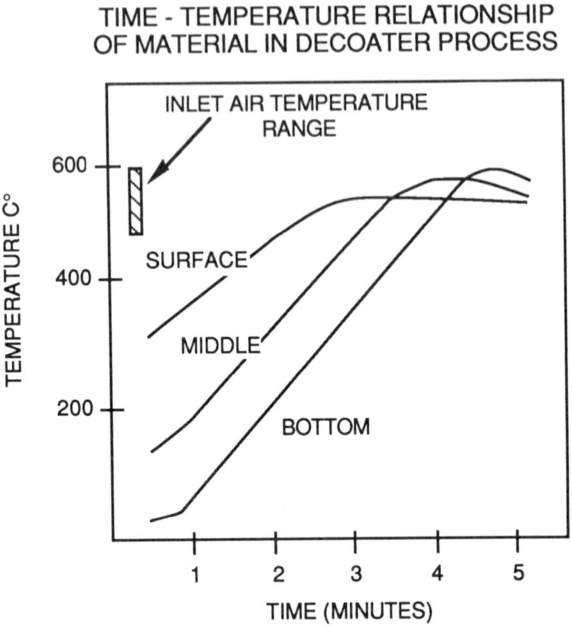

Figure 4

The thin bed process, by improving temperature control, would also reduce melt loss caused by increased oxidation within the decoater due to localized overheating. Material loss is of greater importance than energy efficiency.

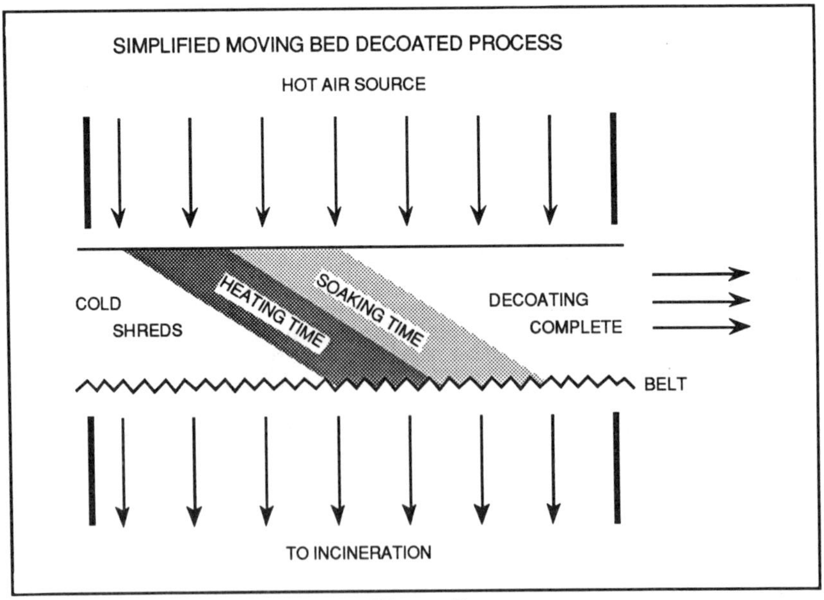

Figure 5

Figure 5 shows a simplified drawing of the decoater process showing the movement of the decoated shreds, the soaking zone where the volatisation of the residual carbon occurs, and the heating zone where the shreds are heated to temperature. The residence time in the decoater is on the order of 4.5 minutes.

The Oswego pilot plant was operational in 1981. After one year of successful operation the process was evaluated to determine the design parameters for the full scale plant to be built at the Greensboro, Georgia Recycling installation. The Oswego process was found to routinely operate at 3.6 tonnes/h (8 000 lb/h) with the decoater itself able to operate at 6.4 tonnes/h (14 000 lb/h). The decoater rate was determined by using preshredded material as the shredding equipment installed at the Oswego plant limited the plant output.

The Greensboro decoater was operational in 1983. This plant had an initial capacity of 55 000 tonnes/y (60 000 USt/y). Continuous process improvements, including optimizing the operation of the decoater, has increased the plant throughput to 75 000 t/y. A third generation recycling plant has been built in Berea, Kentucky. This plant was opened in October, 1989 and reached design production rate of 110,000 tonnes/y in January of 1990. This plant consists of two parallel lines with capability of transferring material from one line to the other as the need arises.

Basic Concept of the Alcan Decoater Technology

The aluminum shreds on a wire mesh belt form a horizontal packed bed. The process is similar to that used for sintering iron ore or nickel suphide pellets. The heat transfer within the packed bed occurs according to conventional packed bed theory, with the process heat moving through the bed as a front.

The thin bed as specified (75 mm thickness) is a means to prevent overheating of the bottom layer of the bed material.

The parameters controlling the process are;

- air flow rate
- process air temperature to the top surface
- metal throughput rate (bed travel speed)
- bed thickness
- residence time

The process flow sheet is shown in figure 6.

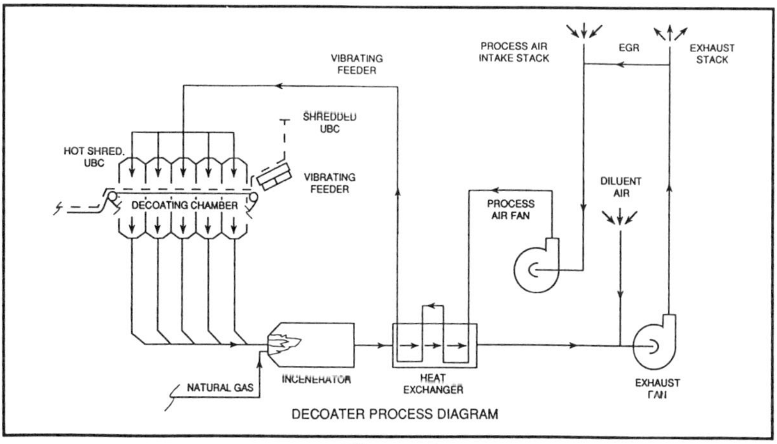

Figure 6

The Alcan process uses a down flow air path. This means that the hot process air first contacts the bed material and not the support mesh screen. The mesh screen is therefore heated by energy surplus to that required for heating the bed material and a lower air input temperature can be utilized. As the mass of the wire mesh support can be greater than the mass of bed material the down flow design is important for the overall process efficiency.

Ambient air is drawn through a flow control device and then through the inlet air blower (process air). Controlled amounts of air passes through the tubes of a heat exchanger. On the shell side of the exchanger pass the exhaust gases after incineration. The incoming process air is tempered with exhaust gases from the process (after incineration and the heat exchanger) to ensure the incoming process air has a bulk temperature above the dew point of common acid gases. In the heat exchanger the process air is heated to 500 - 600 °C after which it passes to the decoater entrance. The process air is divided equally between the five zones of the decoater in the process air plenum. From the plenum, the hot air is pulled through the thin packed bed and the wire mesh support by a negative draft created by the exhaust gas induced draft blower. After the decoater, the hot gases containing the volatile compounds are passed into the incinerator where the fumes are ignited with a grid style burner. The incinerator is maintained at a minimum of 705 °C to ensure the thorough destruction of the organic carbon compounds to CO_2. After incineration the hot gases pass through the shell side of the heat exchanger where they release a substantial quantity of their heat to the incoming process air. The temperature of the waste gases after the heat exchanger are on the order of 300 °C.

Performance of the Decoater System

Ideally a decoater system should result in:

- no volatiles to the atmosphere
- no carbon residue remaining on the treated material
- no evidence of overheating or melting

Energy Balance of Decoater Process

The prototype decoater system energy use is shown in figure 7. A Sankey diagram for the process is shown in figure 8. Energy absorbed by the aluminum shreds represents 50% of the energy supplied by the fuel. Additional heat energy is generated in the process from the combustion of the volatile organic compounds in the incinerator. This process uses natural gas at the rate of 1 050 MJ/tonne of aluminum shreds treated.

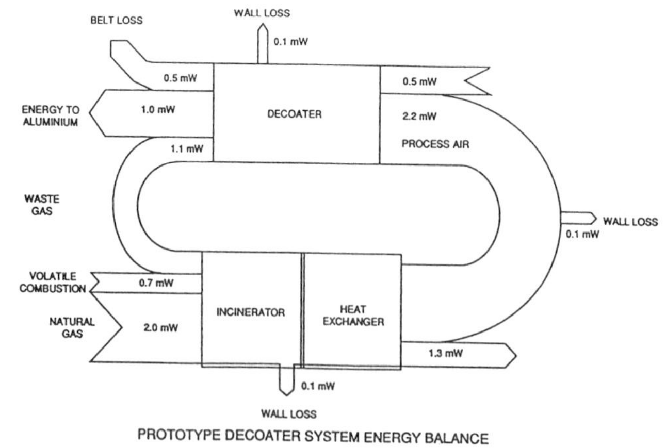

PROTOTYPE DECOATER SYSTEM ENERGY BALANCE

Figure 7

The energy efficiency of the process is determined by the process air flow rates used. mathematical modelling of the decoater process determined that energy use is lowest with a minimum quantity of process air used for a given throughput. Too low of a process air flow rate however results in insufficent process for absorbing the energy generated in the exothermic reaction in the packed bed. For this reason the air flow rate is optimized to minimize energy use while assuring that the Decoating Process is stable and does not result in excessive heating of the particles at the bottom of the thin bed.

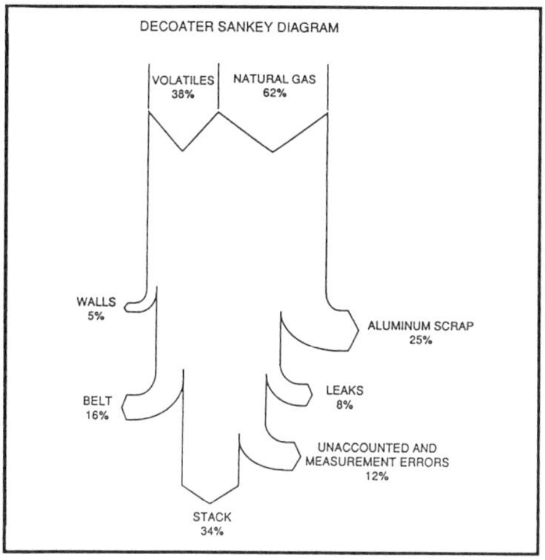

Figure 8

<u>Greensboro Recycling Center Description</u>

The Greensboro recycling center consists of a shredding and decoating system for UBC, two sidewell melters, and a casting furnace with a DC Casting Machine. (see figure 9).

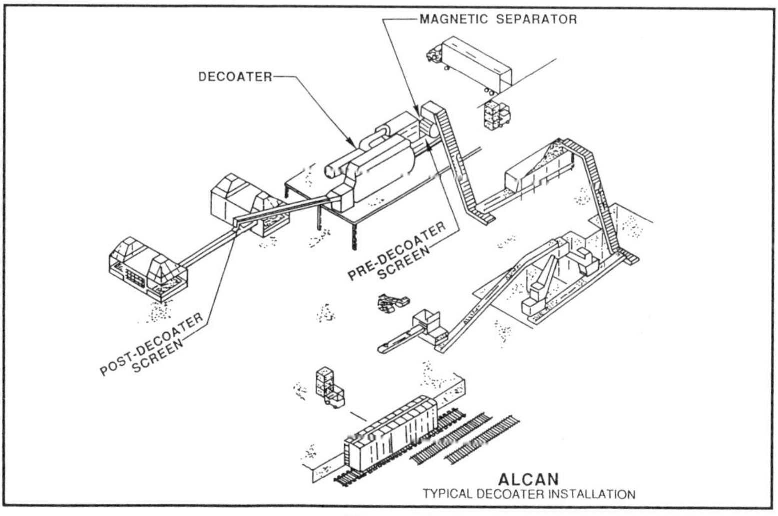

Figure 9

The Decoater is a variable throughput device allowing operation between a range of 5 tonnes (11 000 lb/h) to 11.8 tonnes/h (26 000 lb/h). Through the use of control algorithms installed in multi-loop controllers, the operator simply "dials" the desired feed rate. All parameters are automatically calculated and set to provide efficient decoating at the desired rate. These multi-loop controllers are the means for measuring, displaying, and controlling all analog (variable) signals integral to the process parameters. A programmable logic controller (PLC) provides all of the discrete control for sequencing of material handling conveyors, the decoater fans and alarm annunciation for operator information.

Coated aluminum (UBC) shreds are supplied on a continuoùs basis to a surge hopper with a retention capacity of 9.0 tonnes (20 000 lb).

The shreds are discharged form the surge hopper by means of a 1.8 m wide conveyor belt equipped with a belt weigh scale.

The weighed material is elevated with an inclined conveyor and then passed over a rotating fixed magnet to remove any tramp magnetic material included in the aluminum shreds.

The material is discharged onto a vibratory screen to remove dirt and fines. From the vibrating screen the material is transferred directly onto the wire mesh belt of the decoater. The 1.8 m wide stainless steel wire mesh belt is supported on the inlet side and inside the oven by a herring-bone type grate for improved tracking and to allow for minimal restriction to the air flow through the belt. The belt return is outside the decoating oven and is supported by idler rollers. The belt return is enclosed to minimize heat losses from the belt.

The hot process air enters the decoater heating chamber through a top hood designed to give uniform air distribution over the whole bed area. The lower part of the decoater housing, which forms the gas collection chamber is composed of five exhaust openings, and five cleaning/inspection doors to provide five similar modules making up the windbox chamber.

The process air flowing to the decoater's hood is controlled for temperature and flow rate. Waste gases leaving the decoater windbox chambers are bound for the temperature controlled incinerator prior to passage across the heat exchanger tube bundles and out the exhaust stack. The volume of these waste gases is controlled via the exhaust gas fan whose speed is determined by the static pressure within the hood of the decoater oven.

The Decoating Process makes extensive use of variable frequency drives in order to control air flow rate, air temperature, metal throughput, bed thickness/residence time and exhaust gas volume/hood pressure. One major feature is the control system which allows the operator to set the desired material feed rate to the melting furnace after which the controller sets, monitors and regulates all the operating variables including conveyor speeds, decoater air flows, and temperatures.

From a maintenance point of view the decoater consists basically of two fans, a burner, a heat exchanger, and a wire mesh belt. All of the equipment is well developed and is of "off-the-shelf" design. The areas requiring maintenance are: adjusting the belt tracking, the replacing of conveyor bearings, and periodic replacement of the wire mesh belt.

Environmental Controls and Emissions

The Alcan decoater is equipped with an incinerator which consists of a natural gas fired grid burner followed by an incinerator chamber. The temperature in the incinerator is maintained at a minimum temperature of 705 °C (1 300 °F) to control volatile organic carbon (VOC) and carbon monoxide (CO) emissions. In the incinerator the majority of the volatilized organic compounds from the pyrolised paints are converted to carbon dioxide (CO_2). Typical emissions are outlined in the following table.

Table I

Emission type	**Rate** (kg/tonne UBC)
CO	0.073
VOC	0.041
Particulites	0.041

The emission type of primary concern - the CO - is controlled by the incinerator temperature and is never allowed to exceed 50 ppm in the exhaust gas.

Decoater By-products

There are three by-products streams resulting from the decoater process: irony material, pre-decoater fines, and post decoater fines. All three of these materials are sold as by-products to third party metal reclaimers. The irony material is sold to recover the iron material for reclamation. Pre-decoating screening is performed to remove as much dirt and fines before the decoater to prevent blinding of the wire mesh screen. Finely shredded UBC is therefore found in the fines resulting in a product with a saleable aluminum content. Post decoater fines are also sold as they contain 10 - 40% recoverable aluminum.

Maintenance and Operating Costs

The equipment included in this section ranges from the outlet of the surge hopper to the vibrating conveyor discharging to the post decoater screen. The costs quoted include the decoater and associated handling equipment. Total maintenance costs per year including one new stainless steel wire mesh belt ($40 000 US) are $4.81/tonne UBC throughput. Non-annual maintenance includes the replacement of the following items:

Item	**Replacement Cycle**	**Cost ($US)**
Grid burner	2 years	10 000
Hot side heat exchange tube bundle	7 years	95 000
Cold side heat exchange tube bundle	5 years	37 000
Herring bone belt support table	5 years	65 000

The electrical power and natural gas costs based on a 70% annual utilization rate is $0.88/tonne and $4.44/tonne respectively. Energy costs are directly related to utilization rate as the fans and incinerator burner operate even when the decoater is idling. To minimize these costs sufficient quantities of UBC shredded material must be available as well as sufficient melting and casting capacity to ensure continuous operation of the decoater.

Conclusions

Alcan has developed a proven, environmentally sound technology for the decoating of UBC material. To date, in excess of 398 000 tonnes of UBC has been processed with the four operating decoaters. A fifth decoater unit is now under construction in the United Kingdom.

References

1. N. P. Fitzpatrick, R. E. G. Lee, U.S. patent #4,654,088.

2. A. L. Stewart, J. G. McCubbin, "Melting Aluminum and Its Alloys", Light Metals Age, December 1977, pp 13-15.

A STATE-OF-THE-ART UBC RECYCLING FACILITY

IMSAMET'S IDAHO PLANT

JOE WOJCIECHOWSKI AND W.F. FUNDINE

PLANT MANAGER
IMSAMET, INC.
POST FALLS, ID

INDUSTRIAL ENGINEER
KAISER ALUMINUM
SPOKANE, WA

ABSTRACT

During the past few years, recycling of aluminum Used Beverage Cans (UBC's) has become an integral part of the United States aluminum industry. As a result, UBC recycling has developed to the point that competitive market factors control the collection, distribution and processing of UBC's. At IMSAMET's recently commissioned UBC plant in Hauser Lake, Idaho, state-of-the-art equipment has been installed to ensure best possible recovery of high quality aluminum from the UBC's at the lowest fuel consumption in an environmentally clean operation. The paper describes plant features and operation. The paper describes plant features and operating practices that make this 140 million pound per year plant amongst the leaders in UBC recycling.

**Second International Symposium -
Recycling of Metals and Engineered Materials
Edited by Jan H. L. van Linden,
Donald L. Stewart, Jr., and Yogeshwar Sahai
The Minerals, Metals & Materials Society, 1990**

The UBC recycling facility at Post Falls (Hauser Lake), Idaho was conceived as a result of a contract between Kaiser Aluminum and Chemical Corporation and International Mill Service, Inc. to toll process UBC's and other canstock-related materials for the client, Kaiser.

International Mill Service has been in business for over 50 years, providing recycling services, waste handling, and capital-avoidance work for the metals industry.

Kaiser-Trentwood management recognized the potential savings from this concept by delivery of hot metal, improved data acquisition, and re-direction of capital for continued internal sophistication.

Geography

The Plant is located in western Kootenai County, Idaho, on a small prairie bordered north, south, and east by mountains. Soil is well drained, with excellent compaction. Beneath the Plant (150 feet) is the Rathdrum aquifer, a protected, sole-source, water supply for the entire Spokane area.

Climate

The area receives about 27 inches of annual precipitation, mainly winter snows. Temperatures are relatively mild, being affected by Pacific weather, and protected from arctic blasts by the Bitterroot Mountains to the east.

Demography

The area is the major population center in the Inland Northwest, supported by Kaiser Trentwood and Mead works, timber and mining industries, tourism, and various smaller manufacturing and support groups. The populace is mixed between old-line residents and recent immigrants from the West Coast and Mid-west, attracted by the lifestyle, hiking, fishing, hunting and skiing. There are very strong environmental protection groups and sentiments.

The Plant was intentionally located in Idaho to take advantage of various state tax situations.

The Plant maintains non-union status, pays competitive wages and benefits, plus offers a worker incentive program based upon Plant profitability.

Utilities

The Plant is serviced by the Burlington Northern Railroad, has easy access to the Interstate Highway System, and all-weather highways to the Trentwood Plant (12 miles door-to-door). Natural gas and electricity are supplied by Washington Water Power Company, and the Plant is self-contained for water and sanitation.

Plant-General

The facility was designed to process and melt UBC as the primary feedstock. It was also set up to receive and process class scrap, plus light gauge rolling mill scrap. Maximum recovery, chemistry, good environmental protection, and economy of operation, plus, the supply of end product in molten state, were the primary goals. Also considered essential is a data acquisition and evaluation program to determine fuel efficiencies, moisture losses, fine losses, chemistries and overall recovery of the various types of materials, including drosses.

As prevously alluded, the Plant is in an environmentally sensitive area, seated over the sole source of water for the area, and surrounded by tourism related businesses. It was therefore imperative that all preventive measures be taken to protect the air, soil, and water quality and also present a not-unattractive appearance.

Physical Description

In order to meet the aforementioned criteria, the facility was designed to process materials in the following manner and sequence:

A. Order verification and receipt,
B. Weight check,
C. Inspection and acceptance (or rejection),
D. Inventory and schedule,
E. Pre-break,
F. Magnetic separation,
G. Fines screening,
H. Shred,
I. Magnetic separation,
J. Delacquer and weigh,
K. Melt, and finally,
L. Deliver molten or sow aluminum.

(Drosses are weighed and direct-charged into rotary furnaces, described later).

The main UBC reclamation building is 44,000 square feet, with a 50 foot eave height to accomodate the shredding and delacquering systems. It contains both truck and rail unloading areas and has approximately 12,000 square feet allocated to material storage. The bulk of the inventory is stored directly outside the process area, for ease of access.

The equipment was sized to process 20,000 pounds per hour of UBC, class scrap, and light gauge rolling mill scrap, in any combination.

The pre-delacquering, or cold-processing system consists of an apron feed conveyor to a 60-60 single rotor primary impact crusher, four head pulley and cross-belt magnetic separators strategically located to protect both chemistry and equipment, a vibrating dirt screen which reduces knife wear and enhances chemistry, flux usage, and recoveries, and a shredder to provide uniform feedstock to the delacquerer. The shredder utilizes a trommel, or rotary classifier to insure optimum density (10 to 12 pounds per cubic foot) without excessive fines generation. All of this equipment was supplied to specification by Universal-Hammermills, Inc. of Cedar Rapids, Iowa.

The shredded cans are delivered to the delacquering system by a computer-controlled weigh-belt system, insuring steady feed rates at the operator's command.

The delacquering system is a rotary kiln, unlined, and 40 feet in length. The heat source is external and counter-flow to the UBC directional movement. Cans are introduced to the kiln through a set of airlocks to maintain a low (8%) level of oxygen, thus retarding oxidation of aluminum and preventing premature ignition of the volatilized lacquer. The cans tumble through the drum (and hot gases and vapors) for approximately 14 minutes, during which time the organic coatings are pyrolized. The cans are then discharged, once again through a double airlock system, and onto a proprietary, continuous weighing system and on to the furnaces.

It must be noted that the weighing device was internally developed to provide accurate information on moisture and other delacquering losses. Demonstrated accuracy of this system is better than the other certified weighing systems throughout the Plant, or .03%. Difficulties encountered in the development were (but not limited to) mechanical vibrations, electrical frequency interferences, and constant heat of the weighed material and surrounding environment. Cans exit and are weighed at approximately 975° F.

Hot combustible gases are delivered to the combustor by means of a variable speed fan, where they are ignited by a natural gas burner system. The combustor was designed to provide adequate residence time to insure complete incineration of the hydrocarbons prior to discharge into the atmosphere.

Excess products of combustion are discharged through a stack; system pressures are maintained by a computer controlled damper. The hot (low oxygen) gases needed to sustain the delacquering process are cooled to desired inlet temperature by passing through a recuperator system. The recuperator also provides pre-heated combustion air to the burner. All controls are automated to respond to predetermined setpoints. Also, all controls (and setpoints) are provided with manual overrides, allowing for changing feedstock conditions or requirements, plus providing an additional safety factor in the event of equipment malfunctions.

The two melting furnaces are single well, open hearth, gas fired, reverbatory units, each with 150,000 pounds capacity and 28,000,000 BTU/hour burner systems. The furnaces are positive pressure-controlled to insure low metal oxidation losses in the combustion chamber and also to maximize heat utilization. They are provided with forced-air floor cooling to prevent premature refractory failure, and are direct-controlled by metal temperature. They also have roof, stack, and floor thermocouples as additional safeguards against over-firing. Metal circulation is maintained by (Metaullics) molten metal pumps (M-42, electric drives). Adequate metal circulation is mandatory, and allows for burner settings (at stoichiometric) lightly impinging upon the surface of the metal bath.

The delacquering system and the furnaces were provided to specifications by Gillespie and Powers, of St. Louis, Missouri.

Cans and other feedstocks are submerged in the wells by use of either a vortexer or a puddler system. A standard 50-50 NaCl-KCl flux (with 2% cryolite) is used while charging materials. Flux amounts vary in direct proportion to the dirtiness of the feedstock, and will range from 1/2 to 6%.

Drosses are skimmed by forklift into large surface area skim pans, thus quickly cooling and minimizing thermite losses. The fluxed (black) dross is endothermic, provided that adequate flux is used. Interior (white) dross generation is of a relatively low volume, and is cooled upon contact with the skim pan.

Metal from the melters is tapped either into over-the-road crucibles mounted on special trailers, or into low-profile 1000 pound sow molds, of Kaiser's design. The crucibles are also of Kaiser's development, so that equipment will interface at both ends of the line. Normal trip time, tap to tap on a single unit can be as low as one hour. (12 miles portal to portal.)

Pollution Controls

The Plant is serviced by baghouses, selected for efficiency and compatability with the anticipated particulates generated. The cold-process system is served by a 32,000 cfm pulse-air type baghouse, with pickup points at each dust source. The furnaces are served by a 60,000 cfm shaker-type baghouse (bags are limestone-treated). This unit also has a minor pickup point at the delacquering unit to handle the anticipated fugitive emissions during system startups and shutdowns.

In addition to the air cleaning equipment, the facility has ridge vent and stack emission monitors (opacity meters) and continuous recording devices. Records are kept available for governmental environmental regulatory agencies.

Soil and groundwater protection is provided by strict adherence to operating rules, especially limiting all dross and flux storage to be conducted indoors and on pavement. The plant does not use process water nor does it conduct any waste fluid operations (other than sanitary sewage).

The aquifer has a unidirectional flow under the plant. Test wells are located, penetrating the aquifer upstream and downstream of the plant. Weekly samples are taken, recording the findings for agency scrutiny, and also providing data protecting the company from liability from any upstream pollution. Chlorides, being the fastest acting and moving contaminants, are especially monitored.

The dross processing facility consists of a 38,000 square foot building which houses two rotary salt furnaces, each with a 24,000 pound filled capacity, and a 45,000 pound reverb type holding furnace. All drosses, fluxes, and saltcakes are handled indoors. The furnaces are side-tapped for metal, with a continuous spent-salt overflow.

Metal is tapped periodically into the holding furnace. The holder will accumulate up to 1 1/2 crucible truckloads or will pour into sow molds.

The dross mill is serviced by an 80,000 cfm shaker-type baghouse to help insure maintenance of a pollution-free environment. The use of a shaker-type house is necessary due to the hygroscopic nature of the particulate. Pulse-air systems are recognized to work also, but require more baghouse heating, especially during damp seasons. The particulate collected by this unit is recycled through the rotary furnaces, as it is simply volatilized and recrystalized flux. (The bags are <u>not</u> lime-coated.)

Design and demonstrated capacity of this system is +4,200,000 pounds per month. This capacity was required to service the drosses generated by the UBC facility plus the (white) drosses generated by the customer's cast-house.

Operations

At least equal in importance to proper design and layout is the continued proper operation and management of the endeavor. A proper allegory is that 'anyone with the money can purchase a sophisticated aircraft, but money can't fly it.'

Teamwork and dedication, while sounding trite and overworked, are necessary for ongoing successful operations; maximizing recovery, optimizing chemistry, minimizing all costs, and maintaining the environment.

Each load is weighed, inspected, tested, and data acquired throughout the process. Records are kept by date and vendor pertaining to amounts, contaminants, moisture, load integrity, etc. These records serve to facilitate scheduling, improve operations, and provide help in supplier quality improvements.

Shift, operator, and machine records are kept to improve and standardize operations and equipment reliability. Total records are kept on flux usage, dross generation, and chemistry. The furnaces are sampled at a minimum of every 20,000 pounds of charge, thus identifying excursions of elements before they become harmful, plus identifying the source of contamination. The chemistry of the heat or load is provided to Kaiser prior to the arrival of the metal, thus aiding scheduling and timely utilization. (Chemistry is obtained by spectography; specifically an A.R.L. 3460 series quantometer.)

Operations are enhanced by continuing formal and informal education of the workforce, by a contractual payment schedule based in part by recovery bonuses, and by a worker-incentive system based upon profitability.

Integrity is insured by the fact that Kaiser is the sole owner of all metal units throughout the process, and has audit rights on the operation. (The facility does not process for outside clients.)

Recovery data from UBC operations is proprietary and confidential, but consistently exceeds industrial norms.

Acknowledgements

Engineering, construction, and permitting were done with McGraw Construction Company, a sister company to IMSAMET, under the parent company, Envirosource.

IMSAMET recognizes the quality delivery of goods promised by McGraw and the major sub-contractors, Gillespie and Powers and Universal-Hammermills. The initial groundbreaking was September 6, 1988, and the first heat of metal was tapped mid-March, 1989. Construction was conducted through the worst winter in northern Idaho in 20 years.

Startup was highly successful, surpassing contractual obligations and even the most optimistic forecasts. Capital cost was $14,000,000 for the facility, "turn-key".

TRANSIENT, THREE-DIMENSIONAL BRIQUETTE

PREHEATER MODEL

Patricia A. Wells and Ray D. Peterson

Manufacturing Technology Laboratory
Reynolds Metals Company
P. O. Box 1200, Sheffield, Alabama, USA

Abstract

Preheating and delacquering of lightweight, loose aluminum scraps such as UBC (Used Beverage Containers) and turnings have been shown to increase metal recovery and furnace throughput during the reclamation of these materials. However, a large portion of the light scraps are compressed to form bales or briquettes. Normally these briquettes are not preheated before reclamation resulting in lower metal recoveries.

A dynamic model was developed to estimate the temperature profiles that occur in scrap aluminum briquettes during the preheating/delacquering process. Values of the heat transfer coefficient and thermal conductivity were calculated from individual experimental briquette heat-up tests in conjunction with a transient, three-dimensional heat transfer simulation for rectangular blocks. The single briquette model was expanded into a staged heater model to simulate industrial practice. The effects of briquette geometry and air velocity were examined.

Second International Symposium -
Recycling of Metals and Engineered Materials
Edited by Jan H. L. van Linden,
Donald L. Stewart, Jr., and Yogeshwar Sahai
The Minerals, Metals & Materials Society, 1990

Introduction

The reclamation process has become a significant contributor to the production of aluminum in recent years. Energy costs have increased so that it is now less expensive to remelt existing aluminum rather than produce it from alumina. However, the production efficiency of any reclamation plant will vary with the properties of the scrap metal feed. Currently, large quantities of recycled aluminum are comprised of Used Beverage Containers (UBC). To maintain efficient production in the reclamation furnace, it is desirable to have clean, dry UBC. One method of accomplishing this requirement is through preheating or delacquering. This process involves heating and holding the UBC at an elevated temperature to remove impurities, but not starting the melting process.

Delacquering removes several non-metallic components, the first being moisture and residues. Drying a charge may prevent explosions from occurring and reduce the amount of hydrogen absorbed into the aluminum melt. Furthermore, dirt particles should be less likely to cling to the scrap as the moisture is vaporized. Delacquering also burns away wood chips, plastics, and the paints/lacquers on the can surface. These materials contaminate the metal and may react with the aluminum to increase oxidation.

Since most lightweight scraps tend to float on the aluminum melt surface thus increasing oxidation, it is a common practice to bale these materials. This produces a more dense scrap and makes the material easier to transfer within the plant. This investigation was limited to experiments involving Alloys 3004, 5182 or a combination of the two. Various plants have differing terminologies for baled scrap; however, in this report it will be referred to as briquettes.

The continuous emphasis to reduce costs and increase the efficiency of aluminum production has prompted investigations concerning the exact benefits of preheating briquettes prior to charging. Our goal was to observe the effectiveness of a laboratory scale delacquering unit and model the heating profile trends of individual briquettes. With this information an overall model could be developed to describe the various stages or layers of briquettes in an industrial delacquering unit. For this study, a preheater is assumed to operate by forcing heated air across and through a mass of briquettes.

Experimental

The experimental temperature profiles for each briquette tested were determined utilizing the preheater design shown in Figure 1. The system consisted of a blower which forced air into a chest furnace containing four baffles constructed of firebrick. The baffles increased the residence time within the furnace thus increasing the heat transferred to the air. The heated air was directed into an insulated steel chamber, which minimized the thermal losses to the environment, and then was distributed evenly. The test briquette was placed within the steel chamber just above the baffle plate prior to starting the air flow. Fumes from the system were exhausted to the atmosphere. The air velocity was calculated from pitot tube measurements taken at the exhaust. This value was then used to calculate the velocity in the insulated container.

Seven thermocouples were located throughout each briquette to track the local heating trends. Four were spot-welded in the center of the four briquette surfaces: top, bottom, left and right sides. Three were positioned internally as shown in Figure 2. The air temperatures entering the furnace, within the steel chamber and the exhaust were also measured. Temperature readings were recorded in 1 minute intervals using a datalogger.

Four types of briquettes were investigated: Decorated 3004, Clear 3004, Clear 5182 and Decorated UBC. Each was obtained from a local reclamation plant, but no information of their origin or reclamation condition was acquired. Three experiments were run on each briquette: one at the blower's full velocity, one at half the blower's velocity, and one replicate to check reproducibility. Each test was continued until the briquette center temperature stabilized. In general, this required 2.5 to 3.0 hours, but

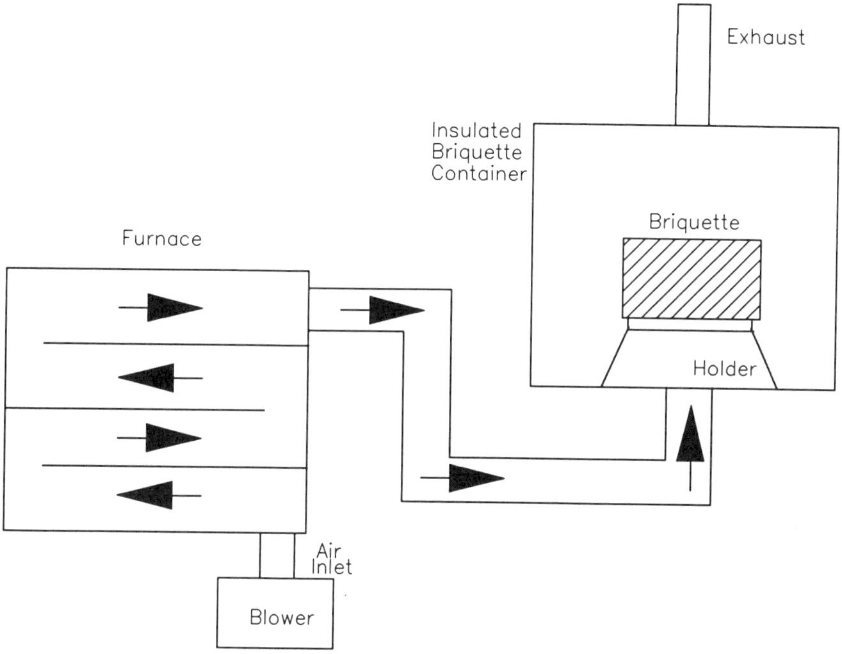

Figure 1 - Briquette preheater schematic.

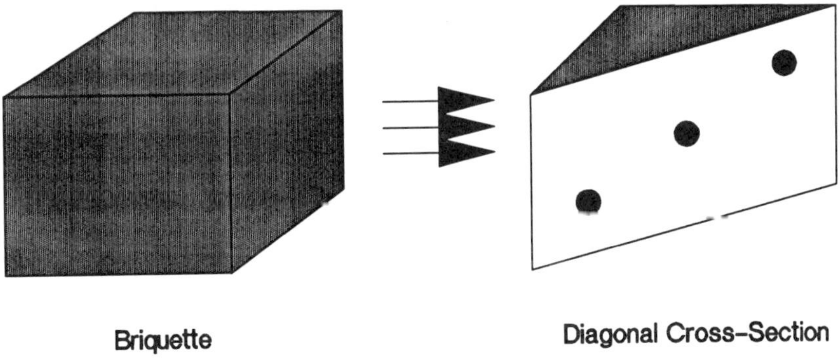

Figure 2 - Internal temperature measurement positions.

did vary with the size of the briquette. Prior to each experiment, the weight and dimensions of each briquette were recorded so that the density of the block and any weight losses due to the delacquering process could be calculated (See Table I).

Table I Briquette Specifications

Material	Test	Weight (kg)	Briquette Dimensions (cm^3)	Average Density (g/cm^3)
3004 Clear	1	21.252	31.8 x 31.8 x 26.7	0.7905
	2	21.067		
	3	21.054		
3004 Decorated	1	17.775	30.5 x 31.8 x 17.8	1.0330
	2	17.596		
	3	17.599		
5182 Clear	1	17.807	28.9 x 33.7 x 30.5	0.9094
	2	17.723		
	3	17.713		
UBC	1	17.371	50.8 x 33.7 x 16.5	0.6154
	2	16.951		
	3	16.892		

Theory

For any solid material placed in the cross flow of a fluid, heat transfer will occur in a two-step process. Initially, by forced convection, the air stream heated the briquette surface. This produced a temperature gradient between the surface and interior regions; thus, resulting in heat transfer to the briquette's center. Since briquettes are composed of discrete pieces of scrap aluminum compressed to form a rectangular bale, voids will exist within the block. Consequently, the thermal conductivity which is dependent upon a material's composition and makeup should be much less than a solid block of aluminum. Conversely, heat transfer coefficients should be a function of the air velocity.

As the briquettes were heated from room temperature, a period of time elapsed before the block and surrounding environment reached thermal equilibrium. This transient heating process accounts for the changes in internal energy occurring within the fixed volume. Hence, to model the heating trends we are interested in the temperatures at various points within the briquette at various times. Prior to investigating the system, it is impossible to make any simplifying assumptions as to the rate determining heat transfer step. Initial model development must include effects for both convection and conduction heat transfer.

For heat transfer in a single dimension, the general differential equation for conduction of heat in a solid can be written as:

$$\frac{\partial T}{\partial t} \qquad \alpha \frac{\partial^2 T}{\partial x^2} \tag{1}$$

where		
	α	= thermal diffusivity = $k/\rho C$.
	T	= temperature
	t	= time
	x	= length
	k	= thermal conductivity
	ρ	= density
	C	= heat capacity

The three boundary conditions for an infinite plate as shown in Figure 3 are:

$T(x,0) = T$, initial condition is a uniform temperature (2)

$$\frac{\partial T(0,t)}{\partial x} = 0,$$ symmetry of geometry (3)

$$\frac{\partial T(L,t)}{\partial x} + \frac{h}{k}[T(L,t) - T_f] = 0,$$ equating the heat flux of the surface to the rate of heat transfer to the fluid (4)

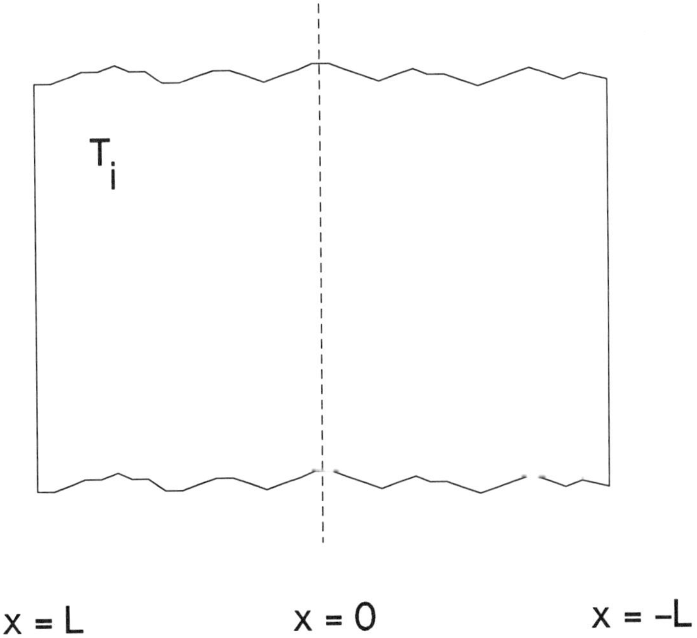

Figure 3 - Briquette geometric configuration.

Equation (1) can be solved by separation of variables as shown in reference (1). The final solution is:

$$\frac{T - T_f}{T_i - T_f} = 2\sum_{n=1}^{\infty} \frac{\sin(\lambda_n L)}{\lambda_n L + \sin(\lambda_n L)\cos(\lambda_n L)} \exp(-\lambda_n^2 \alpha t)\cos(\lambda_n x) \tag{5}$$

or,

$$\frac{\Theta}{\Theta_f} = 2\sum_{n=1}^{\infty} \frac{\sin(\lambda_n L)}{\lambda_n L + \sin(\lambda_n L)\cos(\lambda_n L)} \exp(-F_o)\cos(x/L) \tag{6}$$

where λ_n are the eigenvalues from the equation

$$\cot\lambda_n\ L = \frac{1}{(h/k)L}(\lambda_n L) = \frac{1}{B_i}(\lambda_n L) \tag{7}$$

The four dimensionless groups of variables are:

1. Relative Temperature $= \dfrac{T-T_f}{T_i-T_f} = \dfrac{\Theta}{\Theta_f}$

2. Fourier Number (F_o) $= \dfrac{\alpha t}{L^2}$

3. Biot Number (B_i) $= \dfrac{hL}{k}$

4. Relative Position $= \dfrac{x}{L}$

Since Equation (6) converges rapidly under general conditions, only the first ten or so eigenvalues are needed. The eigenvalues are determined by solving the non-linear transcendental function of Equation (7).

Equation (6) gives the temperature in an infinite plate as a function of the relative position, x/L. To calculate the temperature in a finite body such as a parallelepiped, the law of superposition is used. Equation 6 is used for each of the three axes with the appropriate dimensions used for each case.

$$T(x,y,z,t) = T(x,t) * T(y,t) * T(z,t) \tag{8}$$

A computer program to simulate the heat-up of a block based on the above equations was written in BASIC. For a given set of conditions (i.e., block size, position in the block, thermal conductivity, heat transfer coefficient, density, etc.), the transient temperatures could be calculated and stored into a data file for later analysis or plotting. This program was the basis for further programs used to estimate parameters from experimental data and to simulate the heat-up of a load of briquettes.

Empirical relationships have been found which relate the heat transfer coefficient to velocity and a number of other properties of the fluid and the immersed body (2). For submerged cubes, the proposed equation is:

$$h = \frac{k_f}{d} C' (Re)^n Pr^{1/3} \quad (9)$$

where
- Re = Reynolds Number (= du/v)
- u = fluid velocity,
- v = kinematic viscosity of fluid,
- Pr = Prandtl Number = 0.680 for air,
- C' = Constant,
- k_f = thermal conductivity of the fluid, and
- d = cross-sectional thickness of cube.
- n = 0.675, constant for square geometry

The above correlation for an average heat transfer coefficient is based on experimental data for liquids and gases. Using the estimated heat transfer coefficient (h), obtained from the experimental data, the unknown constants, C' and n, occurring in equation 9 could be calculated and an expression for the thermal conductivity (h) developed.

Results

Experimental Data

The temperature results for each series of tests were recorded and graphed; however, only one set will be discussed in detail due to the similarities in the analyses. The experiment chosen for this example was the third test with the 3004 clear briquette at an air velocity of 0.401 m/sec. Three graphs were produced for each test. First in Figure 4, the air temperatures introduced into the furnace and within the system are shown. The ambient air was a constant 30° C. The furnace was then capable of heating and maintaining the air temperature entering the preheating chamber to a maximum of about 340° C, for a system differential of 310° C. Controlling the temperature was more difficult in tests where the air's velocity was at a maximum since the large volume of flow into the furnace was great enough to induce a cooling effect. Therefore, as the test proceeded, the temperatures had a tendency to decrease slightly. Typically, this decrease was 25°C or less, but for the sake of modeling we assumed the temperature to remain constant.

The two remaining graphs illustrated actual briquette temperatures. As expected, the bottom surface thermocouple heated to a maximum first and then was followed by the top and sides. The left and right sides appeared to heat at similar rates as indicated by the comparable temperature trends. Both reached a stable point between the top and bottom temperatures. If given enough time, these four lines would likely reach a common temperature indicating complete equilibrium. The internal thermocouples had similar heating trends as the left and right sides. In general, there was a rapid temperature increase and then a gradual approach to the equilibrium point. The lower portion of the briquette heated first followed by the upper section and then finally the center.

Although each experiment was different, the manner in which the briquettes behaved was very similar. After 5 minutes, the bottom and top surfaces increased in temperature to within 10 and 20% of their maximum, respectively. The right and left surfaces and inner portions of the block closest to the surface were slower to heat. For the first hour, there was at least 10 minutes of lag between the time that the right surface and center reached the same temperature.

Estimation of Process Parameters

It was an easy task to measure the extensive properties of any of the briquettes. Such properties included the size, weight, and density of the blocks, and the temperature measuring positions within the block. Also measured were the "input" test parameters such as air velocity, and temperature. The individual temperatures which made the heat-

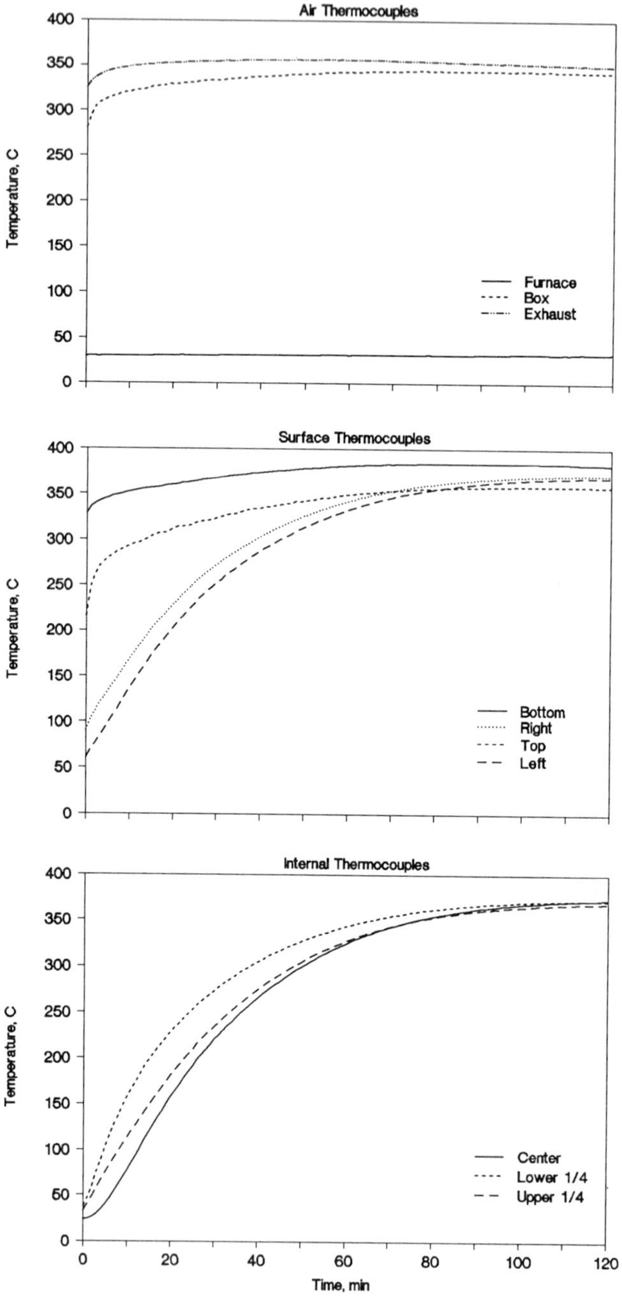

Figure 4 - Experimental data for 3004 clear.

up curves for the various positions within the block were readily collected during the tests. What could not be calculated or directly measured were the two intensive properties of heat transfer coefficient (h) and the thermal conductivity (k). However, with a knowledge of all the earlier mentioned parameters and the heating curves produced during the test, it was possible to estimate h and k.

As stated earlier, some simplifying assumptions must be made to make modelling possible. These include overlooking some variations in h and k. It is assumed that each surface contributes proportionally to heat the interior of the briquette. However, as the surface thermocouples show in Figure 4 this is not exactly true. But, in general an average or overall heat transfer coefficient can be determined. Also, since the briquette may not be homogeneous throughout, the thermal conductivity may be effected. For example, if the briquette was compacted with more force along one axis, then the void in that dimension would be reduced; thus, inducing a preferential heat transfer direction.

For this test work, an error minimization technique was used to estimate the "best" values of h and k for each test. The heating curve at the briquette's center point consisted of roughly 150 data point pairs. The previously discussed BASIC program was given all of the measured intensive and input test data from the experimental test of a briquette. Then estimates for the heat transfer coefficient and the thermal conductivity were input to the program. The program was executed and an output file produced of the simulated heating curve. The square of the difference of the test temperature and the estimated temperature was then calculated for each minute of the test. These squares of the error were then summed for a total measure of the degree of fit between the model (with its estimated parameters) and the experimental data. This value was called the Sum of the Square of the Errors (SSE). The two curves could also be plotted to visually examine the degree of fit.

A more rigorous test was used to determine the global best fit. Since two independent parameters were involved, a response surface was calculated by making the heat transfer coefficient, thermal conductivity, and SSE the x-, y-,and z-axes, respectively, of a three-dimensional system. The goal of the parameter estimation technique was to minimize the SSE. For each experimental test, a contour plot of the response surface was generated as various combinations of h and k were examined. As the number of points increased, it became easier to determine the minimum SSE and the best fit. Generally, the fit between the experimental and computer generated heat curves was quite good as can be seen in Figure 5. For the case of the second test on the 3004 clear briquette, the best fit parameters were determined to be 6.4 W/m-°C for the thermal conductivity and 42 W/m^2-°C for the heat transfer coefficient. Parameters were estimated to the nearest tenth of a unit and are shown in Table II.

The heat transfer coefficient, h, should vary with air velocity and briquette size since these two parameters influence the amount of heat transfer from the air to the briquette surface. Therefore, for any one briquette the h value should be constant for similar air velocities. Furthermore, an increase or decrease in air flow should likewise result in an increase or decrease in h. A comparison of the estimated h values indicates this concept held true in all but two cases. The h value for the third test with the 3004 decorated was lower than expected compared with the previous two tests. Also, during the first test with the UBC, the heat curve experienced an unusual arrest which may have affected the results. However, each of these contradictions occurred with decorated briquettes. Hence, the presence of the paints, lacquers, oils and other contaminants that burned away during the first run may have caused the h values to differ from the predicted estimates.

Conversely, thermal conductivity is a property that is strictly dependent upon the type and condition of a material. Therefore, for an individual briquette the k value should not change. A thermal conductivity of 228 W/m-°C has been reported in the literature for a solid block of pure aluminum. In our tests, these values ranged from 2.4 to 6.6 W/m-°C indicating that the briquettes are much less conductive than solid aluminum. As already mentioned, this is due to the voids within the compacted blocks. In comparing the estimated k values, there appears to be good reproducibility (+/- 0.4) except for the first test with the 5182 clear. In this case, the oils and other volatiles on the briquette surface may have resulted in some heat generation which would cause the test measurement to be shifted upward.

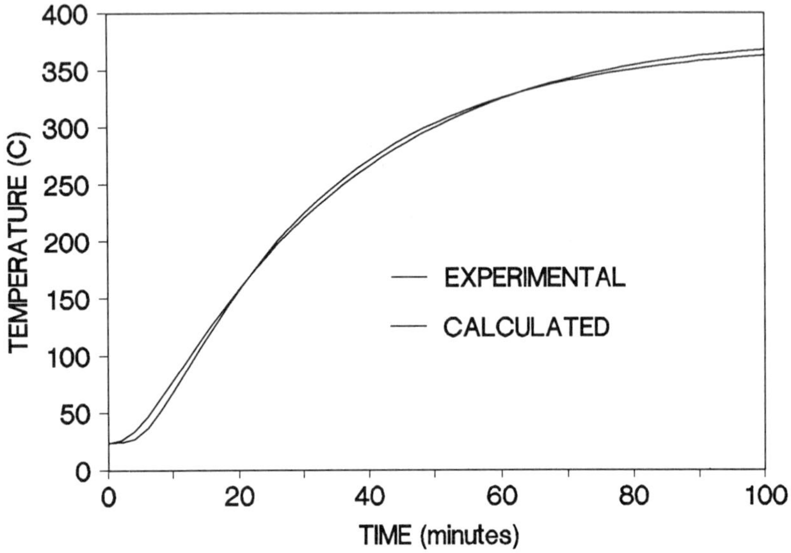

Figure 5 - Comparison of experimental and calculated temperature data for 3004 clear.

Table II Test Results

Material	Test	Velocity (m/sec)	Final Temp. (°C)	HTC h (W/m^2-°C)	Cond. k (W/m-°C)	Parameter Fit (MSE)**
3004	1	0.445	285	-Trial-	-Trial-	
Clear	2	0.444	275	42	6.4	3.01
	3	0.401	375	32	6.4	4.71
3004	1	0.408	365	30	6.6	4.56
Decorated	2	0.477	375	32	6.3	5.82
	3	0.432	410	28	6.0	5.68
5182	1	0.398	428	40	3.4	9.51
Clear	2	0.466	357	58	2.5	5.73
	3	0.405	400	48	2.4	5.07
UBC	1	0.476	400*	68	2.4	13.87
	2	0.424	420	72	2.4	6.72
	3	0.444	365	76	2.8	5.18

** Mean sum of the errors = $(SSE)^{1/2}$

Any conclusions drawn from this study should be based primarily on the second and third tests. For the most part, the first test on a block was a success; however, there is a distinct possibility that the results were skewed because of the presence of various contaminants. By putting more emphasis on the latter two tests, any effect from these components could be avoided.

Modeling Heat-up of Briquette Loads

The data from the latter two preheat tests for each briquette material was used to develop the heat transfer coefficient equation . With the experimentally determined value of h, the constant, C', could be calculated. The average of the two C' values was taken as the value to be used in Equation (9). Table III presents the C' values, the equations developed from the data to use in the full load modelling, and their range of validity (velocity in m/sec and Reynolds number) based on experimental tests.

Table III Modeling Equations

Material	k (W/m-°C)	Expression for h (W/m^2-°C)	Range for Vel. & Re (m/sec)
3004 Clear	6.4	1.3658 $k_f/d\ R^{0.675}\ Pr^{1/3}$	0.401 - 0.444 [2871 - 3179]
3004 Deco.	6.3	1.3054 $k_f/d\ R^{0.675}\ Pr^{1/3}$	0.408 - 0.477 [2213 - 2588]
5182 Clear	2.8	1.7682 $k_f/d\ R^{0.675}\ Pr^{1/3}$	0.398 -.0.466 [1724 - 2019]
UBC	2.5	2.9678 $k_f/d\ R^{0.675}\ Pr^{1/3}$	0.424 - 0.476 [2996 - 3363]

With the data presented in this table, modeling of the heating trends of briquettes at different conditions could now be simulated with a reasonable degree of accuracy.

Staged Heater Concept

In an industrial practice, one method of preheating the briquettes would be to randomly dump them in a steel mesh basket and then place the basket in a tight fitting heating chamber. The internal briquette mass would be externally heated by hot air forced through the chamber by a blower. Then the air is returned to a burner chamber for reheating. After the briquettes are heated to some predetermined temperature (or time), they are removed and charged to the furnace.

The heating process of a single briquette can be precisely defined with Equation (6) and the derived h and k parameters. For a jumbled mass of briquettes it is not so easily defined unless some simplifying assumptions are made. Figure 6 shows a schematic of the staged heater concept. The assumption is made that a jumbled mass of briquettes will heat in a similar manner to an ordered and layered mass of briquettes with the same bulk density, hence the "Staged Heater." The total number of briquettes is evenly divided into n layers with an equivalent height to the randomly ordered mass. The air coming into the first stage will have a known velocity and temperature. Depending upon the spacing

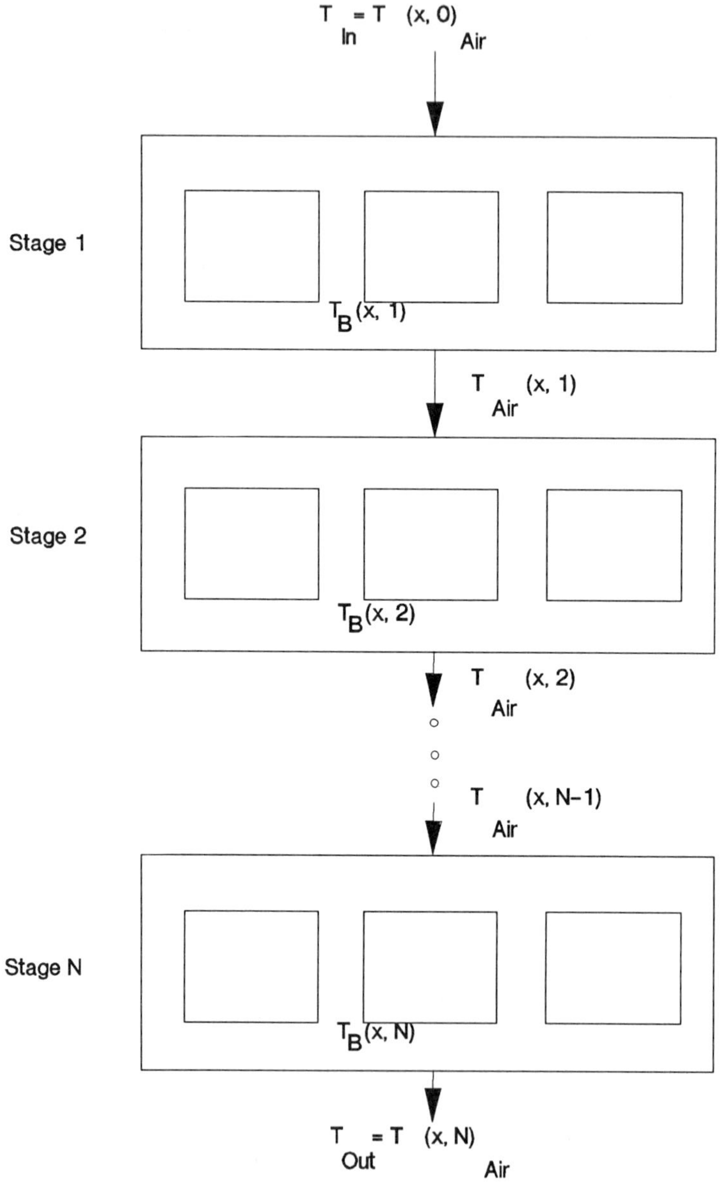

Figure 6 - Staged heater schematic.

between briquettes, a superficial velocity through that layer can be determined. This, in turn, allows determination of the heat transfer coefficient. All briquettes in a single layer are assumed to behave the same; hence, the temperature in any briquette can be modeled. The temperature of the air leaving that layer can be determined based on the amount of heat lost in raising the temperature of all of the briquettes in that layer. This exit temperature becomes the new entry temperature for the next layer or stage. In a similar manner, the temperature in each of the stages can be estimated and the temperature of the jumbled mass can be modeled. The staged heater modeling was completed by modifying the original computer program to accommodate up to ten layers and to use the air from layer (n-1) as the input to layer (n).

An example of the staged heater calculation will be made using a load of 125 of 3004 clear briquettes weighing about 2,656 kg. This load of briquettes, stacked 5 x 5 x 5, would be 1.58 m long x 1.58 m wide x 1.33 m high if close packed . To have a 30% void fraction in the bed, the base would have to be 1.89 m on a side. To handle a single load of briquettes, the basket would need to be 1.89 m long x 1.89 m wide x 1.52 m high. The volume of the basket would be 5.443 m^3. If we assume that there will be no air leakage around the side of the basket, the void area provides an active volume of 1.398 m^3. The cross-sectional area available for air movement is 1.049 m^2

For an initial example of the model, the same velocity as in Test No. 2 of the 3004 Clear will be used (u = 0.44 m/sec).

For this velocity and open cross-sectional area, the volumetric flow rate is 1677 m^3/hr. Under these conditions of velocity and material, the thermal conductivity (k) = 6.4 W/m-°C and the heat transfer coefficient (h) = 38.1 W/m^2-°C. The density of these briquettes is 0.7905 g/cm^3. An inlet air temperature of 500°C will be assumed. For this example, air is blown from the top down, but this is only a convention chosen for this example and could be the opposite way.

Figure 7 plots the heating curve for the center of a representative briquette in both the top and bottom layer of the stack. The heating trends of the two layers are nearly

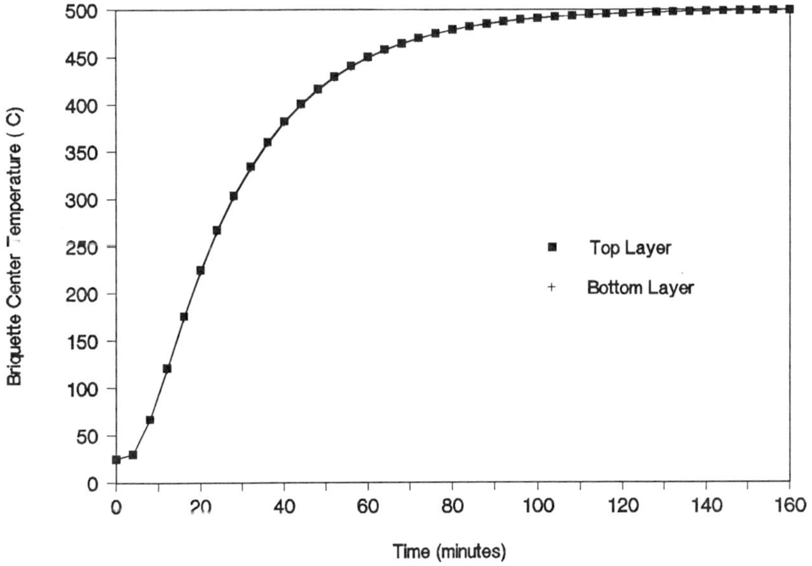

Figure 7 - Comparison of center temperatures for briquette in the top and bottom layers.

identical. Both briquette centers had reached 90% of the temperature change (453 °C) within 55 minutes of startup. The centers were at temperature within two hours. It should be remembered that the center is the last part of the briquette to heat up. The other portions will reach equilibrium much sooner. Since, the temperature difference between the top and bottom layers is not great, then the assumption of treating the jumbled mass of briquettes as a stacked arrangement is probably valid.

Figure 8 follows the temperature of the air exiting from the top and bottom layers of the stack. Note the change in the y-axis scale. The air from the top layer drops by only 1 °C at the most. The temperature in the exit air drops by 6 °C initially, but rapidly regains its temperature to the 500 °C level. These curves indicate the poor utilization of the heat content of the air. This is indicative of poor heat transfer from the air to the briquette.

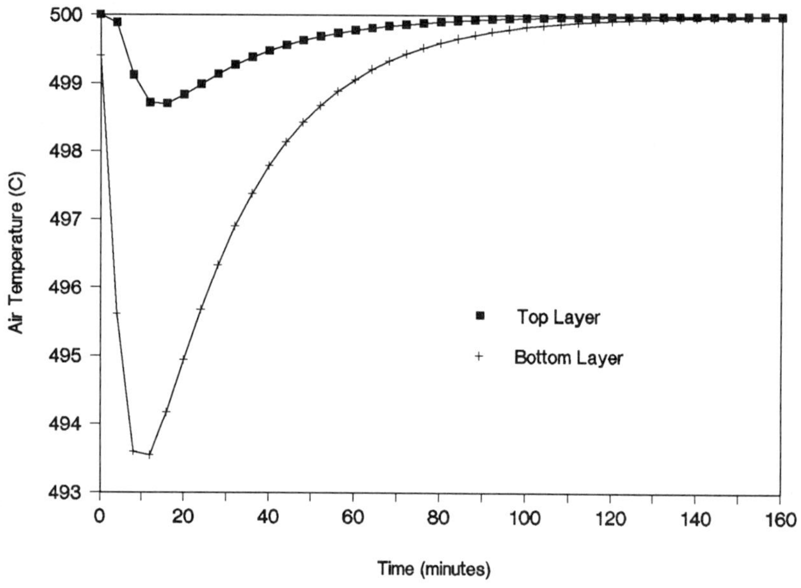

Figure 8 - Air temperature evaluation exiting from top and bottom layers in a stack.

Figure 9 presents the heating curves for briquettes in the bottom layer at three different volumetric air flow rates of 1341, 1677 and 2012 m^3/hr. The upper and lower values are +20% and -20% of the original flow rate, respectively. The influence of air velocity in this range is minimal. The curves vary by only a few minutes.

The effect of using the same heating system on a different material was investigated by looking at the heat-up of a 2336 kg load of UBC briquettes. As shown in Table II, these briquettes have a different shape; they are less cubic. Using the same basket as before, 135 briquettes could be placed in the basket by stacking them three wide and five long in nine layers. This would give a 30% void fraction again. Using the same volumetric flow rate as in the initial case (1677 m^3/hr), the parameters would be 73.7 W/m=2-°C for the heat transfer coefficient, and 2.5 W/m-°C for the thermal conductivity.

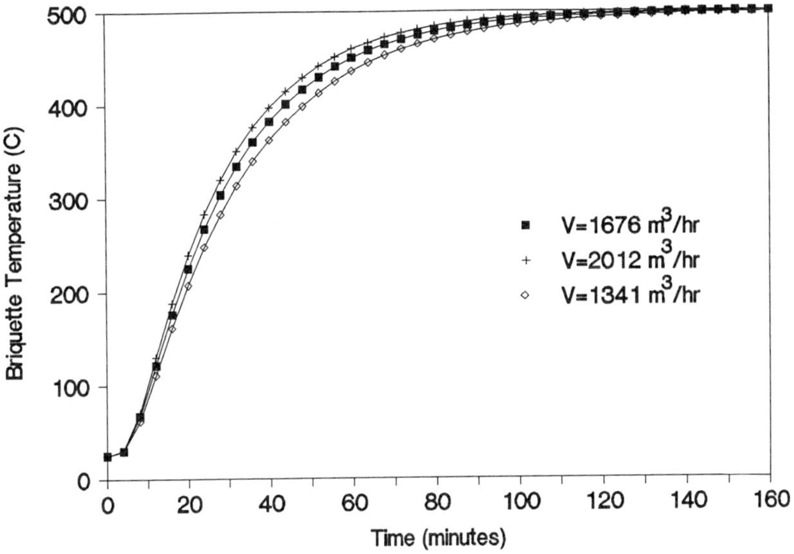

Figure 9 - Comparison of briquette center temperatures at different volumetric air flow rates.

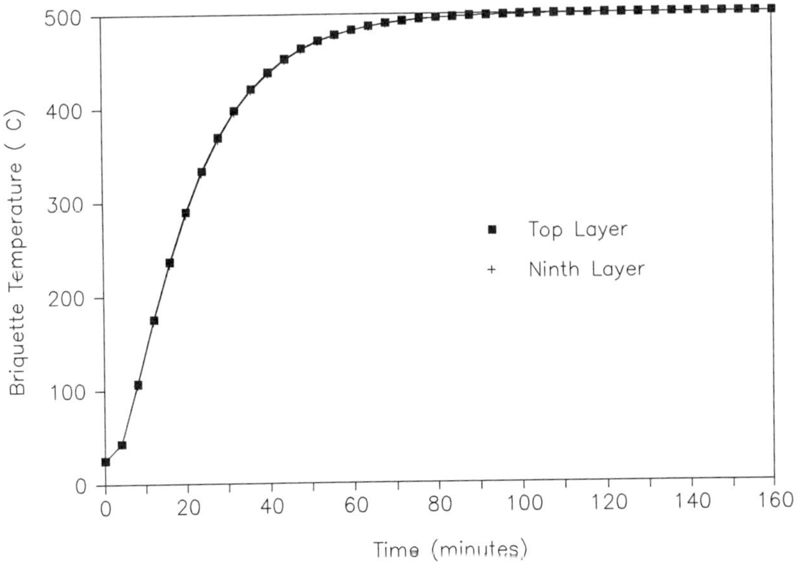

Figure 10 - Comparison of center temperature for UBC briquettes in the first and ninth layers of a stack.

the more cubic 3004 Clear briquettes. This increased rate is primarily due to the shorter distance to the center of the briquette. This material reached 90% of its target value in just 44 minutes as opposed to the 55 minutes for the other briquettes.

Conclusions

1. The experimental unit's design and method of operation adequately supplied the data necessary to model a single briquette.

2. Thermal conductivity and heat transfer coefficient values were estimated and a model developed to predict the heating trends of a single briquette. The thermal conductivity of a briquette is significantly less than a block of pure aluminum.

3. A "staged heater" model was developed to predict the heating profiles of briquettes stacked in a basket.

4. The proposed model will be verified when the preheater is implemented.

5. The model can be used in the design of an industrial preheater.

References

1. Geiger, G. H. and Poirier, D. R., Transport Phenomena in Metallurgy, Addison-Wesley Publishing Co., Reading, MA, 1973, pp. 290-324.

2. Holman, J. P., Heat Transfer, McGraw-Hill Inc., New York, NY, 1981, pp. 240-250.

An integrated aluminium scrap delacquering and melting system meeting strong environmental requirements.

Wolfgang Trosch,

President of Metallwerk Jacobs GmBH

Gelsenkirchen - Germany

It has been necessary for Jacobs, a major aluminium recycling plant in West Germany to devise a fully integrated scrap preparation and melting system to meet stringent environmental standards. The preparation system, designed by Thermcon and installed and operating at Jacobs since 1986, consists of a scrap preheater, and a shredding/delacquering kiln with an afterburner. Recuperators preheat combustion air. The melting furnace operates with both a metal circulation pump and a transfer pump which feeds a melting well. A scrap submergence system, SAMS, is employed to submerge the scrap.

The great benefits of the system include significant reduction of gaseous emissions, the melting of low grades of scrap, and the elimination of salt fluxes usually used during melting and their subsequent disposal problems.

Second International Symposium
Recycling of Metals and Engineered Materials
Edited by Jan H. L. van Linden,
Donald L. Stewart, Jr., and Yogeshwar Sahai
The Minerals, Metals & Materials Society, 1990

1. INTRODUCTION

Metallwerk Jacobs GmbH was founded in 1936 by Mr. Hermann Jacobs. Purpose of the plant was to produce aluminium alloys for die casting. In 1972 Klöckner & Co, one of Germany's biggest industrial groups, took over Metallwerk Jacobs and has since invested in the modernization of the plant and increased its capacity.

The production now is 74 million lbs a year, the annual sales approx. 150 million DM.

The delivery program today, covers aluminium for deoxidation purposes, aluminium for die-casting foundries as well as alloying elements.

Metallwerk Jacobs is situated in the harbour area of the city of Gelsenkirchen and has excellent access to the German Interstate system as well as an own rail connection and immediate access to the harbour and the river Rhine.

Because of the location of the plant in the centre of the Ruhr-area, the majority of the production is deox product material for the steel industry.

The raw materials are very diverse and consist of every thinkable type of aluminium scrap plus drosses.

The main input materials are :

- Sheetscrap, contaminated with i.g. iron, laquer and rubber
- Profiles (extrusions), contaminated with laquer and thermal barriers
- Foil scrap, both laquered and plain
- Low copper swarf and chips
- pretreated alloyed-swarf
- Sheet and cast scrap, pretreated in a heavy media plant
- Low copper drosses
- Drosses from die casting operations

The above materials are processed in two rotary furnaces, one dry hearth furnace and a reverbatory furnace with integrated scrap preheating.

Aluminium scrap is increasingly recycled. Secondary smelters therefore play a correspondingly more important role in the supply of aluminium to the market.

In the Federal Republic of Germany 14 secondary smelters are in operation with a total capacity in 1988 of approx. 530,000 metric tons. The consumption of recycled aluminium is steadily growing (figure 1).

Secondary smelters are processing a wide variety of scrap such as production scrap from rolling mills, extrusion plants, shredded scrap, castings, baled scrap, sheet, profile, mixed scrap, chips, swarf and dross.

Generally, the above scrap types were, in the past, melted down in rotary furnaces under salt without any particular pretreatment, exceptions were certain clean materials which were molten in reverbatory or induction furnaces.

Most furnaces have melting capacities between 4400 and 8800 lbs per hour.

The flue gas of the melting furnaces used to be contaminated with gaseous chlorine and fluorine containing compounds, SO_2, No_x as well as dust and organic elements.

The variety of these elements, the growing environmental awareness, more stringent rules and laws as well as the decreasing quality of the available scrap have lead to different approaches by the secondary melters to deal with these circumstances and to continue its important role in he recycling of aluminium.

To reduce the quantity of organics in the flue gas from the furnaces, scrap is treated in thermal systems or mechancial preparation plants, such as the swarf driers, separation plants of plastics from bottle caps as well as the delacquering of shredded scrap.

Drosses with aluminium contents of 30 to 50% are processed to achieve a concentrate with a metal content of 60 to 70%, enabling melting of this concentrate under salt in rotary furnaces.

To clean flue gasses from melting furnaces the dry separation system has become the most prominent device. The emissions are, with the help of these dry.separation systems, maintained below the present maximum levels as set in the rules and laws.

The dry separation systems eliminate the need for waste water and sludge treatment facilities, connected with a wet scrubbing system, however the amount of residues to be deposited is higher because of the application of lime - $Ca\ (OH)_2$ - as an absorption agent.

In view of the changing nature of the secondary melting operation in 1980 Metallwerk Jacobs embarked on an R & D program towards the conception of an reverbaratory furnace with integrated delacquering and preheating and melting facilities which should be able to present an answer to the new set of circumstances secondaries were faced with and to be prepared and equipped for the future.

Target of the project was, to reduce the emissions of chloride and fluoride fumes, organic elements, as well as sulfer dioxide by the application of primary pretreatment processes.

By maximizing the amount of heat recovery the flue gas quantity was to be reduced; as a major design goal is was decided to try to reduce the specific quantity of flue gases to be vented to atmosphere by 50% compared to the known melting methods such as the rotary salt furnace, whilst still being able to use the at that time already existing modern filter plant, at the same time however the plant should comply with the new German environmental regulations (TA-Luft - figure 2).

A further requirement for the new plant was that it should be able to melt the materials without the usage of salt.

To optimize the fuel efficiency as well as to reduce the flue gas volume, the scrap as well as the combustion air should be preheated.

The solution was found to center around a reverbatory furnace with separate melting well, a new derivative from the hitherto well know side well furnace.

When the project was completed and the plant was in full operation, it was found that the emissions of gaseous and organic chlorine/fluor elements of the Gelsenkirchen secondary melter were reduced by 8,800 lbs per year; the reduction of sulphur dioxide emission was a spectacular 13,500 lbs per year whilst maintaining production output at the same level as in previous years.

2. DESCRIPTION OF THE INTEGRATED DELACQUERING, PREHEATING AND MELTING PLANT (figure 3)

2.1 General

The described plant, as mentioned earlier, has been designed to melt shredded aluminium scrap without the application of salt to the process. Presumption for this type of process therefore was, to delacquer and preheat the scràp.

The plant consists of two major parts :

a) the delacquering kiln with combustion chamber and flue gas recirculation loop

The shredded aluminium scrap is charged from a charge-hopper to a conveyor belt. The conveyor belt transports the scrap to the double flap valve at the charge end of the delacquering kiln.

Whilst travelling through the kiln, the scrap is preheated up to 932 °F (500 °C), it is then charged into the special melting well where two graphite stirrers (shrouded augers) submerge the scrap in the liquid metal.

In the delacquering kiln, under controlled conditions, fumes are generated which, with the help of the hot gas reciruclation fan are moved into the combustion chamber where the fumes are incinerated at temperatures between 1562 - 1652 °F (850 - 900 °C).

Hereafter the gasses pass through a radiant stack type recuperator which cools the gases down to a temperature of approx. 1202 °F (650 °C) which ensures a stable inlet temperature of the flue gas into the delacquering kiln.

To enhance the fuel efficiency of the system, the waste gas from the melting furnace is fed into the combustion chamber; excess flue gasses are continuously discharged to the filter system and subsequently vented to atmosphere via the stack.

On their way to the filter system, the flue gases pass through an second recuperator, the heat recovered in this recuperator is used to preheat the combustion air of both the burners on the melting furnace and the burner on the combustion chamber.

b) The melting furnace with special melting well

The melting furnace has a bath contents of approx. 110,000 lbs (50 metric tons). At a rate of 154,000 lbs per hour liquid metal from the melting furnace is transfered via a Metaullics transfer pump into the special melting well. At the end of this special melting well the liquid metal, cooled down by the charged shredded scrap, flows back into the melting furnace via an overflow launder.

Because of the ideal location and positioning of the burners a optimal flue gas pattern of the flue gasses in the furnace hearth is achieved, resulting in maximum heat transfer onto the liquid bath.

A sophisticated fuel/air ratio control system, integrated in the computer control system of the plant, maintains a constant air fuel ratio, further enhancing the fuel efficiency.

The computer based control system of the plant controls and regulates the quantity of flue gases introduced into the delacquering section of the plant. Pressure variations in the system are eliminated with the help of air cooled dampers, which are similar to the damper which maintains a slight positive pressure in the melting furnace it self, the result of the furnace pressure system for the melting furnace is that the ingress of cold air is prevented and the dwell time of the flue gasses in the furnace optimized.

The hearth of the furnace can be easily cleaned via two large cleaning doors, which are flush with the side walls of the hearth, enabling easy access to the total surface area.

2.2 Description of the plant components

2.2.1 Steelstructure of the furnace

The steelstructure of the furnace consist of heavy mild steel sheets and cold rolled profiles. Two large cleaning/

charging openings with electrically operated vertical doors and watercooled doorframes are mounted on the furnace hearth.

In one of the end walls of the furnace the burner installation which consists out of two medium velocity burners, is installed.

The furnace is further equipped with an enclosed charge well for bales etc. and a pump well. Separate, at the side of the furnace, the special melting well is located in which the actual melting of the shredded scrap takes place, via the two Metaullics graphite stirrers.

2.2.2 Burner installation

The burner installation, suitable for firing light oil as well as natural gas, consists of two medium velocity burners and one combustion air fan. The burners are designed to use preheated combustion air upto 842 °F (450 °C). Heat recovery is effected via a combination of two radiant stack type recuperators.

2.2.3 Control system of the plant

The electrical system, together with the computer based controls, are installed in a large "walk in" style panel.

The system consist of a Texas Instruments PLC system and CVU system, the operater interface is a factory hardened keyboard.

The following data are continuously shown on video monitors (figure 4) and recorded on printers :

- each process parameter of the delacquering system such as:
  - temperatures
  - feed rates
  - oxygen levels
  - kiln speed
- Burner capacity
- Fuel consumption
- Roof temperature of melting furnace
- Flue gas temperatures
- Combustion air temperature
- Opening times of cleaning/charging doors
- Bath temperatures
- all failures according to the first out principle
- operating time

2.2.4 Furnace pressure control

The pressure differences in the furnace hearth are measured in the middle of the roof and transmitted to the pressure control loop in the PLC system. From the pressure control loop a signal is transmitted to the positioner of the hydraulic actuator of the furnace damper which immediately adapts the position of the damper to maintain the furnace pressure in accordance with the setpoint. The fast reaction to changing pressures ensure a very stable pressure in the furnace which forms the basis of stable and reliable process conditions.

2.2.5 Recuperators

Two radiant stack type recuperators are working in tandem. The first recuperator is installed in the flue gas recirculation loop, where its main purpose is to control and cool the flue gas temperature in the recirculation loop. Cold air is supplied to this recuperator by the combustion air fan. The outlet of the combustion air, which is moderately preheated in this first recuperator is connected to the second stack type radiant recuperator which is located in the exhaust flue gas duct to the filter plant. In this second recuperator the combustion air is preheated upto 842 °F (450 °C).

2.2.6 Vortex generators

To submerge, hence, melt the shredded aluminium scrap, the special melting well is equipped with a sophisticated installation to drive two graphite stirrers (shrouded augers).

This feature of the plant is one of the major innovations of the system. The design of the vortex generators enables the efficient melting of a wide variety of scrap.

2.2.7 Fume exhaust hood over the special melting well

Although hardly any fumes are generated in the special melting well, the well is equipped with a refractory lined fume exhaust hood, just to be able to cope with the emissions of the occasional large chunk of rubber or wood which has not been totally eliminated in the delacquering kiln.

2.2.8 Rotary delacquering and preheating kiln

This kiln consists of a double wall structure; in between the double wall the kiln is insulated with ceramic fibres. To enhance the heat transfer onto the scrap in the kiln, the kiln is equipped with scrap lifters along its total length.

2.2.9 Vibratory feeder at kiln charge end

This unit consists of a totally enclosed vibratory feeder with a rectangular trough and a sieve section made of perforated stainless steel. This sieve section serves to removes to fines before the scrap is charged to the kiln.

2.2.10 Duct around kiln charge end

Around the kiln charge end, a refractory lined duct is installed through which the flue gases are guided into the afterburner.

2.2.11 Flue gas ducts

The delacquering loop, apart from the kiln and the combustion chamber, the recuperator and the hot gas recirculation fan, consists of refractory lined steel ducts. The refractory lining of these ducts is carefully selected to minimize the heat loss of the system.

2.2.12 Double flap valves

Heavy duty double flap valves are mounted both at the kiln charge and discharge end as well as in the duct below the sieve section of the kiln, application of these heavy duty double flap vales ensure that the atmosphere in the delacquering kiln can be very closely controlled which not only contributes to the efficiency of the system but also to its safety.

2.2.13 Combustion chamber

The combustion chamber performs a highly important role in the system. It incinerates the fumes off the delacquering kiln and provides extra energy used in the delacquering loop. As mentioned earlier, the waste gas from the melting furnace is introduced into this combustion chamber and is combined applied to the delacquering loop system. A single burner is installed on the combustion chamber, fed with preheated combustion air, just as the two burners on the melting furnace.

The combustion chamber is internally equipped with refractory baffles to enhance the incineration effect and efficiency.

2.2.14 Secondary air supply system

To ensure a proper functioning of the combustion chamber a secondary air supply system is installed around the burner mounted on the combustion chamber. Secondary air is supplied in relation to the temperatures and oxygen levels measured in the system.

2.2.15 Radiant stack type recuperator in the delacquering loop

As mentioned before, to control the temperature in the delacquering loop without the need for water dust systems or other complicated means of controlling, a radiant stack type recuperator is installed in the delacquering loop. This device has proven to be tremendously reliable and contributes significantly to the success of the plant.

2.2.16 Hot gas recirculation fan

The hot gas recirculation fan, with specially designed and constructed impeller, recirculates the necessary flue gas quantity needed to delacquer and preheat the scrap in the delacquering kiln, in the flue gas loop system.

3. PROCESS DESCRIPTION

First we will discuss the delacquering system. Via the charge hopper and conveyor belt the scrap is charged into the rotary delacquering kiln. This conveyor is designed and installed in such a way, enabling an evenly automatic charging of scrap to the kiln, this means that the angle of the conveyor is selected such that excess scrap continuously rolls back over the rubber belt pocket sections.

In the kiln the scrap is transported with the help of lifters and preheated upto 932 °F (500 °C). The first section of the kiln is a sieve, in this part the scrap is preheated up to 248 °F (120 °C) and small particles and fines leave the kiln through the sieve. The rotation speed of the kiln can be freely adjusted via a frequency convertor. Matching the rotation speed of the kiln as well as the inlet temperatures to the specific conditions of the various types of scrap processed enables running the operation with maximum efficiency or yield.

It is quite possible to delacquer scrap types with up to 10% contaminants such as plastics, rubber and laquor. Via the double flap valve at the kiln discharge end, the scrap leaves the delacquering system at a temperatur up to 932 °F (500 °C). It should be mentioned here, that the oxygen level in the delacquering loop is between 3 and 5%. These values are automatically maintained, resulting in the prevention of oxydation of the aluminium scrap during the delacquering process.

The oxygen level is controlled via the secondary air fan of the combustion chamber which has the capacity to supplete 1765 ft^3 per minute.

The delacquering process in the kiln can be watched via a water cooled high temperature camera.

As stated earlier, the waste gas from the melting furnace is added into the combustion chamber, contributing to the fuel efficiency of the system. The operation temperature of the combustion chamber is at present 1652 °F (900 °C). However in view of anticipated even stricter rules in Germany in connection with the dioxin problem, this temperature will in the near future be increased to 2192 °F (1200 °C), this also implies that the capacity of the recuperator in the delacquering loop must be increased to deal with these higher temperatures.

The fumes generated in the delacquering kiln are fed into the combustion chamber where they are incinerated. Because of the changing flue gas velocities in this combustion chamber, caused by the differently spaced refractory baffles, solid particles are separated from the flue gases and can be removed from the system via cleaning doors, located at the various refractory baffles in the combustion chamber.

The radiant recuperator which controls the temperature in the delacquering loop, has performed extremely well and does not need any maintenance.

Via this recuperator and the recirculation fan, the flue gases enter into the kiln after passing a damper which allows only the required flue gas quantity to really enter into this kiln.

Excess flue gas is diverted to the filter system and stack, this filter system is a dry absorption and cleaning system where lime is used as an absorption agent.

The lime binds the Fluorides and Chlorides as follows :

$Ca\ (OH)_2 + 2\ HCl = CaCl_2 + 2\ H_2O$

$Ca\ (OH)_2 + 2\ HF = CaF_2 + 2\ H_2O$

(Figure 5)

Bag filters are used to collect the reaction product particulate.

3.2 Melting furnace and vortex melting system

The melting system discussed is designed to meet all know requirements, both with respect to efficient production and enviromental protection.

The melting furnace as described is an advanced, state-of-the-art melting machine, designed to enable melting, apart from the delacquered an preheated scrap, of a wide range of materials, such as ingot, bulky scrap, and baled scrap, in a highly efficient manner.

Heat transfer onto and in the bath is enhanced by the use of a Metaullics submerged circulation pump. The carefully designed geometry of the furnace ensures that recirculation of the total bath takes place, which is highly important to achieve an optimum heat transfer, this is of paramount importance to maintain the maximum melting rate of the furnace.

Actually, the furnace has three possibilties to charge materials. Clean ingot type of scrap as well as heavy gauge material such as T-bar, sows and heavy gauge material can be charged directly into the main hearth, via the two large access doors.

Baled clean or contaminated scrap can be charged into the charge-well which is connected directly to the furnace. Apart from the fact that this charge well is totally enclosed, this charge well is a fairly conventional feature, found on many of the well known reverbatory furnaces.

The most unique feature of the system presented to you today is the special melting well where delacquered shredded scrap can be automatically melted down with the help of an uniquely configured scrap submergence system.

Scrap submergence systems are not new, however, the system we, in close cooperation with Metaullics Systems, have come up with eliminates all of the complications associated with such systems and enable operation of an integrated delacquering and melting plant on an continuous basis.

The scrap submergence system is attached to a special melting well, this well is only connected to the main furnace via the molten metal supply which is achieved with the help of a Metaullics M-30 transfer pump and an overflow launder at the end of the special melting well.

This arrangement offers the unique possibility of continuous melting irrespective of the metal level in the furnace itself. (obviously a minium metal level in the furnace must be present to allow operation of the pumps and to provide the heat, necessary for melting in this special melting well).

As the vortex melting system, contrary to most other systems found, consists of two stirrers, the surface of the special melting well is kept clean, virtually automatically as any dross generated at the initial submergence of the scrap at the first stirrer, is submerged again at the second stirrer and eventually flows over to the charge well, where it can be skimmed of, again without influencing or even interrupting the melting process in the special melting well!

The drive system for the scrap submergence system is equally advanced as the overall concept, it enables adjustment of the positioning of the vortex-stirrers in the X, Y and Z axis, independently of each other! Positioning of the stirrers is accomplished via the keyboard of the furnace operating system.

This feature has been found to be extremely useful; the exact positioning both with respect to the place of the stirrers in relation to each other and in relation to the walls of the special melting well as well as the depth with which the vortex stirrers are operating below the surface of the molten metal are important to optimize the overall performance of the vortex system.

Also the possibility to lift the stirrers easy out of the bath makes cleaning of the stirrers an easy job to do, in our experience cleaning is already effective by just moving the stirrers up and down a couple of times which will quickly free them completely of any build up which occasionally may occur.

The stirrers are controlled via frequency converters, this means that the speed of the stirrers can be exactly set to match the required melting rate and to avoid excessive ingress of air into the vortex which would - as a consequence - increase melt loss.

The special melting well is not equipped with its own heating system. Experience has shown that it is possible to drain the special melting well over the weekend, just leave it as it is and to start operation on Monday morning without the need to preheat the special melting well prior to introducing liquid metal via the transfer pump.

Having the separate arrangement with both an ordinary charge well for bales and the like and the special melting well offers a tremendous flexibility to the operator, an optimum choice can be made with respect to the charge mix.

Although the special melting well arrangement requires minimum maintenance, such maintenace can be performed efficiently and comfortably as the equipment, being separated from the furnace is extremely well accessible, which results in a high uptime of the equipment. Also during maintenance of the special well system, charging in the charge well can continue and production rates can be maintained.

4. CONCLUSIONS

The advanced melting system as outlined to you today present tremendous advantages and opportunities to the industry.

It is a universal melting system which can handle shredded scrap of almost any nature via its delacquering system.

At the moment Metallwerk Jacobs is engaged in a R & D program for the melting of swarf without the use of salt. With the addition of a chlorine fluxing lance in the recirculation pump it will become possible to produce alloys for example for the automobile industry because demagging will be possible and scrap containing copper and silicon can be melted with the help of the vortex system.

The plant meets the present stringent German enviromental laws, and is capable of dealing with problems such as halogen compound emissions or dioxin (figure 6).

The realization of this project shows that the seconday industry in close cooperation with specialised equipment manufactuerers can come up with systems which can contribute both to the environment and to a responsible way of dealing with our planets resources.

5. ACKNOWLEDGEMENTS

At this time I want to thank both Dr. Neff of Metaullics, Solon, Ohio and Mr. De Groot, president of Thermcon Ovens B.V. in the Netherlands for their assistance when preparing this paper.

Thank you for your kind attention.

ANNUAL PRODUCTION OF THE GERMAN SEKUNDARY MELTER

(figure 1)

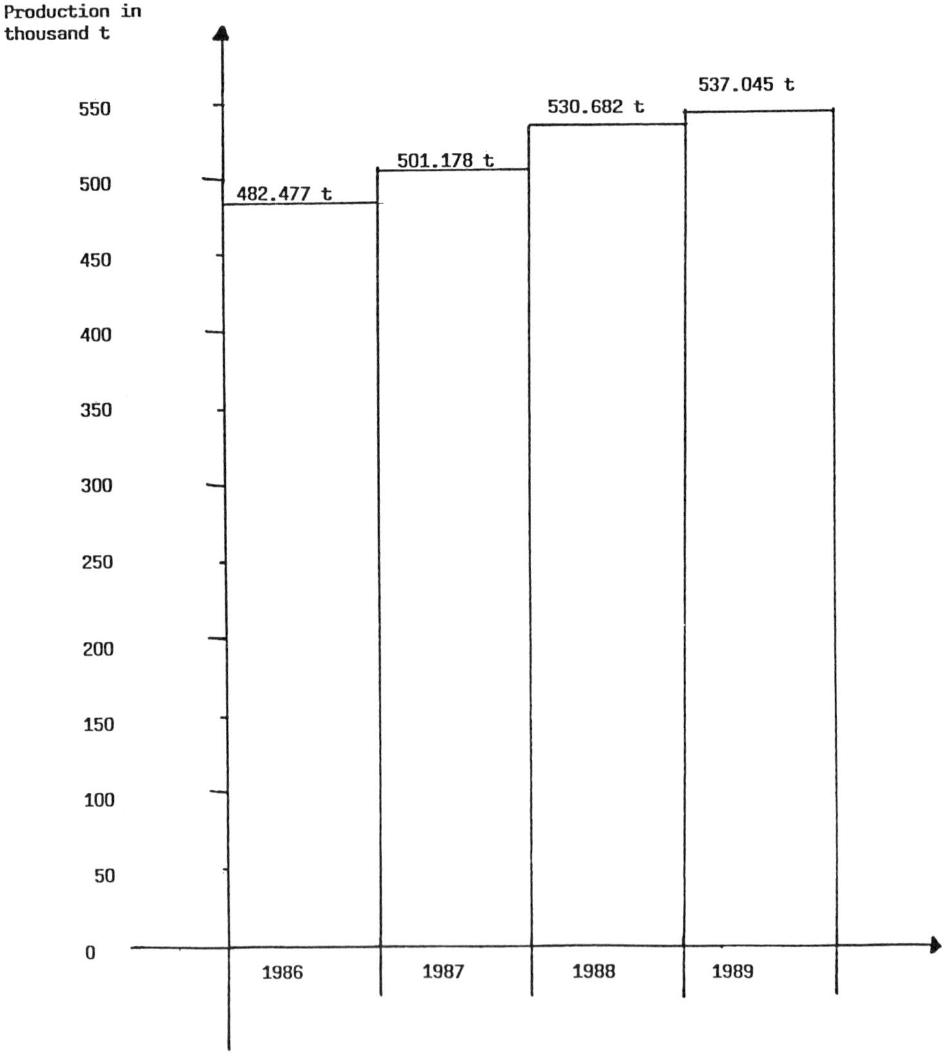

(figure 2)

ENVIROMENTAL AIR CONTROL REQUIREMENTS (TA LUFT)

Sulfurdioxide	SO_2	0,5 mg/m^3
organics - measured as total C		50,o mg/m^3
	NO_x	0,5 mg/m^3
	CO	100,o mg/m^3
effluent organic	Cl complexes	30,o mg/m^3
effluent organic	Fe complexes	5,o mg/m^3
	Cl_2	3,o mg/m^3

(figure 3)

ALUMINIUM UMSCHMELZANLAGE MIT SCHROTTENTLACKUNCKUNG UND SCHROTTVORWÄRMUNG

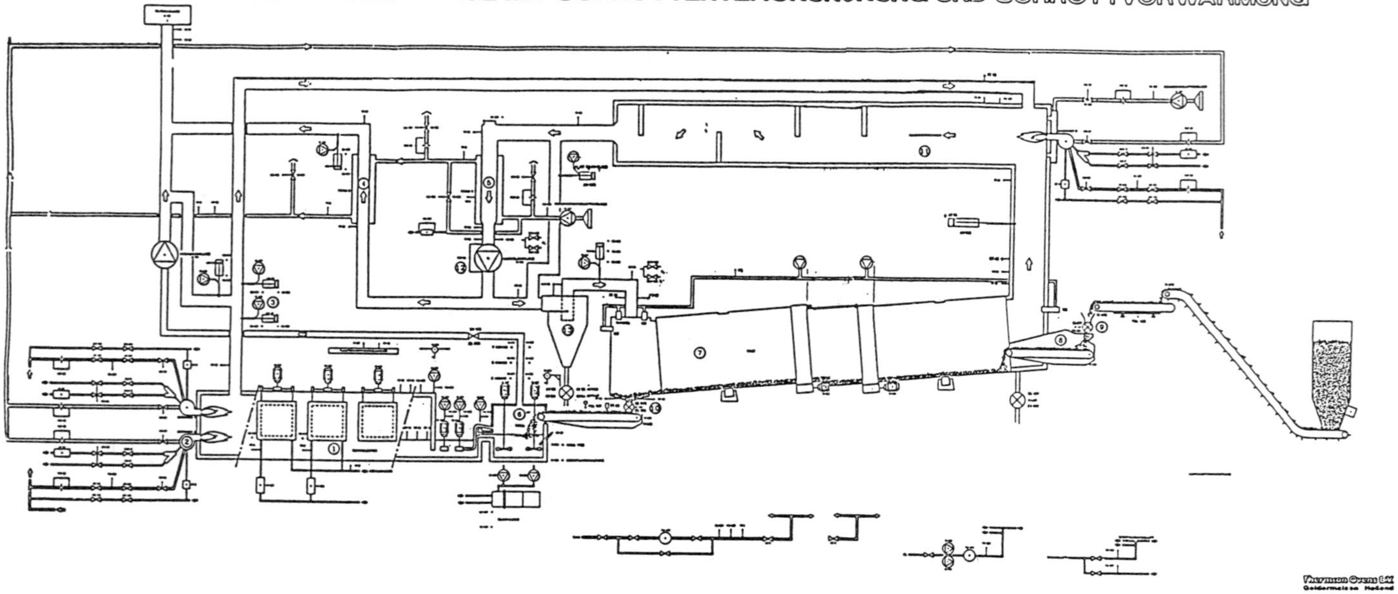

(figure 4)

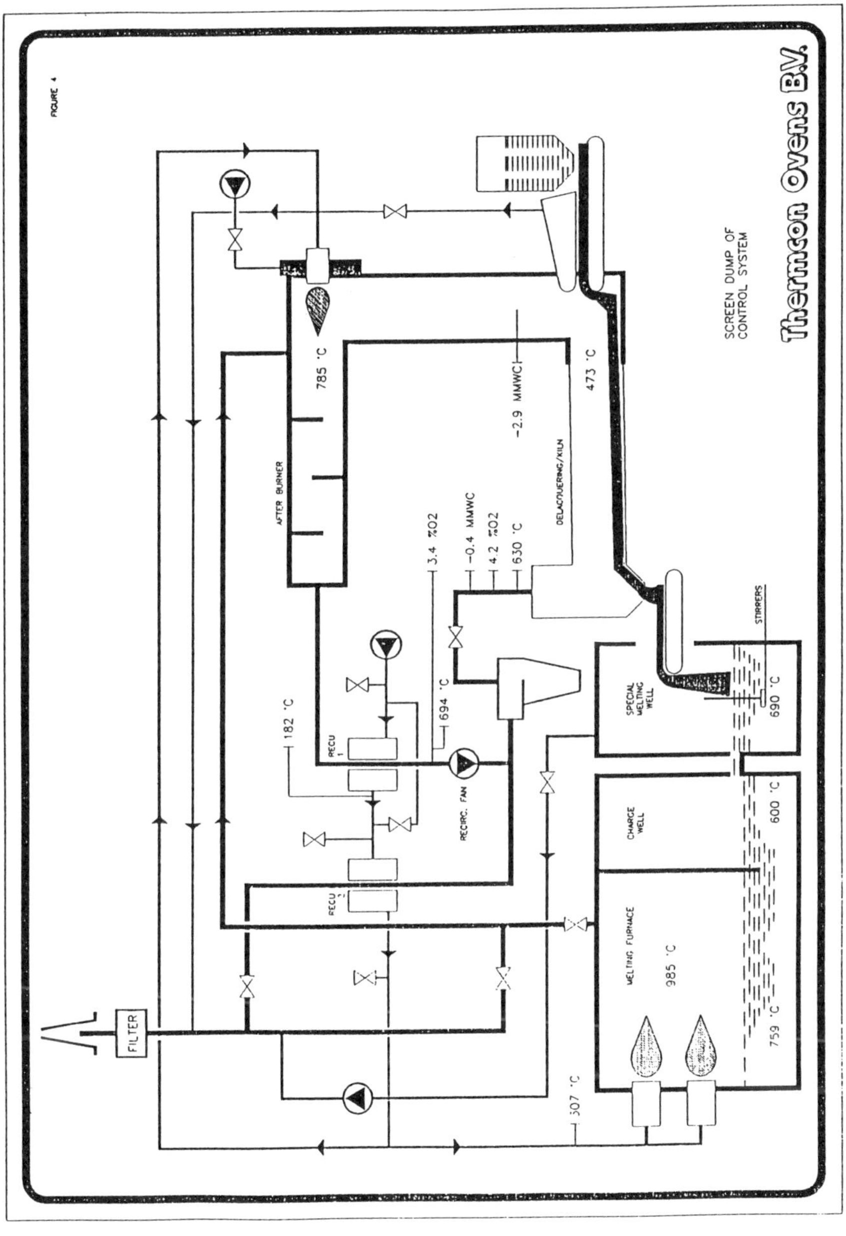
FIGURE 4
Thermcon Ovens B.V.
SCREEN DUMP OF
CONTROL SYSTEM
785 °C
AFTER BURNER
-2.9 MMWC
473 °C
DELACQUERING/KILN
3.4 %O2
-0.4 MMWC
4.2 %O2
630 °C
694 °C
182 °C
RECU 1
RECIRC. FAN
RECU 2
STIRRERS
SPECIAL MELTING WELL
690 °C
CHARGE WELL
600 °C
MELTING FURNACE
985 °C
759 °C
FILTER
307 °C

(figure 5)

$$Ca(OH)_2 + 2\,HCl = CaCl_2 + 2\,H_2O$$

$$Ca(OH)_2 + 2\,HF = CaF_2 + 2\,H_2O$$

COMPARISON OF EMISSIONS ROTARY FURNACE/NEW PLANT (figure 6)

ELEMENT	EMISSIONS-FACTORS FURNACE HOLDING FURNACE	NEW PLANT kg/t	DIFFERENCE %	EMISSION REDUCTION 12.ooo t kg/a
Dust	0,1	0,03	70	840
Chloride Cl^-	0,1	0,03	70	840
Fluoride F^-	0,001	0,0005	50	6
Nitrogen Oxyde	0,856	0,32	63	6.500
Sulphur Oxyde	1,o	0,31	69	8.300
Total Carbon	0,546	0,046	90	5.900
Carbon Oxyde	3,2	0,062	98	37.600

CLOSING THE RECYCLING LOOP FOR STEEL CANS

Gregory L. Crawford, Director of Recycling
Steel Can Recycling Institute

Steel cans have quietly done an admirable job as a container package for years at low cost and without environmental impact, because of their degradability and modest energy usage. In recent years, the steel industry came to recognize that recycling of steel cans was a worthy and achievable mission that deserved priority attention.

With this in mind, the Steel Can Recycling Institute (SCRI) was formed in late 1988 to promote and sustain steel can recycling throughout the United States, or, in the context of this symposium, to "close the loop." While this mission is simply stated, its execution involves an array of complex systems of solid waste and municipal priorities and the integration of public and private sector activities to accomplish rapidly developing and changing goals. This paper will explain how the recycling loop is being closed for steel cans.

Recycling is a relatively new topic to many people. And until recently, those entering the recycling arena for post-consumer food and beverage cans have been pioneers. While recycling receives great media attention, it is only one of the solid waste solutions, and the broader solid waste problem itself is a dominant environmental issue impacting the American culture and economy. By closing the loop, steel cans will be a very welcome and, indeed, fundamental part of solving the problem.

The steel can is 100% recyclable, as are all steel products. And the steel industry has been recycling its products routinely in the steelmaking process—and will assuredly continue to do so. One may say that the recycling loop for other steel scrap has been an active practice for years, and new steel cans will be able to take their rightful place.

Second International Symposium -
Recycling of Metals and Engineered Materials
Edited by Jan H. L. van Linden,
Donald L. Stewart, Jr., and Yogeshwar Sahai
The Minerals, Metals & Materials Society, 1990

Steel Industry Recycling History

The steelmaking process has included the use of ferrous scrap for a least a century. The steel industry, in fact, has recycled well over 50% of its products for over 50 years—an achievement that has gone virtually unnoticed by society as a whole. In fact, those in the steel industry who know all about the enormous scrap usage have taken the process for granted, unaware of the general public's thirst for such information.

With the current environmental emphasis on solid waste issues, two points become abundantly clear: that the steel industry must inform the public regarding its remarkable contribution to alleviating a substantial portion of the nation's landfill input, and that the steel industry must recycle one of its most visible consumer products, the steel can.

Over the last decade, the steel industry has kept one trillion pounds of steel scrap out of landfill by recycling. On an annual basis, this is the equivalent of recycling roughly one third of the amount of all municipal solid waste landfilled in the U.S. each year, and this process has been going on since the turn of the century. The public and legislators must be informed of this ongoing, highly praiseworthy achievement.

But with the nation's current focus on the municipal solid waste stream, the steel industry knows that it cannot relax in the comfort of its overall 66% recycling rate. Through the formation of the Steel Can Recycling Institute, the industry is aggressively working to close the loop that will divert its post-consumer steel products from the solid waste stream before they reach the landfill. Steel food and beverage cans, and other post-consumer ferrous, equal roughly 5-7% of the post-consumer solid waste stream—and generally, over half of this percentage consists of steel cans. Already, steel cans and other post-consumer ferrous items are being successfully diverted from landfill sites.

Solid Waste Issues: A New Focus

The post-consumer solid waste issues have caused the steel industry to take action, but change is also required on the part of consumers, municipal leaders and legislators. The change must be made through a broad focus on solid waste issues and landfill reduction. Until several years ago, recycling was conducted by the environmentally-motivated portion of society or by those who wanted to receive cash, primarily by recycling aluminum cans. The aluminum industry, in developing its used beverage can recycling program, deserves credit for doing an excellent job. But its focus was on "cash for cans" and litter avoidance. And rightly so. Aluminum cans are very useful to the industry in the making of new cans, and aluminum beverage cans take many years to degrade; they can be a very serious litter problem.

But today, while aluminum cans still offer a fair price at buyback centers and while litter is still an important issue, the environmental focus has broadened to encompass municipal solid waste. These solid waste issues involve all of the private sector companies that produce materials for consumer packaging.

However, in some areas of the country, especially those in which recycling has not become a community issue, a small fraction of the citizenry still evaluates recycling on a personal profit basis: "How much money will I get for my effort?" Recycling will not effectively reduce the post-consumer solid waste stream until all citizens take a longer look at the current and future effects of their disposal habits and consider recycling as a serious and routine part of their daily lives that accrues benefits other than direct financial reward.

On a broader level, it is critical that communities do the same. Ideally, a portion of a community's recycling program expenses will be offset by revenues, both large and small, from the sale of recyclables. But such programs are not necessarily "profitable," since the significant operating costs and amortized capital costs may not be fully compensated by the revenues. A very simple cost-benefit analysis would result in the following:

- The costs are the dollars spent on a program, for operations and overhead.

- The benefits are a reduction in post-consumer solid waste, extension of landfill life, cost-avoidance in tipping fees, and revenue generated from the sale of recyclables.

Given the situation some communities face—limited landfill space, and no other locations willing to take the garbage—there are no choices. For those communities that have not yet faced this situation, an added benefit of recycling is the "crisis avoidance" of the problems encountered by communities that implemented solid waste solutions too late to have the luxury of choice.

Finally, for those communities that do not accept and implement a holistic approach to recycling, it may be done for them. Over 20% of the states in the U.S. have adopted mandatory solid waste reduction laws, some with strict punitive measures. So a final benefit accrued by communities that implement recycling programs voluntarily is that they have time to plan, focus and develop the most comprehensive programs that work within their own individual systems.

Changes in Steel Industry Technology and Scrap Generation

Change will also be required of industry; in fact, in many areas, it has been required. Several of the states that have implemented solid waste legislation have also included restrictions and recycling percentages that apply to packaging. These directly affect the private sector, from materials manufacturers to packagers.

Fortunately for the steel industry, change has been the watchword of the last decade, some of it with surprisingly beneficial, but unintentional results.

The steel industry has collectively spent over ten billion dollars in the last decade to modernize production facilities and technologies. Continuous casting, improved gauge and profile controls, better control of coating weights, and increased uniformity of mechanical properties have not only created a higher quality product, they have created a situation in which less scrap is produced in the mills.

Similar advancements have been made in the steel fabricating processes—such as can making and auto part stamping. These changes, too, have resulted in greater efficiency, meaning that there is less scrap coming off the shop floors.

So changes in both steelmaking and fabricating, directly related to product improvement efforts, have also resulted in a reduction of scrap availability. From 1974 to 1988, the percentage of self-generated scrap consumed by steel mills declined from 60.5 percent of total scrap consumption to 28 percent. To better illustrate the significance of this event, it is important to outline the three scrap resources used by the steel industry. They consist of "home" scrap, which is a by-product of the steelmaking process; "prompt" scrap, which is a by-product of the fabricating process; and "obsolete" scrap, consisting of steel products that are no longer used. As home and prompt scrap resources have declined, the industry has had to purchase increasing amounts of obsolete scrap to ensure that it has a steady supply of scrap for steelmaking.

And the trend of change did not stop there. Two types of steelmaking furnaces became most prevalent over the last decade, making the former use of the open hearth furnace almost completely obsolete. Basic oxygen furnaces, which consume from 20-30 percent scrap in the steelmaking process, and electric arc furnaces, which consume virtually 100% scrap, are now the primary furnace types. Their use has increased overall external scrap consumption. The mini-mill facilities, which utilize electric arc furnaces, represent about 40% of all U.S. steel production. This means that an increasing proportion of steel production requires larger volumes of scrap.

These events have lead the steel industry to seek additional sources of scrap, and an important, high-quality source has been found in the post-consumer waste stream—in the steel can. The steel industry has changed in both technology and outlook. Its aggressive environmental posture, in conjunction with its need for outside scrap resources have helped it to become a broad and steady end market for the post-consumer steel can.

A further change, specifically induced by the industry's increasing use of the steel can as a scrap resource, has been the development of modern methodology related to the use of tinned products. The American Iron & Steel Institute's Task Force for the Utilization of Steel Can Scrap has developed an algorithm that effectively helps steel mills determine desirable proportions of tinned and detinned material to be included in the steelmaking process for various product specifications. This algorithm permits the steel industry to knowledgeably employ recycled steel cans as an economically desirable material in steelmaking, which allows for increased use of steel can scrap.

Steel Can Recycling Infrastructure Development

In order to close the loop, the steel industry has to develop the path that the cans take from the hands of the individual citizen all the way to the mill, a path which is called the recycling infrastructure. Development of the infrastructure requires direct contact with each link in the chain, from the communities and sources of steel can scrap, through intermediate processors to the end markets.

The SCRI has focused on infrastructure development by placing recycling representatives throughout the United States whose function is to systematically contact and work with municipal officials, recyclers, waste haulers, county officials, and intermediate processors to ensure that steel can recycling is implemented promptly and effectively.

Sources of Steel Can Scrap

The increasingly strong demand for recycled steel cans matches the abundant supply. For recycling to work effectively, each source of recycled steel cans must be examined in terms of how it fits into the recycling infrastructure. Post-consumer sources of steel can scrap include individual, household, group, institutional, commercial, and industrial users of food, beverage and other steel packages. These are the initial sources—the first link, so to speak.

Collectors and Secondary Processors of Steel Can Scrap

The second link in the infrastructure consists of the collection and processing operations, such as independent recyclers, ferrous scrap dealers, municipalities conducting curbside and other recycling programs, operators collecting steel cans from specialty sources (such as the reverse vending machines that are placed in convenience stores, educational institutions and public buildings), and operators/owners of resource recovery facilities (often publicly owned and privately operated).

Each of these entities may sell steel cans directly to a mill or may seek another private or public entity as an intermediate processor. The largest volumes will be collected and processed by major municipalities and resource recovery facilities which will almost always deal directly with the end market.

Increasingly, ferrous scrap dealers are becoming high volume collectors and processors because they are serving as an intermediate market buyer of post-consumer steel cans from municipalities, and of smaller quantities from their own retail buyback businesses. Ferrous scrap dealers have been partners with the steel industry for decades. They have collected, processed and shipped the traditional forms of steel scrap to the mills for recycling, in quantities that would surprise most people. They are now building their business opportunities by purchasing recycled steel cans regionally. Thus they are a vital link in the infrastructure. With over 1,500 locations throughout the country, sale to ferrous scrap dealers helps local sellers of steel cans decrease their shipping distances and eliminate processing concerns.

STEEL CAN ADVANTAGES

The steel can is now a highly visible product to the steel industry and to environmentalists, consumers and legislators, who recognize it as a commonly used consumer package that must be recycled. Steel food, beverage and general purpose cans offer distinct advantages—as packages and as recyclable items.

The Steel Can as a Package

Steel cans have changed over the last decade; generally, far less metal is used in the can bodies, so they are lighter, but equally durable and of better quality than ever. And the lighter tin coating now used still protects the can, as well as its contents, but conserves valuable resources of tin. Strength, tamper-resistance, convenience and attractive appearance add to the cans' use and marketability, but the steel can also has inherent recycling advantages.

Steel Can Recycling Methods

Steel cans are recyclable through any and all of the current and developing methods of recycling: curbside collection, resource recovery, and drop-off or buy-back centers. One reason that they work so well in each of these programs is due to their magnetic property. Recycling involves separation—or sorting—usually at a collection facility. Most separation processes are still done by hand, in the absence of sophisticated mechanical equipment...except where the steel can is concerned. Its magnetic properties allow it to be automatically magnetically separated from other recyclables, as well as from post-consumer solid waste.

Curbside Recycling Programs

Curbside programs are becoming prevalent throughout the United States as an effective recycling method. Curbside collection follows the traditional form of refuse pick-up, conforming to a pattern that is part of the traditional suburban and in many cases, urban lifestyle.

Again, steel cans offer a special advantage to curbside programs, due to their magnetic property. Generally, curbside programs either require residents to commingle the recyclables, placing all recyclables together in one bin, or to separate each recyclable into its own bin.

Pick-ups from a commingled curbside program are delivered to a material recovery facility where steel cans are easily and effectively separated from other recyclables through the use of magnetic separation. Recyclables are unloaded onto a conveyor for processing. The belt passes under an overhead magnetic separator conveyor. Steel cans are automatically removed and fed into their own bin for further processing, usually baling. This process can be used for all steel cans: food and beverage, as well as household paint and aerosol cans that have been properly prepared and emptied for recycling. Non-magnetic material continues through the conveyor system, for further separation, usually by hand.

The automatic separation of steel cans helps to relieve the labor volume at these facilities because, when steel cans are removed from the other recyclables, there is one less recyclable to be sorted through during the hand-sorting process. It has been documented that, among material recovery facilities, steel food cans are handled in 97% of these facilities, sharing this position only with clear glass containers.

For curbside programs that require consumers to separate each item, all metal cans (steel and aluminum) may still be commingled in one bin. At the material recovery facility, the steel and aluminum cans will be magnetically separated. Essentially one action prepares two recyclables. The advantages of this option are that consumers are relieved of sorting their metal cans, so participation may be higher; one less compartment is required in the collection truck; and the metal cans will be accurately separated for their respective markets.

Resource Recovery Facilities

Resource recovery facilities are unique because they are repositories for all post-consumer solid waste, not just the recyclables. And there are generally two types of resource recovery facility: some create fuel from the waste for use at electrical generating facilities (refuse-derived fuel or RDF); and others create energy from the burn directly at the facility (waste-to-energy or WTE). Some of these facilities separate post-consumer steel products—including steel cans—from waste through front-end or post-burn magnetic separation. This is truly "automatic" recycling that enjoys virtually perfect "participation" by all citizens.

A new EPA proposal may mandate that resource recovery facilities must separate 25% of the reusable components of municipal trash for recycling by 1992. One way in which these facilities will reach this goal is by separating post-consumer ferrous materials at the front-end of the intake process. This will recover as much as 7% of the municipal solid waste received at resource recovery facilities. About 50 - 65% of this ferrous material will be post-consumer steel cans, with the balance consisting of steel items such as tools, small appliances and toys.

Voluntary Drop-off and Buyback Centers

At community voluntary drop-off or buyback centers, magnetic separation also plays a strong role. Consumers often bring mixed steel and aluminum beverage cans, and magnetic separation helps prepare both for market. Cans are fed through the intake of small magnetic separators, through which steel and aluminum cans are discharged into separate bins. The trend toward multi-commodity recycling at these centers means that more are accepting steel food and beverage cans, as well as other materials.

The Market Spectrum

As explained earlier, steel recycling is an inherent aspect of the steelmaking process. Therefore, there is a steady and broad market for recycled steel cans. If every steel can used in this country were collected, the steel industry could recycle them as part of its scrap mix today.

Furthermore, recycled steel cans aren't just used to make new steel cans. The recycling loop is multi-faceted, to everyone's benefit. Steel cans are used as scrap in the production of many types of new steel products because the base metal in a steel can is one of the highest quality types of steel made. While recycling has often been described as a closed "loop" process, steel can and other steel recycling provides many options in the total loop. It is really a continuous cycle or series of multiple loops, through which the material is reused again and again.

The importance of this quality for steel cans becomes more evident when markets are considered. If a product can be recycled only into one or two new products, its recycling depends solely on the commercial market of the product into which it will be recycled. If that market declines or fluctuates for any reason, then market capacity for the recycled products declines as well. But a steel can may be recycled into steel that will be used to make a car, new construction materials, tools, toys, appliances, and of course, steel cans themselves. The breadth of steel products indicates the broad spectrum of the recycling end market options for steel can scrap.

Recycled Content

Steel cans already contain a significant amount of recycled steel. Steel can sheet is produced in integrated mills, using basic oxygen furnaces, so each can contains an average of 25% recycled steel. Recycled content is important because it demonstrates that steel scrap is already an inherent ingredient of the steel can. The steel can, which both contains recycled steel and is recycled, will comfortably meet the mandatory environmental requirements of a state, and more importantly, will meet the true environmental needs of the country.

Degradability

Finally, while the SCRI's efforts are focused on ensuring that no steel cans will be littered or landfilled, realistically, that goal will not be reached tomorrow. Fortunately, the steel can is completely degradable, being reduced to benign iron oxides when exposed to the elements, or degrading more slowly in landfill. The can simply disappears, rather than remaining as a blight on the landscape. This is one more reason that steel cans are environmentally compatible in every respect.

COMPLETING THE CYCLE

End Markets

The end market role for recycled steel cans is shared by detinning companies and steel mills. They purchase steel cans directly from the collector, as with a municipality; or through a secondary processor, such as the ferrous scrap dealer.

Detinning companies have had a long term partnership with the steel industry. They purchase tinned steel products and remove the tin, selling the detinned steel to steel mills and the recycled tin to its appropriate markets. The major detinning companies have expanded facilities and opened additional collection locations over the last 12 months to accommodate the increasing influx of post-consumer steel cans.

There are over 120 steel mill melting locations throughout the country, many of which are now seeking sources of recycled steel cans. While they purchase steel cans directly from the collectors and secondary processors, they also purchase them from detinning companies. These options provide steel mills with a good mix of tinned and detinned scrap, providing a steady flow of material for steelmaking at the right price and quality.

Finally, iron and steel foundries are an additional end market target for the sale of steel can scrap. While the foundry market is currently under development, it represents a potential additional user of post-consumer steel can scrap. Iron and steel foundries typically use 30 to 40 percent purchased scrap mix, in addition to the same proportion of self-generated scrap, to achieve their final product. Approximately 2000 such foundries exist throughout the country, and can represent excellent local markets at substantial transportation savings.

End Market Specifications

Steel cans, and all recyclables, need to be prepared for sale to end markets. Specifications for steel cans vary slightly among end market buyers, and sellers are always advised to contact the individual buyers to determine specifications and transportation options. Steel can scrap may include aluminum lids (found on steel beverage cans), but generally it may not include non-metallics or nonferrous metals, except those used in can construction. General specifications are as follows:

1. Baled can scrap for steel companies should be 2' x 2' x 2' (or 3') in size, with a density of 75 to 80 pounds per cubic foot. Bale integrity must be maintained during shipping and mill magnetic handling. Cans may be baled without removal of paper labels, but bales must be free of nonmetallic materials such as water, plastic, wood, debris, etc.

2. Densified (biscuit) can scrap for steel companies should be stacked and banded into bundles, with a density of 75 to 80 pounds per cubic foot. Bundle weight subject to negotiation.

3. Baled can scrap for detinning companies may be of varied dimensions. Density should be nominally 30 pounds per cubic foot, with higher density subject to negotiation. Wire or other steel banding is acceptable.

4. Loose cans (whole or flattened) are acceptable, subject to negotiation.

5. Shredded cans (loose or baled) are acceptable, subject to negotiation.

STEEL CAN RECYCLING: CLOSING THE LOOP AND MINIMIZING SOLID WASTE

Recycling has become an important solid waste solution, and it involves a broad spectrum of participants, from individual citizens to end markets. Steel cans, made of the recycled material, offer real advantages to food and beverage companies, packagers and municipalities, all of which have a great interest in resolving municipal solid waste problems. And steel cans provide an advantage to the steel industry as well, both as a prominent source of scrap for steelmaking and as an immediately recyclable product that allows the industry to fulfill and maintain its environmental commitment. Infrastructure development and public education, with the help of the SCRI and its many industry partners, will close the recycling loop for steel cans.

FERROUS METALS

Session Chairman:
George St. Pierre

Electrochemical Dezincing of Scrap Galvanized Steel for Recycling

Tianlu Huang, Thomas J. O'Keefe, John L. Watson,

Department of Metallurgical Engineering
University of Missouri-Rolla

Abstract

With the significant increase in the use of galvanized steel, the presence of zinc in scrap fed to steel furnaces is now recognized as a potential processing problem. In addition the zinc present in the scrap constitutes a significant metal resource that should be recovered. Thus this research examines the potential of removing zinc from scrap and reject galvanized steel, prior to recycling as steel furnace feed.

The testwork investigated electrochemical processes, and their control, for the dissolution of zinc from galvanized steel in hydrochloric and sulfuric acids. It is shown that zinc can be effectively removed, and that the process can be controlled through electrochemical monitoring of the cell potential. The rate of zinc dissolution is demonstrated to be a function of acid strength, temperature and agitation, but the most dramatic increase resulted from the use of a platinum auxiliary electrode. The auxiliary cathode provided an alternative site for hydrogen evolution, and permitted improved conditions for zinc dissolution.

Second International Symposium -
Recycling of Metals and Engineered Materials
Edited by Jan H. L. van Linden,
Donald L. Stewart, Jr., and Yogeshwar Sahai
The Minerals, Metals & Materials Society, 1990

Introduction

The practice of using scrap iron and steel as a steelmaking furnace feed has increased dramatically in recent times, and the proportion of scrap in a furnace charge is frequently in the range of 35% to 100%. In addition, due to the increased use of galvanized steel, especially in the automotive industry, the presence of zinc in the scrap furnace charge has become significant. The majority of the zinc is vaporized in the furnace and is ultimately recovered as an oxide from the furnace off-gases. These furnace dusts or sludges are high in iron, and can be recycled, provided the zinc and certain other contaminant levels are controlled. The increased use of galvanized scrap has made the recycling of furnace dusts and sludges to blast furnaces very difficult (1), and zinc has been considered to be the cause of numerous blast furnace problems (2), such as refractory failure and scaffolding. The presence of zinc can also complicate the operation of the steelmaking process in terms of waste product disposal (3), which is rapidly becoming a major problem for the steel industry.

A further point of importance, when considering the use of galvanized scrap as furnace feed, is that the U.S. presently imports approximately 75% of the zinc it consumes (4). Hence the recovery of zinc from waste products is becoming increasingly important in these times of huge trade deficits. Currently research efforts are concentrating on recovering zinc from the furnace dusts and sludges by pyro- and hyrdometallurgical routes (5,6), but an alternative may be to prevent the zinc entering the furnace with the scrap. It therefore appears appropriate to examine the possibility of removing the zinc from the galvanized scrap prior to charging it to the furnace. In this way the steelmaking process will not be contaminated by zinc, and the zinc will be recovered for reuse. It is estimated (7) that for an annual U.S. production rate of 100 million tonnes of steel, sufficient galvanized scrap will have been used to generate 40,000 tonnes of zinc in the form of dusts and sludges. In addition to scrap, there is also considerable potential in recovering zinc from reject galvanized steel products.

To date little information is available relating to the removal of a zinc coating from steel sheet, strip or wire. It would appear that a leach in an acid solution would be effective, but it would be necessary to carefully monitor the leaching process to permit the dissolution of zinc without dissolving iron. It is probably more critical to prevent iron dissolution than to ensure complete zinc dissolution, as the subsequent recovery of zinc from solution is difficult in the presence of iron, while a small amount of zinc on recycled scrap could be tolerated by the steelmaking process.

This research project was initiated to investigate the influence of various process parameters on the electrochemical dissolution of zinc from galvanized sheet.

Theory

The removal of zinc from scrap galvanized strip by acid leaching is essentially a controlled, rapid corrosion process. The overall zinc dissolution reaction is :

$$Zn + 2H^+ \rightarrow Zn^{2+} + H_2 \qquad (1)$$

The electrodes of the corrosion cell will be anodic and cathodic areas of the scrap strip, with zinc dissolving at the former and hydrogen being evolved at the latter.

The corrosion of zinc will proceed at a rate determined partially by the intersection of the anodic and cathodic polarization curves as shown simplistically in Fig. 1 for zinc and iron. The positions and slopes of

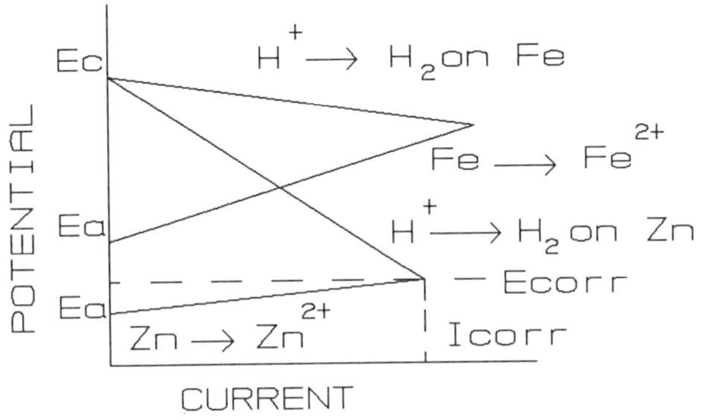

Figure 1 - Corrosion Plot of Zinc and Iron in Acid Solution.

these curves are determined primarily by the kinetic parameters or overpotentials for the reactions. In most cases the hydrogen reduction reaction is likely to be the rate controlling step. Other variables affecting the corrosion rate of zinc will be the ratio of anode and cathode areas, the ionic concentrations of zinc and hydrogen in the electrolyte, solution chemistry, the degree of agitation to provide mass transport, and the temperature of the system.

Experimental

Equipment

The starting material for the tests was galvanized steel strip, which had a 2-3 micron thick zinc coating, with a surface density of $20 g/m^2$. The leach system was contained in a small glass reactor and utilized a magnetic stir bar, and a stirrer/hot plate. The galvanized scrap sheet, with exposed surface area of $1 cm^2$, and a calomel electrode were linked externally through a universal meter, such that the cell mixed potential could be monitored during the leach.

In some later tests, a platinum auxiliary electrode, with an area varying from 4 to 52 cm^2, was attached to the galvanized strip.

Procedure

At the beginning of a leach test the cell potential was measured against the calomel reference electrode. As the leach proceeded the potential became more positive as the zinc was dissolved, and then increased rapidly as the final zinc was removed from the steel surface. A typical potential plot for dissolution in sulfuric acid is shown in Fig. 2, and it can be seen that the difference between the initial and final potentials approximates to the difference between the zinc and iron potentials. The zinc dissolution time was calculated from the potential plot as illustrated in Fig. 2.

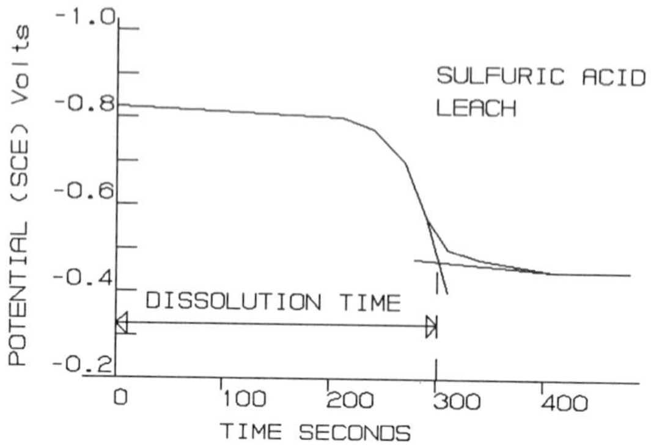

Figure 2 - Potential Time Plot for Galvanized Scrap Leaching in Sulfuric Acid

Design

The zinc dissolution time was determined as a function of a number of factors, at three levels, in an advanced statistically designed experimentation procedure. This orthogonal test design utilizes factor levels which are evenly spaced, to permit a well distributed, balanced design, which can significantly reduce the number of tests required compared to a traditional factorially designed test. The results from the tests can be subjected to analysis of variance, and statistically significant results may then be identified.

Since any zinc solution produced by the electrochemical leaching of galvanized steel scrap will be subjected to a zinc recovery treatment, it is appropriate to utilize acids for which information is available with regard to metal recovery. Thus hydrochloric and sulfuric acids were used in this research. Obviously sulfuric acid would be preferred, due to the ease of zinc electrowinning from sulfate solutions, but there is little information on the kinetics and selectivity of either acid with respect to dezincing of galvanized material.

A) Hydrochloric Acid. A four factor, three level orthogonal design was selected to investigate the use of hydrochloric acid. Table I presents the factors and levels for the experimentation, which used a $L_9(3^4)$ orthogonal design with each of the nine tests being repeated four times.

Table I. Test Conditions for Hydrochloric Acid Leach

Factor	1	2	3
[HCl] g/l	100	150	200
[Zn] g/l	125	75	25
Agitation	none	weak	strong
Temperature°C	45	35	25

B) Sulfuric Acid. To investigate the performance of sulfuric acid, two sets of tests were undertaken. The first set utilized five factors to identify the relative significance of the main dissolution variables, while the second set examined in depth two of the main factors, as identified by the first set of tests.

1 Five Factor Test.- For the first sulfuric acid run, the orthogonal test used a five factor, three level design ($L_{18}(3^7)$) to examine the zinc dissolution time. This test design was carried out twice varying the electrode configuration as follows:

a) No auxiliary electrode
b) With a platinum auxiliary electrode attached to the test piece.

Table II details the factors and levels utilized in the two 18 run tests with a single repeat.

Table II Test Conditions for 5 Factor Sulfuric Acid Leach

Factor	1	2	3
$[H_2SO_4]$ g/l	196	392	588
[Zn] g/l	25	50	75
Agitation	none	weak	strong
Temperature°C	55	40	25
Sample preheat°C	95	25	60

2 Effect of Platinum Electrode on Two Factor Test. To further detail the dependence of the zinc dissolution time on acid strength and temperature, a two factor, three level $L_9(3^4)$ orthogonal test was run with single replicates, using a fixed zinc solution concentration (25 g/l) with no stirring or preheat. Again this test design was undertaken twice varying the electrode configuration as follows:

a) No platinum auxiliary electrode
b) With a platinum auxiliary electrode attached to the test piece.

Table III presents the factors and levels for this test.

Table III Test Conditions for 2 Factor Sulfuric Acid Leach

Factor	1	2	3
$[H_2SO_4]$ g/l	196	392	588
Temperature°C	25	45	65

3 Platinum Electrode Area Test. To examine the effect of the platinum auxiliary electrode area on the dissolution of zinc from the galvanized strip, a two factor test was designed using the levels given in Table IV.

Table IV Test Conditions for Pt Area Sulfuric Acid Leach Test

Factor	1	2	3	4
$[H_2SO4]$ g/l	39	59	78	98
Pt Area cm^2	52	36	20	4

The remaining three factors from Test 1 were fixed at 45°C, with no zinc ions in the leach solution and no stirring.

IV Electron Microscopy

To permit an insight into the mechanism of zinc dissolution, selected specimens were examined under the SEM, at various stages of dissolution. This examination was undertaken to reveal both pictorial and analytical information relating to the dissolution.

Results and Discussion

A) Hydrochloric Acid Leach

The 36 tests produced dissolution times of 16 to 430 seconds, and a statistical analysis (analysis of variance) of these results is presented in Table V. These data illustrate that all the factors are significant

Table V - Analysis of Variance for Hydrochloric Acid Dissolution with No Pt Electrode ($F_{0.01}(2,27) = 5.44$)

Factor	Mean Square	F Statistic	Significance 99% Level
[HCl]	140086	129.2	Very High
[Zinc]	159655	147.2	Very High
Agitation	17521	16.2	Slight
Temperature	77338	71.3	High

Minimum Time = 16.5 secs
Maximum Time = 430.5 secs
Mean Time = 147.1 secs

with respect to the reaction time. However the dissolution time for the zinc coating can be seen to be largely controlled by acid strength and zinc concentration in the leach solution. The effect of zinc ions in the solution may not be simple, due to the influence of zinc ions on the dissolution reaction and on the hydrogen overpotential. As expected the acid strength simply increased the reaction rate, and hence a higher acid concentration resulted in a shorter dissolution time. Agitation had only a slight effect on zinc dissolution indicating that diffusion was not limiting in this case, while the effect of temperature was probably to enhance the reaction rate.

B) Sulfuric Acid Leach

1a) Five Factor Test without Pt Electrode. The 18 tests produced dissolution times of 58 to 1058 seconds, and these results indicate that for approximately the same molarity, sulfuric acid is less effective in leaching zinc then the hydrochloric acid, with the mean leaching time being more than double. A statistical analysis of the sulfuric acid leach results is presented in Table VI, and it can be seen that the major parameters favoring zinc dissolution, under the conditions tested, are acid strength and temperature, while agitation and zinc content are much less significant. This illustrates that the dissolution is not hindered by the presence of zinc ions in solution, and that the rate controlling step in the dissolution is not diffusion of reactants and products to and from the

Table VI - Analysis of Variance for Sulfuric Acid Dissolution with No Platinum Electrode (5 factor-$F_{0.01}(2,25)$ = 5.57)

Factor	Mean Square	F Statistic	Significance 99% Level
$[H_2SO_4]$	514224	83.07	Very High
Temperature	410206	66.26	Very High
Agitation	47092	7.61	Slight
Zn Content	43159	6.97	Slight
Preheating	154918	25.02	Moderate

Minimum Time = 76.5 sec
Maximum Time = 1058.0 sec
Mean Time = 304.3 sec

anode strip. The effect of temperature is far more significant than for hydrochloric acid, and it appears that the mechanism for dissolution differs for the two acids. The tolerance of sulfuric acid to zinc ions in the leach solution, and its pronounced dependence on temperature, will be advantageous when considering the potential use of sulfuric and hydrochloric acids.

1b) Five Factor Test with Pt Electrode. The 18 tests produced dissolution times of 23 to 129 seconds, which are significantly lower than the times without the auxiliary electrode. A statistical analysis of these results is presented in Table VII, where it can be seen that sulfuric acid concentration and temperature are highly significant.

Table VII - Analysis of Variance for Sulfuric Acid Dissolution with Platinum Electrode (5 factor-$F_{0.01}(2,25)$ = 5.57)

Factor	Mean Square	F Statistic	Significance 99% Level
$[H_2SO_4]$	6364	42.62	High
Temperature	6247	41.84	High
Agitation	161	1.07	None
Zn Content	145	0.97	None
Preheating	414	2.77	None

Minimum Time = 23.5 sec
Maximum Time = 129.0 sec
Mean Time = 53.0 sec

The platinum auxiliary electrode permits the cathodic reaction to occur at a lower overpotential, and the corrosion current increases due to this modification of the cathodic curve, as shown in Fig. 3. Essentially the platinum electrode is enabling hydrogen evolution to occur away from the strip anode, such that the zinc dissolution is less disrupted by the bubbling.

The dissolution times were thus reduced by a factor of ten at the longer times, to a factor of two for the shorter times. These dramatic reductions suggest that the use of an auxiliary electrode can be very beneficial in this process and provide considerable time savings and economic benefits. It would probably be worthwhile in the future to further investigate the effect of the nature and form of the auxiliary electrode materials on the

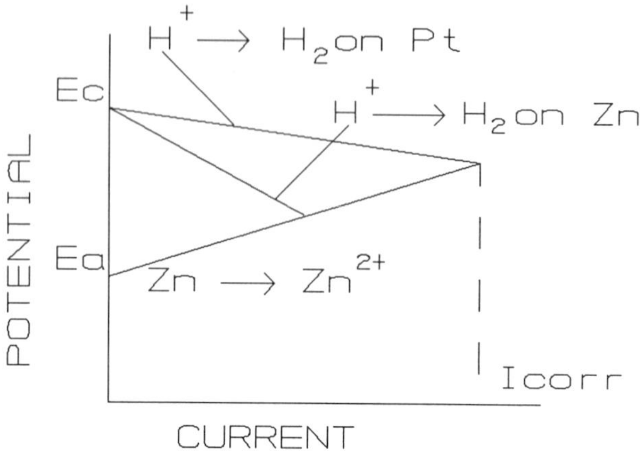

Figure 3 - Corrosion Plot for Zinc in Sulfuric Acid with an Auxiliary Platinum Electrode

dissolution of zinc.

<u>2a) Two Factor Test without Pt Electrode.</u> The nine run test design permitted a detailed examination of the dependence of the dissolution time on acid strength and temperature, and the results are given in Table VIII.

Table VIII Analysis of Variance for Sulfuric Acid Dissolution with No Platinum Electrode (2 factor-$F_{0.01}(2,13)=6.70$)

Factor	Mean Square	F Statistic	Significance 99% Level
$[H_2SO_4]$	363666	64.5	High
Temperature	74494	14.0	Moderate

Minimum Time = 33.5 sec
Maximum Time = 776.0 sec
Mean Time = 271.2 sec

The dissolution times varied from 33 to 776 seconds, and it is was found that high acid strength (588g/l) and temperature (65°C) provided the best conditions for a rapid dissolution of the zinc coating. However the times are excessive when the acid strength is low (196g/l), even at the higher temperature. This illustrates that acid concentration is of prime importance in zinc dissolution from galvanized steel scrap.

The use of a polynomial regression enabled the following equation to be determined to permit the estimation of dissolution time (secs) in terms of acid concentration ([A]-moles) and temperature (T-°C) for these conditions:

$$\text{Time} = 1719 - 413\,[A] + 24.2\,[A]^2 - 14.3\,T + 2.2\,[A].T \qquad (2)$$

The standard error of this equation is 27, and an examination of the data reveals that all experimental points fall within two standard errors of the corresponding estimate.

2b) Two Factor Test with Pt Electrode. The influence of the platinum auxiliary electrode is apparent from the results given in Table IX. The

Table IX Analysis of Variance for Sulfuric Acid Dissolution with Platinum Electrode (2 factor-$F_{0.01}(2,13)=6.70$)

Factor	Mean Square	F Statistic	Significance 99% Level
$[H_2SO_4]$	1236	37.8	High
Temperature	708	52.2	High
	Minimum Time =	18.0 sec	
	Maximum Time =	85.5 sec	
	Mean Time =	41.7 sec	

dissolution times for this test range from 18 to 86 seconds, and therefore the effect of the platinum electrode is again to dramatically increase the dissolution rate for zinc, with the mean dissolution time decreasing from 271 to 42 seconds. It is shown that both acid strength and temperature are highly significant with the auxiliary electrode, but it appears that the presence of the platinum electrode itself is the most significant factor.

The equation to estimate the dissolution time for these data is :

$$\text{Time} = 138 - 7.0\,[A] - 2.5\,T + 0.02\,T^2 \qquad (3)$$

To isolate the actual effect of the platinum electrode, the ratio (k) of the dissolution times, with and without the auxiliary electrode, was calculated for each test, and Figs. 4 and 5 illustrate the variation of this ratio with acid concentration and temperature respectively. It is apparent from Figs. 4 and 5 that the platinum electrode is five times more effective in reducing dissolution times at low acid concentrations, while the temperature effect is small at all acid concentrations. Thus it would

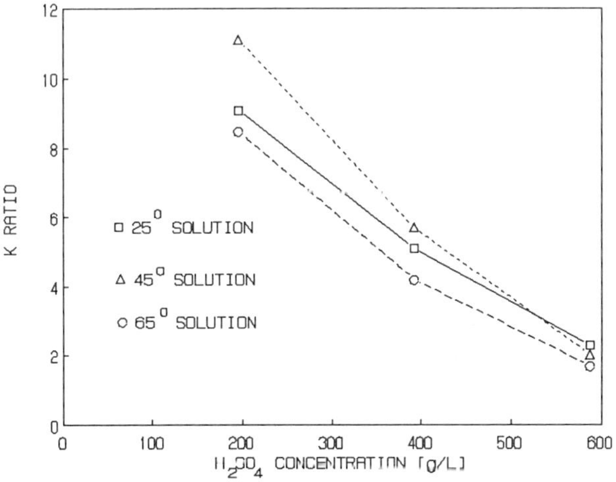

Figure 4 - Effect of Platinum Electrode on K Ratio with respect to Acid Concentration at Constant Temperature.

appear that the use of a platinum auxiliary electrode at ambient temperature with acid concentration of 200g/l is adequate for effective zinc removal.

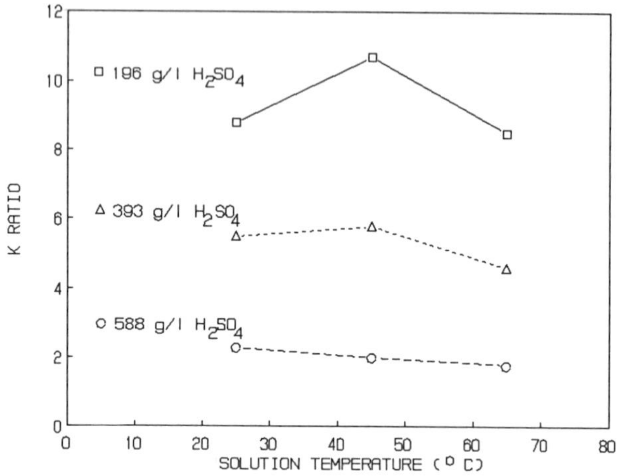

Figure 5 - Effect of Platinum Electrode on K Ratio with respect to Solution Temperature at Constant Acid Concentration

3 Platinum Electrode Area Test. The range of dissolution response times varied from 70 seconds to 255 seconds, and the statistical analysis of the results is given in Table X. It can be seen that the effect

Table X - Analysis of Variance for Sulfuric Acid Dissolution and Pt Electrode Area ($F_{0.01}(3,25) = 4.68$)

Factor	Mean Square	F Statistic	Significance 99% Level
$[H_2SO_4]$	23362	139.5	Very High
Pt Area	3341	20.0	Moderate

Minimum Time = 61.0 sec
Maximum Time = 254.5 sec
Mean Time = 122.5 sec

of acid concentration is more significant than platinum electrode area, but it is apparent from Fig. 6 that the area is particularly effective in decreasing dissolution times at lower acid concentrations. Under the conditions of high platinum area and acid concentrations of greater than 59g/l the dissolution of zinc can effectively be recovered at rates in excess of 12g/m^2/min.

An analysis of the potential measurements prior to and after the dissolution illustrates that the platinum electrode area decreases the initial potential markedly, and the final potential to a lesser extent. This shows the effect of the auxiliary electrode in removing the hydrogen

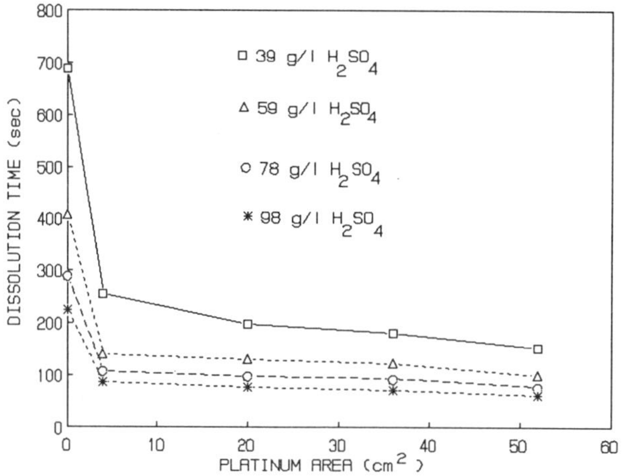

Figure 6 - Effect of Pt Electrode Area on Zn Dissolution Time with Acid Concentrations

overpotential, but there is no obvious relationship between the potentials and the dissolution time.

Electron Microscopy.

To illustrate the pattern of zinc removal from the galvanized strip, several specimen were examined using a scanning electron microscope. Figure 7 illustrates an untreated surface which can be seen to be electro-

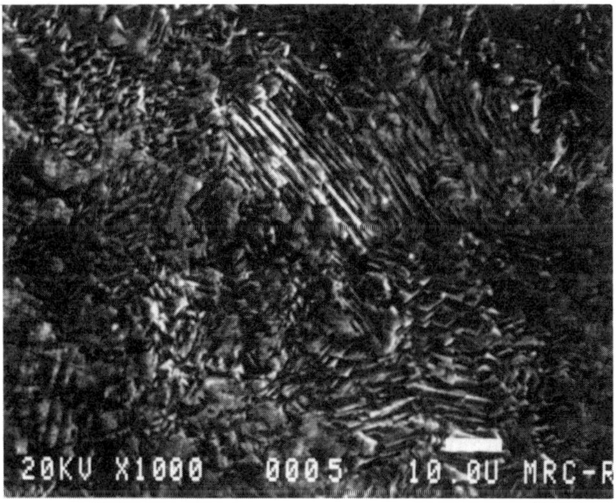

Figure 7 - SEM Image of Untreated Galvanized Sheet (X1000)

galvanized zinc. A partially treated surface is shown in Figure 8, where

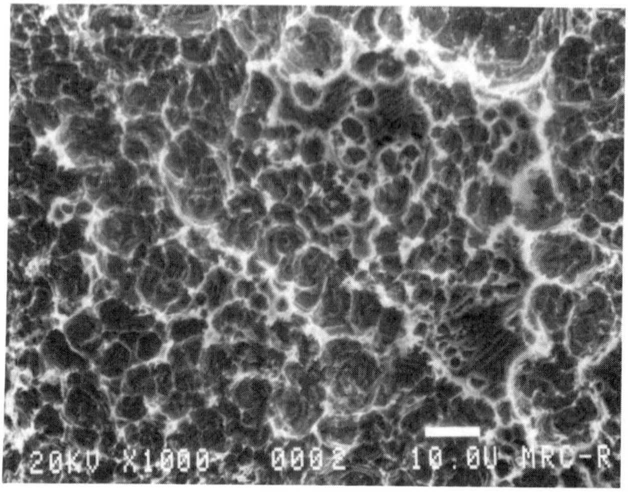

Figure 8 - SEM Image of Partially Treated Galvanized Sheet Surface(X1000)

the morphology of the dissolving zinc surface can be seen to exhibit hexagonal etch pits. Figure 9 illustrates a more complete stage of zinc

Figure 9 - SEM Image of Treated Galvanized Sheet Surface (X1000)

dissolution with large areas of the steel sample completely denuded of zinc, while other areas are still protected. Zinc was generally removed from the center of the test area outwards. Typically the dezincing treatment gave excellent zinc removal, as seen in Figure 10, where the steel surface is exhibiting some charging but less than 1% zinc.

Conclusions

A comparison of hydrochloric and sulfuric acids revealed that the former was more effective in dissolving zinc from the galvanized strip, but was also more susceptible to inhibition by zinc ions in solution. Acid concentration was shown to have the major effect on dissolution time for both acids, but the introduction of a platinum electrode also caused dramatic reductions in these times. Table XI illustrates the effect of

Table XI Comparison of Reaction Times for Sulfuric Acid Concentration and Platinum Area

Acid [g/l]	196	392	588	39	98
Pt Area cm^2	0 / 4	0 / 4	0 / 4	4 / 52	4 / 52
Reaction Time (Seconds)	1058/129	255/47	204/25	255/153	87/61

sulfuric acid concentration and platinum electrode area on the reaction time, with all other factors essentially constant. It can be seen that platinum mesh and strip have drastically reduced the dissolution times, probably by permitting the cathode reaction to occur at a lower overpotential, thus increasing the corrosion current. The effect of platinum area was most beneficial at low acid concentrations and, as seen in Table XI, a dissolution time less than 100 seconds was attainable with sulfuric acid concentrations below one molar. This equates to a zinc dissolution rate of greater than $12 g/m^2/min$.

It is now important to produce acid leach solutions for electrowinning, and to determine extraction efficiencies as a function of the zinc and iron recoveries from the scrap. In addition the economics of a galvanized scrap recovery process must be examined and possible sources of such scrap identified and quantified.

Acknowledgements

The authors wish to acknowledge financial support for the research from the Generic Mineral Technology Center for Mineral Industry Waste Treatment and Recovery at Reno, Nevada (USBM Project No. NV 89557 -0047).

References

1 D.R.Fosnacht, "Recycling Ferrous Steel Plant Fines-State of the Art" Disposal, Recycling and Recovery of Electric Arc Furnace Exhaust Dust, Iron and Steel Society Inc., Warrendale, PA, pp 15-19, 1987.

2 P.C. Chaubal et al., "Sulphation and Removal of Zinc from Electric Steelmaking Furnace Flue Dusts", Ironmaking and Steelmaking, Vol. 9, No. 6, pp 258-266, 1982.

3 L.E.Price, "Tensions Mount in the EAF Dust Bowl", 33 Metal Producing, pp 38-41, Feb., 1986

4 U.S. Bureau of Mines, "Mineral Commodities Summaries", 1987.

5 A.E.Morris et al, "Treatment Options for Carbon Steel Electric Arc Furnace Dust", Disposal, Recycling and Recovery of Electric Furnace Exhaust Dust, Iron and Steel Society, Warrendale, PA, pp 35-50, 1987.

6 D.J. Fray, "Low-Temperature Treatment of Dust from Electric Arc Furnaces", Trans. I.M.M., C95, pp 55-58 1986.

7 Laboratory for Environmental Research Report, "Feasibility of Recycling Zinc Bearing Dusts and Sludges Generated by the Northwest Indiana Steel Industry", Indiana University Northwest, Gary, IN, 1986.

MANUFACTURE OF MINERAL WOOL PRODUCTS

FROM SLAG BY-PRODUCTS

W. F. Porter, J. A. Watson, R. S. Blancett

USG CORPORATION
700 N. Highway 45
Libertyville, IL 60048

Mineral slag waste from the production of steel in blast furnace operations can be converted for use in a variety of applications.

Specific cost performance aspects of mineral fiber derived from slag make it a critical raw material in several major industries. This paper discusses slag chemistry, conversion methods, end use applications, and the present regulatory environment.

Second International Symposium -
Recycling of Metals and Engineered Materials
Edited by Jan H. L. van Linden,
Donald L. Stewart, Jr., and Yogeshwar Sahai
The Minerals, Metals & Materials Society, 1990

Introduction

The foundry industry produces a waste stream of slag material through blast furnace operations that can be converted to fiber for use as a raw material in many applications worldwide.

As a man-made mineral fiber, it has a unique cost-performance relationship that makes it a preferred raw material primarily in industrial applications, as a building material, and in agricultural applications. The ability to recycle the slag waste stream into usable products provides many benefits to modern society, including reductions in landfill requirements.

History

The earliest documented production of slag wool can be traced to Wales in 1840. In the United States the initial primitive production of slag wool began in 1869.

Sources indicate the first commercially successful plant was in Manchester, England and started operating about 1885 while the first U.S. slag wool plant began in Cleveland, Ohio in 1888.

Successful large scale commercial production of rock wool began in the USA in Alexandria, IN in approximately 1897. A chemical engineer named C. C. Hall discovered that a local deposit of an argillaceous limestone could be melted and fashioned into a commercially viable wool. This geological formation had the correct flow properties in the molten state to be blown into wool and was readily available for the production of rock wool. Accordingly, the early rock wool industry tended to center in Indiana.

In Europe production of mineral wool began with slags of various types and continued until the 1940's. Shortly after World War II most of the European plants began using rock rather than slag. At the present, with few exceptions, most European manufacturers continue to use basaltic rock.

In the USA the production of rock wool dominated from about 1900 into the early 1940's. Between 1935 and 1943 the rock wool plants in the USA converted to blast furnace slag, a waste product in the production of pig iron made by using a blast furnace(1). This was mainly due to greater melting efficiency.

Since the mid-1940's the slag wool industry in the United States has relied entirely on blast furnace slag plus small amounts of additives. Slag accounts for 70 to 90 percent of the weight of the materials that make the slag wool.

Melting

Slag and rock wools are composed of calcium-magnesium-aluminum silicate glass. To obtain the desired compositions, various slag and rock raw materials are placed in a cupola along with coke, the fuel that supplies the heat, and melted. Alternately, the slag can be melted in electric or gas heated furnaces. A majority of melters worldwide operate with coke as the fuel source. In manufacturing rock wool a mixture of various natural and/or synthetic rock sources are blended together to produce the desired chemical and physical properties.

In slag wool manufacturing, typically one raw material is the main component and other raw materials are added to make up for some deficit in that raw material. If, for example, the main component is too rich in acid oxides, such as silica, limestone or a slag rich in calcium oxide is added.

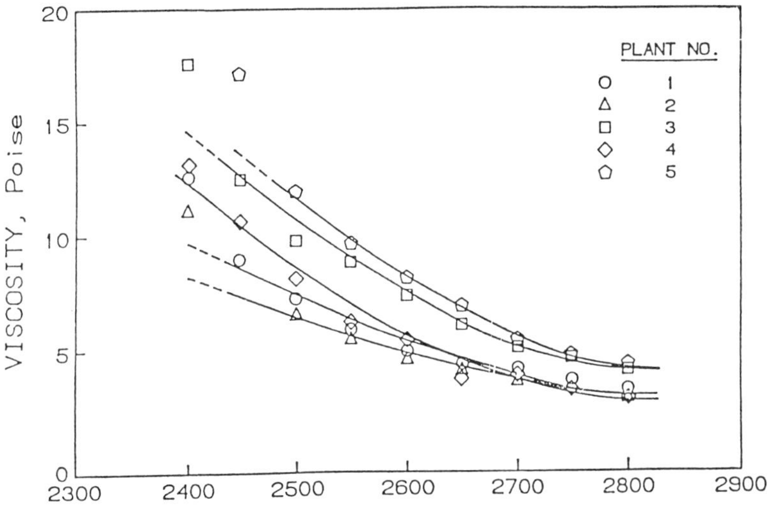

Figure 1 - Typical viscosity curves

Other ingredients must be added to produce a melt with sufficient uniformity that it can be made into a reasonably uniform slag wool. To accomplish this wool producers use additives that effect the viscosity and surface tension of the molten material, important to efficient fiberizing processes. Typical operating viscosity curves are illustrated in Figure 1.

Sizing of raw materials is of prime importance for the smooth operation of the cupola. For proper operation the bed of the cupola must remain "open" i.e. it must allow the free passage of combustion air and the products of combustion. The best operation is obtained when the slag or rock has been well screened and washed. For good operation in a 54 inch diameter cupola the slag or rock size should be about 2 inch by 4 inch in size.

For purposes of broadly predicting how a given slag will perform, it is convenient to calculate and acid/base ratio for the slag. In this calculation, silica and alumina are considered acids and all other components bases. The formula then becomes:

$$\frac{\%\ SiO_2 + Al_2O_3}{100 - (\%\ SiO_2 + \%\ Al_2O_3)} = A/B\ Ratio$$

Generally speaking, the higher the silica and alumina (the higher the A/B) the more viscous the melt. The A/B ratio of the melt should be greater than 1.0 and when the A/B ratio goes above 1.3 the melt becomes quite viscous and difficult to handle. Various additives are used to alter the A/B ratio of the melt. For example, with a slag that is low in silica and alumina, the A/B ratio can be increased by adding river gravel, sandstone, shale, quartz, tile chips, broken brick or basalt. Generally, additives will amount to only about 10% of the charge. Typical slag compositions are illustrated in Figure 2.

Cupola type melters are normally operated at 2500-2700°F. Typical thermal efficiencies of a coke-fired cupola are 33 percent whereas electric melters can operate at 75 percent efficiency.

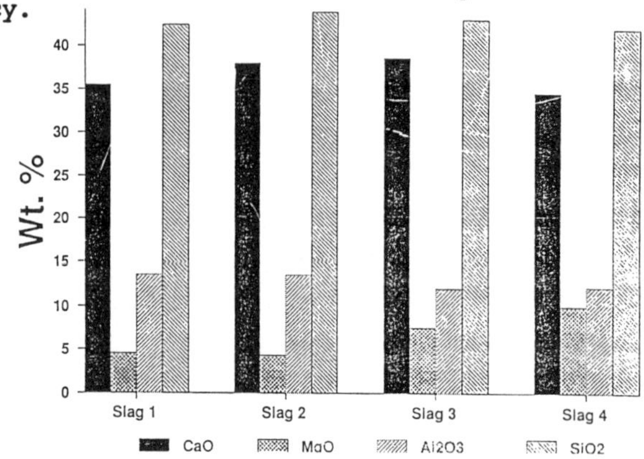

Figure 2 - Typical slag compositions

Fiberization

After the raw materials are melted together, the rock or slag wool is produced by using one of two basic types of spinning processes. In one process, the molten material is conveyed down a trough and onto the outside of the rim of a distributor wheel. While some material is spun off into the air, most of the molten material transferred to either one or more larger wheels which are located in close proximity to the first wheel and are spinning in opposite directions. The majority of the fiber produced in this process is generated from these larger wheels. Additional wheels can be added below the primary wheels in an attempt to fiberize additional portions of the molten material.

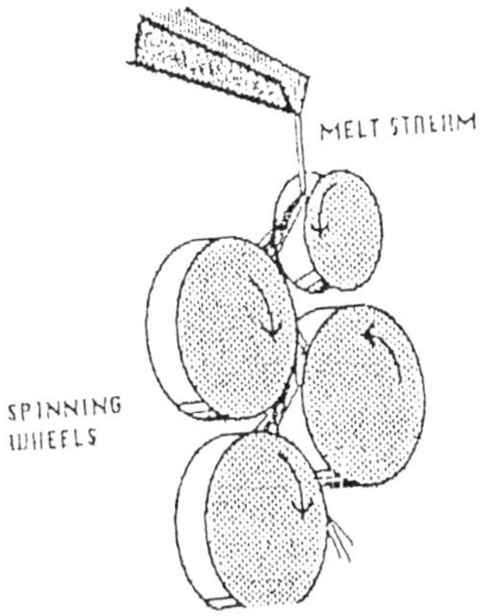

Figure 3 - Cascade or mechanical spinning process

The second method of producing slag wool from the molten material utilizes a single, dish shaped wheel onto which the molten material is placed (as illustrated in Figure 4). This is known as the Downey method. Please note that with this process the wheel can be mounted either with the axis in the verticle or the horizontal planes, and there are reports in the literature of it being used with an inclined axis. The primary function of this disk is to spread the molten material and break it into streams which are forced off the lip of the wheel by centrifugal force. Air or steam jets are directed parallel to the axle, adjacent to the edge of the wheel and they do two things: one, they force the material being spun off the wheel to change direction and go into the blow chamber; and more importantly, they shatter the streams of molten material out into finer streams of material that then freeze and form the wool.

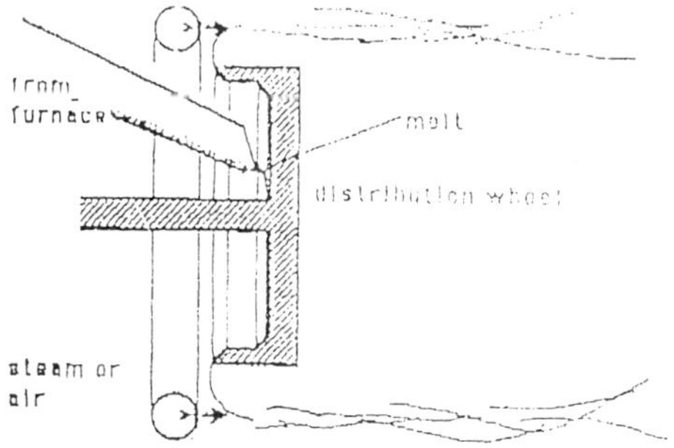

Figure 4 - Single Wheel Spinner

Collection & Fabrication

Once in the collecting chamber the wool is formed into a mat or blanket on a moving screen and is then conveyed either to the blanket making machine or to a bagger or a baler. The lines on which batts are made contain equipment which sprays the phenolic resin binder directly onto the wool as it is made. The mat containing the phonolic resin is conveyed through a curing oven where a high temperature polymerization takes place. The cured blanket is then cooled and the bonded blanket is cut to size. Paper or foil may be adhesively applied on one or more sides and the finished product packaged for shipment (See Figure 5).

The binders used are typically phenolic resins and the percentage of resin to the final weight of the bonded slag or rock wool varies from 1% to 10% depending upon the application.

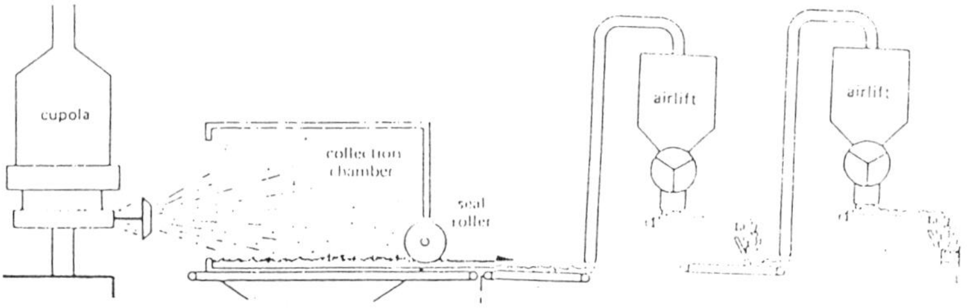

Figure 5 - Collection/Fabrication

Applications

Most of the ceiling tile used in the USA in commercial and residential buildings is made with slag wool which has surface treatments on it when shipped to the ceiling tile manufacturing plant. Mineral fiber acoustical ceiling products are made via casting or water felting processes. The total US market for these materials is two billion square feet of which roughly 90 percent are manufactured with slag wool as the primary raw material component. The slag wool comprises 40-85 percent by weight of the finished tile.

Mineral wool blankets and loose fill products are used in a variety of thermal and acoustical insulation applications. Mineral wool blankets are used as thermal insulation in commercial curtain wall and column fire protection materials where fire endurance ratings are required. Mineral wool exhibits low thermal shrinkage and high melting point, making it ideal as a fire resistive material. It can be formed into complex shapes for use as pipe and vessel insulation for high temperature and cryogenic applications. It also has good sound absorption characteristics and is used routinely in commercial interior wall constructions where speach privacy between closed offices is essential. Mineral fiber loose fill insulation has limited sales in residential applications in North America.

In Europe, however, mineral fiber enjoys a much stronger market share position with competitive fiberglass products.

Mineral fibers are also used in industrial and agricultural markets. In friction applications such as automatic brake linings, they have been a suitable to asbestos fibers, although shot content is a limiting factor. In Europe, manufacturers have developed a mineral wool as a peatmoss extender and growing media for seedlings, tomatoes, cucumbers, and other vegetables and flowers. The hygrophonic exhibit at Disney World Florida, illustrates the use of mineral wool blankets in this application.

Regulatory Activities

Mineral Fibers derived from slag have been the subject of extensive research along with other man-made mineral fibers (MMMF's) relative to health affects on humans. Epidemiological studies to-date have been inconclusive. The International Agency for Research on Cancer (IARC) has classified mineral wool as a Category 2B or "possibly carcenogenic to humans" and additional research is in progress. The large domestic producers of MMMF through the Thermal Insulation Manufacturer's Association (TIMA) have instituted voluntary product labelling programs.

Summary

Mineral fiber products derived from the use of blast furnace slag waste have provided a significant value to the present day society. These products are the raw material of choice, based on cost performance, for several major industries worldwide. They will continue to play a central role in these industries well into the 21st century.

Practical Applications of Eddy Current Separators in the Scrap Recycling Industry

Wijnand L. Dalmijn

Delft University of Technology
Department of Raw Materials Processing
PO Box 5028
2600 GA Delft
The Netherlands

SUMMARY

Eddy-current separators have been developed since the start of this century. The magnetic field applied in these separators was either generated by electricity or by permanent magnets. Not until recently have the generated magnetic fields been sufficiently strong to induce large enough forces in metal particles for practical applications in the scrap metal industry.

At the moment eddy current separators based on permanent magnets or electromagnetic fields are in use for scrap processing. Most of the eddy current separators are used for metal non-metal separation in the size range of 4 mm to 100 mm.

INTRODUCTION

In an eddy current separator, repulsive forces are exerted on electrically conductive non-ferrous particles. In this way these particles can be separated from solid material mixtures. The first eddy current separators were introduced in 1889 for the extraction of gold from sand deposits (1). Recently, separators have been designed for the recovery of high valued, non-ferrous metals from waste (2,3).

Eddy currents are induced in the conductive particles as a result of time dependent changes of a magnetic field. The repulsive forces are caused by the interaction between the eddy currents and the magnetic field. The forces have the direction of the changing magnetic field. The magnitude of the forces depends on the specific conductivity, mass, size and shape of the particles, and on the intensity and distribution of the magnetic field.

Time dependent magnetic fields can be generated in different ways. In linear motors, the magnetic fields are generated by electromagnets with alternating currents. The repulsive forces can also be generated by moving the particles through an inhomogeneous, static magnetic field or through the fields of permanent magnets mounted on rotating discs,

Second International Symposium -
Recycling of Metals and Engineered Materials
Edited by Jan H. L. van Linden,
Donald L. Stewart, Jr., and Yogeshwar Sahai
The Minerals, Metals & Materials Society, 1990

cylinders or other shaped surfaces.

In order to design an eddy current separator, it is necessary to make at least an estimation of the exerted forces. However, the theoretical calculation of these forces is difficult. The repulsive forces depend on the spatial distribution of the magnetic field and on the induced eddy currents. The eddy currents can only be calculated by resolving complex differential equations (2,4). Recent interest in the recovery of non-ferrous metals from waste was the main stimulus for the further development of eddy current separators. The sizes of the particles in these material mixtures are typically larger than those in earlier ore processing. These larger particles are much more strongly deflected in an eddy current separator and thus easier to recover.

During the 1960's, a research program on eddy current separation was carried out at Vanderbilt University (5). This program included a lot of theoretical work, such as the determination of the relationship between the repulsive forces and both the particle dimensions and the applied magnetic field. Several separators were developed and one of the devices, called the Pulsort, has been in operation on refuse fractions in Gallatin, Tennessee.

The operation of a linear motor is similar in principle to that of a stator in a rotating induction motor. The significant difference is that the configuration is flat rather than cylindrical (fig. 1).

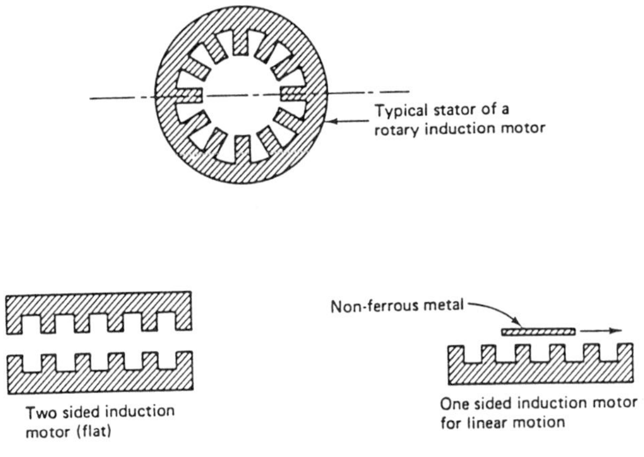

Fig. 1

A linear motor generates a sine wave of magnetic field intensity which travels down the length of the motor. The mixture of conductive and non-conductive particles passes on a conducting belt over a linear motor. For generating time-dependent fields a linear motor needs a high power consumption. In addition to this the magnetic field intensity decreases strongly as a function of the distance of the poles of the linear motor.

The inhomogeneity of the field distribution can disturb the separation results. Recently, new linear motors have been developed which generate magnetic fields with sufficient large intensities at distances in the order of 60 mm to the poles. Itkin has tested a linear motor device for the recovery of aluminium from automobile scrap in comparison with other separation techniques (6). The best results were obtained with a grinding-screening method. Other devices for eddy-current separators are based on the application of high quality, strong permanent magnets made of barium and strontium ferrite, samarium cobalt, or more recently neodynium iron. These magnets can generate strong fields which previously could only have been produced by electromagnets. This means a considerable reduction in the costs of operation of separators and simpler construction. On the other hand, permanent magnets with large magnetization intensities are generally very expensive.
In 1975, Schloemann (7) introduced the ramp separator.

In 1979, Dalmijn, Voskuyl and Roorda (8) presented the Vertical Eddy Current Separator (VECS). The VECS consists of two vertical, parallel, mild steel walls with magnetic strips with alternating polarities mounted on the insides of the walls. The strips are oriented at an angle of 45° with respect to the undisturbed fall direction. The particles fall through the channel between the tow arrays of strips. The exerted forces due to magnetic field variations and eddy currents cause the deflection of conductive particles. The operation is not disturbed by rolling and non-sliding particles as in the ramp separator.

With the quality factor or apparatus constant Q which was developed at the Delft University of Technology, the magnetic configuration can be optimized (4). Thus, optimum forces can be exerted on conductive particles. As a result the performance of the VECS was considerably improved (9). In Rotating Disc Separators (RDS) and cylinder separators, the exerted forces can be much stronger than in a VECS. An RDS consists of permanent magnets mounted on rotating mild steel discs. The time dependent field acting on the particles between the discs is caused by the rotation of the discs in combination with the motion of particles. The magnitude of the repulsive forces can be estimated by the use of the calculations (4).

EDDY CURRENT SEPARATORS

It was not until recently that the induced forces in metal particles by the magnetic field in an eddy current separator have been sufficiently large enough for practical use.
According to the main separation criteria (σ/ρ) for eddy current separators in table 1, a separation between non metals and metals or one metal from other metals should be possible.

TABLE 1

MAIN SEPARATION CRITERIA (σ/ρ) FOR EDDY CURRENT SEPARATORS

MATERIAL	(σ/ρ) x 10^3 m^2/Ω kg
Aluminium	13.1
Copper	6.6
Zinc	2.4
Brass	1.7
Lead	0.4
Stainless steel	0.0
Plastic/glass	0.0

In table 1 σ is the electrical conductivity and ρ the specific density.

For practical applications most metals are alloyed which influences, the density and the conductivity. The values for different metals and their alloys presented in figure 2 indicates that even when ignore the shape and size differences, a separation between metals is almost possible. Therefore, the main application for eddy current separators is limited to the separation of metals from non-metals. The eddy current separators can be broadly divided into two groups, one based on permanent magnets and the other based on an electromagnetic field.
The following separators belong to the first group of separators.

Permanent magnets:

Stationary permanent magnets:

Rampseparator	Raytheon, U.S.A.
Vertical symmetric separator	TU Delft, The Netherlands
Belt deceleration separator	MG Frankfurt, B.R.D.

Moving permanent magnets:

Rotating disc separator	TU Delft, The Netherlands
Rotating cylinder separator	Lindemann, Steinert, BRD. Huron Valley, USA.

Electromagnetic field:

Linear motor - single side	Bird, U.K. Coreco Milwaukee, USA. Feichter Halver, BRD.
Linear motor - two sides	Combusiton Power Comp., USA
Coil	TU Delft, The Netherlands Carpco, USA KWU, BRD
Linear motor - head pulley	Lindemann, BRD

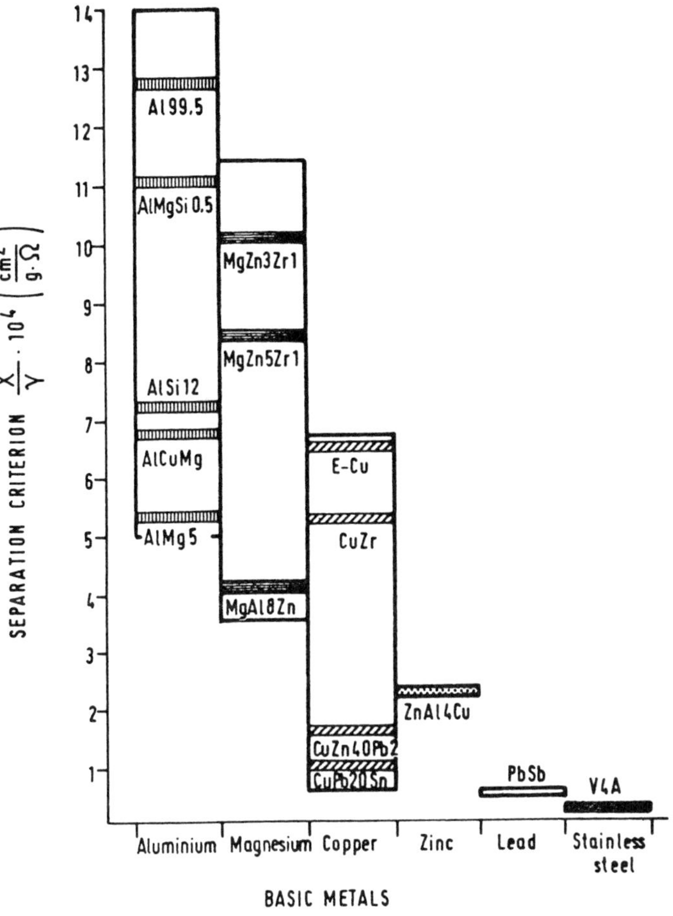

Fig. 2 Separation criterion of metallic construction materials

PERMANENT MAGNETS

The induced forces in eddy current separators with stationary permanent magnets are smaller than those in the moving permanent magnet systems and for this reason their application is very limited. On top of this, all systems based on this principle are very sensitive for magnetic material in the feed. The combination of these factors is the reason that very few systems have been successful in operation although the basic design is simple.

The fast moving magnet systems have been more successful because they are less influenced by magnetic material and induced forces in the metal particles are comparable with those induced by linear motors. A rotating disc separator (figure 3) has been modified by Weiss - K.P. Vertriebs GmbH and used for different types of materials.

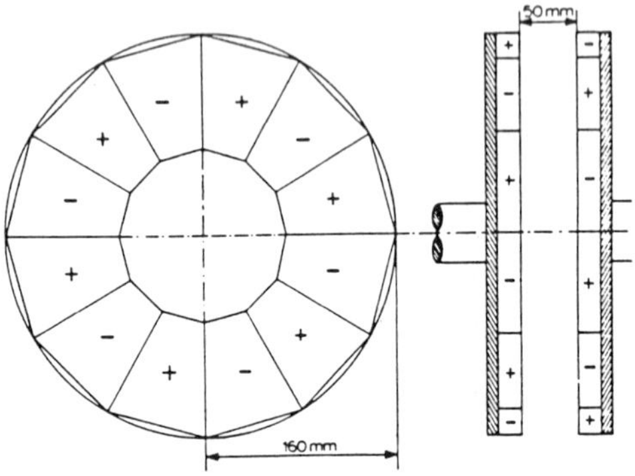

Fig. 3 Schematic view of a Rotating Disc Separator

The results of the pilot plant were excellent at 3000 r.p.m. with a maximum capacity of 1.5 ton/h and a width between the discs of 60 mm with aluminium-stone mixture, shredder fines, a mixture of stainless steel and aluminium turnings and the unmeltables of a selective melting Weiss furnace.

The rotating head pulley separator with permanent magnets of Lindemann and Steinert are similar in concept but different in design, figure 5.

For an aluminium stone mixture from the second stage H.M.S. from car shredder residue in the size range of 6 - 100 mm the results of the Steinert and the Lindemann separator are impressive, with grades and recoveries from the metals exceeding 98 percent depending on the setting of the splitter. The capacity of these separators is density and size dependent because for optimal results a mono layer of feed material on the belt is necessary. These eddy current separators are also in use for the separation of rubber and plastic from magnesium and sheet aluminium in the reject fraction generated by rising current separators in HMS car scrap plants.

Both separators can be used for the separation of metal non metal mixtures with a capacity of 2 - 8 ton/hour.

Lindemann has recently installed an eddy current separator with permanent magnets for the recovery of non ferrous metals from incinerator residue in The Netherlands, figures 4, 5

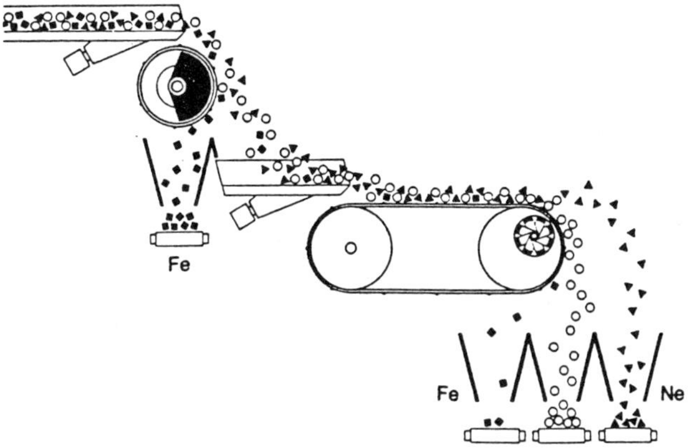

Fig. 4

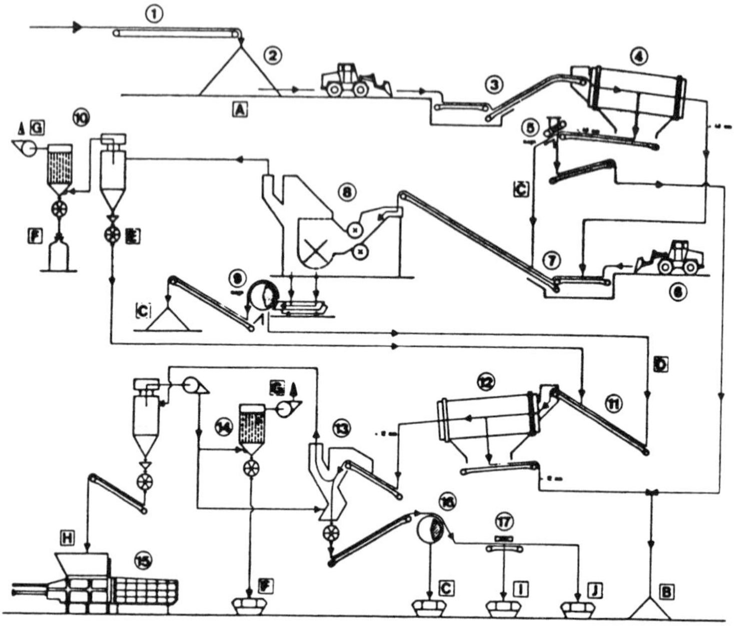

Slag beneficiation

1) Slag withdrawal
2) Intermediate slag storage
3) Rotary screen feeding
4) Rotary screen
5) Magnetic belt classifier
6) Coarse scrap feeding
7) Transfer to crusher
8) Crusher
9) Magnetic drum separator
10) Dust precipitating plant
11) Rotary drum feeding
12) Rotary drum
13) Air classifier
14) Dust precipitating plant
15) Container press
16) Magnetic drum separator
17) Non-ferrous metal separator

A Incineration residues
B Slag
C Ferrous metals
D Non-magnetic material
E Light materials
F Fine dust
G Exhaust air
H Non-burnt material
I Non-ferrous material
J Solid materials

Traitement des scories

1 sortie de scories
2 entrepôt intermédiaire des scories
3 alimentation du cribletambour
4 crible-tambour
5 séparateur magnétique monté au dessus de la courroie
6 alimentation grosse ferraille
7 transport vers le broyeur
8 broyeur
9 séparateur magnétique à tambour
10 installation de dépoussiérage
11 alimentation crible-tambour
12 crible-tambour
13 séparateur à vent
14 installation de dépoussiérage
15 presse container
16 séparateur magnétique à tambour
17 séparateur métaux non ferreux

A résidus de combustion
B scories
C métaux Fe
D produits non magnétiques
E produits légers
F poussiére fine
G air d'échappement
H produits non consumés
I métaux non ferreux
J substances dures

Fig. 5

ELECTROMAGNETIC FIELD

The original linear motor designed by Occidental Research for the aluminium can recovery is still marketed by Coreco, Milwaukee, figure 6. One of the systems has been in operation by Eumet North America, Chicago, Illonois and recently replaced by a Lindemann electrodynamic system because of its superior performance.

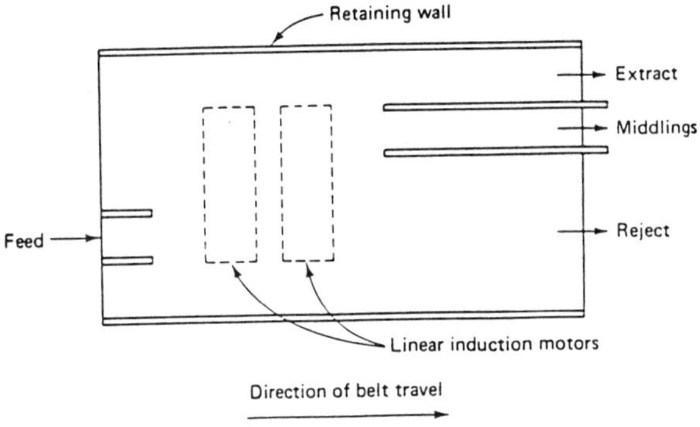

Fig. 6 Typical plan view of linear induction motor system for aluminium recovery.

The Coreco systems operated by Prolerized Schiabo Neu Comp. have also been shut down due to high metal losses in the reject fraction. Furthermore, the Cotswold separator developed by the Bird group, figure 6, with one motor, is less refined with regard to metal recoveries and energy consumption than the Lindemann electrodynamic system.

The metal removal which is perpendicular to the feed direction is one of the important reasons for the lower metal recoveries compared to systems separating into the feed direction.

The linear motor system designed by Mr. Feichter at Halver has one linear motor under a 45° angle in the direction of the feed under a 1000 mm wide horizontal belt. Total power consumption of the system depends on size and type of the material, table 2 (11).

TABLE 2

Size (mm)	Cap. ton per hour	Power consumption KW
2 - 10	1.5	150
10 - 30	2.5	100
20 - 50	3.0	50

The linear motor is cooled by a combination of water and air. Although the power consumption is high compared to all the other systems, it is also that effective that it will recover metal values from the reject from all the other eddy current systems in operation in the non ferrous car shredder residue processing. At low frequencies such as 50 or 60 Hz it is advantageous to use two linear motors on top of each other over a feed belt, figure 7, to separate non ferrous metals out of a feed stock.

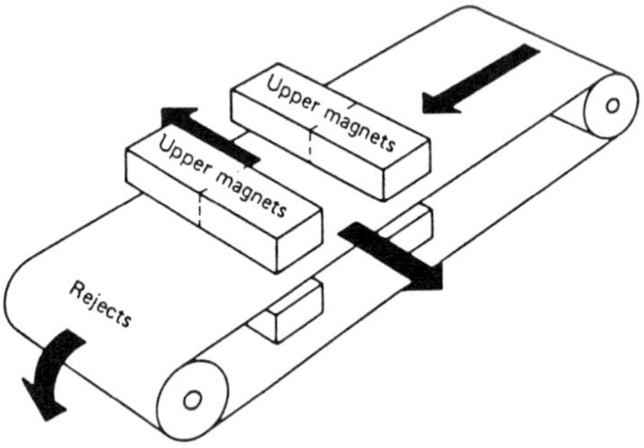

Fig. 7 Typical cross section of double magnet system for aluminium recovery

Combustion Power Comp. designed a machine based on this principle with water colling for the separation of aluminium from household refuse which has not been successful so far.

A continuous high frequency magnetic field or a pulsed magnetic field can be generated with a coil. The design from KWU, the pulsort, is a combination of the older Carpco and TU Delft design and has found no practical application so far in the recovery of non ferrous metals from car shredder residue.

The design from Lindemann with the 40 KW linear motor in a head pulley, figure 8, which has been very successful so far in the car shredder residue processing industry. Depending on the throughput and the layout of the plant, one to three Lindemann separators are used to concentrate magnesium and aluminium sheet from the reject from the first concentration step of the rising current separator in a secondary non ferrous metal concentration plant.

Also the reject from the first HMS step still has metal values which can be recovered with a linear motor. After the second HMS the aluminium fraction still contains from 10 to 40 percent glass, stones, insolated copper wires and other non metallics which can be removed with the linear motor.

The grade and recovery of the different metal products range from 96% to 99%.

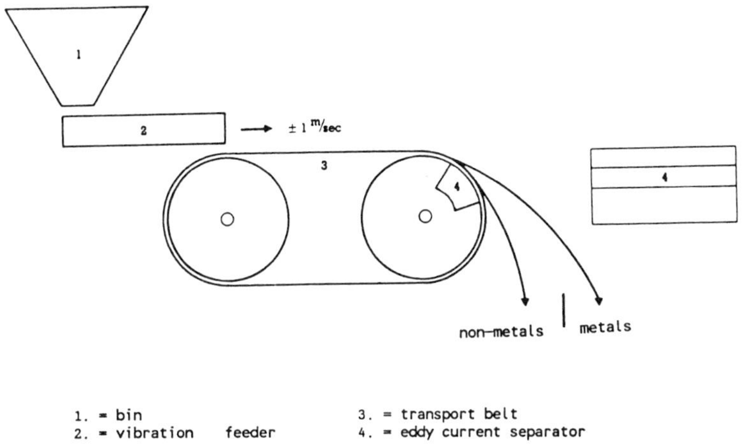

Fig. 8 System Lindemann water cooled eddy current

Due to the large variations in the feed to the scrap plants it is very difficult to get more accurate data. The Lindemann eddy current linear motor separators are also used in the new Lindemann dry separation system for car shredder residue. In this plant a combination of screening, shredding, air classification, air tables and linear motor technology is used to obtain a clean aluminium product and the heavy non ferrous metals.

CONCLUSIONS

Eddy current separators of different designs are available at the moment for the concentration of metals from the non metals in waste streams. In some cases a metal from metal separation is possible.

New magnetic materials will increase the applicability of eddy current separators based on permanent magnets.

REFERENCES

1. Edison, T.A., 1889. US Patent 44, 317.
2. Schloemann, E., 1979. Eddy current separation methods. In: R.J. Wakeman (Ed.), Progress in Filtration and Separation, Vol. 1, Elsevier, Amsterdam, The Netherlands, 346 pp.
3. Kercher, S.A. and Webb, M., 1982. Scrap processing by eddy current separation techniques. Resources and Conservation, 8 (1): 61 - 79.
4. Van der Valk, H.J.L. et al. Eddy current separation by permanent magnets. Part 1, theory. Resource and Conservation.
5. Roos, C.E. (Ed.), 1976. Non ferrous metal sorting at Vanderbilt University. Final report to the Solid Waste Management Office, Environmental Protection Agency, Vanderbilt University, Nashville, TN, 240 pp.
6. Itkin, V., 1986. Benefication of aluminium automobile scrap. Can. Inst. Min. Bull., (March): 93 - 95.
7. Schloemann, E., 1975. Separation of nonmagnetic metals from solid waste by permanent magnets. J. Appl. Phys., 46(11): 5012 - 5029.
8. Dalmijn, W.L., Voskuijl, W.P.H. and Roorda, H.J. 1979. Low energy separation of non ferrous metals by eddy current techniques. In: K.J. Thome-Kozmiensky6 (Ed.), Recycling Berlin '79; Proceedings of the International Recycling Congress held in Berlin, October 1979. E. Freitag-Verlag für Umwelttechnik, Berlin, Vol. 2, pp. 930-935.
9. Dalmijn, W.L. and Van der Valk, H.J.L. 1987. Trennung von Ne-Metallen aus vermischte Material mittels Permanentmagneten in einem vertikalen Wirbelstromscheider. Neuw Hutte 32(7): 273 - 279.
10. Braam, B.C. Dalmijn, W.L. and Duyvesteyn, W.P.C. 1985. Recovery of non ferrous metals from waste using an eddy current separator with rotating permanent samarium-cobalt magnets. In: P.R. Tayor, H.Y. Sohn and N. Jarret (Eds.), Proceedings of the International Symposium on Recycle and Secondary Recovery of Metals and Fall Extractive and Process Metallurgy Meeting, held in Ft. Lauderdale, Fl, 1 - 4 December 1985. The Metallurgical Society Inc., Warrendale, PA, pp. 641 - 654.

11. Rohstoff Rundschau 17, 1986. Page 550.

AUTOMOBILE SHREDDER RESIDUE -

THE PROBLEM AND POTENTIAL SOLUTIONS

Robert J. Schmitt

Center for Materials Production
Carnegie Mellon Research Institute
4400 Fifth Avenue
Pittsburgh, Pennsylvania 15213

Abstract

Shredding of automobiles produces about 3 million tons of shredder residue or "auto fluff" per year. Auto fluff consists of rubber, plastic, fiber, glass, dirt and small pieces of metal. A study was undertaken by the EPRI Center for Materials Production to identify: 1) processes available for disposing of fluff, and 2) research underway on potential disposal methods. It was concluded from this study that if auto fluff were to be declared hazardous, chemical fixation methods are available to treat fluff to allow disposal in landfills. However, other potential disposal methods that can reduce the volume of fluff to be stabilized and/or take advantage of its high heat value are reviewed and include 1) chemical fixation and landfill; 2) screening to produce fuel and landfill cover; and 3) incineration technology such as the rotary kiln, fluidized bed burning, gasification, plasma melting and submerged arc melting. Also potential R&D projects are suggested.

Second International Symposium -
Recycling of Metals and Engineered Materials
Edited by Jan H. L. van Linden,
Donald L. Stewart, Jr., and Yogeshwar Sahai
The Minerals, Metals & Materials Society, 1990

Automobile Shredder Residue Problem

Scrap automobiles, appliances, and other steel objects such as sheets and light structurals are processed by the steel shredding industry to produce a fragmented steel scrap for sale to the electric arc furnace steelmaker. Old automobiles and trucks make up the largest portion of the scrap processed, and it is estimated that as many as 8 to 10 million cars and trucks are shredded annually. Prior to shredding, tires, batteries, radiators, and gasoline tanks are usually removed. At the auto shredder, the scrap material is fed to a hammermill type shredder which chops and tears the old automobiles into small pieces which are conveyed through a sorting system (1). The steel or ferrous metal pieces are separated magnetically from the nonferrous metals (aluminum, copper, etc.) and piled for shipment to steelmakers. The low density or light material often referred to as "fluff" is collected by cyclone air separation. Although limited in practice, some auto shredders use wet shredding where water is sprayed on the scrap during shredding. This is done to prevent explosions from solvents remaining in the automobile, possibly from old paint cans left in the trunk. Although the water is recycled it picks up contaminants and represents a separate disposal problem.

About 200 automobile shredders are in operation throughout the United States producing about 2.5 to 3 million tons of auto fluff per year. Most automobile shredders are powered by electrical motors and power requirements range from 1000 hp (750 kW) to 6000 hp (4470 kW) depending upon size of unit (2). Electrical consumption reportedly is in the range of 15 to 25 kWh per ton.

Fluff is comprised of plastic, glass, rubber, fiber, some metal and dirt. A typical fluff composition is shown in Figure 1. The metals in fluff vary but can be aluminum, copper, cadmium, zinc, iron, etc. The presence of hazardous metals such as cadmium (corrosion resistant coating for fasteners and seat cover color enhancer) present potential disposal problems. Fluff is not considered a hazardous material by the U.S. Environmental Protection Agency (EPA) at this time except in California (cadmium restrictions) and thus is being disposed of primarily in landfills (3). Reportedly, depending upon location in the United States, disposal costs can range from $12 to $100 per ton, not including shipping charges. The concern on the part of the auto shredder is that disposal costs are rising, landfills are closing and some states like Massachusetts are imposing stringent environmental regulations on shredder operations. For example, Massachusetts now requires that a shredder operator shut down

Figure 1 - Typical Auto Fluff Composition*

Sponge & Foam	-	19.3%
Fabric & Batting	-	33.3%
Plastics	-	22.2%
Metal $\geq$ 12 Mesh	-	6.1% (Copper-0.6%, Zinc-5.0%)
Glass, Sand and Dirt Plus Metal < 12 Mesh	-	19.1% (Lead-0.6%, AL-7% and Iron-6%)

*Provided by Puremet Corporation

if the fluff produced contains more than 7 percent of oily substances. Oily liquids come from the engine, transmission, differential and the antifreeze. Polychlorinated biphenyls (PCBs) probably coming from capacitors in appliances is also presenting serious problems for auto shredders. Residue containing greater than 50 ppm of PCBs can no longer be dumped in ordinary landfills but must be disposed of in special landfills or incinerators.

The fluff disposal picture is one of confusion for the auto shredder not knowing what direction government regulation will take in the future. However, auto shredders believe they have about 2 to 5 years to find suitable alternatives to disposing of fluff in landfills or face possibly being shut down.

The purpose of this study conducted by the Center for Materials Production (CMP) was to review information on processes presently available for disposing of fluff, discuss research underway on potential disposal methods where information is available and identify potential processes that require research and development and could benefit from collaborative support from industry and the Electric Power Research Institute (EPRI).

Potential Disposal Methods

A number of organizations are working on developing novel methods for disposing of auto fluff or adapting present refuse disposal processes to accommodate this waste product. The various methods can be classified in several categories described below. Included under each category is a brief summary of specific research on various methods where information was available on the process.

Chemical Fixation and Landfill

Chemical fixation is a non-recovery alternative to secure landfills by which hazardous wastes can be converted into an environmentally nonhazardous material. This method is currently one of several processes being used to treat electric arc furnace dust, which is classified as hazardous by EPA (4). In addition, this technology has been in use for many years for stabilizing mill tailings for mining backfill as well as a means of treating liquid wastes containing metallic hazardous products. Such a process could be used to stabilize auto fluff for disposal in a landfill. In this process the waste is mixed with another material to produce a chemical form that is more resistant to leaching when placed in the ground. Materials used for this process include asphalt, silicates, pozzalanic cements and portland cements. Reportedly, the Schnitzer Steel Shredding Company in Oakland, California, uses a polysilicate treatment (referred to as the "K20" process) to prevent hazardous metal leaching thus permitting disposal in a conventional class 3 landfill. Cost of this treatment is estimated at $20 per ton. Companies actively promoting chemical fixation technology include Chemfix Technology, Inc. (5), Stablex (6), New Materials Technology (7), and Bethlehem Steel (Superdetox) (8).

Screening to Produce Fuel and Landfill Cover

The Midwest Steel Company (auto shredder) and the University of Wisconsin have developed a process whereby they separate auto fluff into two products referred to as "Mill Cover" and "Mill Fuel", (9). Mill cover reportedly can be used as a substitute for the daily soil cover in a landfill. Mill fuel can either be sold or consumed on site to generate steam, heat, or electricity. These products are produced by processing auto fluff through a rotary screen trommel, comprised of two sections, one having a screen with 1-inch holes and the other having a screen with 2-inch holes. The material passing through the two screens becomes the mill cover and contains the majority of the recoverable metals. However, the ferrous and nonferrous metals are either magnetically or hand separated from the mill cover. The balance of the material over 2 inches is the mill fuel. The developers of the process claim that there are many potential markets for the mill fuel with paper mills that currently burn sawdust and wood chips as prime candidates. Also, it is suggested that the shredder operator may also want to consider using the product along with other fuels for cogeneration which could help reduce the operating cost of the operation. It is also claimed that mill cover offers advantages over soil as a daily landfill cover and they include better barrier between waste and animals, insects, and birds; absorbs odors better, thus less smell; more stable surface on which to operate vehicles, and reduces blowing dust. Although, at the present time, mill cover could be used on landfills except in California, future concern would be the hazardous metals in the less than 2-inch material concentrating and exceeding any future, more stringent EPA requirements.

Incineration Technology

An industry-wide research and development study has been conducted by EnerGroup, Inc. to evaluate the energy value of automobile shredder residue (10). The study was conducted for the U.S. Department of Energy, Idaho Operations Office, in cooperation with the Metal Scrap Research and Education Foundation (MSREF) which is the research group of the Institute of Scrap Iron and Steel Industry. The study showed that combustible materials make up approximately 50 percent of auto fluff weight and 80 percent of its volume. The study also reported that based on actual statistical analysis of fluff from twelve different shredder sites the mean heating value of fluff was 5400 Btu/lb. Further, the fuel value of fluff can be enhanced by removing the fines. Recognizing the high heating value of fluff, considerable research has been done on means for incinerating the material to produce steam or electricity. Various incineration methods have been tested but all have experienced problems because of the nature of fluff. The problems range from incomplete combustion and materials handling to production of ash that because of the presence of hazardous metals such as lead, cadmium and chromium, may be classified as a toxic waste by EPA.

Although there are no known commercial combustion plants for incinerating fluff in the United States, the EnerGroup study identified several generic types of combustion systems for fluff and are as follows:

1. Circulating fluidized beds
2. Refractory lined and waterwall rotary kiln combustors
3. Gasification systems
4. High temperature slagging combustors
5. Traveling grate

Of these methods, rotary kiln, fluidized bed burning and high temperature gasification appear to have the greatest potential for success. Although not mentioned in the study, plasma and submerged arc melting also appear to be promising disposal methods. A brief review of each of these methods is provided below.

Rotary Kiln Process

As part of the EnerGroup study reported on above, they selected a rotary kiln to test the waste fuel performance of fluff (11). The rotating action of a rotary kiln maximizes the exposure of the burning material to the hot combustion gases and thus is a highly efficient means for incinerating waste. The test was conducted at Industronics, Inc. in South Windsor, Connecticut, using a rotary kiln rated at 2 million Btu per hour and equipped with a secondary combustor, waste heat boiler, baghouse, and packed tower scrubber. Approximately 3.5 tons of fluff were incinerated over a period of three days and was comprised of equal volumes of as-received fluff from eleven different shredder facilities. The fluff mixture was fed into the rotary kiln by means of a tapered auger at rates ranging from 200 to more than 400 lbs per hour.

The results of the test burn showed that approximately 94 percent of the combustibles in fluff were burned and the fluff reduced 55 percent by weight and 80 percent by volume. The resulting ash was reported to have an average moisture content of 1 percent, volatile matter 6 percent and contained 25 percent by weight of iron. Chemical analysis of an ash sample revealed substantial amounts of lead primarily as lead chloride. Composite samples of the ash were analyzed using the EPA extraction procedure toxicity test and were shown to exceed 5.0 mg/l of lead indicating that ash from incineration of fluff has the potential to be classified as a hazardous waste. Other heavy metals were present in the ash such as cadmium, chromium and barium but substantially below hazardous concentrations. Air emissions were monitored during the test burn with the exception of toxic organics, and the report concludes that emissions from burning fluff can be controlled to within regulatory limits with existing technology.

Fluidized Bed Combustion

Based on results of extensive pilot plant testing, Energy Products of Idaho (EPI) developed a unique fluidized bed combustion system for auto shredder residue (12). A schematic of the pilot plant is shown in Figure 2, the fluidized bed contained nine square feet of bed area and 25 feet of freeboard above the fluidizing nozzles. In the process, fluff is fed through the feed tube and combustion is accomplished with a special fluidized cone arrangement. A particle size of less than six inches had to be maintained to prevent plugging of the feed tube. Combustion products then pass into a refractory cyclone and the ash captured can either be removed from the system or injected back into the fluidized combustor. Flue gases can exit the cyclone or part of the flow can be pulled through a heat exchanger and baghouse.

EPI processed three truckloads of fluff screened to three-inch minus through the pilot plant. Environmental testing was conducted for organic particulate and heavy metal emissions. During this test program, continuous emission monitors recorded O_2, Co, So_2 and No_x emissions. Limestone was added with the fuel on a continual basis because of high concentrations of glass and the potential for acid gas emissions.

EPI reports that the emission test data are proprietary. The author assumes the tests were successful as the study has led to a proposal by EPI to a major auto shredder to design, build, and operate a 7.5 megawatt electric power plant utilizing a fluidized bed combustor to burn auto fluff (13). Reportedly, plant capacity would be about 72,000 tons of auto fluff per year.

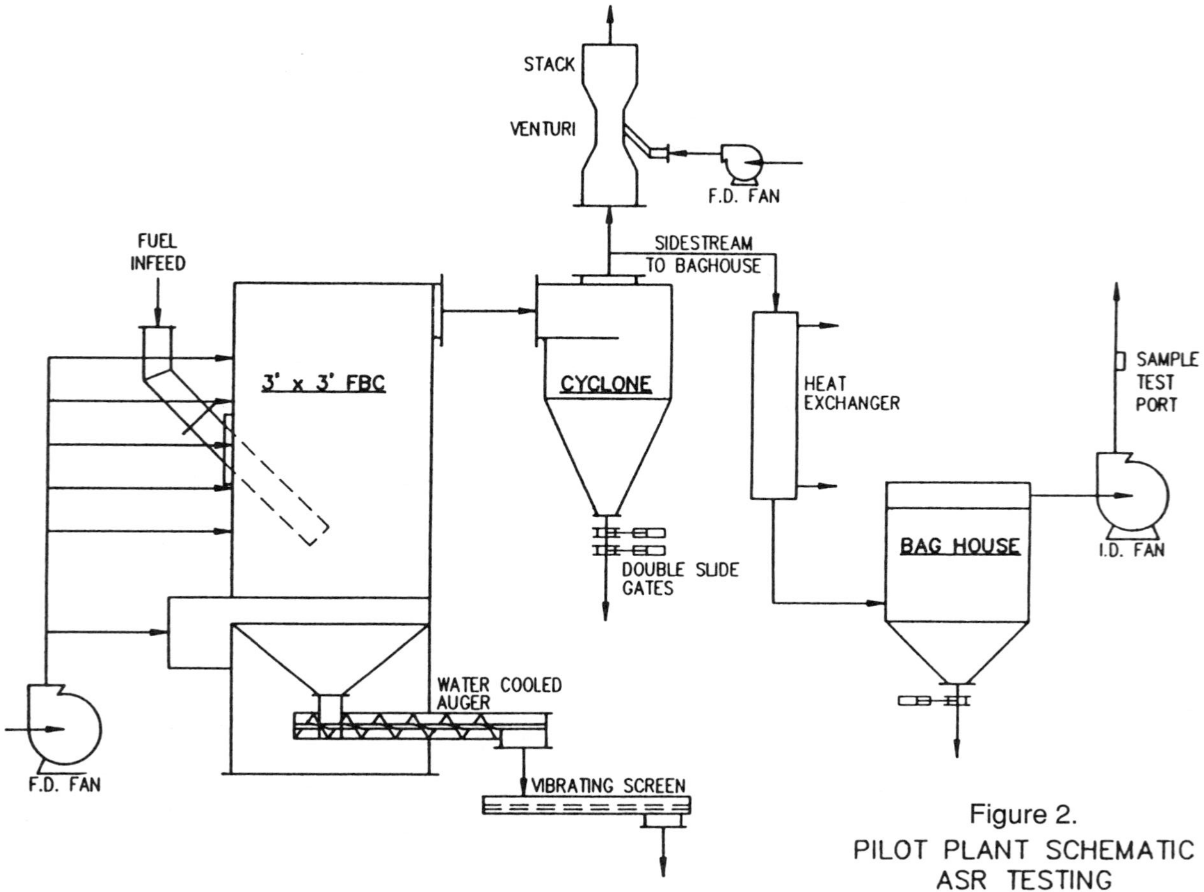

Figure 2.
PILOT PLANT SCHEMATIC
ASR TESTING

High Temperature Gasification Process

Voest-Alpine has developed a patented process for high-temperature gasification of waste (14). The process referred to as HTV converts waste such as auto fluff into clean fuel gas and a mineralic slag granulate which can be used as a building material. In the process fuel such as oil and used solvents are burned in a multi-fuel burner which maintains the temperature necessary for gasification in the reaction chamber, Figure 3. Solid and/or a pasty waste material is fed directly into the gasification chamber and coke is fed from the top into the shaft. The hot gases from the multi-burner combust the waste and coke reaching temperatures of about 3000F (1600C). The shell of the reactor is cooled by forced circulation of boiler feed water. The combustion gases laden with dust pass into a cyclone where most of the dust particles are separated from the raw gas and returned to the reactor. In the next step the sensible heat from the raw gas is exchanged with cooling water to produce hot water or steam. The gas is then processes through a purification plant where dust and chloride and sulfur compounds are removed to produce a clean fuel gas. The slag melted in the combustion chamber flows into a water bath where it is cooled and granulated. A series of gasification trials with waste oil and hazardous waste have been conducted with the HTV process in Linz, Austria. A synthetic hazardous waste comprised of PCBs; industrial wastes; medical wastes; fuel oil containing mercury, cadmium, and lead, etc., was processed. Reportedly, emission data revealed only traces of hydrocarbons and very low No_x levels in the HTV clean gas. The heavy metals in the slag were non-elutable. For specific details of the process, contact Voest-Alpine.

Voest-Alpine has also conducted tests (15) with their HTV-demonstration plant to determine its suitability for processing auto fluff. A four-day demonstration test run was conducted wherein shredder residue enriched with plastics such as ABS, vinylester-SMC, polycarbonate, and polyurethane was fed to the HTV plant. For this test shredder residue and plastic granules were agglomerated and made into briquettes. Both briquettes and crushed plastic parts were fed into the reactor and burned. Reportedly, the HTV gasification process satisfactorily incinerated the auto shredder residue and produced a clean fuel gas and stabile slag.

Based on one year's test operation of the HTV gasification pilot process, Voest-Alpine is presently offering this technology for sale worldwide.

Plasma Incineration

Plasma can produce temperatures in excess of 10,000°C and consequently has been considered as a means for processing hazardous waste. Although some research has been conducted on treating waste by heating with plasma, apparently no work has been done with auto fluff. Plasma can be applied in two ways to heat waste. With direct heating material is passed through the plasma flame, whereas with indirect heating the plasma is used to create a bath of molten material. An excellent review of these methods and experiences with plasma to treat hazardous wastes has been prepared by Laurel J. Staley of the U.S. Environmental Protection Agency, Cincinnati, Ohio, and is entitled "Hazardous Waste Decontamination With Plasma Reactors." (16).

Reportedly, the indirect heating method is particularly suited for treating solid waste which is combined with fluxing agents. The high temperature of the plasma combusts the waste material, volatilizing the organic waste contaminants, which are thermally destructed by the high

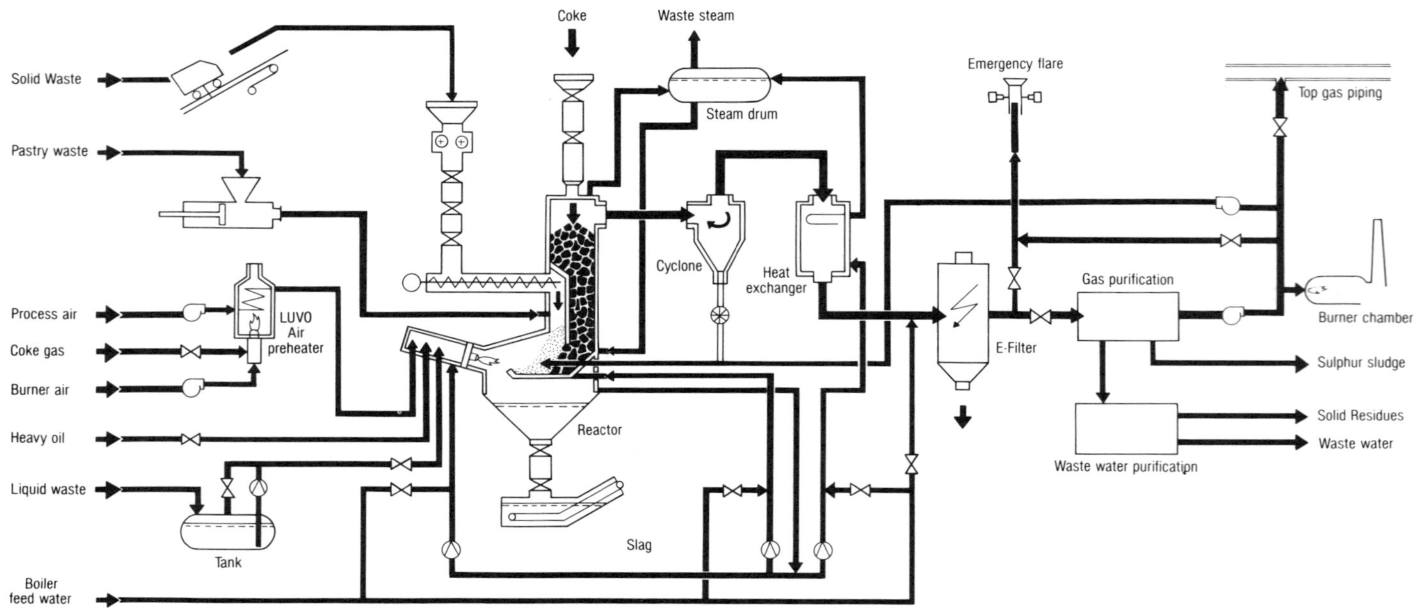

Figure 3. VOEST-ALPINE High-temperature gasification

temperatures. The molten material solidifies into a vitrified mass encapsulating any heavy metals present. The vitrified mass is non-leachable and thus suitable for disposal in a landfill.

Retech, Inc. of Ukiah, California, has developed an indirect plasma heating process referred to as the Plasma Centrifugal Reactor (PCR) which has potential for decomposing and stabilizing auto fluff. Briefly, the PCR design incorporates a stationary plasma torch with a revolving tub inside a sealed reaction chamber (17). The plasma torch heats waste material and fluxing agents producing a molten mass. The centrifugal force imposed upon the molten mass moves it to the outside of the chamber, where mixing of new feed material is accomplished. The effluent gases are subject to a treatment system and a vitrified slag is produced.

The U.S. Department of Energy (DOE) has installed a Retech PCR unit adjacent to their magnetohydrodynamics facility in Butte, Montana, which is being operated by MSE/CDIF to conduct tests for DOE and EPA. Reportedly, MSE/CDIF is interested in conducting tests with auto fluff (18).

The plasma-fired cupola developed by the Electric Power Research Institute and Westinghouse Electric Corporation has a plasma-fired torch fitted to the bottom of the furnace (19). In this application the plasma torch provides energy for iron melting independent of coke combustion primarily used in the conventional cupola. Since this development, the plasma technology group at Westinghouse has applied their developmental efforts pertaining to the plasma cupola to treatment of landfill material. The results of tests to-date indicate that the plasma cupola is capable of vitrifying landfill-type material without the need of elaborate pretreatment of the material. Westinghouse personnel envision that the plasma process would not only combust the fluff but produce a glassy slag encapsulating any hazardous metals and thus be suitable for disposal in a landfill.

Westinghouse has proposed to demonstrate the feasibility of processing automotive shredder residue in the plasma cupola located in the Westinghouse Environmental Service's Plasma Center in Madison, Pennsylvania. A two-heat test is proposed to determine the effects of material composition, feed methods and rates, and energy requirements. For more information, contact Westinghouse Electric Corporation, Madison, Pennsylvania.

Submerged Arc Melting

Cookson Group, PLC in the U.K. reports that auto fluff can be combusted in a submerged arc furnace (20), see Figure 4. In the process, which is called deTOX$_{TM}$, the heavy metals in the fluff are melted with a fluxing agent and encapsulated in the vitrified slag and are non-leachable. For details of the process, contact Dunston Ceramics Limited.

Proposed Methods for Disposal of Hazardous Ash From Incineration Processes

Incineration processes such as rotary kilns and fluidized beds offer promise for combusting auto fluff. However, the ash from the process can contain heavy metals such as cadmium, lead, etc., and thus potentially can be declared a hazardous waste which would require further processing before being disposed of in a landfill.

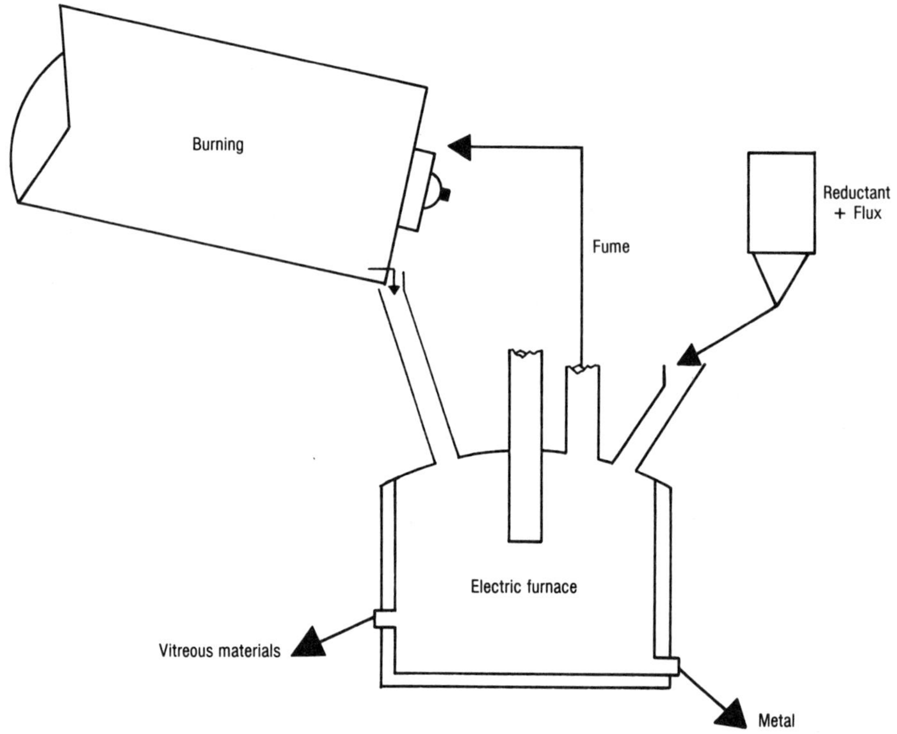

Figure 4. Waste material processing by deTOX™ process

Several processes appear promising for treating ash to make it suitable for landfill and are briefly described below.

Puremet Process

Puremet Corporation of Chattanooga, Tennessee, has developed a hydrometallurgical process for the removal of copper and other metals such as cadmium, zinc, nickel, etc., from steel scrap. Puremet personnel believe their process would be applicable to removal of metals from ash resulting from auto fluff combustion.

A schematic of the patented process is shown in Figure 5 as it pertains to removal of nonferrous metals from steel scrap. Briefly, in this process, scrap from shredded automobiles, appliances, etc., from which the iron and steel have been magnetically separated is fed into a carbon steel reaction tank. A cuprous ammonium sulfate solution is added to dissolve the copper, nickel, zinc, etc. The solution is recycled from the electrowinning cells, and the only heat added is that picked up in the electrowinning process. The reaction time varies depending upon the scrap mix; pure copper dissolves rapidly. Stainless steel scrap, aluminum, and zinc die castings are unaffected in the process and are removed and sold for scrap.

The metal-rich solution then goes to the electrowinning cells where copper, zinc, nickel, etc., are recovered. Also, the metal-rich solution can be sent to a refrigerated tank where copper salts are crystallized, then filtered and dried and can be sold to chemical firms. Puremet claims that all hazardous metals will be recovered and that the remaining residue will meet EPA requirements for disposal in a landfill.

With regard to handling auto fluff ash, Puremet envisions simply processing the ash through the process much like scrap. For further details, contact Puremet.

deTOX$_{TM}$ Process

The Dunston Ceramics Limited deTOX$_{TM}$ Process referred to earlier reportedly is not only suitable for combusting auto fluff and producing a vitrified product but is particularly suited for processing ash from incineration of auto fluff to produce a vitrified product which encapsulates the heavy metals, see Figure 6. The glass/ceramic type product produced can be converted to tiles, bricks, etc., and used in commercial applications. An alloy product stream is possible from some raw materials, thus optimizing value recovery. A pilot-demonstration unit is under construction in the United Kingdom. Processing rates of concentrated raw materials range from 500 to 6000 lbs per hour or more.

Related Research Activities

Recycling of Auto Shredder Residue

Argonne National Laboratory (ANL) is working on a program sponsored by the U.S. Department of Energy to evaluate various recycling procedures that could be applied to recover value added products from auto shredder residue (21). Reportedly, the main thrust of the program at this stage is on the separation and recycling of the plastics content of fluff. A procedure which is under testing at the present time involves the use of organic solvents to extract the various plastics and separate them from each other. Emphasis is on the plastics which are present in appreciable

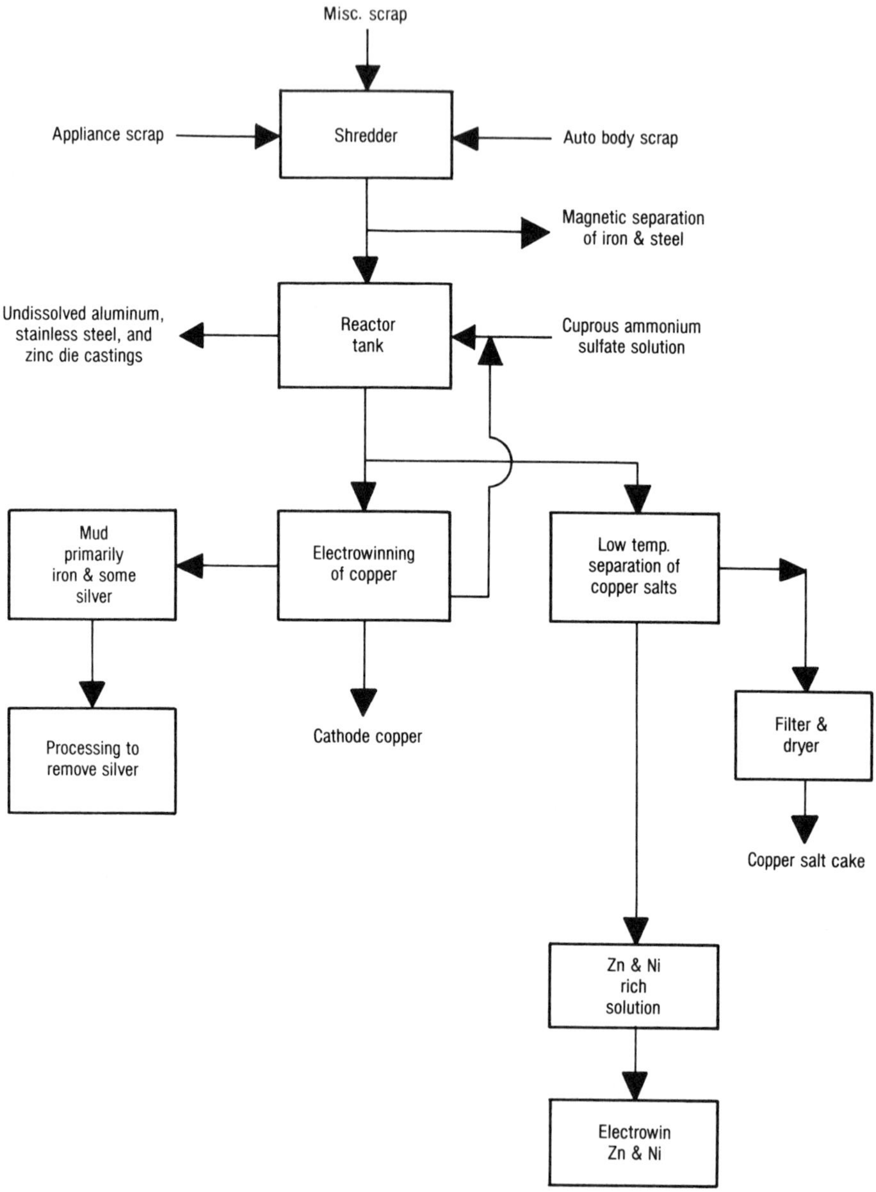

Figure 5. Schematic for puremet process for copper removal from scrap

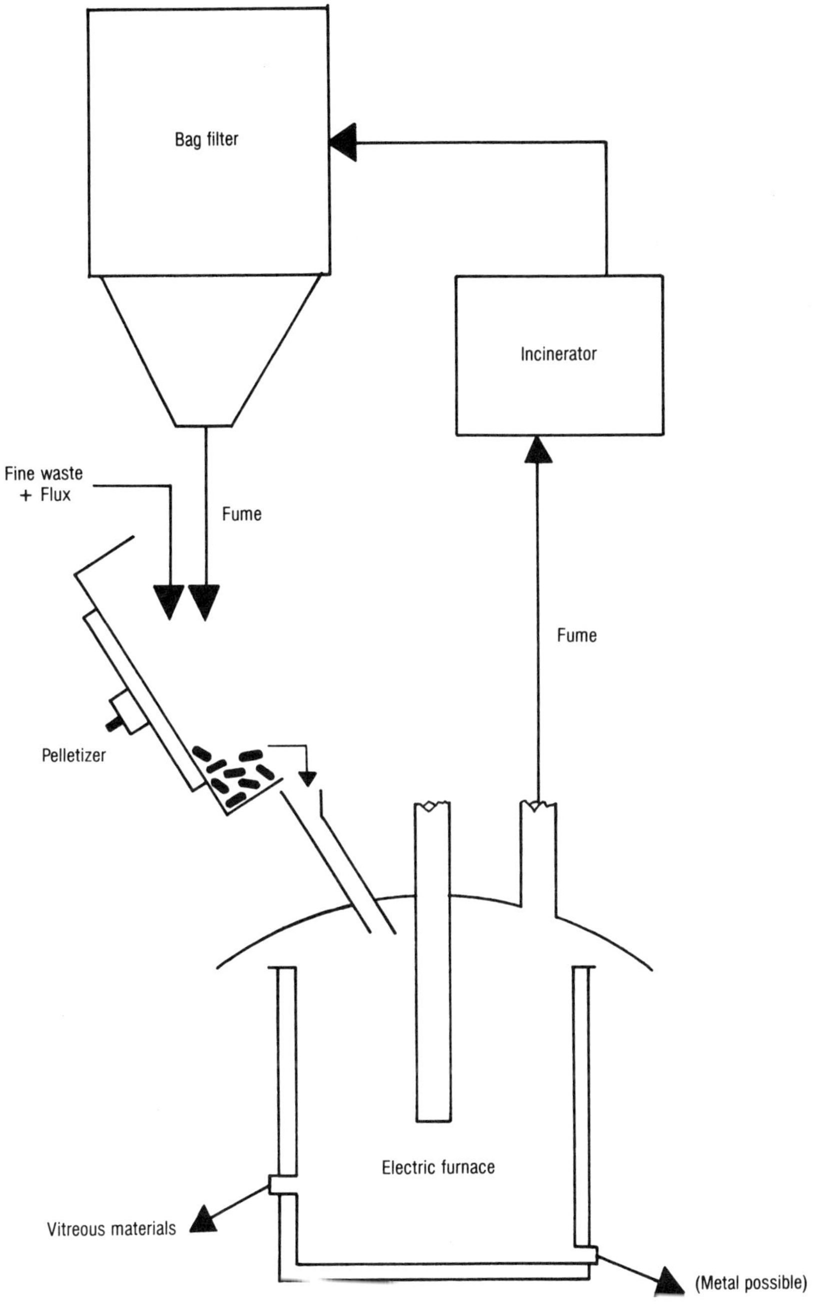

Figure 6. Ash processing by deTOX$_{TM}$ process

quantities in the fluff and for which a growing market exists. These are high and low density polyethylene, polypropylene, polyvinyl chloride, and acrilonitrile-butadiene-styrene (ABS).

Initial experiments showed that a plastics material showing at least two distinct layers could be extracted from fluff. The separation of these layers appears feasible. More work is planned to quantify these initial results. Experiments will also be conducted to identify operating conditions at which fluff can be pyrolized to produce marketable chemicals and/or liquid fuels without having environmentally harmful byproducts. Other chemical processing methods such as hydrogenation may also be evaluated.

Cryogenic Scrap Fragmentation Process

Typically old automobiles are shredded using a hammermill-type shredder. In the process of tearing the automobile apart, the shearing action of the mill tends to smash different metals together. For example, copper wiring may get crimped together with a fragment of steel framing, or a lead tire balancing weight may remain attached to a steel wheel fragment or be smashed onto another steel piece. Magnetic separation used to collect the steel parts will pick up the aforementioned fragments containing copper and lead thereby contaminating the steel pile. Depending upon the scrap dealer they may selectively remove by hand certain parts of the old automobile minimizing mixing of materials. This can minimize the presence of hazardous metals, such as lead and cadmium, that may eventually end up in the auto fluff. Selective removal of parts today typically includes the battery, copper radiator, catalytic converter, tires, gas tank and possibly the engine block. Other possible sources of heavy metal contamination such as lead balancing weights on wheels and cadmium plated parts could be removed by hand but represent an expensive, time-consuming operation.

The use of cryogenics in scrap processing has the potential to produce a cleaner and denser scrap than the conventional process and possibly reduce the volume of fluff produced. The cryogenic scrap fragmentation process was developed by the Belgian scrap processing firm of George et Cie in Liege, Belgium (22). In the process, scrap automobiles are baled and the bales are conveyed through an insulated tunnel where they are cooled to about -7°C by cold nitrogen vapors coming from liquid nitrogen. Exiting through the tunnel, the bales are partially immersed in the liquid nitrogen bath and the temperature of the bale lowered to about -120°C. The frozen bales are then processed through a hammermill where they are fragmented into coin size pieces. Reportedly, nitrogen consumption is estimated to be about 0.3 liter per kilogram of steel scrap produced. Various means for separating the glass, dirt, rubber, plastic and metal once the scrap has been fragmented include air classification, magnetic separation, and water elutriation or density separation (23,24).

Various research organizations have evaluated the cryogenic process and conclude that it shows promise in helping solve the auto fluff problem by making fluff more amenable to separation into various fractions such as dirt, glass, plastics, and rubber and consequently reducing the volume of auto fluff. By reducing the volume of fluff it may help justify the cost of cryogenic treatment. At the present, this process cannot be justified economically because of the high cost of nitrogen.

Possible Research Projects

At the present time, only the state of California has declared auto fluff as a hazardous material requiring treatment of fluff prior to disposal in a landfill. As discussed earlier in the report, chemical fixation methods are available for treating fluff and thereby permitting disposal in a landfill. However, landfill costs are rising and landfills are closing. Further, obtaining permits for new ones is becoming extremely difficult. Incineration in a rotary kiln or fluidized bed will significantly reduce the volume and weight of fluff thereby reducing the amount requiring disposal. However, the ash produced by these methods will undoubtedly be declared a hazardous material by EPA and require treatment. Further, because of the high heat value of fluff, it represents a potential source of energy to produce hot water, steam or electricity. Therefore, it could become an important energy source in the future as our fossil fuel reserves diminish. Consequently, it appears that means other than chemical fixation should be developed for disposing of auto fluff.

The Voest-Alpine HTV process represents a potential method whereby the energy value of fluff can be realized and at the same time the ash from combustion of fluff is made into a non-leachable slag. It appears that this process will require a large capital investment and thus is particularly suited for a regional or central operation whereby fluff from several shredders could be processed. However, shredder operators have expressed the need for a process that they can be installed on site to process fluff on a continuing basis and not have to store or transport the material. It appears that a small plasma reactor could serve this need, and it is recommended that R&D be conducted to develop such a process. The Retech PCR test unit installed at the DOE R&D facility in Butte, Montana, represents a potential means to obtain an evaluation of plasma as a method for disposing of auto fluff. EPRI/CMP is interested in a collaborative project with the shredder operators to evaluate plasma in this application.

In addition, should EPI build their proposed fluidized bed process for processing auto fluff as discussed earlier in this report, this development should be followed closely as it would burn fluff to produce electricity. Also, Puremet's hydrometallurgical process and Dunston's submerged arc furnace process represent potential solutions for treating ash from combustion processes to meet EPA leach test requirements and thus make it suitable for disposal in landfills. The shredder industry should give consideration to supporting projects to demonstrate the viability of these processes.

References

1. Metallic Scrap: The Manufactured Resource, (ISIS) Washington, DC, Institute of Scrap Iron and Steel, Inc.

2. Ibid.

3. An Industrywide Hazardous Waste Identification Study of the Ferrous Scrap Processing Industry," Clayton Environmental Consultants, Inc., for ISIS, November 1980.

4. Donald MacRae and Peter Cowx, Plasma Furnace Treatment of Electric Arc Furnace Dust as Demonstrated by Bethlehem-Tetronics. Center for Metals Production, Pittsburgh, November 1988, Report No. 88-2.

5. Chemfix Technologies, Inc., P.O. Box 1572, Kenner, LA 70063, U.S. Patent 3,837,872, J.R. Connor, "Method of Making Wastes Non-Polluting and Disposable."

6. Stablex Corp., 2 Radnor Corporate Center, Radnor, PA 19087; "Sealosafe$_{SM}$ Process."

7. New Materials Technology Corp., 8709 Arthur Circle, Wichita, KS 67207; "Fujibeton."

8. J. D. Lynn, Technology Group, Bethlehem Steel Corp., Bethlehem, PA 18016; January 15, 1985.

9. K.E. Boerger, H. Samuels Co., and N.R. Braton, Univ. of Wisconsin-Madison, "Mill Fuel and Mill Cover Recycled from Shredder Fluff."

10. W.S. Hubble, I.G. Most, and M.R. Wolman, of EnerGroup, Inc., and S.L. Natof, U.S. Dept. of Energy, Power Generation from Automobile Shredder Waste Fuel: Characterization and System Feasibility, EnerGroup, Incorporated for U.S. Depart. of Energy, Report No. DE-PRO7-871012704, August 1987.

11. Ibid.

12. D.M. Albertson. Development of a Fluidized Bed Combustion System for Auto Shredder Residue, Energy Products of Idaho, presented at 1989 Spring National Meeting of the American Institute of Chemical Engineers.

13. R.D. Smith, Thermo Energy Corporation, M. Murphy, V.P. Sales for Energy Products of Idaho, private communication with author.

14. Dr. P. Freimann, Voest-Alpine, and Dr. G. Starudinger, University of Graz., "Voest-Alpine High Temperature Gasification Process."

15. Dr. P. Freimann, Voest-Alpine, Thermal Utilization of Shredder Residues by High-Temperature Gasification, Report No. 2228B, January 27, 1989.

16. L.J. Staley. Hazardous Waste Decontamination With Plasma Reactors, U.S. Environmental Protection Agency, Cincinnati, OH 45268.

17. R.C. Eschenbach, R.A. Hill, and J.W. Sears. Process Description and Initial Test Results With the Plasma Centrifugal Reactor, Retech, Inc., presented at Forum on Innovative Hazardous Waste Treatment Technologies: Domestic and International, June 19-22, 1989, Atlanta, GA.

18. S. Kujawa, MSE/CDIF, private communication with author, December 6, 1989.

19. Plasma-Fired Cupola Development, Volume 2, Description and Analysis of Test Program and Test Unit, EPRI Report EM-5901, September 1988.

20. N. Pocock, Cookson Group Plc, private communication with author, December 13, 1989.

21. F. Dudek, Argonne National Laboratory, private communication with author, December 8, 1989.

22. J.H. Bilbrey, Jr. and E.G. Valdex. Use of Cryogenics in Scrap Processing, Bureau of Mines, U.S. Department of the Interior, Advances in Cryogenic Engineering, 20, pp. 411-416, 1975.

23. C.J. Chindgren, K.C. Dean, and L. Peterson, Recovery of the Non-Ferrous Metals from Auto Shredder Rejects by Air Classification, Bureau of Mines TPR 31 (April 1971), 11 pp.

24. L.J. Froisland, K.C. Dean, and C.J. Chindgren, Upgrading Junk Auto Shredder Rejects, Bureau of Mines TPR 53 (March 1972), 11 pp.

AUTOMATIC SORTING OF NON-FERROUS METALS

FROM AUTOMOBILE SHREDDERS

Hans-Peter Sattler

Metallgesellschaft Aktiengesellschaft
Reuterweg 14
D-6000 Frankfurt a. Main
F. R. Germany

Abstract

The non-ferrous metals which remain together with rubber, glass, plastics etc. from shreddered automobiles after separation of steel, iron and fluff are till today won by heavy media flotation and handpicking. This is unefficient and expensive. To rise the quality of the recycling products as well as the recycling rate a special sorting technique for these non-ferrous metals was developed. After preseparation of rubber, wires and cables the mixed metals are identified analytically piece by piece and sorted in fractions of aluminium-, magnesium-, zinc- and copper-base alloys, brass and stainless steels. With a detection and sorting rate up to 100 pieces per second and a particle sizes 15 to 65 mm up to 5 t per hour material can be sorted. The purity of the sorted metal fractions is better than 96%.

Second International Symposium -
Recycling of Metals and Engineered Materials
Edited by Jan H. L. van Linden,
Donald L. Stewart, Jr., and Yogeshwar Sahai
The Minerals, Metals & Materials Society, 1990

1. Introduction

Three fractions, ferro-magnetic steel scrap, which can be charged directly in the steelworks, fluff, which must be dumped, and a mixture consisting of light and heavy metals with admixes of rubber, plastics, glass, stones, etc., are produced in shredding of used cars. This last fraction must be further processed to permit recovery and reutilization of the metals which it contains, which is generally accomplished nowadays in heavy media flotation systems (Figure 1). After screening out of the fines on screening drums, mainly at a 13 mm particle size - the minimum particle size which can be processed varies between 10 an 20 mm, according to the type of flotation system - the various constituent components of the material are further separated on the basis of their specific density (1). In the first flotation stage, featuring a heavy medium density of 2.2 g/cm^3 (composed of water and ferrosilicon (FeSi) powder), the majority of non-metallic constituents and any magnesium as well as large-area aluminium sheet present are flotated out in the form of coarse waste. The metallic sink fraction is then separated into flotated aluminium still containing up to 15% extraneous constituents (stones and cables of identical density) and a heavy-metal sink fraction containing up to 10% aluminium in the second stage, using a medium density of 3,4 g/cm^3.

In many cases, the two flotated fractions are aftertreated for recovery of magnesium or to increase aluminium content to above 95% using so-called eddy-current separators, which exploit electrical conductivity to separate the non-ferrous metals from non-metallic materials. 2 to 3% aluminium, however, are lost in this process.

The heavy-metal fraction which sinks in Stage II is entirely handsorted, since all so-called separative melting processes proposed and investigated, in which the differing melting temperatures of the heavy metals (Pb $<$ 330°C, Zn $\sim$ 400°C, Cu $\sim$ 800°C) are used for separation, have failed to gain acceptance for economic and quality reasons.

This processing method for the recycling of non-ferrous metal scrap from car shredders is extremely costintensitive, and not without its problems. The flotation process necessitates complex and expensive environmental protection provisions and achieves only a low-grade scrap quality, due to undesirable extraneous metal contents and carried-over FeSi powder. Handpicking can nowadays be performed economically only in third-world countries; even here, the lack of the necessary expertise leads to the production of only inadequate scrap qualities.

We therefore set ourselves the task of replacing these processing stages with a completely dry and extensively automated system for reycling of shredder product to achieve metal fractions of maximum possible grad purity.

2. Scrap Pretreatment

Before metal sorting, the scrap must firstly be rendered suitable for machine processing, i.e., long, snagging and/or rolling components must be removed. In order to optimize capacity utilization of the sorting machine, it is also desirable to raise the metal content of the feed material by means of the most extensive possible removal of non-metals.

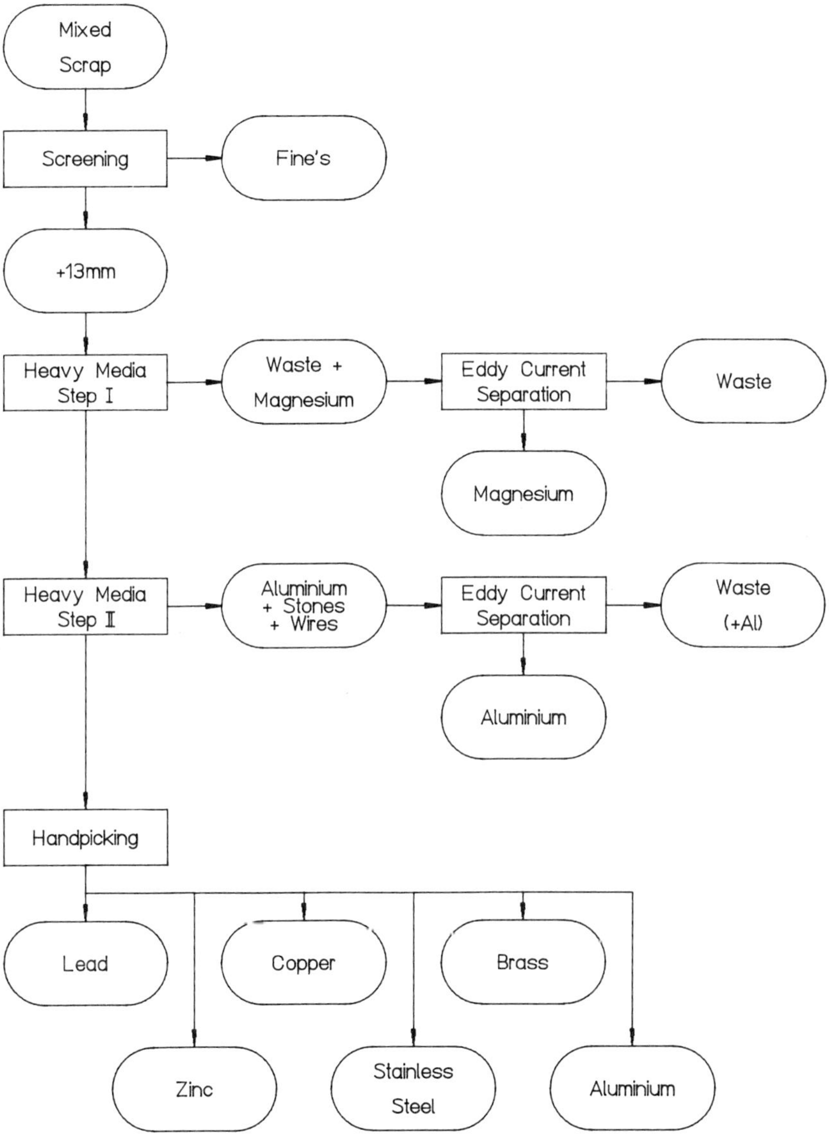

Figure 1 - Idealized processing and product flowchart for currently practised recycling of non-ferrous metals from the automobile shredder

There are various systems which can be used for the latter task, for instance, a rising current separator, using which average metal content can be increased from 50% to 80-85%, or an eddy-current separator, with which metal contents of greater than 97% can be achieved. Both systems, however, have substantial process-engineering disadvantages, in addition to high prime cost. The rising current separator, for instance, entails the problems concomitant with a wet process. In the eddy current separator, up to 15% metals are lost in the waste fraction, principally stainless steel, lead, zinc and high-alloyed copper grades, due to the nature of this process. This means that the waste may contain as much as 40% metals, which is problematic for its subsequent treatment (dumping or incineration).

For this reason, we resolved upon the use of one of our own developments (2), a so-called rubber separator, which functions on the friction principle, i.e., the materials are separated on the basis of their specific coefficients of friction. Such systems make it possible to concentrate the feed materials to over 80% metal, the waste fraction consisting principally of rubber, containing less than 1% metal.

The entire process cycle for pretreatment of scrap for sorting now has the following configuration (Figure 2): firstly, overgrain material and fines are screened out, the fines because excessively small material reduces the throughput capacity of the sorting machine too greatly, and the overgrain because liberation is still incomplete in it. The overgrain material is size-reduced again in a mini-shredder.

Wire and long components such as trim strips, for instance, are then separated, again in one of our own developments, and themselves then further size-reduced.
This operation is followed by the above-mentioned rubber separator, which discharges onto the final pretreatment stage, an inclined belt for removal of rolling constituents. The rolling constituents, principally lock cylinders, seat-belt rollers, etc., are also size-reduced again, and thus made susceptible to sorting. The material pretreated in this way is then further processed on automatic sorting machines and separated according to its basic alloying components into aluminium, copper, brass, magnesium, lead, zinc and stainless steel fractions, and a final waste fraction.

3. Automatic Sorting Machine

The sorting machine consists essentially of three components: a singling mechanism, a detector, and a sorting mechanism (Figure 3).

In the singling mechanism, the scrap particles fed in a mass flow from a storage hopper via vibrating feeders are detached from one another by means of passage through a chicane system and accelerated in stages to 2.50 m/s via parallel runing conveyor belts and to them vertical running guide belts, and, during this process, drawn out to form a chain-like file. The spacing between the individual scrap particles varies between 1 and 10 mm after this treatment.

The scrap thus isolated now passes to the detector (3,4), the centrepiece of the system. Here, sensors which measure electrical conductivity determine whether each individual scrap particle consists of metal or not.

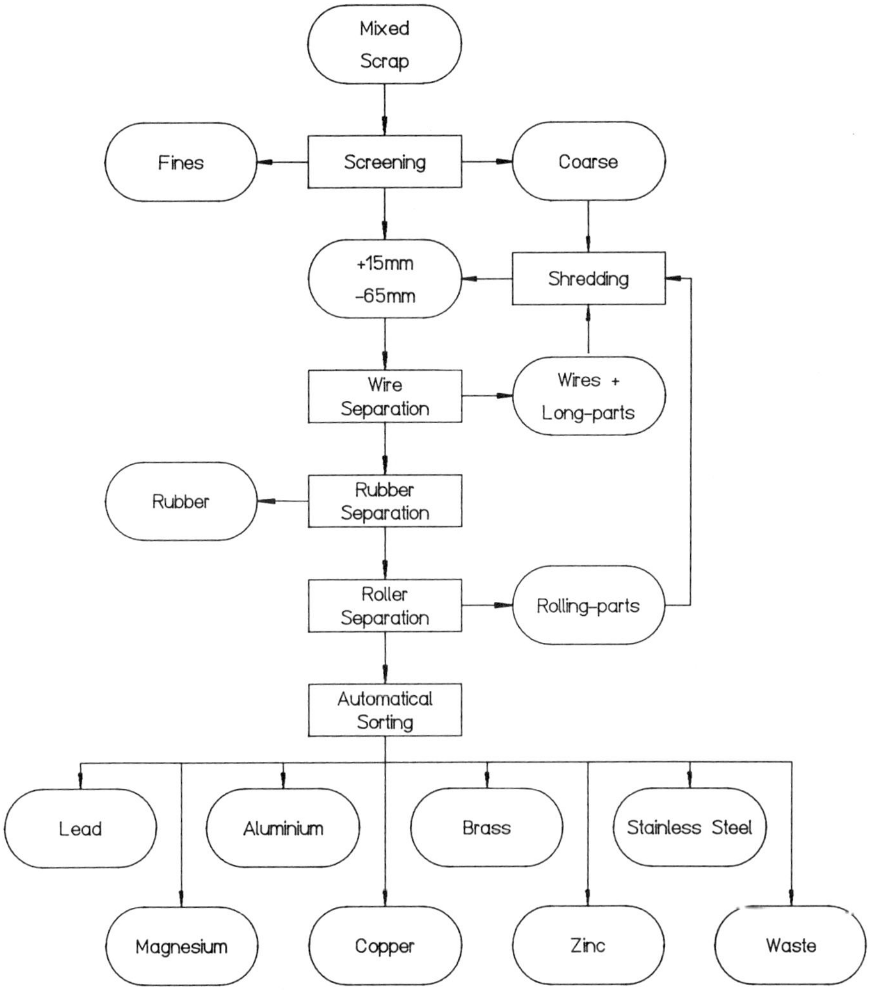

Figure 2 - Pretreatment and automatic sorting of non-ferrous metal fraction from the automobile shredder

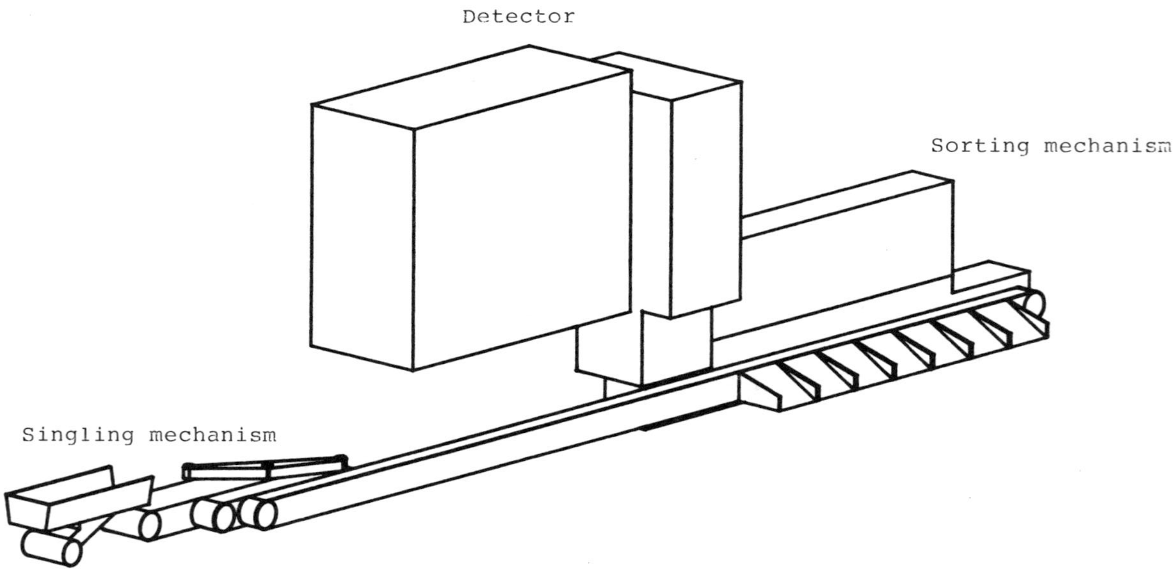

Figure 3 - Diagram of a sorting machine for recycling of non-ferrous metals from the automobil shredder

Only the metals are then analytically identified using the classical atomic emission spectroscopy process. A pulsed laser beam is used for excitation of atomic emission by evaporating an area of approx. 1 x 2 mm on the surface of the scrap to plasma, from which the analytical information is then obtained. In order to avoid misidentifications arising due to surface dirt, etc., the analysis surface on the items of scrap is first cleaned by means of a "cleaning shot" from a second laser head.

The detection process is illustrated diagrammatically in Figure 4:

The arrival of the singled scrap particle in the detection zone of the conveyor belt it detected by a trigger system which activates the pulsed laser. The pulsed laser beam is directed onto the scrap particle by means of mirrors and a focusing lens. The spectral lines required for identification are isolated from the plasma generated by the laser beam by means of spectral detectors with spectral filters mounted in front of them, and converted to sorting signals by the detector electronics and the evaluation computer. During measurement, the detector electronics system ensures that the spectral detector registers the plasma radiation only during a defined of interval of time. This ensures that the background noise of the radiation from the plasma is reduced to a minimum, thus making it possible to better identify the various spectra.

Up to 100 identifation cycles per second, or 360,000 per hour, are at present possible using this system, with an identification accuracy of normal-quality shredder scrap of 98 % or better (3).

In the sorting mechanism, the scrap particles identified are then discharged from the conveyor belt into the corresponding sorting channels by means of pushers. The pushers running parallel to and at the same speed as the conveyor belt are activated on the induction principle by the sorting signal. An activation cycle
(= command + activation + return) takes 3 ms, with 150 activation cycles per second. Differing numbers of the 12.5 mm broad pushers are activated in each sorting cycle, according ot the length of the piece of scrap, which is determined by an optics system. 3 to 5 t per hour non-ferrous metals can be recovered from shredder scrap depending on the feed material. The purity of the sorted metal fractions is better than 96%.

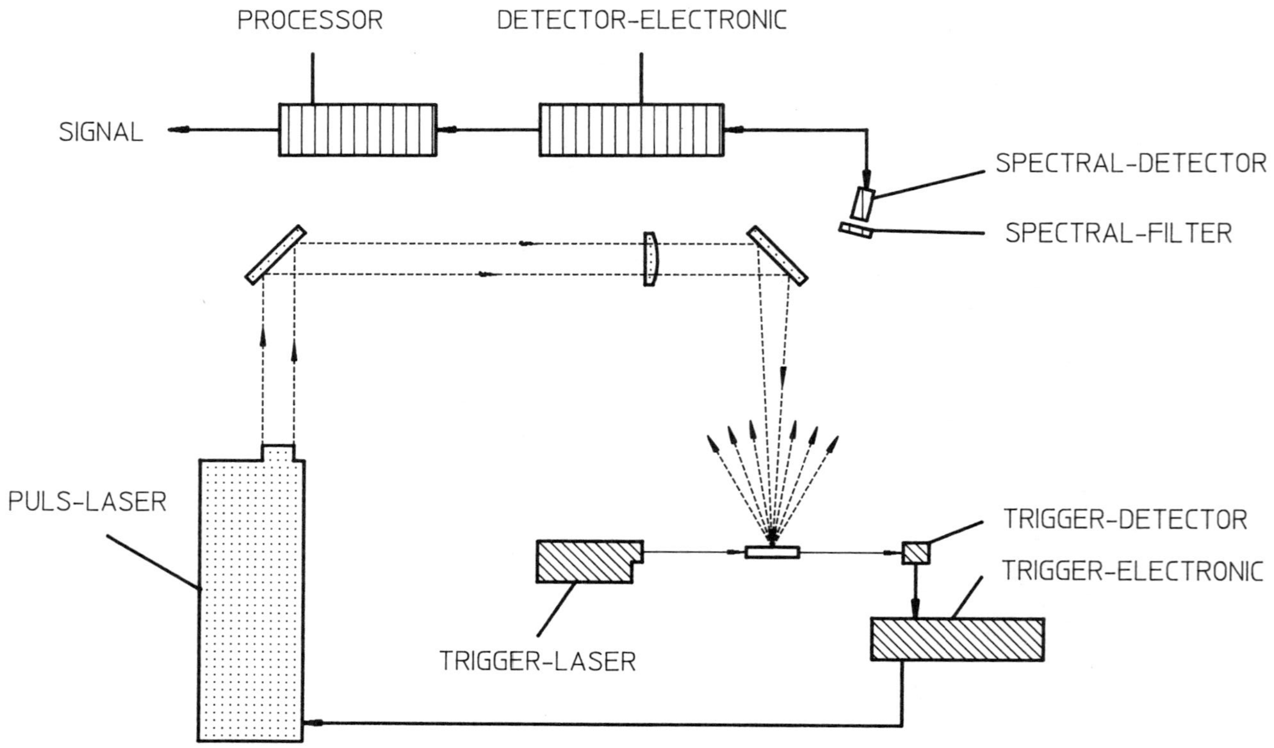

Figure 4 - Diagram of the detection process for analytical identification of shredder scrap

References

1. H.-P. Sattler: "Werkstoffbedingte Recyclingprobleme am Beispiel des Automobils" (Recycling Problems conditional on Materials using the Example of the Automobils), VDI Reports, No. 694 (1988), 349-368.

2. German Patent Application, P 39 37 265.0

3. H.-P. Sattler: "Trockene Abtrennung von NE-Metall-Fraktionen aus Automobil-Shredderschrotten mit mechanischen Sortierverfahren der Erzaufbereitung" (Dry Separation of Non-ferrous Metal Fractions from Automobil Shredder Scrap using Mechanical Sorting Processes from Ore Dressing", in the seminar "Stofftrennung und Stoffanreicherung" (Materials Separation and Materials Concentration), Goslar (1987), p. 89, editor PL Material- und Rohstoff-forschung, Nuclear Research Facility, Jülich.

4. European Patent Application, EP 88 20 10 73.

LEAD - ZINC - TIN

Session Chairman:
John L. Watson

A LABORATORY INVESTIGATION OF THE SODA ASH SMELTING OF LEAD ACID BATTERY RESIDUE

by

C.A. Pickles
Department of Metallurgical Engineering
Queen's University
Kingston, Ontario K7L 3N6

and

J.M. Toguri
Department of Metallurgy & Materials Science
University of Toronto
Toronto, Ontario M5S 1A4

ABSTRACT

A pyrometallurgical process for the recovery of lead from battery residue was investigated on a laboratory scale. The process involved pelletizing the residue with sodium carbonate followed by a carbothermic smelting operation. The optimum conditions for the maximum recovery of lead were determined. These were: a smelting temperature of 950°C, a charcoal addition of three mass percent and a sodium carbonate addition of twenty mass percent. A moisture addition of ten mass percent during pelletizing was found to be sufficient to complete the sulphate-carbonate exchange reaction and to produce a strong green pellet. Sintering of the green pellets prior to reduction was found to have no effect on the lead recovery. However, below 500°C, the strength and abrasion resistance of the pellets decreased with sintering temperature.

INTRODUCTION

Lead production from the secondary lead industry represents a substantial portion of the total lead produced. At the present time about forty percent of the refined lead originates in the recycling industry. Lead-acid batteries account for the majority of the lead generated by the secondary industry. In addition, the majority of recycled lead is employed in battery production.

A typical scrap lead acid battery would contain approximately 32%Pb, 3%PbO, 17%PbO_2 and 36%$PbSO_4$, in addition to the plastic and acid components. The average mass of the finely divided residue from one spent battery is approximately 6kg of which 19% is PbO_2, 60%$PbSO_4$ and 21% metallic Pb.

The current methods for the recovery of lead from lead-acid batteries are based on pyrometallurgical smelting. The various routes can be categorized according to the furnace type and the smelting operations involved. Generally, secondary lead smelting is performed in two stages. In the first, the drained and de-cased scrap is smelted in an oxidizing atmosphere in a reverberatory or rotary furnace. The charge is melted and the antimony and the other alloying elements are oxidized. A litharge slag forms which contains virtually all of the alloying elements and impurities (1,2). The primary bullion contains low concentrations of antimony,

Second International Symposium -
Recycling of Metals and Engineered Materials
Edited by Jan H. L. van Linden,
Donald L. Stewart, Jr., and Yogeshwar Sahai
The Minerals, Metals & Materials Society, 1990

arsenic, and tin. A secondary lead-antimony alloy is produced in stage two which involves reductive smelting of the oxide slag. Reductant and fluxes can be added to the same vessel after removal of the primary lead bullion or the second stage could be performed in a small blast furnace or cupola (2). The final slag can be discarded if the lead content is sufficiently low to be environmentally acceptable. The main advantage of this two-stage process is that two useful lead products are obtained. The product from stage one can easily be refined to high purity lead, while the alloy product requires only moderate refining and alloy adjustment to form a marketable antimonial alloy.

The major disadvantages of the traditional pyrometallurgical processes are the environmental problems associated with the treatment of these sulphur-bearing raw materials. The presence of large quantities of lead sulphate leads to sulphur dioxide emissions during the pyrometallurgical smelting operation. Combined with the presence of finely divided oxides of lead, the control of the sulphur dioxide and lead particulates from a high temperature process is not only difficult but expensive. Containment of airborne lead, as dust and fume, requires extensive gas treatment facilities. It is much simpler and less expensive if the process is compact and air-intake is minimized. The multi-stage nature of the traditional processes is a disadvantage in this respect. Also, the disposal of slag presents a major problem. The addition of soda ash during the smelting of lead acid battery residue in a rotary kiln process has resulted in reduced sulphur dioxide emissions(4). However, due to the presence of sulphides and thiosulphates special slag disposal techniques are required.

The U.S. Bureau of Mines has investigated an electrolytic process for the recycling of lead from scrap batteries (5,6). In this process lead metal is separated from the residue by ball milling, washing and screening. Then the metal is cast into anodes and electrorefined by the Betts process using fluosilicic acid as the electrolyte. The sulphate-oxide-metal residue which remains after removal of the metallic lead is treated in a two-stage leaching operation. Ammonium carbonate, fluosilicic acid and lead powder are employed to solubilize the lead which is recovered by electrowinning.

Recently, Arai and Toguri (7) reported the kinetics of conversion of $PbSO_4$ to $PbCO_3$ using Na_2CO_3 in aqueous solution. This conversion occurred readily at room temperature. Experimentally, the lead carbonate recovery was found to be 98%. The overall reaction for the carbonate leaching of lead sulphate was controlled by the mass transport of carbonate ions through the solution boundary layer.

Over the years, there have been many hydrometallurgical processes proposed (8-14). There are a number of disadvantages to these hydrometallurgical leaching processes which include: a large quantity of water must be handled, there are many steps in the process, there are environmental problems involved with the treatment of the waste liquor (Na_2SO_4 solution) and the carbonate must still be dried and reduced at high temperatures to metallic lead. Thus, it would be advantageous to develop a simplified high temperature process utilizing the same desulphation reaction.

In the present study, a pollution free pyrometallurgical process for the recovery of lead from battery residue was investigated on the laboratory scale. A flow sheet of the process is shown in Figure 1. This process is based on the exchange reaction:

$$Na_2CO_3 + PbSO_4 = PbCO_3 + Na_2SO_4 \qquad (1)$$

followed by the decomposition of the carbonate to the oxide according to:

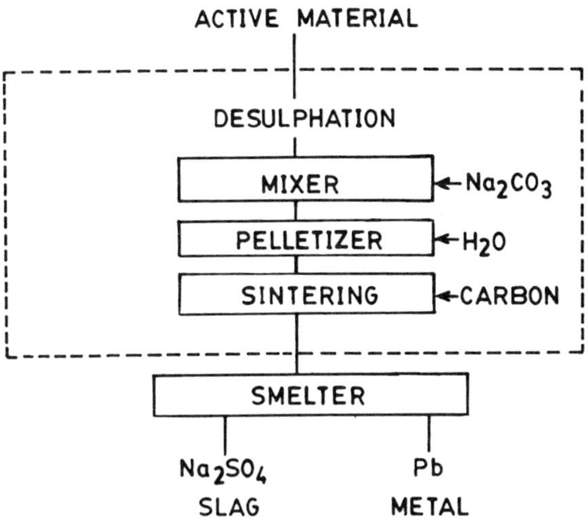

FIGURE 1: Flowsheet of the proposed process for the recovery of lead from lead-acid battery residue.

$$PbCO_3 = PbO + CO_2 \qquad (2)$$

and finally the reduction of PbO to lead as follows:

$$PbO + CO = Pb + CO_2 \qquad (3)$$

For this purpose the metallic lead fraction was separated from the residue by crushing, grinding and sieving. Then, the battery residue was mixed with sodium carbonate and pelletized with moisture. Subsequently, the pellets were smelted with charcoal in a fireclay crucible. The sulphur dioxide emissions are minimal and the sulphur is fixed in the slag as sodium sulphate (15). The lead particulate emissions are reduced since the smelting temperature is low and pellets are employed. Also, the by-product sodium sulphate slag could be a marketable product.

EXPERIMENTAL

Raw Materials

A typical analyses of the battery residue as received is shown in Table I. This material was ground in a ball mill and the metallic lead was separated by sieving the ball mill product at 28 mesh (Tyler). The analyses of the two sieved products are also shown in Table I. The metallic lead content of the minus 28 mesh residue was relatively low at 5.92 mass percent. The sodium carbonate was minus 200 mesh and was Fisher reagent-grade. Fisher activated charcoal in the size range of 6 to 14 mesh was employed as the reducing agent.

Smelting Experiments

The sodium carbonate was mixed with the residue and pelletized with moisture. In some cases, the charcoal was incorporated into the pellet

	Battery Residue	Minus 28 Mesh Material	Plus 28 Mesh Material
Total Pb	--	71.67	90.39
Metallic Pb	32	5.92	78.60
$PbSO_4$	36	61.38	11.64
PbO	3	--	--
PbO_2	17	--	--
Al	--	0.02	0.05
Sb	--	2.80	0.43
As	--	0.51	0.25
Bi	--	0.12	0.009
Cd	--	0.17	0.002
Cu	--	0.11	0.22
Fe	--	0.08	0.63
Ni	--	0.005	0.006
Ag	--	0.001	0.0007
Na	--	0.01	0.06
Sn	--	0.10	0.35
Zn	--	0.16	0.59
Total S	--	6.53	1.45
Sulphide S	--	0.045	0.22

TABLE I: Analyses of Battery Residue in mass %, Minus 28 Mesh (Tyler) Material and Plus 28 Mesh (Tyler) Material.

during the mixing operation. These pellets were either air dried or sintered and subsequently smelted in a fireclay crucible in a muffle furnace. The variables studied included: smelting time, smelting temperature, sodium carbonate addition, and charcoal addition. The lead button was separated, weighed and the lead recovery was determined. Slag and metal samples were analyzed for impurities.

A series of experiments were performed in which the pellets were smelted without a carbon addition. The purpose of these tests was to determine if lead oxide rather than lead could be recovered from the battery residue. In these tests, the sinter temperature was varied between 25°C and 800°C. The lead metal, lead oxide and sodium sulphate phases were analysed for antimony, tin, and arsenic. Also, the sodium sulphate slag was analysed for lead.

Reaction Degree

To measure the degree of reaction between sodium carbonate and lead sulphate, a refractory boat containing about five grams of pelletized battery residue was placed in the hot zone of a resistance furnace. The temperature of the hot zone was maintained at 500 $\pm$ 5°C. Commercial argon, purified by passing it over Ascarite, Drierite and magnesium perchlorate, was continuously swept over the sample. Within thirty minutes, the sample temperature reached 500°C and both the moisture and the carbon dioxide were evolved from the sample. The moisture was absorbed by a column of Drierite while the carbon dioxide was absorbed by Ascarite in a Nesbitt bulb. The degree of reaction was calculated from the ratio of the actual amount of carbon dioxide evolved to the maximum theoretical available amount of carbon dioxide in the pellet. The sample was weighed before and after the experiment and thus the moisture content could be calculated by difference. The efficiency of the Drierite column was tested by placing about 1 gram of water in the boat. It was observed that in this case there was no increase in the weight of the Nesbitt bulb containing the Ascarite. Thus, the weight increase of the bulb could be attributed solely to carbon dioxide absorption. For each experiment, the same gas flowrate was employed and fresh Ascarite was used. The variables studied included: the moisture added during pelletizing, the amount of sodium carbonate added, prior sintering temperature, and time after pelletizing.

In addition, a test was designed to measure the time required for the exchange reaction during pelletizing and the effect of moisture content on the degree of reaction. This test consisted of determining the temperature rise on mixing twenty grams of battery residue with sodium carbonate, adding various amounts of water and vigorously mixing in a beaker. The temperature rise of the mixture was measured with a thermometer.

Melting Behavior

The melting range of the battery residue was determined by melting the residue in a fireclay crucible and allowing it to furnace cool. A chromel-alumel thermocouple, within a fireclay sheath, was placed in the melt for temperature determination.

Pellet Strength

The pellet strength was measured by two methods. In the first, one hundred and fifty grams of pellets were autogenously ground in a ball mill for fifteen minutes. The material from milling was screened on a sieve shaker. Then, the percent retained on 8 mesh was calculated. In the second method, the pellets were sorted into three size ranges: 1 to 3 mm, 4-5 mm and greater than 6 mm. The pellets were individually compressed to rupture on a Chantillon pellet tester. The force required to break the pellet was recorded. For comparison purposes, tests were performed on both pure lead sulphate and battery residue.

Porosity Determination

The average porosity at each sintering temperature was determined as described in the following procedure. Six dried pellets having diameters ranging from 5 to 10 mm were weighed. They were then immersed for 24 hours in ethanol. Using the density of ethanol (0.789 g/cc), the weight increase was employed to calculate the volume of pores. The pellets were approximately spherical and thus the percentage of pore volume was calculated as follows:

$$\% \text{ of pore volume} = \frac{\text{volume of pores}}{\text{volume of pellet}} \times 100\%$$

RESULTS AND DISCUSSION

Smelting Experiments

The cooling curve for the molten battery residue indicated that the residue melts over the temperature range from 915 to 935°C. This is about 50° lower than the liquidus region at the appropriate composition of the published phase diagram for the lead oxide-lead sulphate system. This can be attributed to the presence of other impurity oxides and sulphates in the system which would lower the melting range. In the solid state, the exchange reaction between sodium carbonate and lead sulphate would be expected to proceed very slowly. However, in the presence of liquid the exchange reaction would be accelerated. Since sodium carbonate melts at 851°C and a liquid phase forms in the battery residue at 915°C, high reaction rates would be expected at temperatures over 900°C.

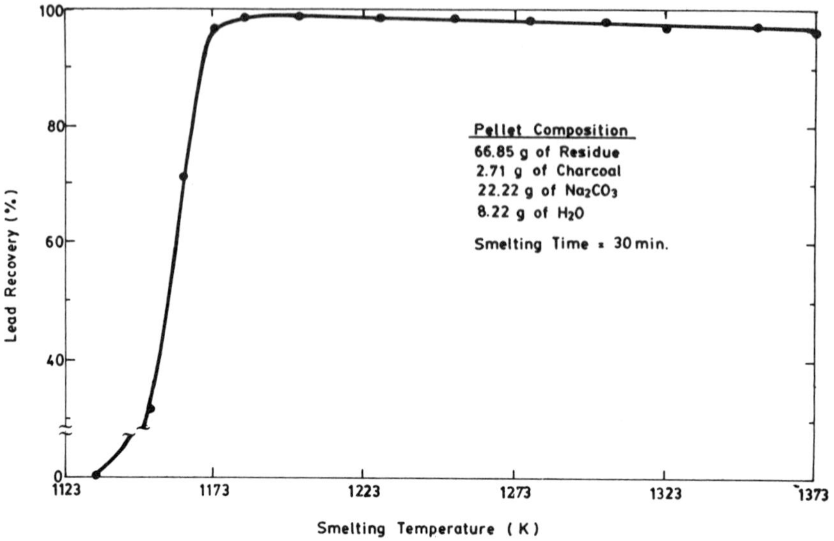

FIGURE 2: The effect of smelting temperature on lead recovery.

The lead recovery as a function of smelting time at 925°C was determined and these results showed that a smelting time of approximately 30 minutes at 925°C was required to achieve good slag-metal separation. At extended smelting times, some lead was oxidized to the slag phase and thus the lead recovery marginally decreased.

The effect of smelting temperature on lead recovery is shown in Figure 2. The smelting time was maintained at 30 minutes. Smelting temperatures of over 900°C were required to achieve good slag-metal separation. Lead recoveries of over ninety eight percent were obtained in the temperature range of 900 to 1000°C. Table II gives the impurity content of the metal and slag phases. The impurity content was not a function of temperature in the range studied. Also, as shown by the slag-metal distribution ratio, a large percentage of the impurities remained in the sodium sulphate phase, particularly barium, antimony, arsenic, nickel and tin. The lead contents of the slags were in the range of one to two mass percent.

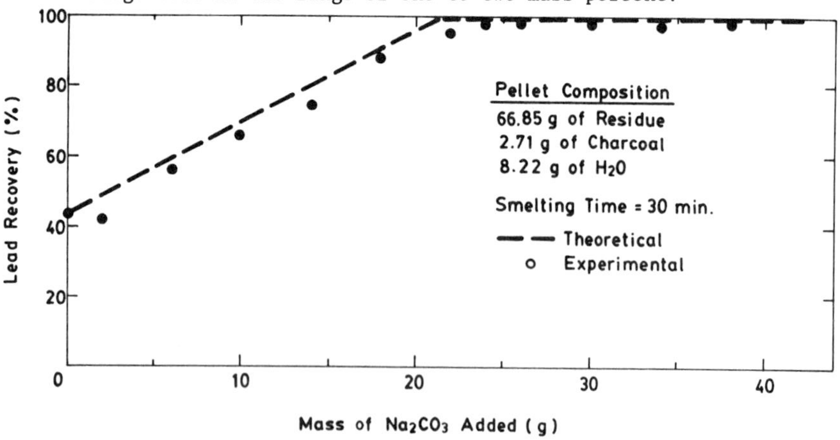

FIGURE 3: The effect of sodium carbonate additions on the recovery of lead.

METAL PHASE ANALYSES

Smelting Temperature	930°C	950°C	970°C	990°C	1010°C
Element					
Ba	.0002	<.0002	<.0002	<.0002	<.0002
Sb	.015	.016	.013	.0082	.011
As	.0047	.0016	.0033	.0016	.0016
Bi	.0081	.0083	.008	.0085	.0085
Cu	.049	.049	.097	.047	.048
Fe	.0045	.0067	.0099	.004	.0038
Ni	.0002	.0002	.0002	.0021	.002
Ag	.0015	.0016	.0016	.0017	.0014
S	.0022	.0016	.0018	.0018	.0018
Sn	<.004	<.004	<.004	<.004	<.004

SLAG PHASE ANALYSES

Smelting Temperature	930°C	950°C	970°C	990°C	1010°C	Slag-Metal Distribution Ratio at 950°C
Element						
Ba	0.145	0.127	0.141	0.147	0.158	725.0
Sb	0.36	0.38	0.36	0.32	0.18	22.5
As	0.20	0.21	0.20	0.20	0.21	125.0
Bi	0.0077	0.010	0.0067	0.0059	0.0076	0.9
Cu	0.0041	0.0046	0.0056	0.0063	0.0036	0.08
Fe	0.015	0.016	0.077	0.12	0.070	2.2
Ni	0.0091	0.0089	0.0086	0.0078	0.0077	45.5
Ag	0.0028	0.0024	0.010	0.0075	0.0018	1.75
S	<0.005	<0.005	<0.005	<0.006	<0.005	3.1
S Total	21.30	21.31	21.11	20.97	22.93	------
Sn	1.10	1.09	0.93	0.96	0.83	275.0
Pb	1.54	1.49	1.21	1.87	1.32	------

TABLE II: Impurity contents (in mass percent) of the metal and slag phases obtained during the smelting tests at the indicated temperatures. Also included is the slag-metal distribution ratio at 950° C.

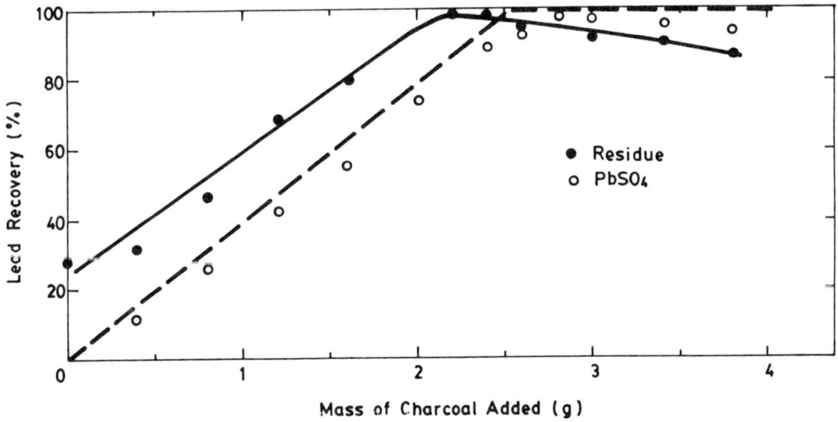

FIGURE 4: The effect of carbon additions on the lead recovery.

Figure 3 shows the effect of sodium carbonate additions on the lead recovery. A sodium carbonate addition of about 21.5 mass percent/100g residue was required to achieve lead recoveries of over 95 percent. This amount is slightly in excess of the stoichiometric requirement as given by the following equation:

$$PbSO_{4(s)} + Na_2CO_{3(s)} + C_{(s)} = Pb_{(l)} + Na_2SO_{4(l)} + CO_{(g)} + CO_{2(g)} \qquad (4)$$

At high sodium carbonate contents, the lead recovery decreased slightly. This can be attributed to the increase in the oxygen potential of the system due to the carbon dioxide in the sodium carbonate. In the absence of sodium carbonate some lead will be recovered since there is metallic lead and lead oxide in the residue. Also some carbothermic reduction of lead sulphate will occur

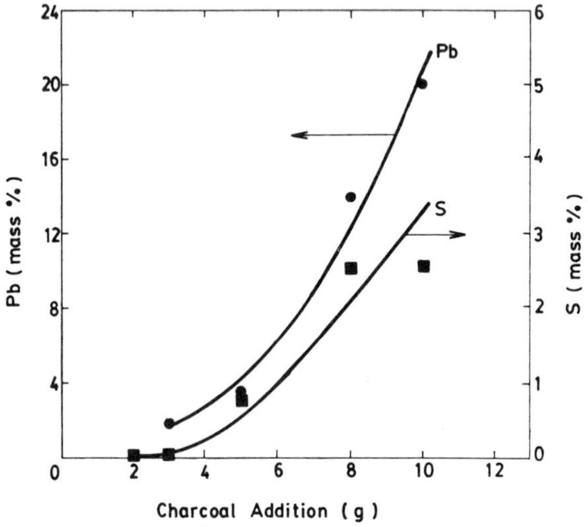

FIGURE 5: The effect of carbon additions on the lead and sulphur contents of the slag.

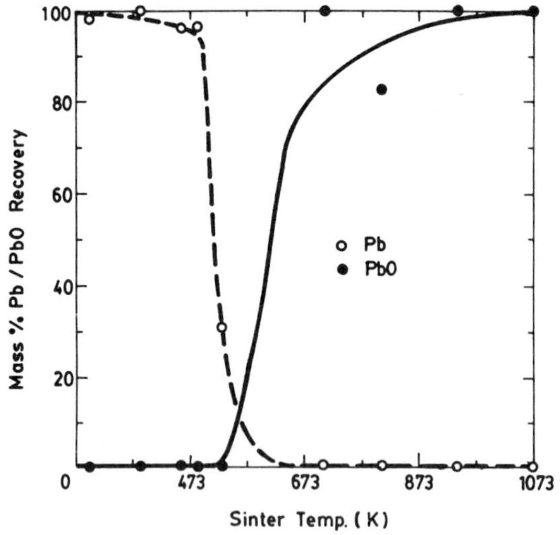

FIGURE 6: Lead and lead oxide recoveries from the smelting tests without a carbon addition.

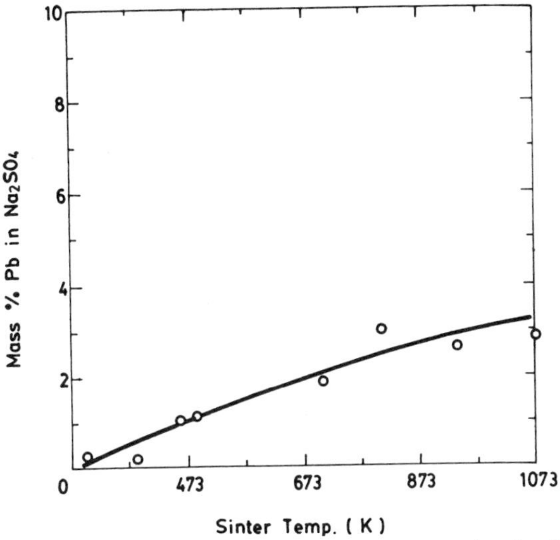

FIGURE 7: The effect of sintering temperature on the lead content of the sodium sulphate slag from the smelting tests without a carbon addition.

The effect of charcoal additions on the lead recovery and the lead and sulphur contents of the slag are shown in Figures 4 and 5 respectively. At high carbon contents, the lead recovery decreased and the lead and sulphur contents of the slag increased. This is due to the increased production of matte under the highly reducing conditions. On the laboratory scale the amount of matte is not sufficient to produce a separate layer and remains entrapped in the slag.

The results for the tests in which carbon was not added are summarized in Figure 6. For sintering temperatures below about 250°C the metal recovery was over ninety-five percent. Above this temperature, the metal recovery decreased rapidly to zero at about 400°C. This can be attributed to the removal of the organics in the residue at sintering temperatures over

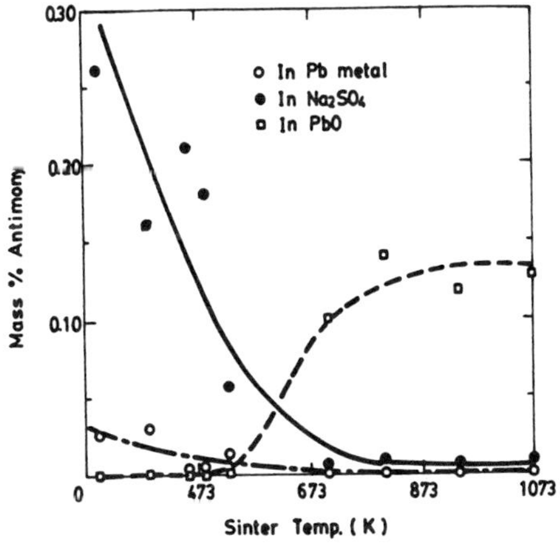

FIGURE 8: The antimony contents of lead metal, lead oxide, and sodium sulphate from the smelting tests without a carbon addition.

200°C. The lead content of the sodium sulphate is shown in Figure 7 and increases with sintering temperature to between 2.5 and 3.0 mass percent for sintering temperatures over 500°C. This is due to the increase in the oxidation potential during smelting at the higher sintering temperatures. The antimony, tin, and arsenic contents of the lead metal, lead oxide, and sodium sulphate slag are shown in Figures 8, 9 and 10 respectively. All three impurities exhibit similar trends. In the lead metal and the sodium sulphate slag, the impurity concentration decreases with increasing sintering temperature. On the other hand, the impurity content of the lead oxide phase increases with sintering temperature. Thus, in this process the sodium sulphate slag is relatively clean and the impurities concentrate in the lead oxide. Since the reducing potential is relatively low, the sulphide sulphur content of the sodium sulphate is less than 0.01 mass percent.

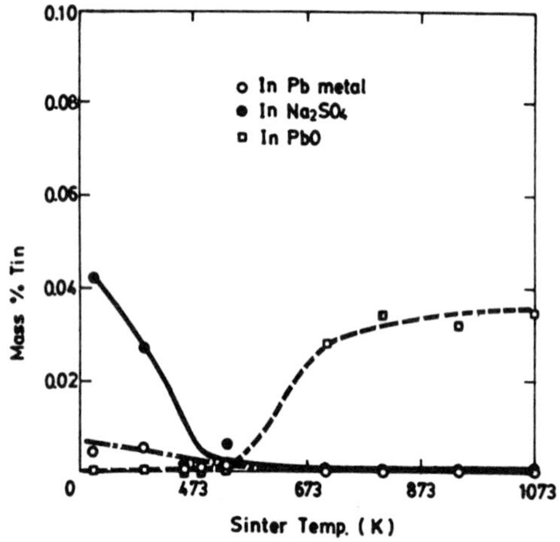

FIGURE 9: The tin contents of lead metal, lead oxide, and sodium sulphate from the smelting tests without a carbon addition.

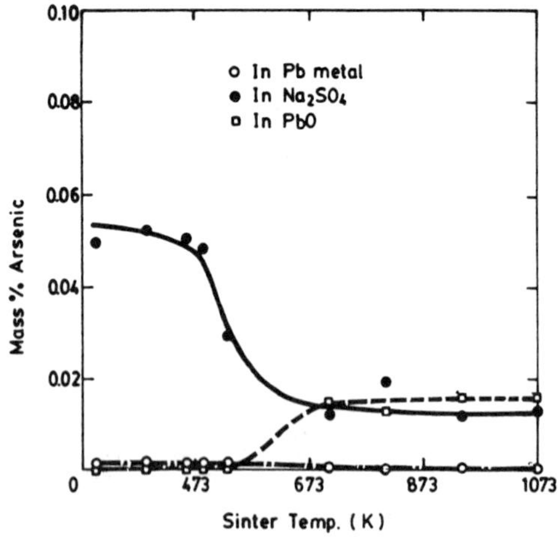

FIGURE 10: The arsenic contents of lead metal, lead oxide and sodium sulphate from the smelting tests without a carbon addition.

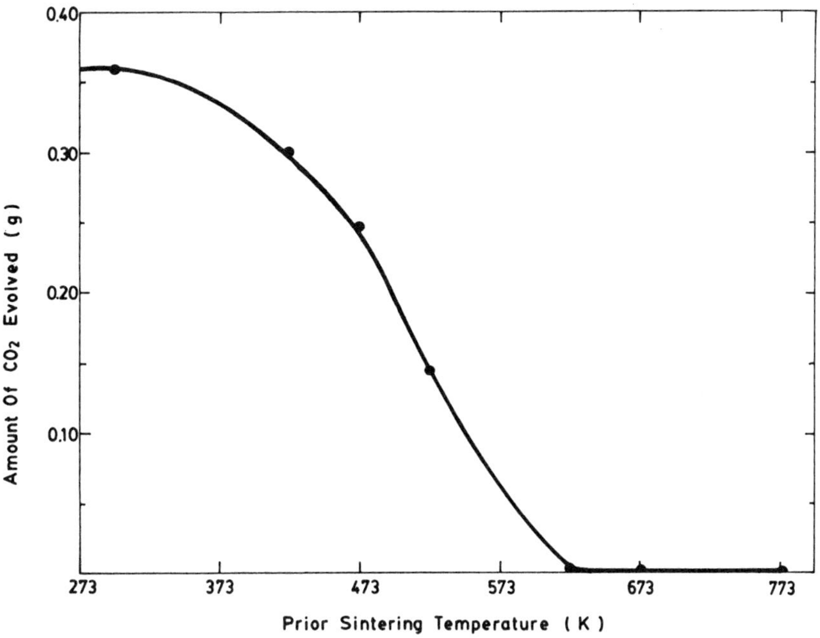

FIGURE 11: The effect of prior sintering temperature on the carbon dioxide evolved.

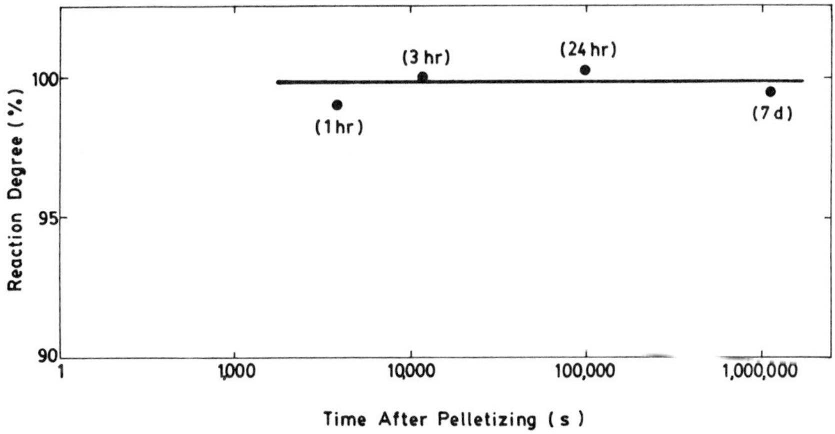

FIGURE 12: The effect of time after pelletizing on the reaction degree.

Reaction Degree

The effect of prior sintering temperature on the amount of carbon dioxide evolved is shown in Figure 11. A sintering temperature of about 350°C was required to remove the majority of the carbon dioxide. This corresponds to the decomposition temperature for lead carbonate which is about 315°C.

The effect of time after pelletizing on the reaction degree is shown in Figure 12. These results demonstrate that the degree of reaction is independent of time and is completed thirty minutes after pelletizing. This indicates that the exchange reaction occurs during the pelletizing operation, which is in agreement with the results of a study of the leaching of lead sulphate in a sodium carbonate solution by Arai and Toguri (7). They found that under the optimum leaching conditions a lead carbonate recovery of ninety-eight percent could be obtained in under thirty minutes at ambient temperatures.

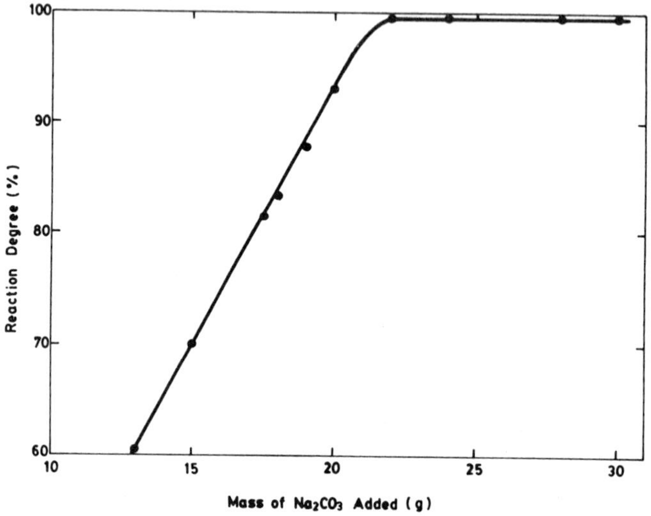

FIGURE 13: The effect of sodium carbonate additions on the reaction degree.

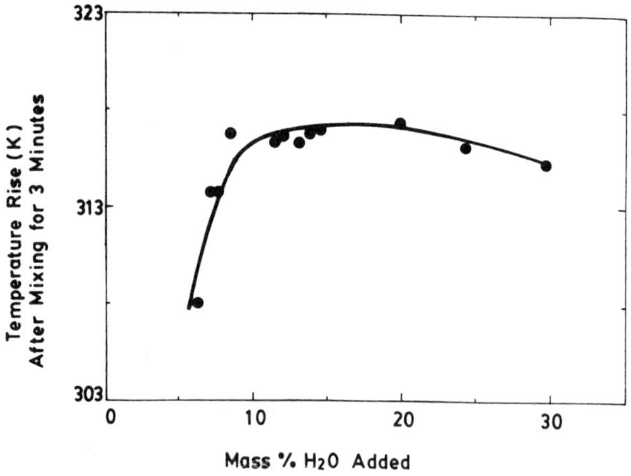

FIGURE 14: The temperature rise of a mixture of sodium carbonate and battery residue as a function of the amount of water added.

Figure 13 shows the reaction degree as a function of the amount of sodium carbonate added. As discussed previously a sodium carbonate addition of about twenty mass percent was required to obtain acceptable recoveries. This value is slightly in excess of the stoichiometric requirement. Also, sodium carbonate additions in excess of the stoichiometric do not generate excess carbon dioxide.

The temperature rise during the mixing of sodium carbonate, battery residue and water as a function of the amount of water added is shown in Figure 14. It can be seen that the maximum temperature rise of 44°C is achieved with a water addition of about ten mass percent. Below this value the exothermic exchange reaction:

$$PbSO_{4(aq)} + Na_2CO_{3(aq)} = Na_2SO_{4(aq)} + PbCO_{3(aq)} \quad (5)$$

is complete. At water additions of over twenty mass percent the amount of water is in excess and only serves to reduce the temperature rise.

Pellet Strength

The effects of drying time and sintering temperature on the percent retained on an 8 mesh screen after autogenous milling are shown in Figures 15 and 16 respectively. The abrasion resistance decreased with drying time. Also, the residue had a lower abrasion resistance than the pure lead sulphate. There was a minimum in the strength at about 500°C. Above this temperature some sintering occurred and thus the strength increased.

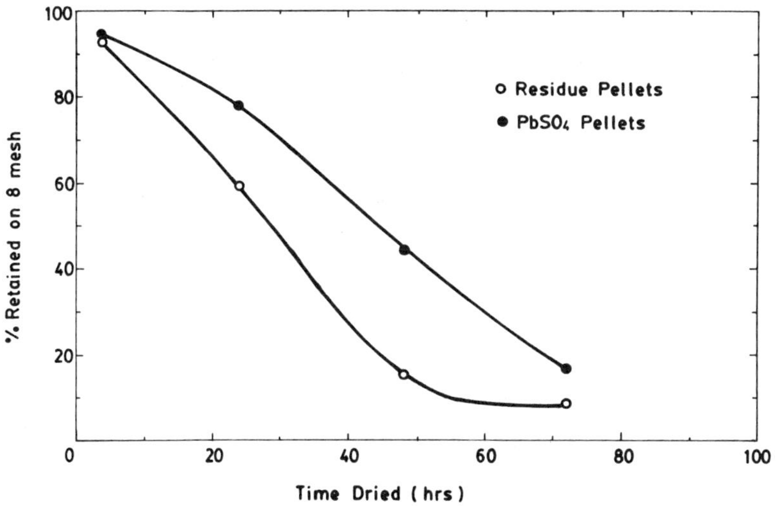

FIGURE 15: The effect of drying time on the amount of material on 8 mesh after autogenous milling for fifteen minutes.

The results for compressive strength demonstrated similar trends. Again, the strength decreased with sintering temperature, reached a minimum at about 500°C and then increased as some sintering occurred. As expected the compressive strength increased with pellet diameter. In general, the compressive strength of the pellets prepared from battery residue was less than that of the pure lead sulphate pellets.

The percent porosity in the pellets is shown in Figure 17. The porosity increased above the decomposition temperature of lead carbonate (315°C) and then decreased after about 550°C. These trends demonstrate that the decomposition of lead carbonate produces porosity in the pellet and this accounts for the decrease in strength with sintering temperature (Figure 16). At high temperatures the strength increases as sintering occurs and consequently the porosity decreases.

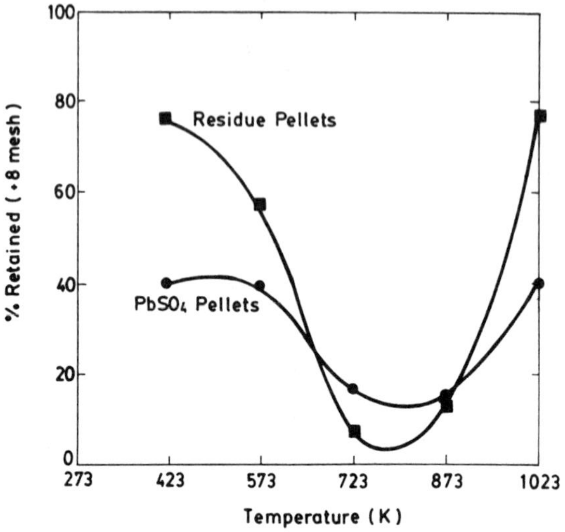

FIGURE 16: The effect of sintering temperature on the amount of material on 8 mesh after autogenous milling for fifteen minutes.

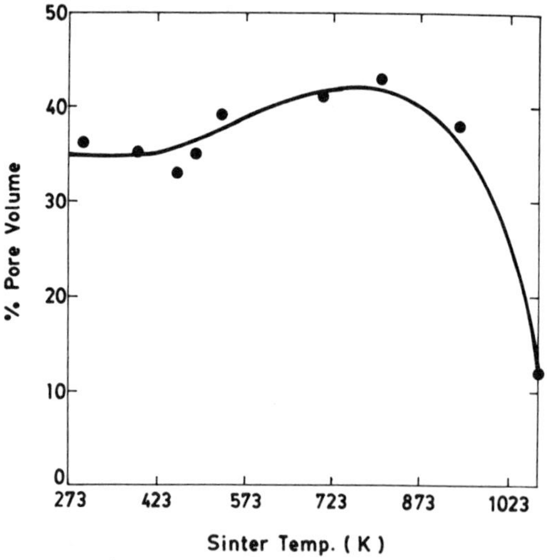

FIGURE 17: The percent pore volume of pellets made from residue as a function of sintering temperature.

CONCLUSIONS

(1) A pyrometallurgical process for the recovery of metallic lead was investigated on the laboratory scale. The overall reaction is:

$$PbSO_{4(s)} + Na_2CO_{3(s)} + C_{(s)} = Pb_{(l)} + Na_2SO_{4(l)} + CO_{(g)} + CO_{2(g)}$$

Lead recoveries of ninety-six to ninety-eight percent were achieved and the lead content of the slags was in the range of one to two mass percent.

(2) The results show that the exchange reaction:

$$PbSO_{4(aq)} + Na_2CO_{3(aq)} = Na_2SO_{4(aq)} + PbCO_{3(aq)}$$

is rapid and occurs in the presence of moisture during the pelletizing operation.

(3) In the absence of carbon and with sintering temperatures over 400°C pure lead oxide and relatively pure sodium sulphate can be produced.

(4) Both the compressive strength and the abrasion resistance of the pellets decreased with sintering temperature up to about 500°C. Above this temperature some sintering occurred and the strength increased.

ACKNOWLEDGEMENTS

The authors would like to thank Mr. R. Hickson, formerly of Canada Metal Co., Toronto for his technical assistance in this work. Also, Canada Metal provided the raw materials and the analyses. Financial support from the Natural Sciences and Engineering Research Council of Canada through the PRAI program is appreciated. The assistance of D. Smith in performing some of the experiments is acknowledged.

REFERENCES

1. R.D. Prengaman "Reverberatory Furnace - Blast Furnace Smelting of Battery Scrap at RSR", Lead-Zinc-Tin '80, TMS-AIME World Symposium on Metallurgy and Environmental Control, 1980, pp. 985-1002.
2. S. Bergsoe and N. Gram, "Lead Smelting, Refining and Pollution", Lead-Zinc-Tin'80, TMS-AIME World Symposium on Metallurgy and Environmental Control, 1980, pp. 1023-1030.
3. L.A. Lyons and D.B. Gillett, "The New Secondary Lead Smelter of SIMS Metal Pty., Ltd., Brooklyn, Victoria, Australia", Proc. Advances in Ext. Met., 1977, London, pp. 105-110.
4. R.C. Egan, M.V. Rao and K.D. Libsch, "Rotary Kiln Smelting of Secondary Lead", Lead-Zinc-Tin '80, TMS-AIME World Symposium on Metallurgy and Environmental Control, 1980, pp. 953-973.
5. E.R. Cole, A.Y. Lee and D.L. Paulson, "Recovery of Lead from Battery Sludge by Electrowinning", J. of Metals, 35, (8), (1983) pp. 42-46.
6. E.R. Cole, A.Y. Lee and D.L. Paulson, "Update on Recovery of Lead from Scrap Batteries", J. of Metals, 37, (2), 1985, pp. 79-83.
7. K. Arai and J.M. Toguri, "Leaching of Lead Sulphate in a Sodium Carbonate Solution", Hydrometallurgy, 12, 1984, pp. 49-59.
8. D.A. Wilson, "A New Sulfur Dioxide-Free Process For Recovering Lead From Battery Scrap", U.S. Bureau of Mines, R.I. 8123 (1976).
9. F.A. Acoveno and T.A. Fuediger, "Process for the Removal of Sulphur from Battery Wreckers Material Using Ammonium Carbonate Solutions", U.S. Pat. 3383348 (1975).
10. M. Yamada, N. Wakamatsu, and S. Nakano, "Lead Carbonate of High Purity from Crude Lead Sulphate", Japan Tokyo Koho, Japan Pat. 744819 (1974).
11. M.E. Elmore and J.K. Klang, "Process for Recycling Junk Lead-Acid Batteries", U.S. Pat. 4118219 (1978).
12. Diamond Engineering Co., Ltd., "DEC Lead Recovery System from Battery Scrap", (1977).
13. H. Soto and J.M. Toguri, "Pollution-Free Hydrometallurgical Process for The Recovery of Lead from Battery Scrap", Recycle and Secondary Recovery of Metals, edited by P.R. Taylor, H.Y. Sohn and N. Jarrett, AIME, Warrendale, Pennsylvania, 1985, pp. 257-273.
14. M. Kato, M. Imai, K. Arai and T. Izaki, "New Hydrometallurgical

Treatment Process of Lead Battery Slime", J. Japan Inst. Metals, Vol. 50, No. 7 (1986) pp. 631-639.
15. C.A. Pickles, D. Smith, V. Tomlinson and J.M. Toguri, "Sulphur Dioxide Emissions from Soda Ash Smelting of Scrap Lead-Acid Battery Residues", International Symposium on Primary and Secondary Lead Processing, 28th CIM Conference Symposia, August 1989, pp. 133-141.

Zinc Fuming from Slags - A Steady State Volatilization Model

V.L. PRABHU and R.G. REDDY

Department of Chemical and Metallurgical Engineering
University of Nevada, Reno
Reno, NV 89557

ABSTRACT

A steady state volatilization model was developed to understand the industrial zinc fuming process. The activity coefficient of ZnO, used in this model was derived by assuming a stepwise equilibrium during the entire fuming cycle. Industrial fuming operation slag composition data were used in this study. The activity coefficient of ZnO as a function of composition of slag and temperature was calculated. The rate of change of zinc in the bath was calculated using our newly developed steady state volatilization model. An excellent agreement between the calculated data and the industrial zinc fuming data was obtained.

Second International Symposium -
Recycling of Metals and Engineered Materials
Edited by Jan H. L. van Linden,
Donald L. Stewart, Jr., and Yogeshwar Sahai
The Minerals, Metals & Materials Society, 1990

INTRODUCTION:

Slag fuming is a batch process in which a reducing mixture of air and pulverized coal is injected into a bath of molten zinciferous slag. The liberated zinc, in vapor form oxidizes in contact with air above the bath. The products pass via a cooling system to a baghouse in which the zinc oxide is collected. Lead and tin also volatilize and collected in the baghouse with zinc.

Metallurgists first became interested in the recovery of zinc from slags around the turn of the twentieth century. But no great attempt was made to study the chemistry of the process, until recently. Bell, etal[1] published the first useful model of the slag fuming process in 1953. They assumed a state of equilibrium between the slag and reduced gas and calculated the activity of zinc oxide which they used for the prediction of the furnace performance. Quarm[2,3] considered that the assumption of equilibrium was unjustified and proposed that the rate of reduction of ZnO indicated a pseudo-first order reaction involving ferrous oxide present in the slag. He argued convincingly, that if the assumption of equilibrium was true, the slope of the zinc vs time curve could be altered at will by changing the CO/CO_2 ratio. But no such effect could be detected experimentally.

Thereafter Kellogg[4,5] came up with the first computer model of the slag fuming proces. Kellogg assumed that the slag-gas equilibrium conditions postulated by Bell etal[1] contributed to the rate control. Kelloggs' method of assuming stepwise equilibrium facilitated the use of a computer. But Kellogg made several assumptions, primarily due to lack of consistent data. Since fuming raters are quite high, γ_{ZnO} should change appreciably during the process. Kellogg assumed ZnO slags obey regular solutions behavior with $\gamma_{ZnO} = 1.85$ at 1500K. This value was chosen to make the model agree better with practical results.

None of the above theories have been able to explain the slag fuming process convincingly[6,7]. In this paper, the steady state volatilization of the product Zn vapor is considered to be the rate controlling step. This procedure has been used with great success to explain the volatilization of Bi from the copper converter by Nagamori and Chaubal[8].

STEADY STATE VOLATILIZATION MODEL:
MODEL DEVELOPMENT:

Assume a system consisting of slag and gas phase under equilibrium conditions. Let $[Zn]_s$ denote the concentration of zinc in the slag phase. Assume further that the concentration of the minor elements is sufficiently low so that the partial pressures of zinc containing gases, P^T_{Zn} is proportional to the concentration of zinc in the slag phase.

Hence,

$$P^T_{Zn} = K_{Zn}\,[Zn]_s \tag{1}$$

where K_{Zn} is a constant.

When the condensed phases are flushed out with a small volume of neutral gas, Δvg, is such a way that the gas will be saturated with zinc, we have

$$P_{Zn}^{T} = \frac{\Delta V_{Zn}}{\Delta V_{Zn} + \Delta V_g} \approx \frac{\Delta V_{Zn}}{\Delta V_g} \qquad (2)$$

where, ΔV_{Zn} is the total volume of zinc containing gases.

The weight of zinc, ΔW_{Zn} grams contained in volume ΔV_{Zn} liters (S.T.P) is given by

$$\Delta W_{Zn} = A_{Zn}\, \Delta V_{Zn}/22.4 \qquad (3)$$

where A_{Zn} is the atomic wt of zinc.

The difference in mass balance of zinc before and after volatilization of ΔW_{Zn} grams may be expressed as

$$\Delta W_{Zn} = -\ 0.01\ \Delta[Zn]s\ Ws \qquad (4)$$

where Ws is weight of the slag.

Substituting (1), (2), (3) into (4)

$$\frac{A_{Zn}}{22.4}\ (\ \Delta V_g\ K_{Zn}\ [Zn]s\ ) = -0.01\ \Delta[Zn]s\ Ws \qquad (5)$$

$$\Delta V_g = -\ \frac{0.224}{A_{Zn}\ K_{Zn}}\ Ws\ \frac{\Delta[Zn]s}{[Zn]s}$$

$$\frac{dV_g}{d[Zn]s} = -\ \frac{0.224\ Ws}{A_{Zn}\ K_{Zn}\ [Zn]s} \qquad (6)$$

Integrating (6) from initial conditions

$[Zn]_s^i$ at $Vg = 0$ to $[Zn]_s^f$ at $V_g = V_g$

$$Vg = \int_i^f \frac{-0.224\ Ws}{K_{Zn}\ A_{Zn}}\ \frac{d[Zn]s}{[Zn]s} = \frac{-0.224}{K_{Zn}\ A_{Zn}}\ Ws\ \ln\left\{\frac{[Zn]_s^f}{[Zn]_s^i}\right\} \qquad (7)$$

Equation (7) can be written as

$$[Zn]_s^i = C_{Zn}\ [Zn]_s^f \qquad (8)$$

$$\text{where } C_{Zn} = \exp\left\{ \frac{K_{Zn}\, A_{Zn}\, V_g}{0.224\, W_s} \right\} \tag{9}$$

and $K^*_{Zn} = K_{Zn}\, A_{Zn}$, is the volatilization constant. This volatilization constant K^*_{Zn} must contain all the important gaseous species $Zn_{(g)}$, $ZnO_{(g)}$ and $ZnS_{(g)}$

EVALUATION OF K^*_{Zn}

Consider the equilibrium reaction

$$ZnO_{(s)} = Zn_{(s)} + 1/2\ O_{2(g)} \tag{10}$$

$$K_{10} = \frac{a_{Zn}\, PO_2^{1/2}}{a_{ZnO(s)}}$$

$$a_{Zn} = K_{10}\, a_{ZnO}\, Po_2^{-1/2}$$

For the equilibrium reaction,

or

$$ZnS_{(g)} = Zn_{(g)} + 1/2\ S_{2(g)} \tag{11}$$

$$K_{11} = \frac{a_{Zn(s)}\, PS_2^{1/2}}{a_{ZnO(s)}}$$

$$a_{ZnS} = \frac{a_{Zn}\, PS_2^{1/2}}{K_{11}}$$

By substituting value of a_{Zn} from (10)

$$a_{ZnS} = \frac{K_{10}\, a_{ZnO}}{K_{11}} \left(\frac{PS_2}{PO_2} \right)^{1/2}$$

consider

$$P_{Zn} = P^o_{Zn}\, a_{Zn}$$

$$P_{ZnO} = P^o_{ZnO}\, a_{ZnO}$$

$$P_{ZnS} = P^o_{ZnS}\, a_{ZnS}$$

and

$$P^T_{Zn} = P_{Zn} + P_{ZnO} + P_{ZnS}$$

$$= P^o_{Zn}\, K_{10}\, a_{ZnO}\, PO_2^{-1/2} + P^o_{ZnO}\, a_{ZnO}$$

$$+ P^o_{ZnS} \frac{K_{10}}{K_{11}}\, a_{ZnO} \left(\frac{Ps_2}{PO_2}\right)^{1/2}$$

$$P^T_{Zn} = a_{ZnO} \left[P^o_{ZnO} + P^o_{ZnS} \frac{K_{10}}{K_{11}} \left(\frac{PS_2}{PO_2}\right)^{1/2} + P^o_{Zn}\, K_{10}\, (PO_2)^{-1/2} \right] \tag{12}$$

From above equation we have

$$K_{Zn} = \frac{P^{T}_{Zn}}{[Zn]_s}$$

or

$$K^{*}_{Zn} = \frac{P^{T}_{Zn}\ A_{Zn}}{[Zn]_s} \qquad (13)$$

Substituting P^{T}_{Zn} from (12) into (13), equation (14) is obtained

$$K^{*}_{Zn} = \frac{a_{ZnO}\ A_{Zn}}{[Zn]_s} \left[P^{o}_{ZnO} + P^{o}_{ZnS} \left(\frac{PS_2}{PO_2}\right)^{1/2} \frac{K_{10}}{K_{11}} + \frac{P^{o}_{Zn}\ K_{10}}{PO_2^{1/2}} \right] \qquad (14)$$

Substituting this value of K^{*}_{Zn} in equation (9) one obtains

$$C_{Zn} = \exp\left[\frac{a_{ZnO}\ A_{Zn}\ V_g}{[Zn]_s\ 0.224\ W_s} \left[P^{o}_{ZnO} + P^{o}_{ZnS}\frac{K_{10}}{K_{11}} \left(\frac{PS_2}{PO_2}\right)^{1/2} + \frac{P^{o}_{Zn}\ K_{10}}{PO_2} \right] \right] \qquad (15)$$

and also $[Zn]^{f}_{s} = C_{Zn}\ [Zn]^{i}_{s}$ (16)

Therefore, knowing C_{Zn} and $[Zn]^{i}_{s}$ at each microstep, $[Zn]^{f}_{s}$ can be calculated.

EVALUATION OF ACTIVITY OF ZnO:

The activity of ZnO in derived here by assuming a stepwise equilibrium process between molten slag and the reducing gases. By knowing the composition of the slag at different times, six simultaneous equations as shown in Table 1, which are assumed to cover all the major reactions, are solved. This gives the partial pressures of the gaseous species present in the slag. Components in the slag other than Zinc oxide are assumed to observe regular solution behavior. Then by knowing the instantaneous slag composition, bath temperature, input rates of coal and air and the rate of zinc removal, the activity of zinc oxide in the slags can be determined.

The loss of zinc, lead and sulfur from the bath, the input of coal ash, and the change in ferrous/ferric ratio in the slag cause a continuous variation in the mass of liquid slag, DST Kg/min. The molar rate of change of zinc (MZN) in the bath in given by:

$$MZN = \frac{WSL \times DZDT}{100 \times 63.57} + \frac{DST \times (\%ZnO)}{100 \times 81.37}\ \text{Kgmole/min} \qquad (17)$$

where, WSL is the estimated liquid bath mass and DZDT is the slope of the %Zinc oxide vs. Time curve.

Similarly, the equation for lead and sulfur are

$$MPB = \frac{WSL \times DLDT}{100 \times 207.19} + \frac{DST \times (\%PbO)}{100 \times 223.19} \qquad (18)$$

and

$$MS = \frac{WSL \times DSDT}{100 \times 32.064} + \frac{DST \times (\%CaS)}{100 \times 72.14} \qquad (19)$$

The rate of change of slag bath mass is given by:

$$DST = MZN \times 81.37 + MPB \times 223.19 + MS \times 16.07 \qquad (20)$$

The decrease in slag man is DST x 10.0, based on the equilibrium constants calculated from Table 1. The partial pressures of the gaseous species CO_2, CO and S are determined as follows:

$$\frac{P_{CO_2}}{P_{CO}} = \frac{K_{(4)} \times a_{Fe_3O_4}}{a_{FeO}} \qquad (21)$$

$$P_{CO} = K_{(1)} \times P_{CO_2}/P_{CO} \qquad (22)$$

$$P_{S_2} = (a_{CaS} \times a_{Fe_3O_4} \times K_{(6)})/(a_{CaO} \times a^3_{FeO}) \qquad (23)$$

The initial total molar gas rate is

$$MGR = CIN + NIN + HIN - MZN \qquad (24)$$

where, CIN; NIN and HIN are the input rates of carbon, Nitrogen and hydrogen.

Then the partial pressure of zinc P_{Zn} can be calculated as

$$P_{Zn} = - \frac{MZN}{MGR} \qquad (25)$$

By combining equation 21 to 25 and reaction (2), the activity of zinc oxide can be obtained by the expression as shown in equation (26).

$$a_{ZnO} = \frac{P_{Zn} \times P_{CO_2}}{K_{(2)} \times P_{CO}} \qquad (26)$$

RESULTS AND DISCUSSIONS:

The model was tested on actual fuming plant data obtained from studies reported by Grant[9]. The operating conditions in the plant and the slag composition at various intervals of time are reported. A FORTRAN program was written to evaluate the activity coefficient of zinc oxide based on the earlier derivations.

The steady state volatilization model is an iterative procedure, thereby facilitating the use of a computer. The calculated values of activity coefficients as function of slag composition were used. The activity coėfficient of zinc oxide as a function of composition of zinc oxide in slag is shown in fig 1.

During the application of the model, steady state conditions were assumed to exist in the bath for 0.1 minute. Given the initial composition of ZnO in the bath, the model predicts the zinc oxide composition during the entire fuming process.

The model results along with the observed composition of zinc oxides are given in fig 2. It is seen that the calculated data is in fair agreement with the industrial data. It must be mentioned here that the same model did not predict as accurately for another set of data from the same plant. During this operation, 50% solid slag was charged initially with molten slag and the model is not formulated to accommodate such derivations. Development of a model to incorporate above parameters is progress.

ACKNOWLEDGEMENTS:

The authors wish to acknowledge the financial support of the United States Bureau of Mines under the Mining and Minerals Research Institute Generic Center program (grant number G1125132-Mineral Industry Waste Treatment and Recovery Generic Center.

REFERENCES

1. Bell, R.C; Turner, G.H; Peters,E: Journal of Metals, 1955, pp 472-477
2. Quarm, T.A.A; Mining Magazine, 1965, vol.113, No 2, pp. 114-121
3. Quarm, T.A.A: Eng. and Min. Journal Jan, 1968, pp. 92-93
4. Kellogg, H.H: Trans-AIME, 1967, vol 239, pp 1439-1449.
5. Kellogg,H.H; Eng. and Min. Jorn 1957, vol 158, No 3. pp. 90-92
6. Ruddle R.W; Inst. of Min and Met London. 1953
7. Richards, A.W; Canad. Met. Quart, 1981, vol 20, No 2, pp. 145-151.
8. Nagamori M.; Chaubal R.C; Met. Trans, 1982, vol 13B, pp. 339-348
9. Grant, R,M: Australia-Japan extractive Metallurgy Symposium, Sydeney, Australia, 1980
10. D.R.Stull and H.Prophet; JANAF Thermochemcal Tables, 2nd ed., NSRDS-NBS 37, Supt. of Documents, Washington, D.C., 1971 and Supplements.

Table 1. Chemical Reactions and Equilibrium Constants(10)

S.No	Reaction	*ln(K) = A/T + B A	B
1.	$CO_{2(s)} + C_{(s)} \rightleftharpoons 2CO_{(g)}$	-20533.4	+20.985
2.	$ZnO_{(s)} + CO_{(g)} \rightleftharpoons Zn_{(g)} + CO_{2(g)}$	-22250.0	+13.919
3.	$\frac{1}{2} S_{2(g)} + 2CO_{2(g)} \rightleftharpoons 2CO_{(g)} + SO_{2(g)}$	+14853.0	-4.423
4.	$Fe_3O_{4(s)} + CO_{(g)} \rightleftharpoons 3FeO_{(s)} + CO_{2(g)}$	-2290.0	+2.612
5.	$Pb_{(g)} + \frac{1}{2} S_{2(g)} \rightleftharpoons PbS_{(g)}$	+15689.0	-4.884
6.	$(CaO)_{(s)} + Fe_3O_{4(s)} \rightleftharpoons CaO_{(s)} + 3FeO_{(s)} + \frac{1}{2} S_{2(s)}$	-25598.0	+12.985

* where K is the equilibrium constant and T is the absolute temperature in degree K

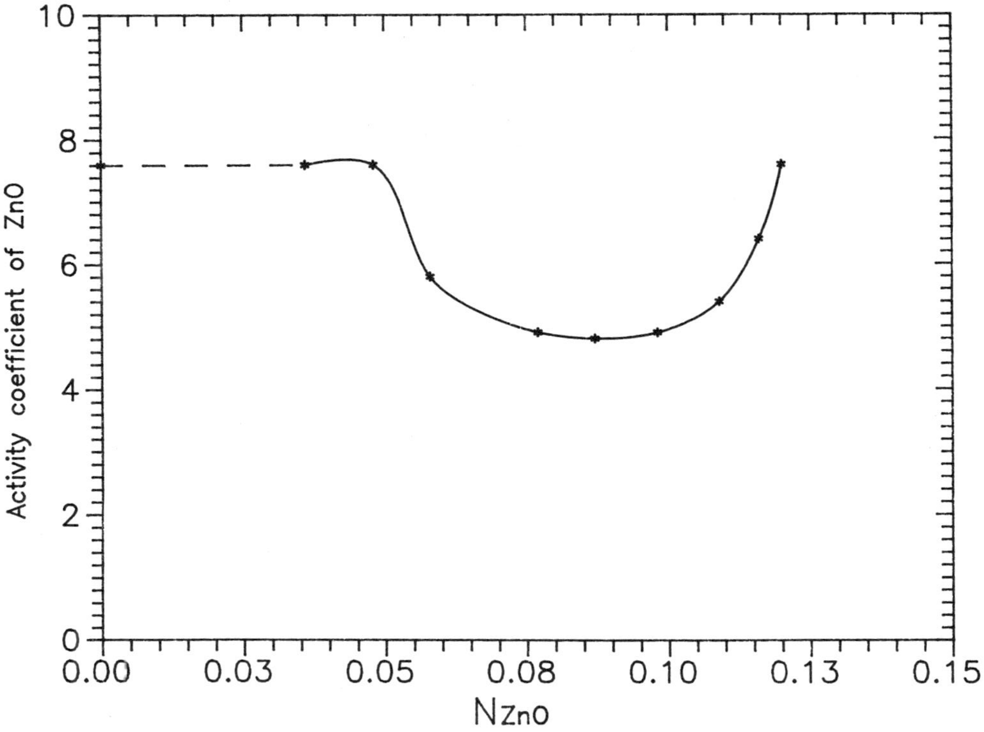

Fig 1. Activity coefficient of ZnO as a function of mole fraction of ZnO in slags.

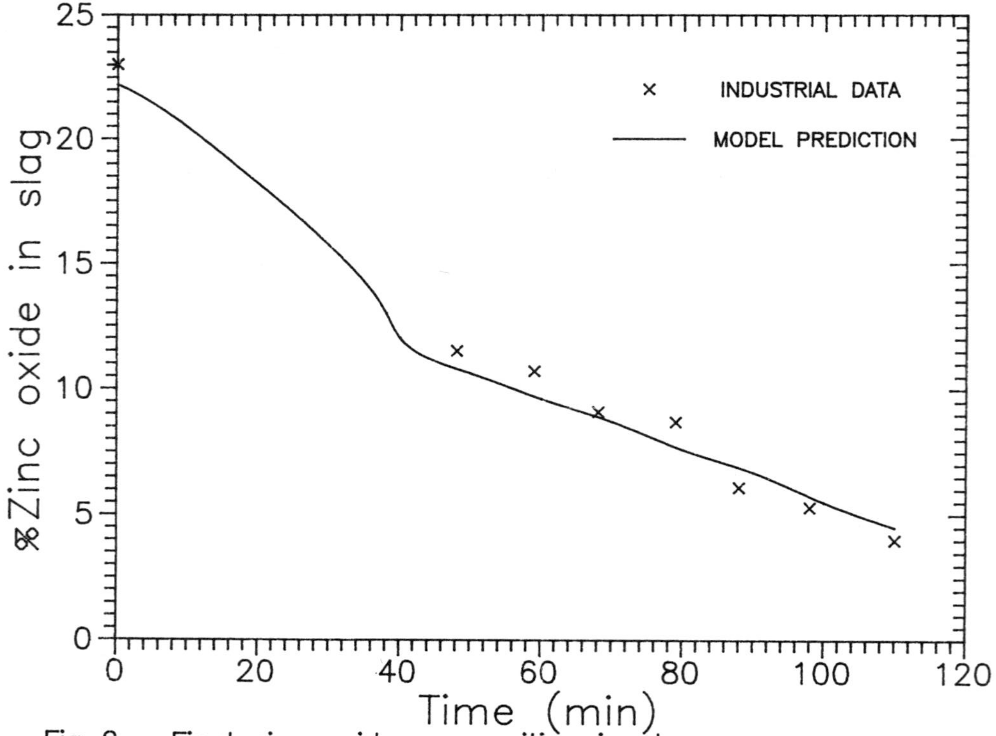

Fig 2. Final zinc oxide composition in slags as a function of time

IMPROVED EFFICIENCY IN THE RECOVERY OF TIN FROM RECYCLED LOW GRADE RESIDUES IN THE CAPPER PASS SULPHIDE FUMING FURNACE

A.M. Strachan and I.A. Linton

Capper Pass and Son Limited
North Ferriby
North Humberside, England

Abstract

Capper Pass and Son Limited has for many years operated sulphide fuming furnaces to recover tin along with other volatile metals from recycled low grade slags, fumes and residues. In 1985 a major development program was commenced to improve the operating efficiency of the Company's 60,000 tpa water-jacketed fumer without interfering with production constraints. The primary objectives of the program were to raise energy and sulphur efficiency and increase the proportion of recycled material which could be incorporated into the furnace charge. This paper describes the factors investigated and highlights the salient improvements achieved during the successful program including a substantial increase in plant throughout which allowed the closure of two smaller refractory lined fuming furnaces and a 50% increase in pyrites efficiency.

Second International Symposium -
Recycling of Metals and Engineered Materials
Edited by Jan H. L. van Linden,
Donald L. Stewart, Jr., and Yogeshwar Sahai
The Minerals, Metals & Materials Society, 1990

Introduction

Capper Pass and Son Limited is a leading processor of complex low grade tin concentrates and recycler of complex low grade tin bearing residues. In addition to tin, feed materials contain varying amounts of lead, antimony, bismuth, copper and precious metals, all of which are recovered in the metal phase. Capper Pass and Son Limited produce 11,000 t.p.a. of electrolytic tin (99.93% min.) 6,000 t.p.a. of electrolytic lead 600 t.p.a. of electrolytic copper and varying amounts of antimony, bismuth, silver, gold and indium. The Capper Pass operation treats 50,000 t.p.a. of raw material, approximately 60% of which is secondary material.

The complex process flowsheet of the installation consists of a sintering operation, to agglomerate the concentrates and fine particle sized residues, followed by a blast furnace to smelt the agglomerated feed along with those residues fed directly into the furnace. The metal product from the blast furnace is dry refined prior to electrorefining to a tin product of minimum purity 99.93% Sn. Higher purity tin is also produced in the refinery. The tin refinery slimes which collect the lead, antimony and precious metals are then smelted in a rotary furnace to an impure lead which is dry refined and then electrorefined. The precious metals are recovered from the lead refinery slimes.

The slag produced from the initial blast furnace smelting operation contains on average 5-7 wt.% tin. This tin is economically recoverable by fuming the slag in a sulphide fuming process. Collected tin, lead and antimony bearing fume is recycled to the sinter plant for agglomeration.

In total there are nine possible entry points for the various types of raw material treated by the Capper Pass smelter including the sintering operation, blast furnace smelting and fuming operations. Each of these possible flow routes is accurately costed and, based upon the raw material assay and physical form, the most cost effective process route can be selected. This costing information is a major factor in assessing the treatment charges levied for a particular material offered for sale to the Company.

The fuming operation has been carried out for many years at the Capper Pass site, initially in refractory lined furnaces and more recently in a water cooled jacketed furnace [1, 2]. The integrity of the water cooled jackets and tuyeres has been substantially developed and improved over several years, both in design detail and in fabrication techniques, and the evolution of a reliable pyrometallurgical plant was obvious fertile ground to critically examine and then improve the efficiency of the entire fuming operation. In 1985 a major development programme was commenced with the prime aims of raising the fuming, energy and sulphur efficiency and allowing a substantial increase in the proportion of purchased low grade residues which could be incorporated into the furnace charge. Because the processing unit was a vital component of the Company's flowsheet, extreme caution and awareness was required at all major stages of the development programme to ensure no adverse effects occurred to the Company's operation.

In total 1,200 cycles of the fuming furnace were scrutinised over a two year period during which numerous parameters were altered and the effects monitored.

This paper describes only some of the major aspects of the operation investigated and a summary of improvements achieved during the development programme.

General Description of the Capper Pass Fuming Furnace

The tin sulphide fuming reaction may be written as:

$$\underset{(Slag)}{SnO} + \underset{(Slag)}{FeS} - \underset{(Gas)}{SnS} + \underset{(Slag)}{FeO} \qquad (1)$$

Tin sulphide, being volatile, fumes out of the slag bath, is oxidised in the flue system to tin oxide and the fume collected.

The Capper Pass fumer is operated on a continuous basis as a batch process. A blend of granulated slag and low grade residues along with a known addition of iron pyrites, are fed via a timed pan conveyor into the furnace through a charge port at the top of the furnace. This charge port also serves as the secondary air inlet to oxidise the fumed tin sulphide to tin oxide prior to cooling and collection. The furnace is fuelled by heavy fuel oil fed through thirty-two water cooled tuyeres. The falling charge is first heated by the rising combustion products and then melts in the fluid slag bath maintained over the tuyeres. The combustion products rising through the fluid bath create a high degree of turbulence encouraging tin fuming.

After approximately 110 tonnes of tin bearing material has been fed to the furnace along with the required addition of iron pyrites, charging is stopped, the oil burning rate is reduced and the tin content of the bath is allowed to fume down to the discard level. This period of the cycle is referred to as the blowdown. This discard level of tin in slag is derived from a small computer program based on oil price, tin price etc., and currently is running at 0.35 wt.% Sn.

When the desired level of tin in slag is reached, the contents of the furnace are tapped in approximately 15 minutes, air cooled and sold as landfill, etc.

Fume and combustion product gases are pulled off the top of the furnace by a 196,000 Acfm induced draught fan, cooled through an evaporative cooling tower and the fume collected in a large baghouse with 37,500m^2 of cloth filtration area. The fume containing the tin, lead, antimony and zinc is collected and wetted to 12% moisture in an intensive mixer to render it suitable for satisfactory handling. Collected fume is recycled to the sinter plant for agglomeration.

The furnace hearth, approximately elliptical in shape, has an area of 14.2 metres^2 surrounded by water cooled mild steel jackets, each 1.95 metres high rising to the feed port at a height 12.6 metres above the hearth. Water cooled tuyeres of 30mm internal diameter are inserted through the lower level of jackets, approximately 150mm above the hearth and penetrate 250mm beyond the internal face of the jacket. The furnace hearth has a single keyed layer of Corhart 104 refractory bricks under the lower ring of jackets which projects 100mm up the inside jacket face. In-board of this refractory brick ring, cast alumina constitutes the remainder of the hearth. The hearth is also water cooled by inserted cooling tubes.

Jacket life is largely dependent upon the position it occupies within the furnace, those occupying the most aggressive area, around the tuyeres and immersed in the slag, now have an average life of 2 - 2.5 years. Tuyere life is 12-15 months and are refurbished merely by replacing the tuyere nose area. Tuyere service life is diligently recorded and nose condition is monitored by X-radiography. Tuyere failure mode is now largely by thinning of the inner curved area of the nose tip.

Slag Bath Temperature Monitoring

It was considered at the start of the development programme that a key factor in improving control of the furnace would be to develop the ability to continuously monitor the slag bath temperature. The use of infra-red techniques was precluded by the high fume density above the bath. Considerable effort was directed to developing a thermocouple sheath material which could both withstand the mechanical and thermal shock of insertion through one of the thirty-two tuyeres, project beyond the tuyere tip and survive in the aggressive conditions prevailing in the bath. Eventually by altering design and materials of construction the average life of the sheath was extended to two months. This was achieved using an iron-chrome alloy (60% Fe - 28% Cr) of wall thickness of 7.3mm. The thermocouple and sheath project 90mm beyond the tuyere tip and are supported by a 75mm surrounding shroud welded on to the end of the tuyere. The reliability of the direct bath measurement was proved by simultaneous measurement using sacrificial dip thermocouples.

The sensitivity of the inserted thermocouple to changes in bath temperature was excellent as illustrated by a typical trace reproduced in Figure 1. The zig-zag response reflects temperature changes resulting from the aliquots of feed material charged to the furnace. As might be expected, temperature variations at the start of the charging cycle are greater than at the end of the charging cycle. The end of the charging cycle is clearly illustrated and the steady temperature rise during the blowdown period is shown. Temperature rise during the blowdown was restricted to 100°C, i.e. a rise from 1100°C to 1200°C.

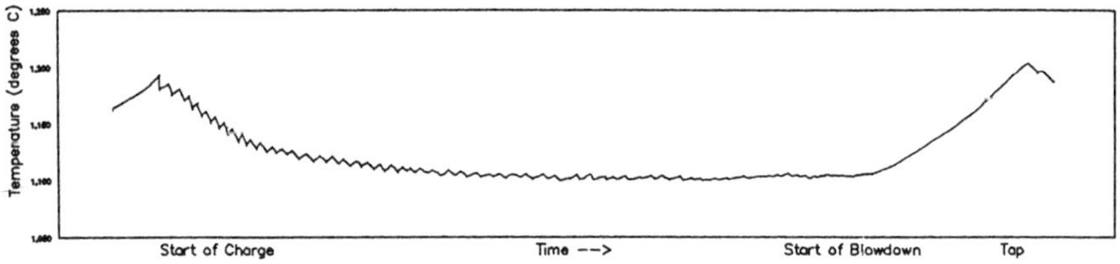

Figure 1: Typical Bath Temperature Profile During Fuming Cycle

The ability to monitor the bath temperature allowed the determination of the most important parameters in the fuming process. However, as one of the major aims of the program was achieved, i.e. a substantially increased proportion of low grade residues in the furnace charge, the reliability of the sheath material decreased. This aspect was investigated and found to be a consequence of the increasing copper content in the bath arising from the residues attacking the Fe/Cr alloy. Little could be done to minimise this problem but efforts are still continuing towards again producing a reliable sheath material as the ability to directly measure the slag temperature is still considered to be a major control benefit.

Efficiency of Pyrites Usage

With some 1,200 furnace fuming cycles investigated during this program it is clearly impractical to reproduce the data obtained. Therefore the data is presented largely in graphical form indicating relationships where established.

A period of four months fuming operation was closely monitored prior to varying any parameter in order to establish a base-line condition. During this observation period the pyrites consumption was 220 kgs/tonne of tin bearing feed which is approximately 3.5 times the theoretical pyrites requirement and represents poor use of the reagent.

The sulphur and tin profiles through the course of several cycles were determined by taking dip samples every hour for subsequent assay. Figure 2 reproduces fairly typical profiles which show a steady increase in sulphur content through the course of the charging cycle up to a maximum at completion of the charging period. (The slight climb of the tin from the discard level to a plateau during this same period will be discussed later in the text.) Clearly there was a considerable excess of sulphur contained in the bath at the commencement of the blowdown period which was merely combusted to sulphur dioxide during this period. Alterations to the amount of pyrites added were made to maintain the sulphur level in slag at around 2 wt % at all times throughout the cycle. Small quantities of pyrites were also added during the blowdown period to sustain this level. Figure 3 shows a typical sulphur profile obtained under the current operating regime where pyrites consumption is 100 kgs/tonne or approximately 1.5 times the theoretical re-quirement. This major improvement in pyrites efficiency represents a substantial increase in the fuming capacity of the furnace. In fact the furnace capacity during this entire investigation was increased from 90 to 110 tonnes of tin bearing charge.

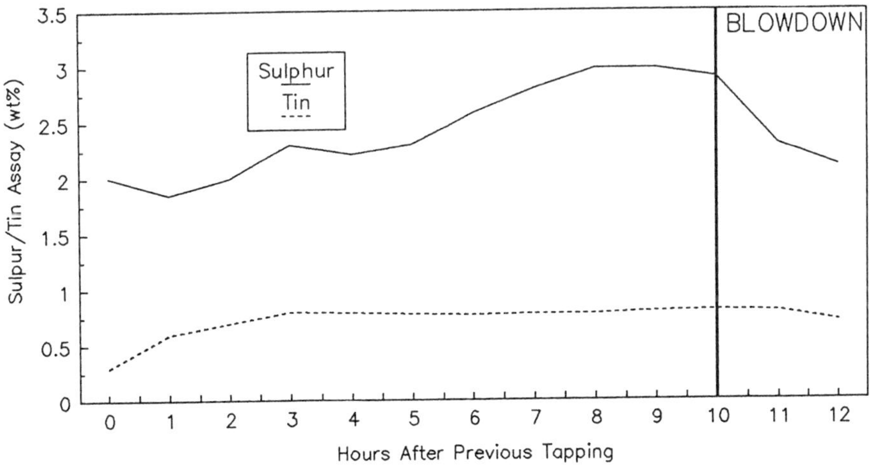

Figure 2: Typical Slag Bath Sulphur and Tin Profiles at the Start of the Program

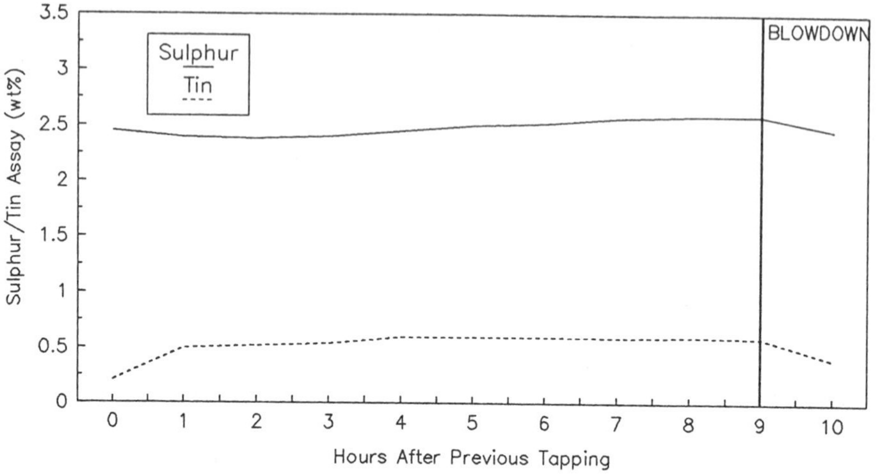

Figure 3: Typical Slag Bath Sulphur and Tin Profiles Under Current Operating Conditions

A series of experiments using the addition of coal to further increase the pyrites efficiency (by reducing the combustion of FeS to FeO) gave no significant improvements and this aspect of the investigation was discontinued.

Although rarely available in a suitable form, other sulphide bearing minerals can and have been used successfully as a substitute for iron pyrites.

Effect of Feed Moisture Content on Charging Rate

Figure 4 shows the furnace charging rate achieved as a function of the moisture content of the feed at a constant oil burning rate and bath temperature. Not surprisingly the charging rate decreases as the moisture content of the feed increases. The observed increase in feed rate with decreasing moisture content of the feed is in good agreement to that which can be calculated, assuming the energy input to evaporate the moisture at 1100^0 is used to melt solid slag.

Increasing the proportion of purchased low grade tin residues (a major aim of the investigation) usually results in an increase in the moisture content of the charge with a consequential increase in cycle charging time and therefore a sensible balance requires to be struck. During the course of the investigation the typical proportion of recycled material incorporated into the furnace charge was increased from 10-15% to 30-35%.

To date the expected return on capital investment and the inevitable hygiene problems has not justified the inclusion of a drier in the feed system.

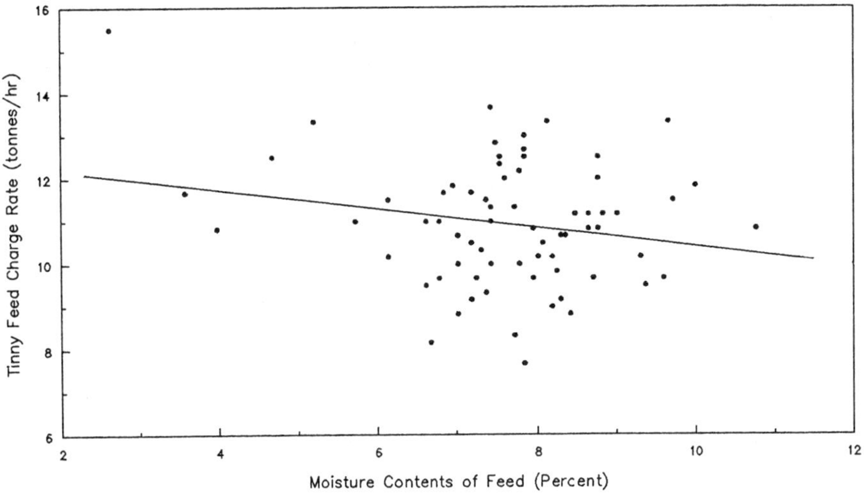

Figure 4: Effect of Moisture on Charging Rate

Tin Fuming Rate

(a) During the Charging Period

Figure 5 shows the substantial effect of increasing the oil burning rate on the furnace charging rate keeping bath temperature and feed moisture constant. The oil rate could not be increased further because of the limitations of the existing screw compressors.

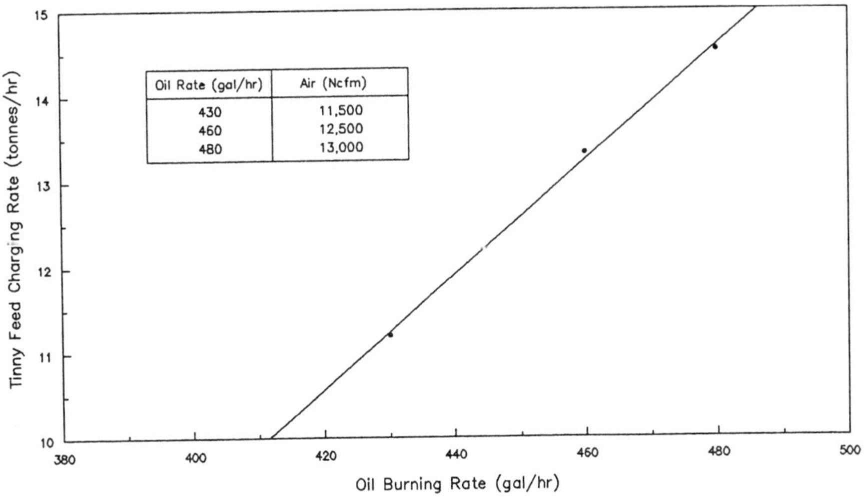

Figure 5: Effect of Oil Burning Rate on Charging Rate

Figures 2 and 3 reproduced earlier in the text to illustrate typical bath sulphur profiles during the course of the investigation also included data on the tin concentration profiles during fuming cycles. The tin level rises relatively quickly from the previous tapping value (0.35 wt %) up to a plateau value sustained throughout the charging period. This phenomenon was observed on all cycles investigated. The two cycles reproduced in Figures 3 and 4 were obtained at the same bath temperature but at very different conditions of contained bath sulphur concentration. It can be concluded that provided a minimum bath sulphur concentration is maintained (1.8 - 2.0 wt % in Capper Pass experience) for a given level of tin in feed the rate of tin fuming is constant during the charging period.

Figure 6 shows the fuming rate (expressed as tonnes of tin fumed per hour) during the plateau period as a function of the tin content of the feed charged to the furnace at constant oil and air rate (430 gal/hr and 11,500 Ncfm) and bath temperature. Clearly the fuming rate increases linearly with increasing tin content of the furnace charge. Implicit in this relationship is the conclusion that under the specific boundary conditions prevailing in the Capper Pass fumer the tin fuming reaction illustrates first order behaviour.

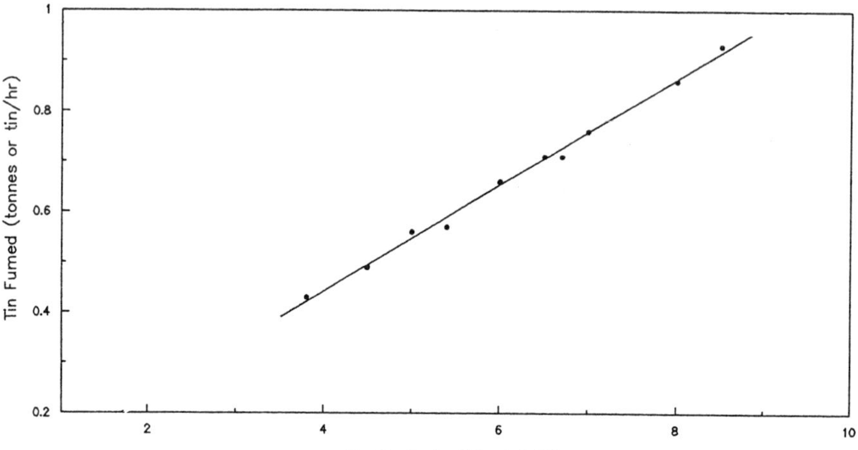

Figure 6: Effect of Tin Content of Feed on Tin Fuming Rate

Considering the concept of fuming intensity defined by Wright[3] as the blowing rate per tonne of slag, clearly this method of comparing fuming furnaces must be used judisciously as the Capper Pass furnace fuming intensity will decrease as the cycle progresses. The Capper Pass fumer intensity could be increased by reducing the weight of charge per cycle but this would serve no useful purpose since the fuming rate has been shown to be constant during charging. In fact the fuming capacity would effectively be reduced due to additional blowdown periods and furnace tappings.

The instantaneous fuming efficiency is defined by Denholm et al[4] as the weight of tin volatilised per unit of sweeping gas blown per percent of tin in the slag bath. At an oil rate of 480 gal/hr and an air rate of 13,000 Ncfm for a typical charge tin concentration of 6.3% Sn the instantaneous fuming efficiency of the Capper Pass fumer computes to a value of 0.042 kg of tin per NM^3 air per percent of tin in the slag bath. This value compares favourably with those values reproduced by Denholm et al[4].

By recalculating the data reproduced by Halsall[2] and Wright[3] the Capper Pass furnace instantaneous fuming efficiency at the start of the investigation was 0.028 kg of tin per NM^3 air per percent tin in the slag bath. Hence an improvement of approximately 50% in fuming efficiency was achieved during the course of this investigation.

(b) <u>During the Blowdown Period</u>

At the end of the charging period additions of tin bearing feed to the furnace ceases, the oil and air rate are decreased and the tin level is allowed to fume down to the discard level. Typical tin and sulphur profiles during the blowdown are included in Figures 2 and 3.

Figure 7 compares the tin level in the furnace charge with the tin level in the bath at the start of the blowdown (effectively the plateau tin value) and again a clear relationship is shown to exist.

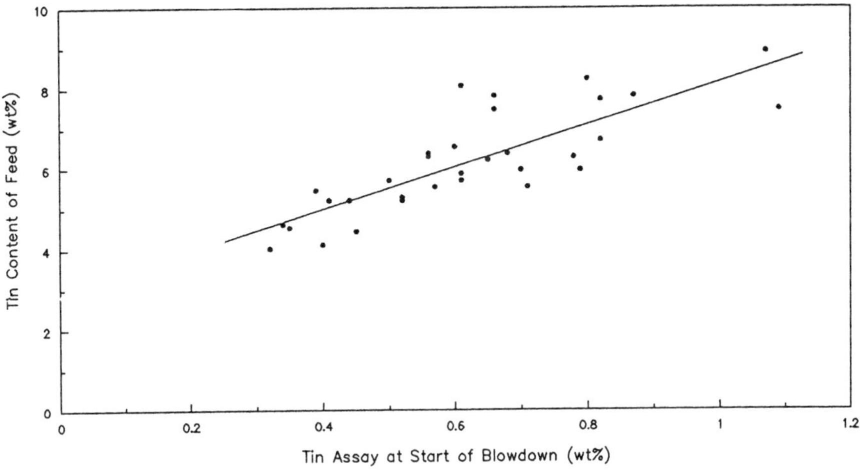

Figure 7: Effect of Tin Assay of Feed Material on Tin Assay at Start of Blowdown

At constant temperature the tin fuming rate decreases according to the first order rate equation. Typical rate constants obtained during blowdown were 0.01 min^{-1}. The value of the rate constant was shown to increase as the temperature was increased. For operational reasons the temperature rise during blowdown was restricted to 100^oC and the slag bath temperature was not permitted to exceed 1200^oC.

<u>Zinc Fuming Rate</u>

Zinc is considered a nuisance element in the Capper Pass circuit by creating a cycling load which then requires fluxing in the blast furnace smelt. Therefore efforts are made to fix the zinc in the fuming furnace slag thereby bleeding zinc from the circuit and reducing the cycling load.

Zinc profiles during a cycle were determined and the concentration was found to be constant throughout the cycle. Fuming of zinc was shown to be strongly temperature dependent, increasing rapidly as bath temperature increased. This was one of the reasons why the temperature rise during blowdown was restricted to 100°C. At the operating temperature of 1100° and under the Capper Pass furnace operating conditions approximately 70% of the charged zinc remains in the final discard slag.

Fuel Oil Additives and Alternative Fuels

The use of fuel oil additive as a combustion catalyst was investigated in an attempt to improve the heavy oil combustion efficiency. A supplier of such additives was consulted for advice and an iron fatty acid salt in an aromatic solvent was recommended as being the one most likely to succeed. A week long trial was conducted using the maximum recommended quantity of additive but no improvements were observed and this line of investigation was terminated.

Gas oil and recovered fuel oil were investigated, the former indicating no benefit. Progress with the use of recovered fuel oil, although economically attractive, was restricted by the availability of sufficient quantities of the oil.

Partial substitution of heavy fuel oil by coal was investigated and some benefit was indicated but only if injected through the tuyeres rather than incorporation within the solid charge. The lack of suitable delivery equipment to the tuyeres precluded further investigation of this aspect. Future fuel price fluctuations will probably dictate when this avenue is worthy of further investigation.

Conclusions

A description of the Capper Pass fuming furnace and its operation has been presented together with a summary of the major improvements achieved during the investigation program. Approximately 1,200 furnace cycles were examined under varying parameters and the most significant observations are reported in graphical representations.

Sulphur profiles in the slag bath during the cycles led to significant improvements in pyrites consumption from 220 kg/tonne of tin bearing feed to 100 kg/tonne.

The tin fuming rate during the charging period has been increased by 50% from 0.028 to 0.042 kg of tin fumed/NM^3/percent tin in bath, a value which compares favourably with rates attributed to other fuming operations.

The proportion of purchased low grade residues incorporated into the fuming furnace charge has been increased from 10-15% to 30-35%.

Tin fuming as the sulphide in the Capper Pass fuming furnace has been shown to exhibit first order reaction behaviour.

The various mathematical relationships established during the investigation have been incorporated into a fuming furnace computer model which is now employed to predict plant capability from available feedstock.

Acknowledgement

The authors gratefully acknowledge the kind permission of the Capper Pass directors to publish the above information.

References

[1] P. HALSALL, The development of a tin sulphide fuming process at Capper Pass Ltd., Advances in Sulphide Smelting, A.I.M.E. Symposium, ed. H.Y. Sohn, D.B. George, A.D. Zunkel, 1983, Vol.2, pp 553-582.

[2] P. HALSALL, Pyrometallurgical concentration of Low Grade Tin Ores, Mineral Processing and Extractive Metallurgy, ed. M.J. Jones, P. Gill, I.M.M. London, p 303 (1984).

[3] P.A. WRIGHT, Mechanism of the tin sulphide fuming process, Trans.Instn.Min.Metall. Vol.95, pp C58-61 (1986).

[4] W.T. DENHOLM, K.A. FOO, N.J.B. POCOCK, 'Tin volatilisation: smaller plants for mines and smelters', Mineral Processing and Extractive Metallurgy, ed. M.J. Jones and P. Gill, I.M.M. London, pp 275-291 (1984).

MATERIAL RECYCLING OF POLYPROPYLENE FROM AUTOMOTIVE BATTERIES

- PROCESS AND EQUIPMENT -

Reinhard Pfaff

Metallgesellschaft Aktiengesellschaft
Reuterweg 14
D-6000 Frankfurt a. Main
F. R. Germany

Abstract

Today's materials recycling has to meet among others the following requirements:

- it has to take care of environment
- it has to pass nearly without residue
- it should create high leveled products with definable duration of life
- it has to be most economic

The presentation refers to the reuse of plastic scrap on the basis of battery waste. Necessary development work for separating, treating and recompounding polypropylene are described. Moreover results of application specific tests are shown, which have been done on the resulting moulding compounds to specify material and processability. Special attention has been paid to the heavy metal contaminants and their effect on the creep strength of the compounds.
Finaly today's and future economic aspects are discussed.

Second International Symposium -
Recycling of Metals and Engineered Materials
Edited by Jan H. L. van Linden,
Donald L. Stewart, Jr., and Yogeshwar Sahai
The Minerals, Metals & Materials Society, 1990

People driving a car are customary not interested in owning the car's battery, they only wish to use the electricity in the battery and return the battery after use.
Therefore, more than 150.000 metric tons per year battery scrap has to be treated today in the Federal Republic of Germany.
Different methods are known to manage that difficult hazardous waste problem. The furhter explanations confined to the aspects of a nearly complete material recycling process which has been realized first in our battery scrap processing plant named BSB-Recycling.

In that plant, which came on stream in june 1984, more than 60.000 metric tons battery scrap a year are treated in several steps.

The first step is called the "TONOLLI-PROCESS" (1), which is installed in several plants in Europe. Used car batteries are disassembled in that step into their components. The picture shows the percentage composition of a starter battery.
Each component is treated in a special way, the further explanations are only directed on the treatment of the polypropylene scrap of the battery housings.
Problem is the purity of the polypropylene chips, which is only 97%. This is the main reason why polypropylene coming from the "TONOLLI-PROCESS" is not suitable as a basic material for high leveled applications.

To solve that problem, we started a R&D project in 1984 by order of the German Ministry for Research and Technology.

The results of the two years lasting laboratory studies and engineering activities culminated in the erection of a pilot plant which allows the automatic treatment of 1.5 metric tons per hour polypropylene scrap.

After passing a metal separator, the preliminarily washed polypropylene chips are subjected to a wet grinding process.
Next, the regrind enters a washing trough, properly dosed and previously dewatered via a worm conveyor. In that washing trough, the mixture of water and polypropylene is properly stirred at the turbulence, necessary for the washing effect.
The cleaned regrind is fed by a worm conveyor into a whirl strainer for mechanical predrying. The material is then blown through a costreaming flow dryer, where the required dryness is achieved by using relative velocity, difference in temperature and dwelling time. For reducing energy consumption during thermal drying, a heat exchanger is installed for using hot exhaust air from other plant sections, so that additional heat only had to be generated electrically for increasing the temperature to 80°C.
Subsequent to the drying process disarranging grinding fluff is removed in a cyclone.
Finally the finished product is stored in silos for further treatment.
With a purity of more than 99,5 per cent, the regrind represents an exellent basic material. How does it differ from the material "fresh from the factory" of the polypropylene manufacturer ?

As the manufactures of battery cases naturally use polypropylene types from various manufacture, this polypropylene basic material consists of a mixture of different molecular composition with different stabilizing concepts.
In addition it is characteristic for plastics, that each processing step during productions, processing and also recycling changes the physical properties of the plastic.

Percentage composition of a factory-fresh 12V 44Ah starter battery with a hard rubber case and with a polypropylene case

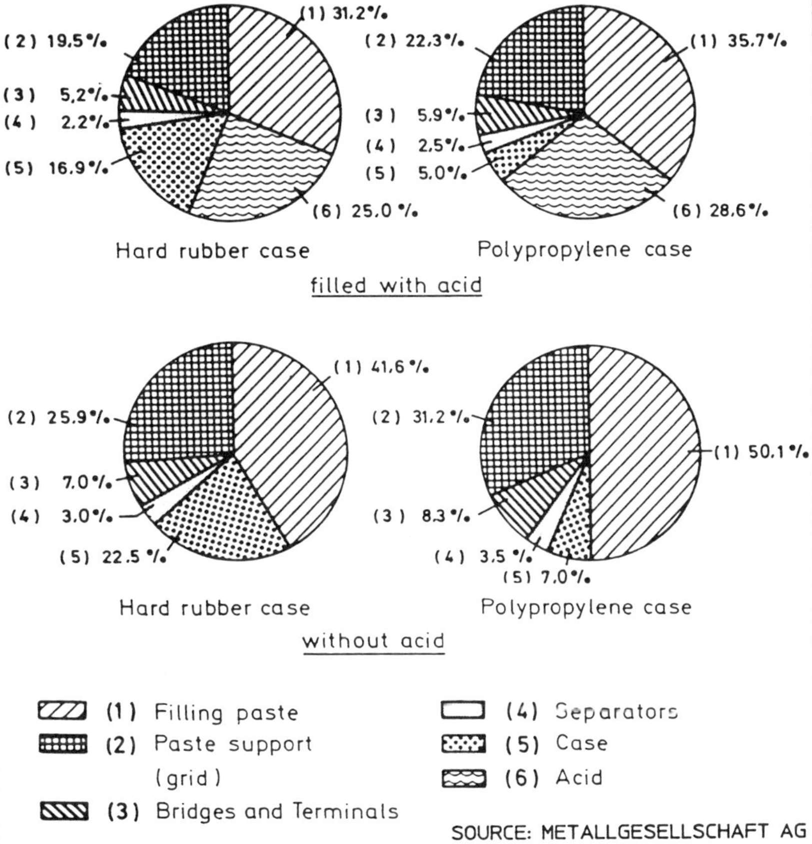

SOURCE: METALLGESELLSCHAFT AG

25.01.1989	STARTER BATTERY	LA

The same applies for their "life cycle" which has passed.
In addition to the thermal load of the plastic during the moulding phase by battery cases production the material is exposed to a considerable extractive stress due to the watery acid and anoxidation during loading process. Particularly the last mentioned process can lead to a drastic reduction of the originally contained stabilizers, because they migrate to a considerable extent in some cases and are extracted from the polymere.

Further on additional impurities are gathered in by collecting and pretreating steps. The picture shows the average-concentration of impurities by 1000 hours production time.

Wooden impurities, which are resulting out of collecting and transporting steps are reaching an average of 150 gramms per metric ton.

Plastics material recycling on a high level contents further steps to get from a nearly pure regrind to a compound with tailor made properties. By skillfully combining antioxidants, which decompose peroxides and desactivate metal traces, compensate for remaining acids and radically destroy them, are compatable with remaining colouring and pigments, are immune against foreign matter and act rapidly at high temperatures during process use. The molecular regression in the recycling material during processing in a melted condition can be prevented to a large extent. For the recovery of a usable regenerate it is thus necessary to carefully coordinate the stabilizer with its damage condition.

Passing a test serie with a pilot plant of more than 1000 hours, we succeeded in developing the necessities of stabilizing and blending to create new polymer products with tailor made properties. Basing on this experience we designed the concept of a new recompounding plant as final step of our Braubach activities.

The pretreated basic polypropylene is directly pneumatically supplied to the storage silos of the compounding plant.
In addition to an extended silo for the regrind there are further silos for storing prematerials like colour additives, blending polymers and fillers. The storage capacity is designed for a one week production.
Feeding and metering of all components are done by pneumatic feeders and gravimetric scales.
For preparing the stabilizer compound an automatic mixing unit is used which is also controlled by microprocessor.
The raw material mixture, being applicated custom made, is gravimetric dosed into a special designed twinscrew kneader. There it will be melted by external heat and internal shearing and mixed with all other components to a homogenous compound, while gaseous products are eliminated by intensiv degassing.
Wooden impurities are continously removed in an especially designed filter unit, the filter process is controlled by melt preasure drop, the change of used filter screens is done automatically (2).
Pellet granulation of the melt and cooling of the pellets are carried out in a waterring pelletizer. The pellets obtained from the waterring pelletizer are either lense shaped or round shaped, depending on the temperature of the waterring and on the viscoelastic properties of the melt. Present R&D activities concentrate on the recovery of the accumulated heat in the waterring to preheat the regrind to save energy in the extrusion process.

COMPOSITION OF FOREIGN SUBSTANCES

PRODUCTION RESULTS (1000 h)

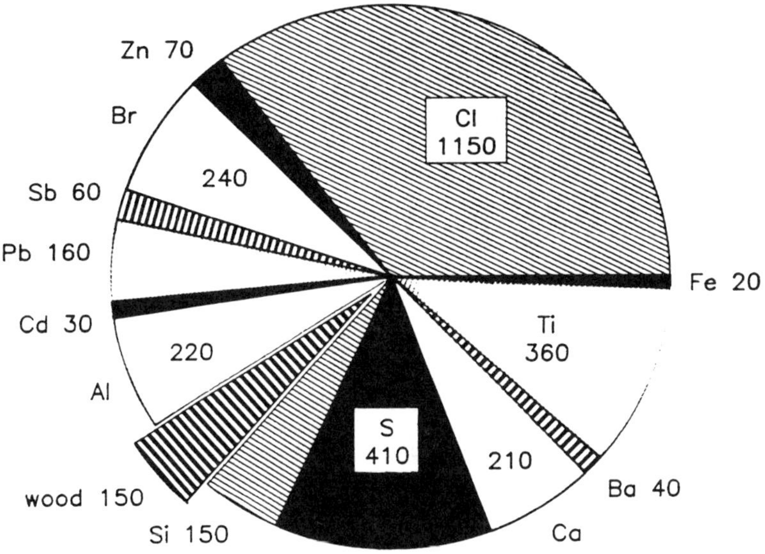

values in g/t

SOURCE: METALLGESELLSCHAFT AG

25.01.1989	BSB-REGRIND	PFF/NUER

The finished products are stored in silos for delivery by silotruck or in big bags.
The heart of the plant, the twinscrew kneader, has been shown at the plastic-exhibiton "K 86" in Düsseldorf. The "WHITE GIANT", as it was called in reports by the expert-press, represents one of the largest and most modern recompounding-units, which has ever been built.

According to the specification, the PP recompounding process is to be equipped with a computer assisted quality assurance (CAQ) system (3),which system has been established at the lowest hierarchy level close to the process (picture).

Pilot-plant tests had been run to establish, which process parameters are relevant for the quality of the finished products.
An automatic data collection system was then installed for processing round about 75 parameters.
By the implementation of that CAQ system, our customers always get the surety of having received just in time inspected commodity following the standards of quality mangement, quality system elements and quality assurance (ISO 9000).
Demonstration and documentation of specified parameters, such as the melt flow index or the output of the stabilizer's gravimetric dosing unit are scope of delivery (picture).
The machinery used in Braubach and the proposed CAQ system together now provide an opportunity of turning recycled storage battery scrap into secondary polypropylene, whose properties are comparable to those of primary polypropylene.
The use of the CAQ system enables us to run the production plant with only three workers a shift under an output of 1.5 metric tons per hour.

The compounds marketed under the trade name SECULENE PP are custom tailored in close cooperation with the client to meet his specified profile of properties.

SECULENE PPT 1220 (picture) represents a low cost product with well balanced properties for universal technical applications.

SECULENE PPT 8027 (picture) is used as an economy-priced TPE modified polypropylene blend with application specific inpact strength for injection moulded fender liners in the automotive industry.

Most important member of our group of talc reinforced polypropylene compounds is SECULENE PPT 1240 TV20 (picture) with a talc content of 20 per cent. It is used for instance in automotive applications like air cleaners or air ducts.

Finally, SEUCLENE PPR 1250 KV40 (picture) is a compound filled with 40 per cent calcium carbonate.

The introduced materials represents only a small cut of our extensive possibilities to meet our clients ideas of economy-priced products just being aimed to their specific requirements.

The described activties are not only a contribution to manage a difficult hazardous problem, they evidently have shown, that it is possible to return plastics from post use durables to the economic cycle.

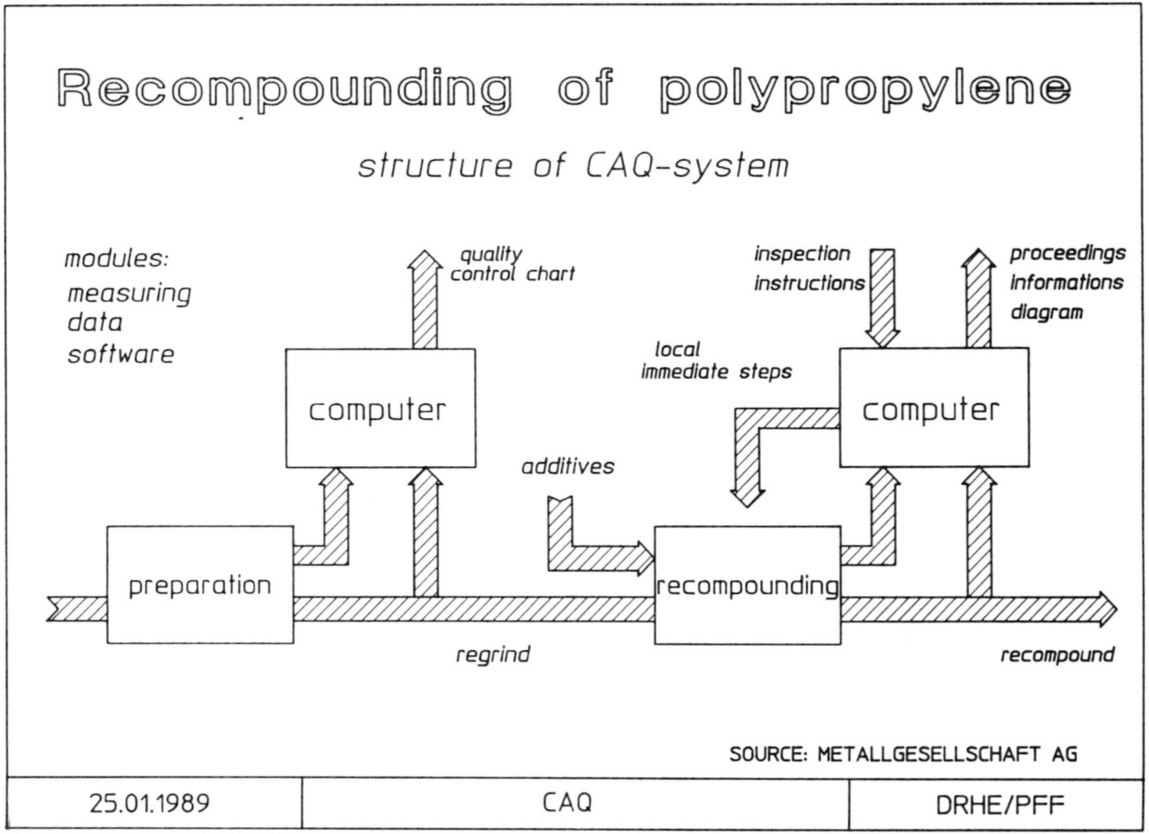
Recompounding of polypropylene
structure of CAQ-system
modules:
measuring
data
software
quality
control chart
computer
preparation
regrind
additives
recompounding
local
immediate steps
inspection
instructions
computer
proceedings
informations
diagram
recompound
SOURCE: METALLGESELLSCHAFT AG
25.01.1989
CAQ
DRHE/PFF

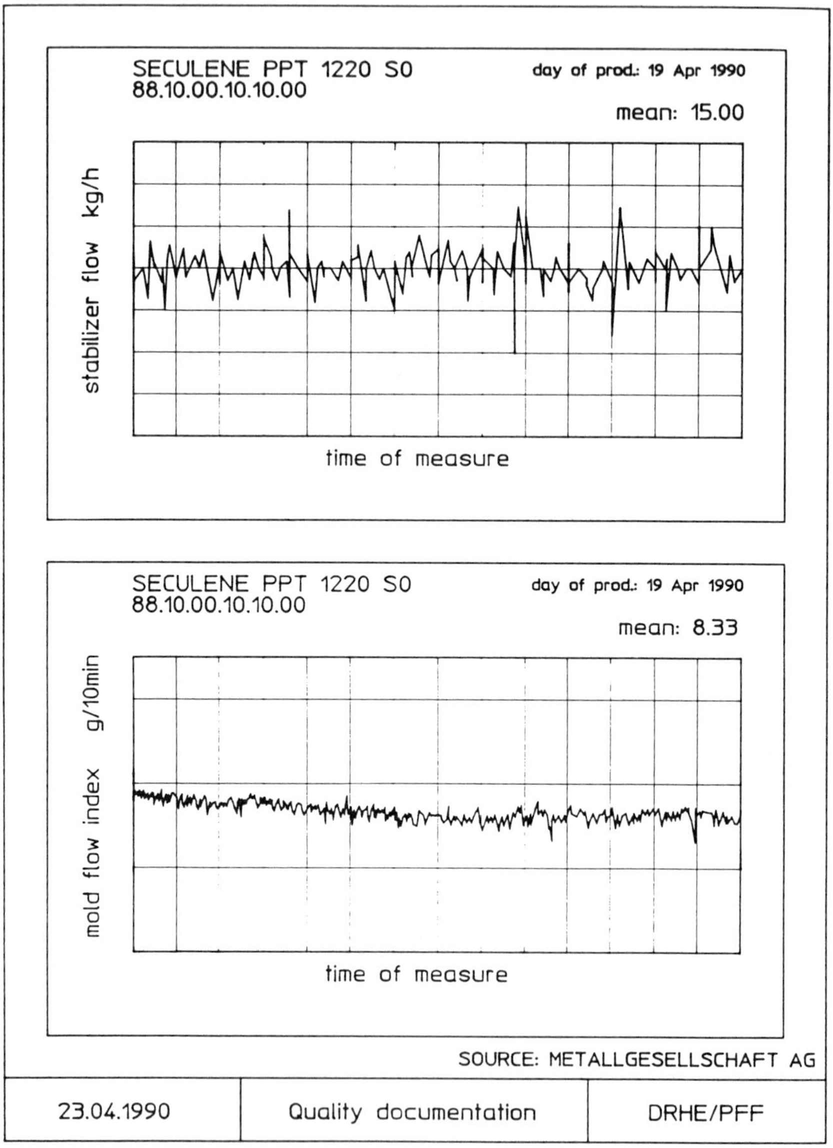
SECULENE PPT 1220 S0
88.10.00.10.10.00
day of prod.: 19 Apr 1990
mean: 15.00
stabilizer flow kg/h
time of measure
SECULENE PPT 1220 S0
88.10.00.10.10.00
day of prod.: 19 Apr 1990
mean: 8.33
mold flow index g/10min
time of measure
SOURCE: METALLGESELLSCHAFT AG
23.04.1990
Quality documentation
DRHE/PFF

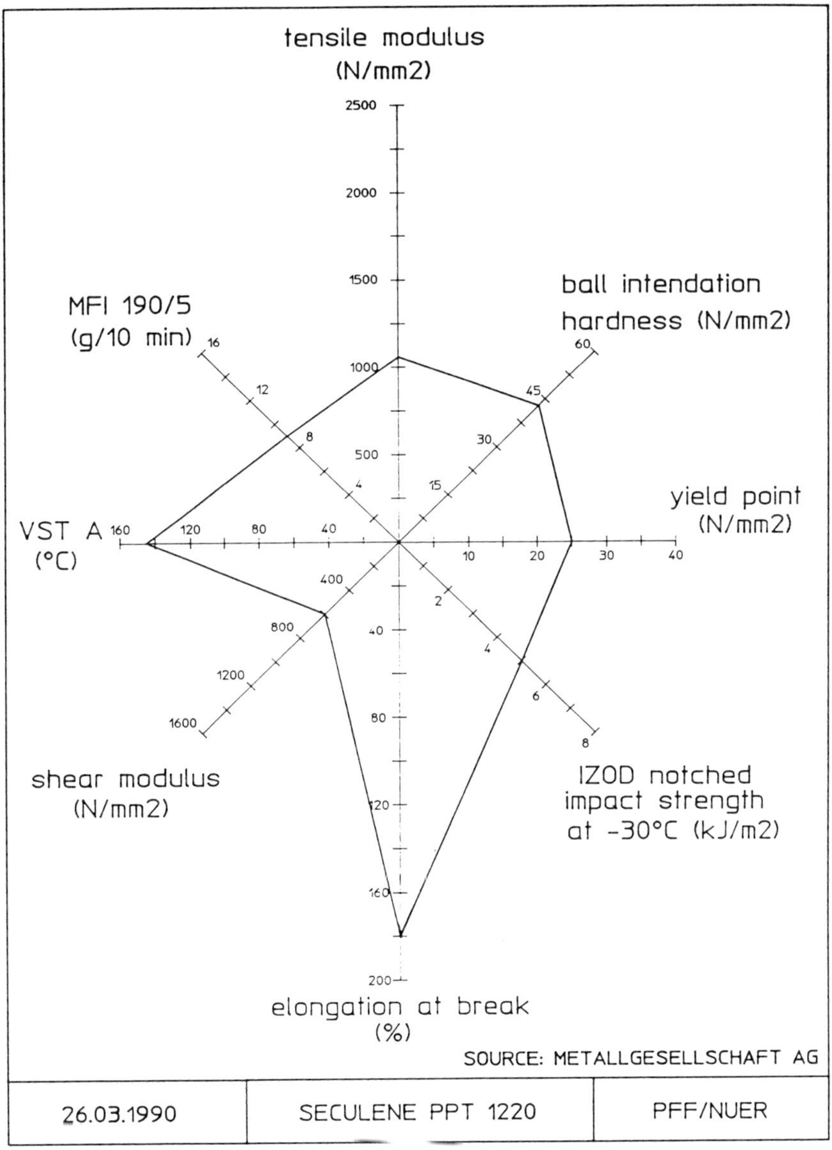
tensile modulus
(N/mm2)
2500
2000
1500
1000
500
ball intendation
hardness (N/mm2)
60
45
30
15
yield point
(N/mm2)
10
20
30
40
IZOD notched
impact strength
at -30°C (kJ/m2)
2
4
6
8
elongation at break
(%)
40
80
120
160
200
shear modulus
(N/mm2)
400
800
1200
1600
VST A
(°C)
160
120
80
40
MFI 190/5
(g/10 min)
16
12
8
4
SOURCE: METALLGESELLSCHAFT AG
26.03.1990
SECULENE PPT 1220
PFF/NUER

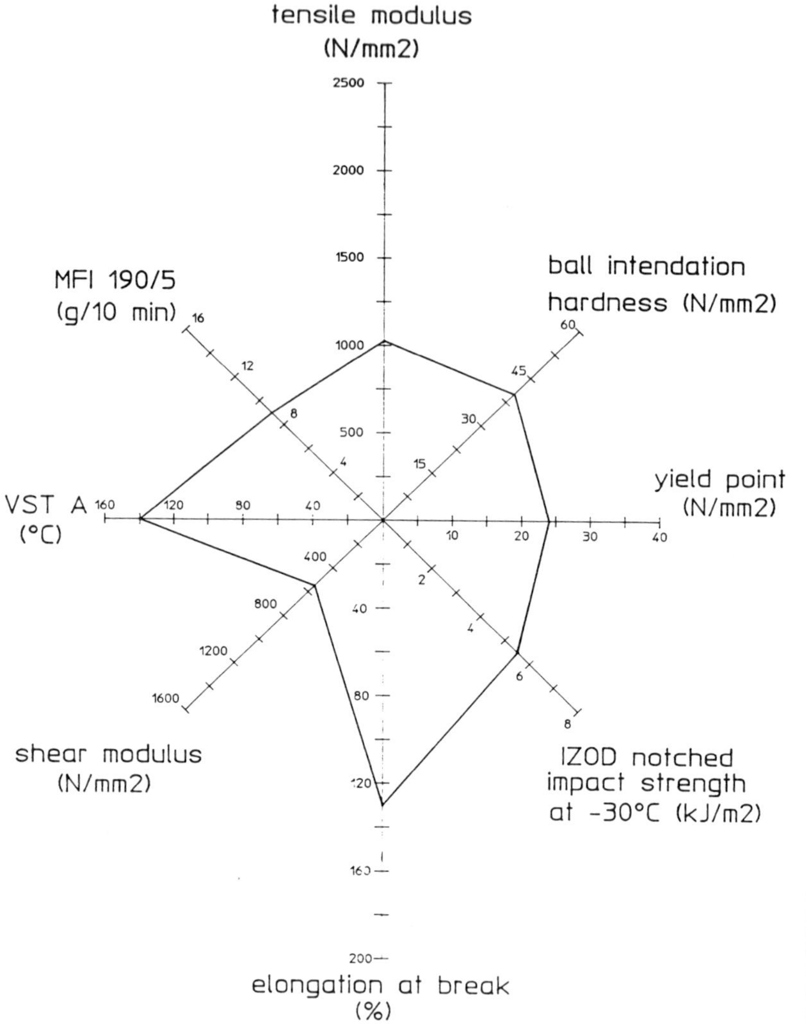

SOURCE: METALLGESELLSCHAFT AG

26.03.1990	SECULENE PPT 8027	PFF/NUER

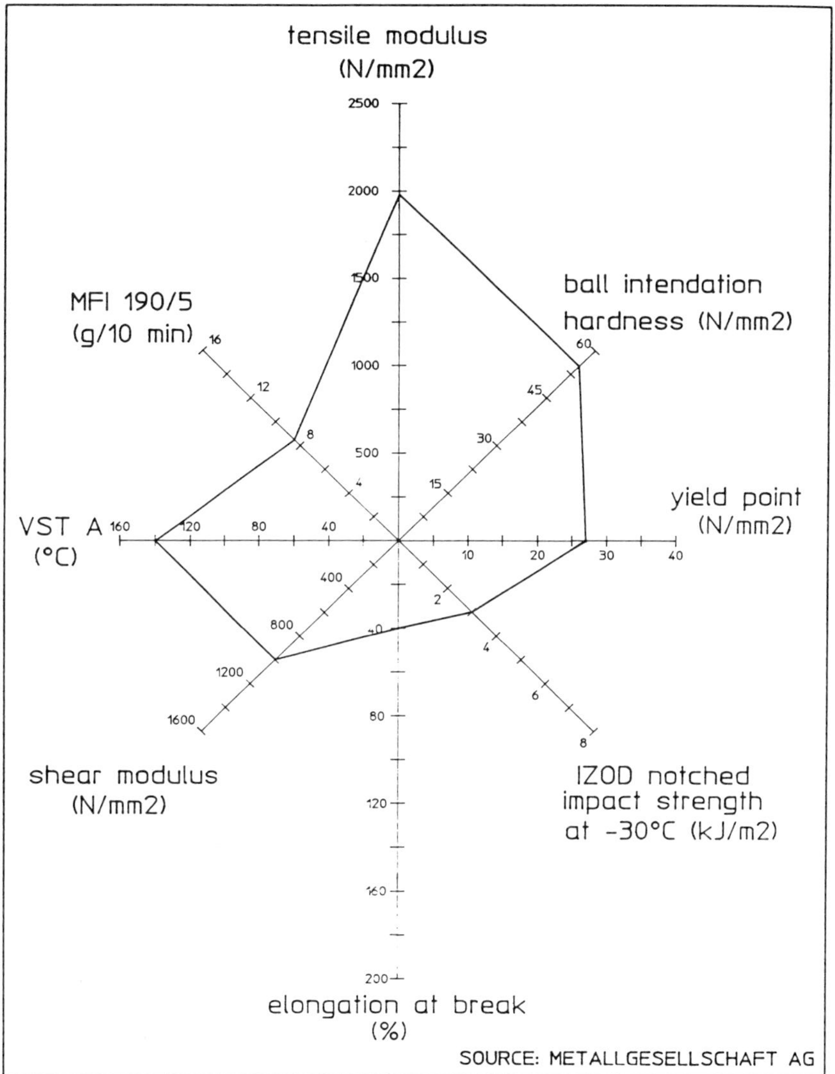

26.03.1990	SECULENE PPR 1240 TV20	PFF/NUER

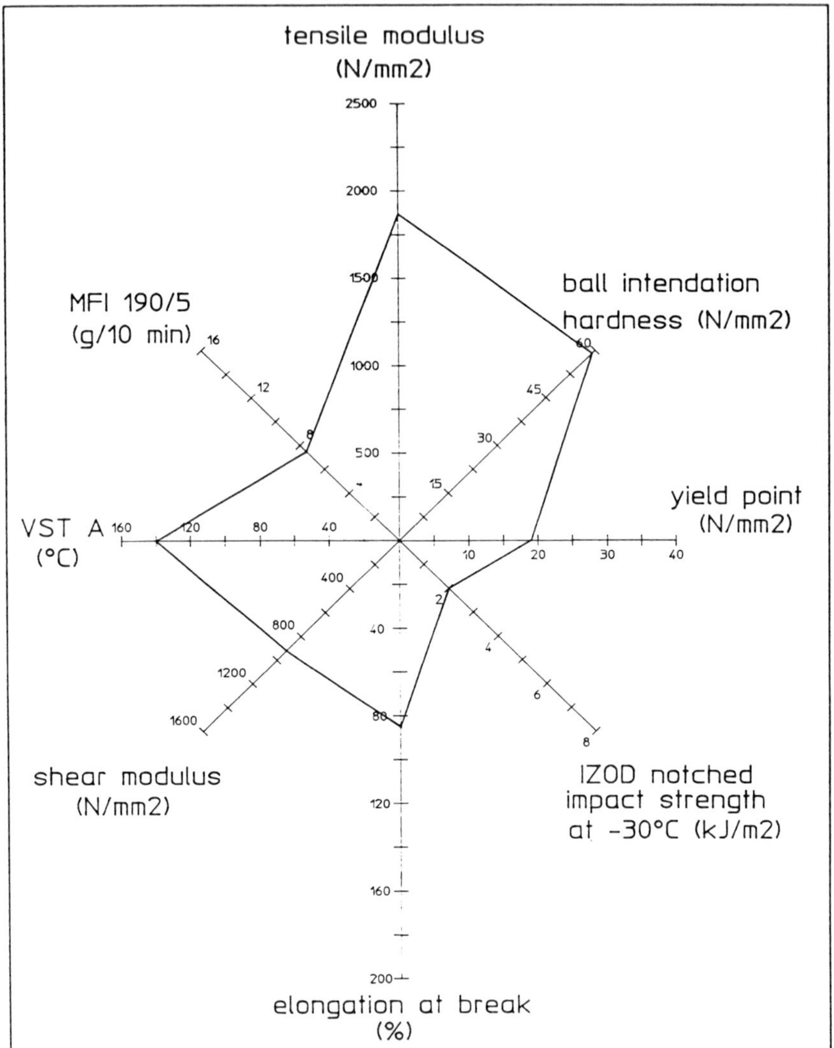

SOURCE: METALLGESELLSCHAFT AG

26.03.1990	SECULENE PPR 1250 KV40	PFF/NUER

References

1. R. Capetti, K.-F. Lamm and R. Pfaff, "Separation of Plastic Waste in the Recycling of Lead-Acid Batteries", 4. International Recycling Congress, Berlin, October 1984.

2. R. Pfaff and D. Gneuß, "Reprocessing of Plastic Waste", Industrial & Produktion Engineering, 11 (1987), 43.

3. K. Heil und R. Pfaff, "Qualitätssicherung bei der Wiedergewinnung von Polypropylen aus gebrauchten Starterbatterien", Vortrag anl. des Fachsymposiums "Prozessleittechnik und Automatisierung für die Rohstofftechnologie", Hannovermesse Industrie, 1988.

ENVIRONMENTAL/HEALTH

Session Chairman:
Yogesh Sahai

Environmental problems in recycling of complex low grade non ferrous metal bearing residues

by Jean GOLDSCHMIDT
President
Jean Goldschmidt International S.A. and Hydrometal S.A., Brussels, Belgium

1. This is not a technical paper, and I must apologize to this audience, composed to a large extent of chemists, metallurgists, engineers and the like, to be unable to tell them many new technical facts about our processes and how we solve environmental problems, be it only because I am not a technician. Yet I believe our experience can be of value to a number of people present because it could well be complementary to theirs.

2. Most people think that recycling of non ferrous metals consists of treatment of metallic scrap by fabricators or secondary smelters and refineries.

However, not all secondary non ferrous metals are collected in a form suitable for use in the furnaces of these industries, in particular when they present themselves as complex mixtures such as fumes, water treatment sludges or production residues from manufacturers of pigments, catalysts and other chemical compounds.

Hydrometal S.A., a Belgian company, was set up some five years ago with the purpose of treating these sorts of residues. After having acquired the wet chemical part of a former electrolytic zinc refinery at Engis near Liège in Belgium, the company invested progressively about $ 3 million in land (15 acres), existing plant and new equipment such as reactors, storage tanks, piping and pumps, filter presses, handling equipment and laboratory instruments.

One very important item we acquired together with the existing plant were the operating permits which enabled us to immediately start treating secondary raw materials. Without these existing permits, it would have taken months and months and much money before industrial operations could have started.

Currently Hydrometal employs about 35 persons and treats around 1,000 tons per month of raw materials acquired mostly under long-term contracts.

The basic philosophy behind Hydrometal's way of operating is different from that of most other industries. Usually, the management decides what to produce, builds a plant accordingly and buys the necessary raw materials.

Second International Symposium -
Recycling of Metals and Engineered Materials
Edited by Jan H. L. van Linden,
Donald L. Stewart, Jr., and Yogeshwar Sahai
The Minerals, Metals & Materials Society, 1990

Hydrometal started from an existing plant and processes raw materials which happen to be available. It then decides which salable products - which will serve as raw material for other plants - can be produced at economically acceptable terms.

The first type of material treated by Hydrometal were flue dusts from copper smelters which contain zinc, lead, and copper as well as silver, gold or tin.

However flue dusts also contain unwelcome elements such as cadmium, chlorine and bromine. These elements are unacceptable to most pyrometallurgical plants, but they can be eliminated by hydrometallurgical means. As to zinc, lead and copper, while there exist several smelters accepting any two of these metals, very few will treat a material containing a mixture of 10 to 20 percent of each of these three metals.

Tin and precious metals are welcome additions to these flue dusts : their recovery is fairly easy, although one has, of course, to be careful to avoid losses, and they are important as means of covering costs.

Another type of raw material which has become very important to Hydrometal are cadmium-bearing residues such as sludges from the water treatment plants of nickel-cadmium battery makers and production residues from cadmium-selenium pigment manufacturers.

The raw materials treated by Hydrometal should be oxidic. Sulphides normally cannot be accepted because we have no furnaces in which to roast them, and processing sulphides by wet chemical means is very expensive and can be applied only to a limited number of sulphidic materials.

Neither can the plant accept metallics, except in very fine particles such as precipitates. The size of all materials must be relatively small because the company has no crushers to reduce big lumps into sizes amenable to wet chemical processing.

There are, of course, materials which we could easily treat at Hydrometal, but not competitively because of our high processing costs and the low value of our finished products. These are clean and relatively simple materials, without impurities such as chlorine and cadmium : brass mill fines, galvanizers ashes, etc.. Other plants can pay much higher prices for such materials and we do not try to compete against these plants because there are sufficient complex materials available for us to treat.

The hydrometallurgical treatment consists essentially in a succession of solid-liquid separations : the materials are submitted to a leaching operation - usually an acid leach - which will put certain elements in solution and leave others insoluble. The insoluble elements are then filtered out.

The filter cake may have to be washed again to rid it of unwelcome impurities (certain soluble compounds crystallize and will remain with the insoluble until washed out). The solutions normally will have to be purified by the addition of zinc dust or aluminium powder, or by changing the pH or temperature. The precipitate is filtered out.

The products of the plant are not refined metals but raw materials for other metallurgical or chemical plants.

Zinc is leaving the plant in the form of zinc sulphate solutions, which assay according to customer specifications. Lead is usually recovered as lead sulphate containing the precious metals and/or the tin, if any, and goes to lead smelters. Copper is shipped to copper smelters in the form of copper cement and cadmium is delivered as sponge or cadmium sulphate solution to producers of refined cadmium metal.

3. Let us now come to the subject of the environment. There are essentially two aspects in our relations with the environment :

- on the one hand our operations are non-polluting;
- on the other hand our activity helps keeping the environment clean.

a) Our operations are non polluting : no neighbour, no water consumer down-stream can complain about any environmental nuisance caused by Hydrometal.

First of all, as we have no furnaces, we do not emit any smokes or hot gases into the atmosphere.

To avoid dust emissions, we specify that raw materials must be supplied either thoroughly wetted - e.g. as filter cakes - or, if they are dry, be delivered in silo trucks which can be unloaded by compressed air direct into one of our 3 large silos, or be packed in big bags which we unload into a special hopper at the bottom of which a vertical knife slits the bag open; the material is then carried by compressed air into one of our silos.

Despite these precautions, we found during the first years of operation that certain quantities of zinc, lead and cadmium were carried into the river by heavy rains. So we are now collecting all rainwaters from the in-plant roads and the storage areas in a decanter and return them together with the plant and laboratory effluents to the head of the process.

We have indeed, because of our shipments of zinc or cadmium sulphate solutions, a negative liquids balance : we ship to our customers about 4 times more solutions than we take in acid, and the difference in liquids is made up by pumping water from the river. Of course, before pumping in this water, we recycle at the head of the process all our liquid effluents from filter washing etc, as well as the rainwater we collect.

b) Our activity helps keeping the environment clean : we can say this because we are returning to the industrial circuit metal bearing residues which otherwise would have landed in a dump.

Being in business for us means not only making money - which so far has fortunately been the case - it also means contributing to keeping the world a little cleaner. The 1000 odd tons of raw materials we are taking in per month may not be much in comparison with household or chemical waste generated, they contain however not unimportant quantities of metals such as lead and cadmium which are among those which received lately the strongest adverse publicity from environmentalists.

Yet it is precisely in this field that we unfortunately do not receive from our authorities all the help we could and should obtain.

Our plant may not be perfect - nobody is perfect, according to Marylin Monroe - and some of the procedures adopted in our plant could perhaps be further improved. We are open to any constructive criticism in this respect.

However when it comes to our positive contribution in keeping the environment clean, we still find all too frequently a prejudice of some of our national or regional authorities who frown at imports of recyclable waste, even by an environmentally clean operation like ours, because they don't like such an activity occurring "in their own back yard". This is a very short sighted view, but, alas, all too frequent.

4. Under these circumstances, what future do we have to expect and what do we have to fight for ?

Currently, European Common Market regulations - or rather directives issued by the EEC authorities which the various European countries are then supposed to confirm by legally binding laws and regulations - make a very opportune distinction between "waste" and "residues of non ferrous metals destined for recycling". The latter, which are really secondary raw materials for the non ferrous metal producing industry, can so far cross European borders accompanied only by a particular form which the importer returns to the authorities of his country. Real waste on the other hand can cross European borders only after prior authorization by the local authorities of the importing country.

Unfortunately certain types of real waste have been illegally moved across European borders under the facilities designed for the secondary non ferrous raw materials, by alledging that they contained non ferrous metals destined for recycling, whereafter they were just dumped and not at all recycled.

The consequence of these abuses are that certain European authorities - regional, national and supranational - want to suppress the facilities reserved so far for non ferrous metal bearing secondary raw materials and submit all imports of recyclable non ferrous metal scrap and residues to prior authorization. This would be an obstacle no metal processor could accept, be it only because all good secondary materials would be gone to countries with easier rules by the time the import permit is delivered and because of the impossibility to hedge against price changes during the period of uncertainty.

Moreover, these same European authorities intend to submit in future all non ferrous recycling operations to the same type of very strict and cumbersome operating permits as pure waste processors.

The effect of such changes in the rules applicable to our type of operation would be economically most detrimental and would eliminate some of the most important incentives to recycle rather than to dump a metal bearing residue.

I have only a vague idea as to the solutions given to these problems in the U.S. As far as I know, you have here an even more legalistic way of solving the problem of cross-border transportation of recyclable waste by listing the materials in detail rather than giving broad definitions.

I understand also that if a positive price is paid for a residue, export from the U.S.is much easier. This however can not always be guaranteed : imagine a material bought under long term contract on basis of a price paid per pound of metal contained, less a treatment charge per ton of material received. Such a contract could result in a positive price at the time it is concluded, but in a negative one after a few months if the price of the metal drops.

The B.I.R. too has suggested for secondary raw materials a definition which incorporates the idea of value, no doubt because most of its members buy and sell their materials on the basis of so much per pound and not under a formula with a treatment charge. Here it is :

"Redundant goods, by-products or residues which have a value and are recycled for further use, are secondary raw materials."

We feel that incorporating in this definition the idea of value as an essential determining factor is wrong and will not contribute to keeping the environment clean, because it excludes too many recyclable residues.

In our opinion, the distinction between a waste and a secondary raw material should be based not on what it is but on what is made of it : if it is converted into a product returning into the industrial circuit, then it is a secondary raw material, whatever its value. Of course a positive value is an indication that it will actually be returned into the industrial circuit (otherwise it would be difficult to pay a price for it) but a positive value is not the only way to prove return into the industrial circuit.

We feel that plants possessing an operating permit covering the conversion of non ferrous metals bearing residues into industrial products (other raw materials, intermediates and finished goods) should be given a blank permit to receive such residues from anywhere in the World, whatever their value, subject only to immediately declaring receipt of these materials and to keeping records permitting subsequent control by local authorities that these residues were effectively used as secondary raw materials.

I am concluding this paper by expressing the hope that all of us who have an interest in these problems will, whenever possible, contribute to influencing their authorities in the right direction.

CAUSES AND PREVENTION OF
MOLTEN ALUMINUM-WATER EXPLOSIONS

PREPARED BY

Seymour G. Epstein
Technical Director

The Aluminum Association, Inc.
900 19th Street, N.W.
Washington, D.C. 20006

Abstract

For many years the aluminum industry has been concerned with controlling molten aluminum-water explosions. The industry has made an extensive effort to gain an understanding of these explosions, the factors that promote them, and means of preventing their occurrence.

More than 40 years ago, the Aluminum Company of America began research to determine the conditions leading to explosions during DC casting and to develop preventive measures. Other aluminum companies in this country and abroad have spent considerable time and effort investigating explosions which have occurred during DC casting and melting of aluminum and searching for and evaluating methods of protection. Through Aluminum Association sponsorship, basic studies of molten aluminum-water reactions and their initiation and prevention have been conducted at leading industry, government and independent research laboratories.

Much has been learned from these efforts, and the results have been used as the basis for widely disseminated guidelines and recommendations. However, findings from a recent Aluminum Association worldwide reporting program clearly show that explosions continue to occur in our industry.

The present state of knowledge on molten aluminum-water explosions is reviewed and the recent findings relating to plant incidents are discussed.

**Second International Symposium -
Recycling of Metals and Engineered Materials
Edited by Jan H. L. van Linden,
Donald L. Stewart, Jr., and Yogeshwar Sahai
The Minerals, Metals & Materials Society, 1990**

Introduction

Molten metal is encountered in nearly every segment of the aluminum industry: reduction; secondary smelting and recycling; continuous and semicontinuous (direct chill or DC) casting; sand, permanent mold and die casting; sow and pig casting; atomizing; and various transfer and transportation operations.

In some operations, water is an integral part of the process and contact between molten metal and water cannot be fully prevented with present technology. In other operations, water and other oxidizers are inadvertently brought into contact with the molten metal. Indeed, there are still processes being employed, albeit on a small scale, where contact is deliberate, such as producing powder by dropping molten metal into water.

Millions of pounds of molten aluminum are handled every day, safely and without incident. However, there are inherent dangers in handling any molten material and explosions still occur in plants throughout the world. Fortunately, even when relatively large amounts of metal are involved, only a small percentage of incidents results in an explosion. Most often steam is harmlessly evolved or perhaps a small “pop” occurs.

It is the violent, unpredictable explosion that is of greatest concern to our industry because of its potential for harm to employees and property damage. There is considerable evidence that water or other oxidizing agents must be trapped between the molten metal surface and a solid surface for an explosion to occur (alloys with lithium or other reactive elements may pose exceptions). However, the exact conditions that lead to a violent explosion rather than a minor blow are not fully understood.

For many years the aluminum industry has been concerned with controlling molten aluminum-water explosions. The industry has made an extensive effort to gain an understanding of these reactions: in particular, the factors that promote them; the mechanisms of their initiation; and means of preventing their occurrence.

The two major projects undertaken by the Aluminum Association are:

1. A world-wide reporting system to obtain details of incidents occurring in aluminum plants.
2. A basic research effort with emphasis on initiation mechanisms and the effects of surfaces.

These are discussed below.

Plant Incidents

Reports on molten aluminum-water explosions date back to the 1940s, although earlier incidents had undoubtedly occurred. One event in particular—an explosion during DC casting with an experimental collector pan designed to catch minor bleed-outs that occurred in 1949, killing one and injuring three others—is thought to have led to the extensive industry research efforts, discussed in the next section.

Early in the 1970s the Association, as part of its research activity, made an effort to collect and review reports on molten metal incidents. Nearly 90 reports were obtained from various sources: 70 involved aluminum; the others were primarily iron and steel, with a few on copper, zinc, and magnesium.

Based on these reports it would seem that, while the minor blows or “pops” are continually occurring, the more violent explosions occur in a cyclical pattern. The “pops” throw metal a short distance and do little, if any, equipment damage; while they may result in burns, no fatalities have been reported for such incidents. The violent explosions have frequently resulted in injuries and equipment damage; in a few instances damage to the building itself has been reported. Figures 1 and 2 depict the reported incidents for the period 1949-1978 and the injuries associated with those events, respectively.

As the figures show, violent explosions with serious consequences were reported for the late 1950s, early 1960s, mid-1960s (1967 was a particularly bad year for the industry), and mid-1970s. Following the single explosion in 1978, no further reports were received for a period of about 5 years. There were those of us who naively believed the problem had been solved by the previous efforts and the situation was under control.

Then in late 1983, reports of violent explosions again began to be received. At that time it as decided to institute a formal reporting program to determine what was happening world-

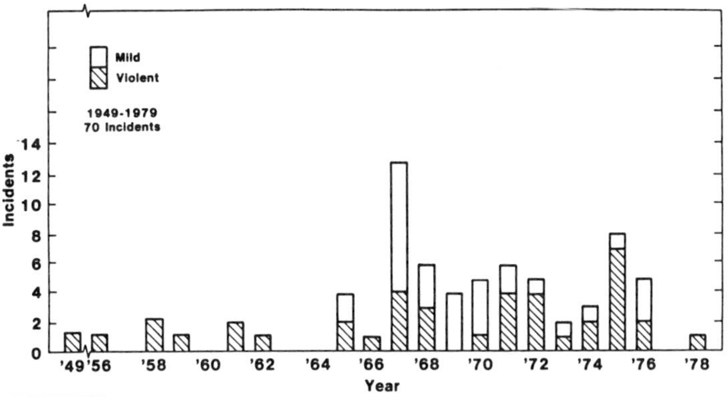

Figure 1.—Incidents Reported During the Period 1949-1978.

wide. The Association, through its Safety Committee, developed a reporting form, shown in Figure 3, to gather the pertinent details of these events. International participation in the reporting was invited. At the time of this writing, more than 150 reporters had agreed to participate. Forms were sent out in mid-1985; 647 reports have been received for incidents that occurred in nearly 20 countries during the period 1980-1989.

A key element of the form is "Explosion Characterization." The reporter is asked to evaluate the explosion by several criteria provided on the form, which are based on members considerable experience in investigating and reporting such events over the years. The ratings not only describe the magnitude of the force, but may also provide valuable insight on the nature of the explosion itself.

A "Force 1" explosion is one that essentially causes no damage and throws metal a small distance. It is thought to be a steam eruption from the rapid expansion (about 1100 time) of the trapped water turning to steam.

A "Force 3" explosion generally causes considerable equipment damage, throws metal a considerable distance, and is often accompanied by an intense light flash. From previous studies, one is led to hypothesize that this explosion is the result of a chemical reaction between finely divided aluminum droplets with air and/or water. In such a reaction, which is still controversial, the energy released would be equivalent to detonating about 3 pounds of TNT for each pound of aluminum fully reacting.

A "Force 2" explosion lies somewhere between, causing some damage and dispersing metal an intermediate distance. It may be a violent steam eruption, a relatively small-scale chemical reaction, or perhaps a combination of the two postulated mechanisms.

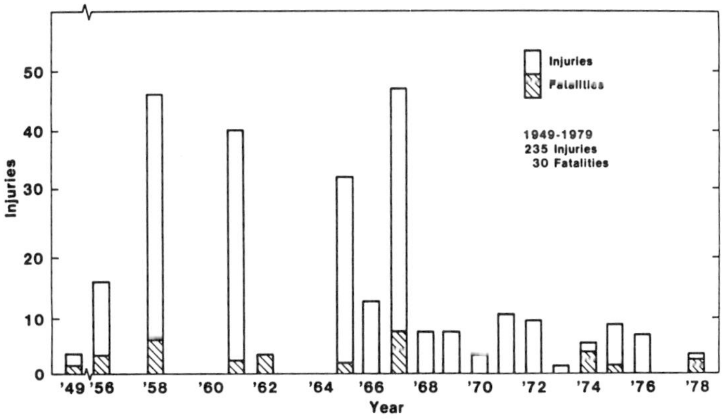

Figure 2.—Reported Injuries for the Period 1949-1978.

ALUMINUM ASSOCIATION
MOLTEN METAL INCIDENT REPORT

Date of Incident ______________________ Type of Plant ______________________
(month/year) (Reduction, Recycling, Rolling, Extrusion, etc.)

Explosion Characterization* (see explanation below): Force 1________ Force 2________ Force 3________

OPERATION:

Charging/Melting Type of Furnace ______________________
(Reverb, Topcharging, Induction, etc.)

Furnace Capacity ____________lbs. % Full ____________Alloy ____________

Metal Temperature ____________ °F Approximate amount of metal involved ____________ lbs.

Materials Charged ______________________

Outside Storage? Yes ☐ No ☐ Preheat? Yes ☐ No ☐ Preheat Time and Temp.: _______Hrs. _______°F

Transfer Type ______________________
(Crucible, Trough, Truck, etc.)

Alloy __________ Temperature __________°F Approximate Amount of Metal Involved __________lbs.

Casting Type of Unit ______________________
(D.C., Continuous, Sand & P.M., etc.)

Alloy __________ Temperature __________°F Approximate Amount of Metal Involved __________lbs.

Stage of Operation: ______________________
(Start-up, Termination, etc.)

Other Describe ______________________

Brief Description of Incident and Results, Including Extent of Injuries: ______________________

Use back of form if more space is needed.

***Explosion characterization**

Force [1]:	Property Damage	- None
	Light	- Minimal
	Sound	- Short Cracking
	Vibration	- Short Sharp
	Metal Dispersion	- < 15' Radius
Force [2]:	Property Damage	- Minor
	Light	- Flash
	Sound	- Loud Report
	Vibration	- Brief Rolling
	Metal Dispersion	- 15-50' Radius
Force [3]:	Property Damage	- Considerable
	Light	- Intense
	Sound	- Painful
	Vibration	- Massive Structural
	Metal Dispersion	- > 50' Radius

Please return to:

S. G. Epstein
Technical Director-
Safety, Health & Environment
Aluminum Association, Inc.
900 19th Street, N.W.
Washington, D.C. 20006

Figure 3.—Aluminum Association Reporting Form.

Of the 647 reports received to date, 436 of the incidents were rated Force 1; 176 were rated Force 2; and 35 were rated Force 3. The distribution by year is shown in Figure 4. Whether the higher numbers shown for recent years are indicative of a trend or whether they merely reflect the initiation date of the program will only be established with time. Since there is not universal participation, the number of incidents occurring world-wide may be substantially higher than actually reported. From personal experience, the data appear to be sufficiently representative to serve as the basis for observations of present conditions in the industry. Although the number of incidents appears to be on the increase, one must bear in mind the following:

- Prior to the present program, instituted in 1985, reports were normally available on only the more violent explosions.
- Ignoring the Force 1 explosions, for which no fatality has ever been reported, the damage and injury per incident are significantly less in this decade than the previous ones.

Based on the present data it may be concluded that, while the industry is still unable to prevent metal-water reactions, it has made progress in controlling the results of these phenomena and in protecting its workers.

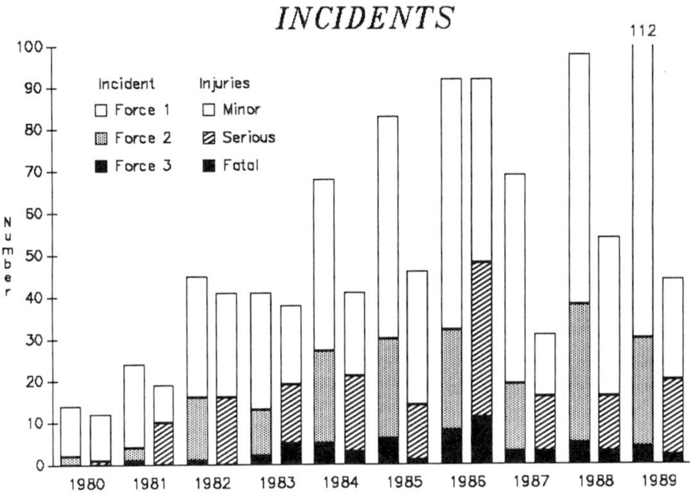

Figure 4.—Incidents Reported for the Period 1980-1989.

Research Efforts

The aluminum industry has made an extensive effort to gain an understanding of molten aluminum-water explosions, the factors that promote them, and means of preventing their occurrence. Over 40 years ago, the Aluminum Company of America (Alcoa) began research to determine the conditions leading to explosions during DC casting and to develop preventive measures. Other organizations, including other aluminum companies in this country and abroad, have spent considerable time and effort investigating explosions which occurred during the DC casting and melting of aluminum and searching for and evaluating methods of protection. Through Aluminum Association sponsorship, basic studies of molten aluminum-water reactions and their initiation and prevention have been conducted at Battelle Memorial Institute, IIT Research Institute, and Alcoa Laboratories. The latest study was initiated at Sandia National Laboratories by the Association with sponsorship by many companies in this country and abroad.
The principal findings of the earlier studies and the objectives of the present effort are discussed below.

Battelle Studies

In 1968 the Aluminum Association funded a project at Battelle Memorial Institute to study mechanisms of initiation and propagation of explosions between molten aluminum and water. Battelle first prepared and submitted an extensive "Review of Knowledge."[1] Experimentation began in September, 1969. First, the Battelle investigators attempted to produce explosions by dropping small amounts, 10 pounds or less, of molten aluminum into water in a steel container under conditions similar to those developed at Alcoa Laboratories. None of the 20 experiments run in this manner was successful.

Next, the quantity of aluminum was increased to nearly 30 pounds. Even with this amount, the investigators still had a difficult time producing an explosion. Only when the steel bottom plate was oxidized in a certain way and metal heated to about 1900°F were explosions produced.

By constructing water containers with Plexiglass sides, the initiation of the explosion within the container could be observed on motion picture film employing high filming speeds. The films showed that the most violent explosions occurred when nearly all the metal had entered the container, sometimes as much as a second after the initial contact of the molten metal with the bottom. They also seemed to show spots of light in the aluminum-water mixture, mainly at the container bottom, on one or two frames immediately preceding the explosion. A violent explosion then produced complete white-out of the film. At a filming speed of 5,000 frames per second, the observable explosion initiation occurred in less than 1 millisecond.

The Battelle investigators also determined that aluminum-water explosions could be initiated by detonating a small explosion charge either within the container or outside the container with the force of the explosion transmitted into the aluminum-water mixture by means of an aluminum bar embedded in the container wall. In every instance, the artificial "trigger" produced an aluminum-water explosion, even with 10 pound quantities of aluminum and with metal temperatures as low as 1350°F. In one experiment in which 10 pounds of aluminum at 1900°F were dropped into a water container with a glass bottom, a length of Pyrofuse was embedded in the container and burned in the aluminum-water mixture, giving off intense heat but no pressure pulse; no explosion resulted.

The instrumentation employed by the Battelle investigators (fast responding thermocouples and pressure gauges in the water container) did not yield meaningful data. Spectrographic analysis of the explosion by-products did reveal the presence of AlO, which is only stable at higher temperatures. Based on this finding and from thermodynamic calculations, the temperature of the explosive reaction was estimated as greater than 3,000°K, supporting the premise of a chemical reaction.

The Battelle studies[2] added considerably to the existing knowledge, but did not provide the basic understanding of molten aluminum-water explosions or their initiation which was sought.

IITRI Studies

During 1973-75 a study of initiating mechanisms for molten aluminum-water explosions was jointly sponsored at IIT Research Institute by the Aluminum Association and the National Institute for Occupational Safety and Health (NIOSH). Although their experimental results were for the most part inconclusive, the IITRI investigators did propose a model for the initiation mechanism. The premise for the model, which is based largely on the work of others, is that liquid water must be present in cavities ("quench pits") of the surface contacted by molten aluminum for initiation to take place. Initially, a vapor film forms which prevents intimate contact between the aluminum and entrapped liquid water. A stimulus is needed, most likely an impact-produced shock, to collapse the vapor film. When this occurs, the resulting hot liquid-cold liquid contact gives rise to spontaneous conversion of liquid water to gaseous water, resulting in a vapor explosion.

The IITRI investigators concluded that flow of molten aluminum over the "quench pits" produced sufficient shock to collapse the vapor film. Although the model postulated by IITRI during the last days of the contract has plausibility, no repeated documentation was developed. Representatives of the Aluminum Association and NIOSH monitoring the research program at IITRI felt that the postulated model was based on inconclusive evidence. In their opinion, further basic research and hard data are needed to fully understand what initiates a violent explosion.

Alcoa Studies

The first research program at Alcoa Laboratories, conducted by George Long and published in 1957, was an empirical study in which molten aluminum was dropped from a crucible into a tank containing water to simulate a bleed-out during DC casting[3]. In all, 880 tests were carried out. The bulk of these tests involved sudden discharge of 50 pounds of commercially pure aluminum through a 3-1/4 inch diameter tap hole into 12 inch x 12 inch x 12 inch steel or concrete containers partially filled with water at ambient temperature.
Long summarized his findings by citing three requirements for producing an aluminum-water explosion:

1. Molten metal in considerable quantity must penetrate to the bottom surface of the water container.
2. A triggering action of some kind must take place on this bottom surface when it is covered by the molten metal.
3. Water depth, temperature, and composition must be proper for the rapid transfer of a large quantity of heat from the metal to the water.

Of particular significance was Long's finding that coatings applied to the steel or concrete water container into which molten aluminum was dropped would prevent explosions.
Subsequently, P. D. Hess and K. J. Brondyke of Alcoa conducted a study[4] to determine more accurately the mechanisms involved in explosions and to develop more efficient ways to prevent them.
In this program, molten aluminum was dropped into water under a variety of conditions. Several organic coatings applied to the inner surfaces of the water container were found to prevent explosions.
Hess and Brondyke concluded that aluminum-water explosions may be prevented in two ways:

1. Maintain sufficient depth water. Deep water probably prevents explosions by sufficiently cooling the metal and reducing its fluidity before it strikes the bottom so that it loses the ability to contact the surface intimately and entrap water.
2. Apply a protective coating. Organic coatings on exposed surfaces probably guard against explosions in two ways: (a) The coatings are decomposed by contact with hot metal; and hot gases are generated which agitate the metal, preventing a restraining crust from forming, and permitting steam to escape, thereby avoiding a pressure buildup. (b) The coating prevents contact between the hot metal and surface rust or other oxides or hydroxides, eliminating a possible source of reaction.

During 1977-79 the Aluminum Association sponsored a study at the Alcoa Laboratories on engineering solutions for the prevention of explosions between molten aluminum and water. Findings[5,6] confirmed the importance of providing proper protective coatings on surfaces which may be struck or contacted by molten aluminum and water.
Since that study Alcoa investigators have confirmed the critical role that mechanical shock can play in explosion initiation[7]. Violent explosions occurred when a 150 pound weight was allowed to impact the side of the pan shortly after metal entered the water. In every case without the application of mechanical shock, the tarset coating prevented the explosion. Additionally, it was reported[8] that molten AlLi alloys dropped into water would repeatedly explode prior to contacting the bottom surface in a tarset coated pan without external mechanical shock.

Sandia Studies

Sandia National Laboratories is a facility under the U.S. Department of Energy. Scientists there have considerable experience in studying and attempting to control steam explosions that could conceivably occur in the nuclear industry by contact of molten materials with aqueous coolants. In September 1985, a proposal was submitted to the Association for a long-range fundamental study of molten aluminum-water reactions. Through funding support by many aluminum companies in this country and abroad, the Association sponsored the first phase of the proposed program: a 2 year study, primarily on surface effects, alloying additions and initiation mechanisms, through small-scale experiments. A final report was distributed to the participants early in 1989; a second phase is now under consideration. A working hypothesis was developed relating explosion probability to surface wettability.

Discussion

When two liquids at different temperatures are suddenly brought into contact the result can be, and often is, a violent reaction. The violence depends considerably on the temperature differential and the chemical reactivity of the substances. In the classic case of molten metal-water explosions, the metal is normally the hot liquid and water the cold liquid. However, severe explosions have occurred involving contact between LNG and water where water is now the "hot" liquid and liquefied natural gas is the cold liquid.

There are those who maintain that all such liquid-liquid interactions are physical phenomena involving such factors as rapid boiling and bubble collapse. Others take issue with this generalization and contend that chemistry plays an important role. Certainly, the chemical activity of aluminum must be considered in assessing aluminum-water interactions.

Aluminum is among the most chemically active metallic elements; its affinity for oxygen is on a par with sodium. If it were not for the inherent nature of the thin oxide film that instantly forms on the metallic aluminum's surface to protect it against further oxidation,[9] aluminum as a metal would be as unusable as sodium. Disturbing the oxide film or exposing fresh surface to the environment invites chemical reaction. If sufficient surface is exposed at one time, a violent reaction might be expected in a manner analogous to a dust explosion.

Although considerable information has been brought to light through previous research efforts, the precise nature of aluminum-water explosions has not been ascertained. Neither are the mechanisms for initiation of these events completely understood. Until a more fundamental understanding can be achieved, the industry cannot be satisfied that optimum controls are in place.

The need for further basic research efforts is seen for several reasons:

1. A better understanding of the mechanisms for initiation and reaction is needed for development of better controls.
2. It has never been known how or why organic coatings afford protection against explosions.
3. In recent experimentation, certain factors such as shock waves and some alloying additions have been found to override the protection afforded by the coatings.
4. Explosions are still occurring in the industry.

Acknowledgements

The author is grateful to Dr. Ron Miller of Alcoa for his technical inputs and review of the manuscript, and to the other members of the Association's Task Force on Molten Metal Explosions: Dr Ernie Ewing of Alcan and Harry Miller of Kaiser, for their helpful discussions and efforts over the years.

References

1. Rengstorff, G.W.P., A.W. Lemmon, Jr., and A.O. Hoffman, Review of Knowledge on Explosions Between Molten Aluminum and Water, Battelle Memorial Institute (1969 April).
2. Lemmon, A.W., Jr., Explosions of Molten Aluminum and Water, Light Metals 1980, Conference Proceedings, The Metallurgical Society of AIME (1980).
3. Long, G., Explosions of Molten Aluminum in Water—Cause and Prevention, Metal Progress, 71, 107-112 (1957).
4. Hess P.D. and K.J. Brondyke, Causes of Molten Aluminum-Water Explosions and Their Prevention, Metal Progress, 95, 93-100 (1969).
5. Hess, P.D., R.E. Miller, W.E. Wahnseidler, and C.N. Cochran, Molten Aluminum-Water Explosions - 1979, Final Report submitted to the Aluminum Association (1979 October 15).
6. Hess, P.D., R.E. Miller, W.E. Wahnseidler, and C.N. Cochran, Molten Aluminum-Water Explosions, Light Metals 1980, Conference Proceedings, The Metallurgical Society of AIME (1980).
7. Miller, R.E., Letter to S.G. Epstein, The Aluminum Association (1980 August 26).
8. Miller, R.E., Letter to S.G. Epstein, The Aluminum Association (1981 October 09).
9. Epstein, S.G., Aluminum and Its Alloys, Aluminum Association Report C-6 (1984).

RECYCLING AND GLOBAL WARMING

J. Szekely and D.R. Forrest

Massachusetts Institute of Technology

Department of Materials Science and Engineering

Cambridge, Massachusetts

The metals and materials industry is responsible for consuming 10-15% of the total energy used in the U.S. The main contributors being steel, aluminum, and paper. Most of the energy used is fossil-fuel-derived and therefore contributes to the total carbon dioxide emitted. Since the energy crisis in the 1970-s, very major improvements have been achieved regarding the specific energy consumption in the steel and in the aluminum industries. However, more extensive recycling promises even greater potential reductions in CO_2 emissions. The technological, economic and societal problems that will have to be overcome will be discussed in the paper.

Second International Symposium -
Recycling of Metals and Engineered Materials
Edited by Jan H. L. van Linden,
Donald L. Stewart, Jr., and Yogeshwar Sahai
The Minerals, Metals & Materials Society, 1990

A MOBILE IN-FLIGHT SUSTAINED SHOCK-WAVE PLASMA REACTOR FOR TREATING ELECTRIC ARC FURNACE DUST

Malcolm T. Hepworth[a], Jozef K. Tylko[b], Marek K. Tylko[c], and S. Vrchota[d]

Abstract

A reactor based on the use of Sustained Shock-Wave Plasma technology has been developed and successfully tested at a laboratory scale for the treatment of electric arc furnace dust. The reactor utilizes a conical, highly diffused, orbiting and pulsating discharge into which mixtures of carbon and electric furnace dust are introduced and treated in-flight. The residence time of such material treated in the plasma ranged typically from 100 to 180 milliseconds. Orbital speeds from 20,000 to 80,000 rpm were employed to accomplish the in-flight reduction and separation. Electronic control was provided, maintaining the orbiting and pulsating functions between six anode segments and a centrally-placed cathode. The unit is capable, thereby, of wide regulation of the height of the discharge, orbital speeds, pulse amplitude, and mark-space ratios. A particularly important feature of this unit is its ability to operate with very low flows of plasma forming gas (argon), which provided both high enthalpy of the ambient discharge and efficient condensation and collection of the volatile fraction in a bag filter.

---

a. Professor of Metallurgy, Mineral Resources Research Center, Department of Civil and Mineral Engineering, University of Minnesota.
b. Adjunct Professor of Plasma Technology, MRRC, University of Minnesota
c. Plasma Engineer, PlasmaTech, Inc.
d. Graduate Research Assistant, MRRC, Department of Civil and Mineral Engineering, University of Minnesota

Second International Symposium
Recycling of Metals and Engineered Materials
Edited by Jan H. L. van Linden,
Donald L. Stewart, Jr., and Yogeshwar Sahai
The Minerals, Metals & Materials Society, 1990

Introduction

Electric Furnace Steelmaking Dust (EFD) fails to pass the EPA standard leach test for lead and cadmium, and in some instances chromium, and therefore is classified as a hazardous waste(1). At present the average cost incurred in shipping and dumping EFD is approximately $250 per ton. Until recently, untreated EFD could be disposed of in specially designated hazardous landfills; however, beginning in August, 1990, only pretreated EFD can be landfilled. In light of these developments, a number of investigators have studied treatment methods to recover the base metal content, primarily zinc, chromium and lead from the other constituents including the Waelz kiln, flame reactor, and other processes, which have been described in detail by McAloon(2). The approach which is described in this study is to employ an in-flight continuous treatment principle developed by J. K. Tylko(3). The goal has been to explore the possibility of creating a low-cost mobile unit which can be moved from site to site instead of transporting the wastes to a central facility.

Prior thermal treatment methods in the field of EFD treatment fall broadly into two categories: those based solely on fossil fuel and those employing plasma technology. In the former approach, reduction occurs either by reaction with natural gas or carbon monoxide. The well-known tendency for the reverse reaction:

$$Zn + CO_2 = ZnO + CO,$$

occurs upon cooling with the characteristic "blue fume" formation which makes the efficiency of the process in terms of conversion to zinc metal difficult and potentially expensive.

The current plasma processes are exemplified by the Tetronics plasma systems known as the Expanded Precessive Plasma. This system has been used successfully in the recovery of platinum from spent catalyst and more recently has been successfully applied to the treatment of EFD using a zinc-splash condenser to collect a metallic zinc product(4).

The current plasma processes employ a plasmatron which is either static (e.g. SKF, Elkem etc.) or mechanically rotated at relatively low speeds (Tetronics). These plasmatrons at the present state of the art have a limited life span requiring frequent replacement. In addition to the high capital cost they have high thermal losses (17 to 20 %) due to the intensive cooling of the arc-constricting parts. In the system developed by Tetronics the generated plasma is employed indirectly, by heating a partly molten starting charge with an impinging jet of plasma. This mode of operation implies not only considerable electrical energy losses in the plasmatron itself, but also necessitates a large, refractory-armored hearth in which the thermal energy derived from the plasma and carbon could be safely distributed. The above implies not only a substantial size and cost of any type of such installation, but also involves lengthy start-up procedures. Furthermore, relatively large quantities of argon are required in order to stabilize the plasma and to transfer heat to the molten charge.

The Sustained Shockwave Plasma (SSP) avoids these major difficulties by replacing the (external) plasmatron with inexpensive graphite electrodes, and by virtue of the electronic controls which orbit and pulsate the discharge, provides a highly turbulent plasma environment requiring only a minimal amount of argon for stabilization.

The SSP reactor used in this work is operated by maintaining a discharge between a central cathode and six equi-spaced radially-disposed anode segments. The so-called base current was maintained between the cathode and all the anode segments. An external d. c. coil was placed above the plane of the anode segments which helped to distribute the base current and provide a highly diffused, conically shaped plasma. The SSP action was developed when each of the anode segments received sequentially current pulses which were superimposed on the base current. It is these current pulses which lead to the development of shock waves. With the presently-used electronic system and computer controls, orbital speeds ranging from 10,000 to 100,000 rpm can be accurately set and maintained. Characteristically, this type of orbiting and pulsating discharge is capable of entraining large quantities of EFD feed which operate with a minimal argon gas flow (Cf. Section Experimental Results below.) Tylko et al.(5,6) have described in detail the special features of the SSP technology elsewhere.

Application of SSP to the Treatment of EFD

In order to treat EFD properly in an SSP reactor, a design was required that provided the separation of volatile metals from the iron product. This was achieved in a laboratory scale device capable of operating up to 35 kW of plasma power. The goal was to produce a recyclable iron product and an upgraded zinc-lead product which is preferably in the metallic form suitable for sending to a refinery.

A cross-section view of the basic reactor design is shown in Figure 1. The plasma section of the reactor consists of a 16-inch diameter stainless steel shell that supports a high alumina refractory liner. This liner consists of a thin corrosion- resistant, impermeable inner layer and an insulating outer layer. Thermocouples were placed throughout the refractory shell to measure the temperature profile. Six 1.125 inch diameter graphite anodes were evenly spaced around the circumference of the plasma section. The anodes are canted down at 25 degrees from the horizontal to keep the arc attachment away from the wall.

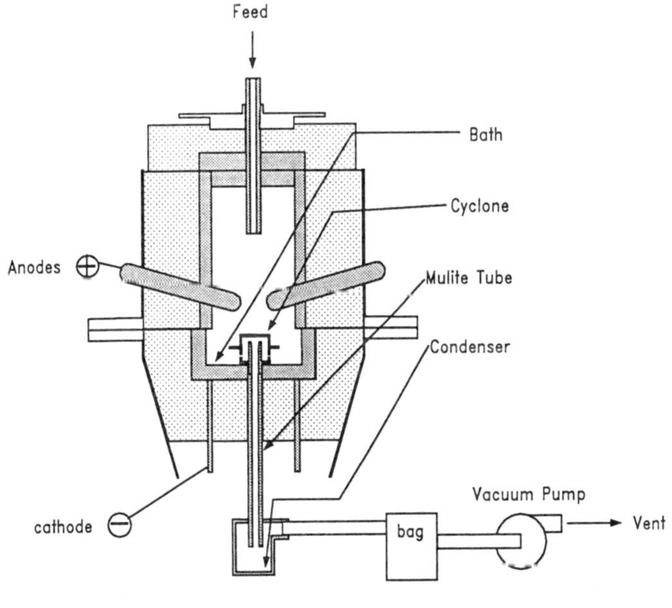

Figure 1. EFD Reactor

Attached to the bottom of the plasma section is an insulated section that contains a graphite bath. This section is connected to the plasma section by a gas-tight flange. An exit tube made of mullite exits the center of the bath and leads to a condenser. Gases leaving the condenser are filtered through a bag to remove any remaining particulates.

To start the reactor, an arc is established between the graphite feed tube and the anodes. The arc is then transferred to the bottom cathode in the center of the bath. This forms an inverted cone of plasma, which easily entrains falling feedstock. This plasma cone is approximately 3 inches high. This allows falling particles a residence time of about 100 milliseconds. The metal oxides are reduced in-flight as they fall through the plasma. Feedstock is fed from a gas-tight vibratory feeder through the graphite feed tube at the top of the reactor. A majority of the iron forms liquid droplets that collect in the bath, while the zinc, lead, and cadmium vaporize and leave the reactor through the exit tube. Most of the iron that is volatilized in the plasma impacts on the walls of the cyclone before it can exit with the other gases.

All experiments were conducted with the excess reactor pressure maintained at +0.1 inches of water gage. This positive pressure was critical to preventing outside oxygen from entering the reactor. The inside wall temperature of the reactor was calculated using a finite element program and the thermocouple data. The wall temperature was maintained above 1200°C by controlling arc power. It was critical to maintain this temperature to prevent zinc condensation on the reactor walls.

The feedstock dispenser was controlled by a computer program which improved the accuracy of feeding. The feedstock used was EFD from a local source assaying 27.15% Fe, 20.04% Zn, 2.42% Pb, 0.12% Cd prior to the addition of 17% carbon in the form of char. This mixture was blended with small amounts of gelatin (<2% by weight) as a binder with enough added moisture to form prills. The feed was passed through 28 mesh Tyler screen and maintained at 105°C to keep it dry.

Experimental Results

Figures 2 and 3 on following pages illustrate results of two representative experiments conducted on EFD using the SSP reactor. The results show excellent volatilization of zinc, lead and cadmium from the iron product and a good purity zinc-lead product. Each figure shows a cross-section of the reactor along with a tabular representation of the distribution of the elements and their location with respect to the reactor and collection system. Tables 1 and 2 are summaries of two representative tests, EFD 21 and EFD 22, respectively, which list operating parameters and elemental mass balances. The overall mass balance is not good because not all the products can be recovered from the refractory and the bath for the relatively small amount of feed (400 gms) to the reactor.

Conclusions

Results from the bench scale SSP reactor are promising. The residual amounts of lead, zinc and cadmium left in the iron product were near the detection limit of the assay equipment (<0.2) and were in a non-leachable form. The zinc-lead product contained 65% zinc and 6% lead. Oxidation of fine metallic products unavoidably occurred prior to analysis. Argon flow rates were typically 5 to 6 SCFH for feed rates of 17 to 20 grams per minute. Upon increasing the feed rate by 150%, no additional argon flow was required indicating that in the range of feed rates studied, the rate of argon flow was not a critical factor in stabilizing the arc. In fact the system operated well with no argon flow

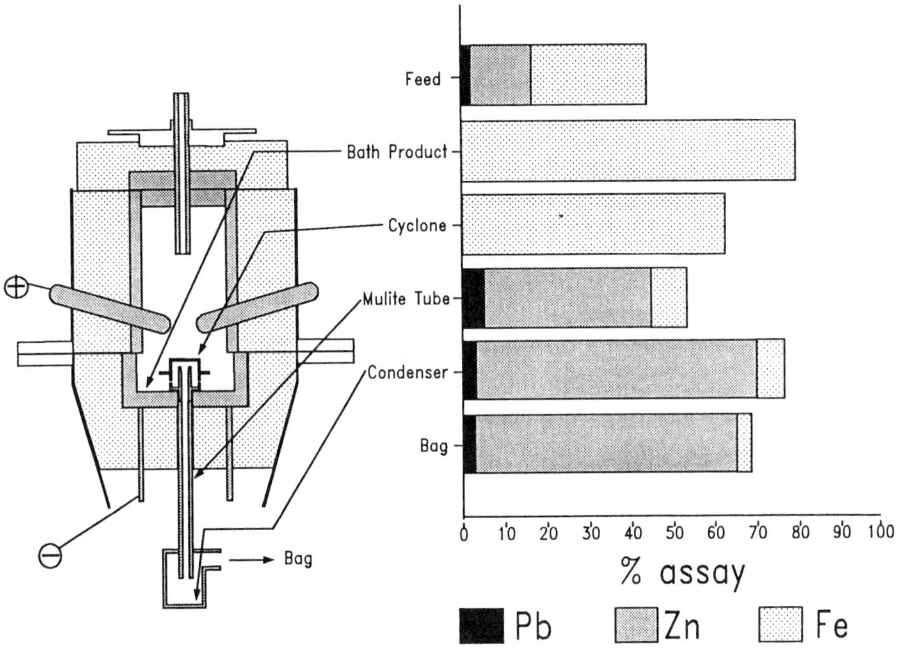

Figure 2. EFD 21

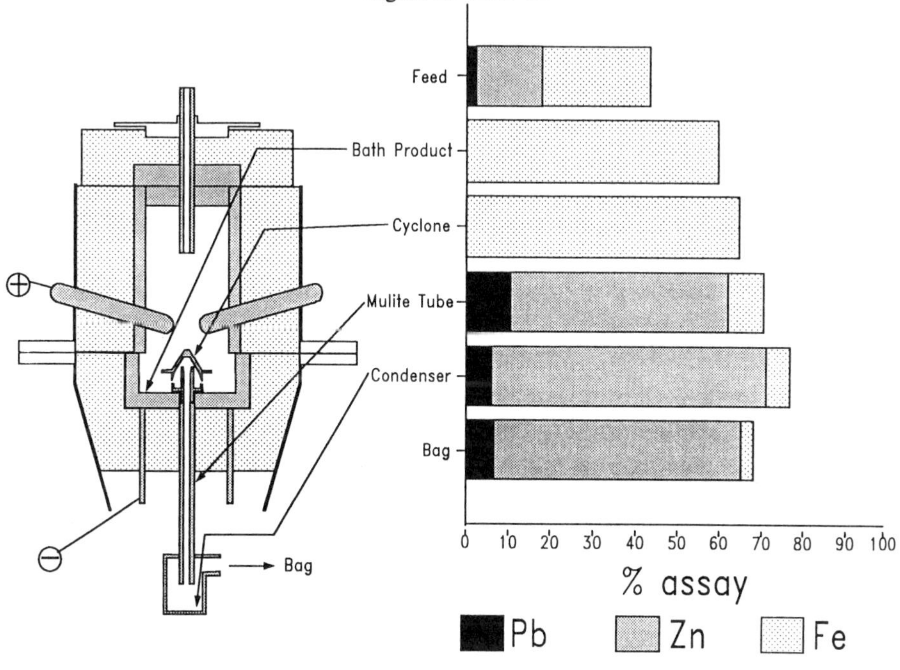

Figure 3. EFD 22

Table 1. Test Results - EFD #21

Run EFD #21

Feedstock 17% Carbon		
Cathode gas	6.0	SCFH
Aux gas	0.0	SCFH
Arc length	2.5	in.
Mass Fed	390.0	gm.
Feed time	23.0	min
Feed Rate	17.0	g/min
Total Plasma time	202.0	min
Arc Voltage	90.0	Volts
Arc Current	125.0	Amps
Aux Voltage	160.0	Volts
Aux Current	65.0	Amps
Power	21.6	KW
Orbital frequency	75000.0	RPM

SAMPLE	Mass g.	Pb%	Zn%	Fe%
Feed	390.0 (100.00)	2.35 (100.00)	15.76 (100.00)	25.11 (100.00)
Bath Product	14.0 (3.59)	0.01 (0.02)	0.01 (0.00)	79.00 (11.29)
Cyclone	3.0 (0.77)	0.02 (0.01)	0.04 (0.00)	62.40 (1.91)
Mulite Tube	28.0 (7.18)	12.60 (38.49)	32.10 (14.62)	8.46 (2.42)
Condenser	34.5 (8.85)	5.01 (18.86)	62.95 (35.33)	4.95 (1.74)
Bag	34.5 (8.85)	6.05 (22.77)	59.95 (33.65)	2.83 (1.00)
Shell	18.0 (4.62)	0.03 (0.06)	0.00 (0.00)	90.60 (16.65)
	0.0 (0.00)	0.00 (0.00)	0.00 (0.00)	0.00 (0.00)
	0.0 (0.00)	0.00 (0.00)	0.00 (0.00)	0.00 (0.00)
	0.0 (0.00)	0.00 (0.00)	0.00 (0.00)	0.00 (0.00)
Total Mass Recovered (% Recovered)	132.0 (33.85)	7.35 (80.21)	51.39 (83.61)	34.29 (35.02)

Table 2. Test Results - EFD #22

Run EFD #22

Feedstock 17% Carbon		
Cathode gas	6.0	SCFH
Aux gas	0.0	SCFH
Arc length	2.5	in.
Mass Fed	330.0	gm.
Feed time	18.5	min
Feed Rate	17.8	g/min
Total Plasma time	150.0	min
Arc Voltage	110.0	Volts
Arc Current	120.0	Amps
Aux Voltage	155.0	Volts
Aux Current	70.0	Amps
Power	24.1	KW
Orbital frequency	75000.0	RPM

SAMPLE	Mass g.	Pb%	Zn%	Fe%
Feed	330.0 (100.00)	2.35 (100.00)	15.76 (100.00)	25.11 (100.00)
Bath Product	9.0 (2.73)	0.02 (0.02)	0.03 (0.01)	59.60 (6.47)
Cyclone	3.2 (0.97)	0.02 (0.01)	0.01 (0.00)	64.40 (2.49)
Mulite Tube	26.1 (7.91)	10.53 (35.44)	51.40 (25.79)	8.23 (2.59)
Condenser	26.6 (8.06)	6.03 (20.68)	64.80 (33.14)	6.13 (1.97)
Bag	22.0 (6.67)	6.81 (19.32)	58.00 (24.53)	3.27 (0.87)
Shell	13.0 (3.94)	0.01 (0.02)	0.01 (0.00)	87.80 (13.77)
	0.0 (0.00)	0.00 (0.00)	0.00 (0.00)	0.00 (0.00)
Total Mass Recovered (% Recovered)	99.9 (30.27)	5.85 (75.49)	43.42 (83.48)	23.34 (28.16)

for short periods, indicating that the system gases (predominantly CO) were sufficient to maintain arc stability. Although the power requirements for this small reactor were high (21 kWh/kg of feed), a scale up of the reactor by at least a factor of ten is required to obtain a real estimate of the power requirements required for steady state operation, where reactor losses become less significant with respect to the reduction reactions. For our small test reactor, heat losses accounted for approximately 90% of the energy input. The major problem now is plugging of the exit tube from the reactor chamber. This phenomenon has been observed by Cowx et al.,(4) who attribute plugging to the presence of zinc halogens. Chemical methods are being explored to surmount this problem. With respect to carbon losses from the electrode, wear was minimal, amounting to approximately 1 gm for 1,000 gms fed. Since the electrodes are commercial-grade graphite, the cost factor of electrode losses is expected to be minimal.

Acknowledgement

The authors wish to express their thanks for the support of the United States Bureau of Mines which funded this research through its Generic Center for Mineral Industry Waste Treatment and Recovery and also the Center staff of the Mackay School of Mines at the University of Nevada, Reno.

References

1. Federal Register 45 No. 98, Section 261.32, pp 33, 124; EFD Hazardous Waste Number: K061.

2. T.P. McAloon, "So Little Time, So Many Choices," Iron and Steelmaking 8/87 pp 14-20.

3. J.K. Tylko, Treatment of Matter in Low Temperature Plasmas, U.S. Patent 4,361,441, 11/30/82; and Tylko, J. K.: Hybrid Plasma Reactor, U. S. Patent 4,801,435, 7/31/89.

4. P. Cowx, et al., "The Processing of Electric Arc Furnace Baghouse Dusts in the Tetronics Plasma Furnace," AIME TMS Annual Meeting, Annaheim, CA, 2/90.

5. J.J. Moore, et al., "In-Flight Plasma Reduction of Mineral Concentrates," International Metallography Conference, 7/21/81.

6. J.K. Tylko and M.T. Hepworth, "In Flight Chromite Reduction in Orbiting and Pulsating Plasmas," Conference ISPC/Tokyo, 1987: Proceedings of the International Union of Pure and Applied Chemistry, Editor: Akashi, K and Kinbara, A.

ALUMINUM:
BYPRODUCT RECOVERY

Session Chairman:
Ray D. Peterson

ALUMINUM DROSS PROCESSING IN THE 90'S

G. J. Kulik
IMSALCO
Goodyear, AZ

J. C. Daley
IMSAMET, Inc.
Litchfield Park, AZ

Abstract

Recovery of aluminum from drosses has a significant economic impact on the United States aluminum industry. Today's modern dross processor uses the latest technology to maximize metal recovery at the lowest cost while complying with all environmental laws and regulations. However, some industry business policies continue to encourage start-up of small local processors who emphasize short term profits at the expense of sound environmental procedures. The paper discusses some of these factors and draws a comparison with practices developed in the steel industry where long term metal recovery arrangements exist with reputable service companies.

Second International Symposium -
Recycling of Metals and Engineered Materials
Edited by Jan H. L. van Linden,
Donald L. Stewart, Jr., and Yogeshwar Sahai
The Minerals, Metals & Materials Society, 1990

Introduction

During a relatively short period in the history of man, aluminum metal has become one of society's most important commodities. In less than a century, aluminum producers have developed equipment and operating techniques to convert naturally-occurring aluminum oxide into pure metal, and metallurgists have learned to alloy aluminum with other metals to give it a limitless number of physical properties. These alloys can be cast, extruded, machined or otherwise formed and worked to produce a large number of consumer products that are often taken for granted in modern societies.

However, a very important aspect in the production of aluminum metal has not been mastered, namely preventing the formation of dross on molten aluminum. Due to its great affinity for oxygen, the metal returns to its oxide state in the presence of even minor amounts of oxygen and this reaction is accelerated at high molten aluminum temperatures. The aluminum oxide has a higher melting temperature and a lower specific gravity than molten aluminum, so aluminum furnace baths are normally covered with a layer of oxide solids. Skimming of this material from the bath invariably results in the coincidental removal of some of the underlying metal.

Metallurgists have also failed to develop a method of converting this "secondary" aluminum oxide back to the metallic state once it has formed. Numerous attempts to develop a practical and economical method of recovering aluminum from this oxide material have been unsuccessful. Consequently, production of aluminum metal is accompanied by the formation of large quantities of aluminum dross.

Types of Aluminum Dross

Webster's Dictionary defines dross as "the scum that forms on the surface of molten metal". Aluminum furnace operators learned many years ago that excessive formation of this scum resulted in high aluminum metal losses, both through formation of the oxide which could never be reduced back to the metal, and the entrapment of aluminum metal during the manual or mechanical removal of the scum from the furnace. Metallurgists also learned that oxide formation could be reduced by the use of a flux that was molten at furnace operating temperatures and would float on the molten aluminum. This flux would act as a barrier between the metal and atmospheric oxygen, reduce oxide formation, lower the production of dross and increase aluminum recovery. Salt and potash have been found to be effective for this purpose, and these chloride salts are widely used in the industry today.

Fluxing techniques and production practices vary widely throughout the industry, resulting in the production of different types of dross.

White Dross

The generation of white dross occurs at primary aluminum smelters, extruding plants, sheet mills, foundries and die casters. The furnaces at these facilities are operated without fluxing, and the dross skimmed from the furnaces is grey or metallic white in color.

White dross as it is skimmed has a very high aluminum metal content. However, oxidation or "thermiting" will occur very rapidly because of the high temperature and the absence of flux. If not controlled, all of the aluminum metal could convert to aluminum oxide. Thus, thermiting has a major impact on the value of the metal recovered from white dross and the amount of waste aluminum oxide generated.

Thermiting can be controlled by use of an oxygen-free environment or mechanical dross coolers that quench the oxidation process. Some producers simply spread the thermiting dross on steel floor plates and cover it with old dross fines.

Depending on the efforts made to control thermiting at the producers' plants, recovery of metal from white dross can be as high as 70% or as low as 15%.

Primary aluminum producers also generate metallic spills and pot bottoms from electrolytic cells. These items are often recycled by dross processors that have furnaces capable of handling large, irregular shaped pieces of metal.

Black Dross

Black dross is generated by secondary aluminum smelters who use open hearth reverberatory furnaces for melting old castings, clippings, turnings and used beverage cans. A salt potash flux is used in the open hearth to reduce the amount of oxidation occurring on the exposed metal. At the high molten metal temperatures, the flux melts and becomes dark colored, hence the name "black" dross.

Because of the flux, thermiting of the aluminum metal is a minor problem. The flux does dilute the dross, however, and the metal recovery is only in the 12%-18% range.

Saltcake

Recovery of aluminum metal from white and black dross is accomplished in molten salt bath rotary furnaces. A salt-potash flux is used in the furnace to maximize the recovery of the metal, and the spent flux is discharged from the furnace either continuously or on a batch basis. This residue, containing 3%-5% metal, is referred to as saltcake and is normally disposed of in landfills.

Saltcake may be economically reprocessed for aluminum metal recovery if the dross processor has the necessary crushing and concentrating equipment. However, the non-metallic portion, including the spent flux, must be landfilled.

Value of Aluminum Dross

Aluminum industry trade groups, government agencies, stock analysts and others publish a wide array of statistics regarding the industry. Information is available on primary metal production, scrap aluminum collection and shipments, exports, imports, can sheet production, UBC collection and recycling rates, and all conceivable aspects of the business. Data is provided monthly, quarterly and annually on orders, production levels and inventories of various mill products, together with graphs and charts showing capacity in use, relative rates of change, and anticipated market effects. Similar information is available for metal prices -- past, current and future.

It is interesting to note that within this mass of data, there is no information on aluminum dross. Metal Statistics reports on the tonnage of Recycled Secondary Ingot (RSI) that is consumed by primary and secondary producers, and Type 380 RSI metal quotations are readily available. However, the quantity of dross produced by the aluminum industry is not reported so the annual tonnage of this material must be computed using other published statistics.

Table I is an estimate of U.S. dross production during 1988 based on typical dross production factors and metal recoveries. These figures indicate that aluminum recovered from dross totals approximately 540 million pounds per year. Assuming an RSI value of $0.60 per pound, dross processing and metal recovery is a $325 million-a-year business.

Table I 1988 Estimated U.S. Aluminum Dross Production (000's Short Tons)

	Metal Prod'd.	Dross Fact. %	Dross Gen'd.	Alum. Rec. %	RSI Rec'd.
Primaries	4,350	2.0	87	40	35
Mills & Extruders	7,850	5.0	393	45	177
Secondaries	1,700	15.0	255	15	38
UBC Recyclers	650	10.0	65	30	20
			800		270

Source: Aluminum Association Statistical Review

Table II shows the impact of dross processing on the U.S. economy. Based on typical industry factors, aluminum recovery from dross is worth $120 million to dross generators, creates 1,250 jobs and supports a capital investment of $67 million. This is clearly a significant segment of the U.S. aluminum industry.

Table II Economic Value of Dross Processing

	$ Million Per Year
RSI Recovered - 270,000 tons	
- 540 million pounds	
RSI Value @ $0.60 per pound	$ 325
Estimated Distribution:	
To Dross Generators	$ 120
To Transporters	45
To Dross Processors	160
	$ 325
Capital Investment	
Based on depreciation of 1.25¢ per pound RSI and 10 year plant life	$ 67
Jobs Created	No. Jobs
Transportation	610
Dross Processing	640
	1,250

Current Dross Treatment Technology

The recovery of aluminum metal from dross is accomplished in molten salt bath rotary furnaces. Some high-grade metallic slabs and pieces may be hand-sorted from the dross and fed to reverberatory furnaces, but the greatest tonnage of dross is charged into rotary furnaces.

Rotary furnace design can vary in terms of size, feeding arrangement, metal tapping and saltcake discharging methods, and burner design. Small entrepreneurs can become dross processors by simply installing a refractory lining in the drum of a discarded concrete delivery truck. The size can range to behemoths 15 ft. diameter and up to 30 ft. long.

Regardless of the dimensions, the principle of all rotary furnaces is identical, namely to melt the salt flux and to coat the aluminum metal particles with the flux to avoid oxidation. The rotation of the furnace on a set of trunnions accomplishes this by forcing the dross beneath the surface of the bath and away from the direct burner flame as quickly as possible. The furnace burners are thus used to heat the exposed refractory which exchanges heat to the bath during the rotation cycle.

Dross operators have developed a number of ways to feed the furnace, ranging from the simple use of front-end loaders with removable chutes, to sophisticated bins, conveyors and slingers. The objective in all cases is to feed the furnace as quickly as possible to minimize refractory cool-down and to improve on-line time.

Metal is tapped from the furnace through a hole located in an end or in the belly of the furnace. The aluminum is laundered into RSI sow molds of various design.

Saltcake can be removed from the furnace in any number of ways. Some furnaces are tilted so saltcake can continuously overflow through an opening in the center at one end of the furnace. Some operators have mounted the furnace on a set of longitudinal rockers so the saltcake can be poured from the furnace quickly by tilting the furnace toward one end. Saltcake can also be tapped from the furnace through a large taphole similar to that used for the metal.

Rotary furnaces can be fired with natural gas or fuel oil. Burners can be fixed, or mounted on swinging brackets if the furnace is fed through the burner port. Off-gases are normally cleaned in baghouses.

Most large dross processors have installed concentrating facilities to upgrade black drosses to a 60%-70% concentrate before being charged to the rotary furnaces. Capital and operating costs for the concentrating plant can be justified on the basis of fuel requirements, flux consumption, saltcake generation and furnace productivity. In fact, metal cannot be economically recovered from black dross without concentrating in most situations.

Concentrating flowsheets vary among processors, but in all cases involve the following basic processes:

Crushing. Black dross is skimmed from reverberatory furnaces into boxes, tubs and containers of a great variety of shapes. The dross freezes into these shapes and the dross processor must be prepared to crush large blocks of dross into pieces no larger than 6 in.-8 in. size. Jaw crushers or impactors are used for this purpose.

Milling. Crushed dross is milled to free aluminum metal from aluminum oxide and salt particles. This can be accomplished in a hammer mill, ball mill or autogenous mill where pieces of large aluminum metal act as the grinding media. Most milling is done dry so baghouses are used to control dust generation. In some situations, wet milling is acceptable and this can result in a higher grade concentrate because the water dissolves the flux clinging to the aluminum particles. Wet milling can produce a brine suitable for subsequent flux recovery and low-chloride mill products suitable for marketing.

Screening. The mill discharge is passed over vibrating or trommel screens to recover the aluminum concentrate and other potentially marketable products. For example, the +12 mesh material will be the concentrate, and the -30 mesh fraction the tailings. The intermediate product may contain 15%-25% aluminum metal and is valuable as an exothermic product. Tailings from wet milling are low in chlorides and suitable for cement manufacturing.

Figure I is a schematic flowsheet showing a typical dross processing operation.

Figure I Dross Processing Schematic Flowsheet

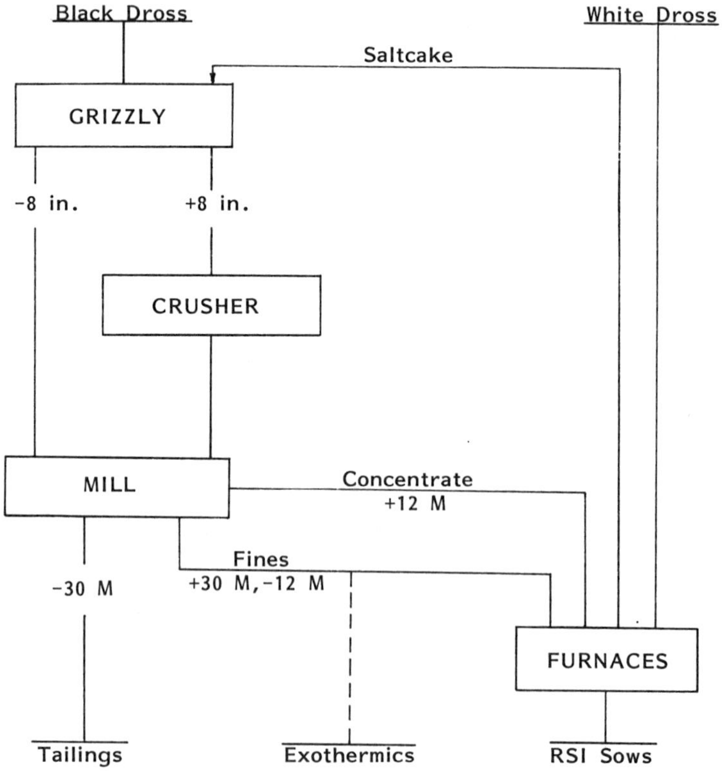

Dross Treatment Developments

Because of current economic and environmental concerns, dross and dross treatment are beginning to draw the attention of research and development groups within and associated with the aluminum industry. Some recent developments are:

Dross Cooling. Mechanical dross coolers are designed to halt the thermiting of white dross by cooling the dross as quickly as possible after being removed from the furnace. This is accomplished by feeding the hot dross through a water-cooled steel drum where quenching will typically improve recovery from 30% to 45%. This economic incentive, plus the reduction of waste aluminum oxide, is being recognized by the industry, and dross coolers will become generally accepted during the next decade.

Flux Recovery. Recovery and recycling of flux from black dross has been attempted on numerous occasions over the last several decades. Salt crystallization plants have been installed to accept brine produced by digesting black dross or saltcake, and the few process problems in recovering flux in this manner have generally been identified and overcome. Unfortunately, salt recovery from dross has not developed in the U.S. because of high capital costs, high operating costs in terms of energy and maintenance caused by chloride corrosion, low value of the recovered flux, and low landfilling costs.

During the 90's, dross producers and processors will turn to flux recovery to reduce the volume to be landfilled, and to make the final product more acceptable to landfills because of its low chloride content. Crystallization and solar evaporation technologies will develop so that processing of all salty drosses will become the accepted standard.

Plasma Arc Technology. Research has shown that white dross can be processed without the use of flux in a rotary furnace fitted with a plasma arc torch, and construction of a full-scale production unit has been announced. This process will avoid the generation of saltcake and make the residue more suitable for low cost landfilling. The high energy cost of the operation is reportedly off-set by economics in pollution control equipment costs and landfilling.

Environmental Concerns

Aluminum dross processing until now has focused almost exclusively on the recovery of aluminum metal values from the dross. Furnace operating and fluxing procedures have been directed toward achieving maximum possible metal recovery at lowest possible costs. Dross concentrating equipment and operating techniques have evolved to the point that black dross with a metal recovery as low as 8% or 9% can be profitably treated, depending on freight cost factors. In some special cases, saltcake with recoveries of only 4% or 5% can be economically processed through concentrating equipment and rotary furnaces.

It is apparent that dross processors extract every pound of aluminum into RSI that can be economically recovered. Some processors find that aluminum-bearing fines can be recovered as a by-product of the concentrating process and marketed as an exothermic product to steelmakers. This material can generate attractive incremental revenues depending on the relative location of the dross plant and exothermic markets.

Unfortunately, dross also contains large quantities of by-product materials that are not recovered today. Table III shows that in a typical year, approximately 850,000 tons of aluminum oxide and flux are generated as dross processing by-products.

Table III Solid Residues from Dross Processing
(000's Short Tons)

	Dross Gen'd.	Flux Used	RSI Rec'd.	Oxide & Flux
Primaries	87	50	35	102
Mills & Extruders	393	215	177	431
Secondaries	255	42	38	259
UBC Recyclers	65	13	20	58
	800	320	270	850

Tests and determinations by industry groups and various federal and state environmental agencies have proven that, with few exceptions, dross by-products are neither toxic nor hazardous. Aluminum oxide was recently removed by the U.S. Environmental Protection Agency from the list of toxic chemicals that must be reported under Sara Title III. Flux is composed of sodium chloride (common table salt) and potassium chloride (potash fertilizer). These are widely used in modern society, and have neither toxic nor hazardous characteristics.

Dross processing generates two types of residues:

1. Aluminum oxide tailings from wet milling of black drosses and saltcake. Because water is used in the milling operation, the soluble salts are dissolved and the resultant tailings contain less than 1% chlorides.

2. Saltcake and saltcake residues from rotary furnaces that contain 40%-50% flux.

Large quantities of these have accumulated at dross processing sites across the country because by-product markets have not yet developed for them. In metropolitan areas, this material is landfilled if it passes the EP Toxicity Test and receives approval from the local regulatory agency as a "special" waste. At some disposal sites where the soluble salts affect ground water quality, mandatory cleanup efforts are underway. In addition, numerous small accumulations of this material are located at many sites in the country where small dross operations sprang up when aluminum prices were strong, and failed when prices dropped. Small processors exit the business as easily as they enter it, often leaving behind a pile of dissolving saltcake which can contaminate local drinking water supplies.

In summary, at a time when the country is emphasizing waste reduction and recycling, the U.S. aluminum industry finds itself producing large quantities of by-product materials that have limited markets. Unless this problem can be solved, aluminum producers' profits from dross will shrink, and for some, dross disposal will become a major cost factor rather than a significant revenue item.

Causes of the Problem

During the growth of the U.S. aluminum industry, some business practices have developed that have hindered the development of by-product markets. Although some small amounts of aluminum oxide tailings are marketed to alum producers for use in water treatment, and some to the cement industry, markets for the bulk of the residues have not developed.

Some aluminum industry practices that have caused the problem are:

Short-Term Dross Arrangements. With few exceptions, dross is sent to processors under short term tolling or sales arrangements. Dross processors are asked to bid on small accumulations of dross at the producer's plant or on anticipated generation of dross for the next quarter.

Small Quantities. Contracts for more than one million pounds of dross rarely occur. Lots as small as one or two truckloads are common.

"As Is" Guarantee. Dross is sold on an "as is" basis. The purchaser of the dross assumes all risk associated with metal recovery and RSI metal prices. The dross buyer is never sure of profitability until the dross has been processed and the metal sold.

Successful High Bid. Dross producers will always sell to the highest bidder (or the lowest bidder in the case of tolling). It is not unusual for a long standing relationship between a producer and processor to be suddenly severed on the basis of a fraction of a cent per pound of dross or a few hundred dollars on a quarter's production. The successful bidder is often unaware of current metal recoveries from the dross or the value of the recovered RSI and may try to back out of the deal. The dross producer then resumes shipments to the original processor, but the interruption is very upsetting to efficient plant operation.

Part-Time Brokers. Most dross generators will sell to anyone interested in submitting a bid. The highest bid is often accepted with little concern for the ultimate disposition of the dross or residual materials. Successful bidders are often part-time speculative brokers who solicit tolling prices from established dross processors after their dross bid has been accepted. It is not unusual, for example, for a dross processor in Arizona to receive a call from a stranger in Miami requesting a quote for dross that he owns in the state of Washington. This material often finds its way to small out-of-the-way dross processors who can offer the best financial arrangement to the broker because environmental control costs are a minor factor.

Steel Industry Practice

The U.S. steel industry has many similarities to the aluminum industry. Steel is produced at 81 integrated and electric-furnace mills in the country. Steelmakers produce large quantities of slag, an oxide by-product, and recovering iron and other metallics from slag is analagous to the recovery of aluminum from dross. However, the annual volume of steel slag generated far exceeds that of dross (Table IV).

Table IV U.S. Steel and Aluminum Production
(000's Short Tons)

	Steel	Aluminum
Annual Production	97,500	14,550
Slag/Dross Generated	18,000	800
Steel/Aluminum Recovered	540	270
Residue Produced	17,460	850
Residue Recycled	17,460	0

Virtually 100% of U.S. steel slag is handled and processed by independent service contractors operating on site under long-term contract arrangements. The development of outside service contracts blossomed during the 1960's when steel producers realized that specialists could handle this work more efficiently and economically than they could themselves. Today, all steel mills are serviced by a handful of relatively large, independent processors who have developed the technical expertise to provide slag processing and metal reclamation services, and have the financial strength to withstand the cyclical nature of the steel industry.

The major benefits provided to the steelmaker by the independent slag processor include:

- The return of the valuable metal units of known chemistry to the steel-making process at a predictable cost.
- The processing of the residual non-metallic material into marketable by-products.
- Capital investment made by the slag processor.
- Environmentally-sound operating and disposal practices.

All steel slag non-metallics, or aggregates, some 17,000,000 tons per year, are marketed to the highway and building construction industries. Because of the long term arrangement with steel producers, today's successful slag processor has been able to invest valuable time and financial resources into identifying and developing these aggregate markets. For example, International Mill Service, a firm that processes slag at over 40 mills in the United States, operates an aggregate research test lab in Pennsylvania. The lab works with the company's marketing group and various state highway deparments to include slag aggregate standards in highway construction specifications.

Service contracts in the steel industry (Exhibit I) have been developed over a long period of time to satisfy the needs of both the independent processor and the steel producer. The contract term, typically 5 to 10 years, assures both parties with continuity of the arrangement and helps the processor obtain low cost financing for his equipment. This investment can be depreciated over a long time period, allowing the processor to pass on lower processing fees to the steel mill.

Exhibit I Typical Steel Slag Service Contract

Term	5-10 years
Slag Ownership	Steel mill owns metal units Aggregate owned by steel mill until shipped to third party
By-Product Sales	Royalty paid to the steelmaker
Fees	Monthly service fee Fee per ton steel produced Fee per ton specification metal reclaimed from the slag
Price Adjustment	Periodic adjustment based on mutually agreed formula and indices

In these contracts, the steel producer always retains ownership of the recovered metals and aggregates. This arrangement together with the fee structure assures the steelmaker that the slag processor is operating in the best interests of the mill. A feeling of mutual trust quickly develops between the two companies and their on-site personnel, and renewal of contracts at the end of the term is usually completed to the mutual satisfaction of both parties.

The negotiated fee structure provides predictability and protection that is fair to both parties. Built-in incentives ensure maximum prices for by-product aggregates. A price adjustment formula provides long-term financial security for the service contractor and allows both parties to avoid renegotiation of prices and fees prior to contract termination.

As a result of these business arrangements, 100% of steel slag in the U.S. is recycled. Moreover, a large number of slag pile accumulations throughout the country have been profitably processed, the metallics recovered and the aggregates marketed.

Aluminum Dross in the 90's

Webster's Dictionary also defines dross as "something that is base, trivial or inferior". To some extent, the aluminum producers have adopted this attitude toward their by-product and consider dross processors as the industry's "scavengers". These attitudes must change if the industry's by-product disposal problems are to be solved during the 90's. Long term service contracts with reputable dross processors can offer the answer.

THE ENGITEC SYSTEM FOR TREATMENT OF SALT SLAG FROM SECONDARY ALUMINUM SMELTERS

Randal M. Reynolds, P.E.
Lake Engineering, Inc.
Atlanta, Georgia

Dott. Ing. Marco Olper
Engitec Impianti, S.p.A.
Milano, Italy

The secondary aluminum industry is under increased pressure from state and federal solid waste agencies to greatly reduce or eliminate the land disposal of its salt slags. This material is generated from rotary furnace processing of dross and scrap aluminum. Engitec Impianti has designed and constructed salt slag recovery systems in Carisio, Italy and Cacapava, Brazil. The Carisio facility, a second generation design, is capable of producing 21,000 m. tons per year of recovered salt from a brine solution stored in old evaporation ponds at a secondary smelting plant. This plant utilizes a single evaporation/crystallization unit with mechanical vapor recompression. The Cacapava plant is designed to produce 8,000 metric tons per year of salt and utilizes a flash evaporator. The flash evaporation technology is easy to operate, provides flexibility, and can be coupled with evaporation/crystallization for larger size systems. A third facility will be completed by late 1990 in Borgofranco, Italy for Alcan Italia. This system represents the latest in the developing technology to address brine purification, elimination of noxious gases, and production of an oxide fraction low in chlorides which enables reuse for addition to cement.

Second International Symposium -
Recycling of Metals and Engineered Materials
Edited by Jan H. L. van Linden,
Donald L. Stewart, Jr., and Yogeshwar Sahai
The Minerals, Metals & Materials Society, 1990

Introduction

The secondary aluminum industry encompasses the recovery of a wide range of aluminum scrap both in form and composition. The scrap is sorted, prepared, blended, melted and refined into a variety of alloys with controlled specifications. In the U.S., the open well reverberatory furnace and the rotary furnace are commonly used for the melting of aluminum scrap. During the melting process, fluxes consisting primarily of NaCl or KCl are used over the molten bath to reduce metal oxidation and to remove impurities from the charge. The salt slag formed contains salt fluxes, some aluminum metallics, aluminum oxide fines, and water reactive aluminum compounds.

The salt slag from a reverberatory furnace may contain 15 to 20 percent aluminum metal and is commonly sent to an off-site rotary furnace operator. The salt slag may be stockpiled or landfilled, depending on the local solid waste regulations. The salt slag from a rotary furnace may contain three to five percent metallic aluminum and is usually disposed of by landfilling. Certain rotary furnace operators crush the rotary furnace slag to recover the metallic aluminum content.

Salt slag, although not yet classified as a hazardous waste, presents its own set of problems concerning land disposal. The slag is very prone to leaching chlorides when in contact with rainfall or groundwater. Also, salt slag in contact with water evolves gases, including mainly hydrogen, ammonia and methane. Smaller amounts of hydrogen sulfide and phosphine may also be evolved. Over the past one to two years, increased regulatory scrutiny has occurred in many states concerning the land disposal of salt slag.

Over the last 13 years, Engitec Impianti, S.p.A. has developed and constructed a salt slag recovery system in Carisio, Italy. This paper describes the process technology evolved from that system. Engitec has also recently completed the installation of a smaller system in Cacapava, Brazil. A third installation is planned in Borgofranco, Italy by late 1990. This paper will discuss the design and operating parameters of both the small and large systems.

Conclusions

Since 1977, Engitec Impianti's salt recovery system has been successfully operated at Carisio, Italy. This plant was completely renovated and modernized beginning in 1988 and includes an evaporator/crystallizer unit with mechanical vapor recompression.

At about the same time as the Carisio renovations took place, Engitec completed pilot testing and patents of a flash evaporation system for smaller capacity plants as well as to serve as a pre-concentration unit for large evaporator/crystallizer systems which includes elimination of noxious gaseous emissions. The flash evaporation system is not only easy to operate but is less capital intensive.

A flash evaporation plant having a production capacity of 60 metric tons per day of salt slags processed would have an operating cost of about $57 per metric ton of salt slag treated. A larger plant (200 tons per day) with a flash evaporation and evaporator/crystallizer would have an operating cost of about $33 per metric ton of slag charged. These costs include the necessary milling costs. The estimated revenues generated including recovery of salt fluxes, aluminum metallics and net savings on landfill disposal would be about $104 per metric ton of salt slag treated. The estimated revenues do not include any sales value for the clean oxide fines.

The process technology to treat salt slags from the secondary aluminum industry in an economical and environmentally sound manner has been demonstrated by Engitec. Over the next one to three years, both primary and secondary aluminum producers in the U.S. should begin taking steps towards alternatives to the land disposal of salt slag materials.

Background/History

Engitec designed and installed the first generation salt recovery system at Carisio, Italy. The facility was designed to process 50 metric tons per day of salt slag from the existing operation. Molten slag was transported in ladles to a large rotary drum. The exterior of the drum was sprayed with recycled water to cool and granulate the slag. The resulting granulated slag was leached and the brine evaporated utilizing waste heat from the rotary furnaces.

In 1987, Engitec commissioned a pilot plant for the flash evaporation system at Novate, Italy. The flash evaporation system had two primary goals: To provide a simple process flowsheet for the recovery of salt slag up to 60 metric tons per day; and secondly, to provide a combustion system to incinerate the gases generated during leaching. The pilot plant is capable of treating three tons per day of salt slag. During a two-year experimental operation period, all the parameters for operating the flash evaporation system were developed and the first plant was designed.

Process Description

Slag Crushing and Screening

The salt slag is fed by conveyor to the primary crushing mill. The crushed slag is then conveyed to a multistage screen where the following fractions are obtained:

A. Crushed fines with negligible amounts of metallic aluminum with the size less than 1.5 millimeters;

B. Crushed material having a size between 1.5 and 10 millimeters; and

C. Crushed material between 10 and 100 millimeters.

Fraction A is sent to the leaching plant for salt brine production. Fraction B is crushed in a secondary mill and then returned to the multistage screen. Fraction C is returned to the primary crushing mill. The oversize aluminum is returned to the furnace.

Industrial ventilation is provided at critical transfer points and at each of the crushing mills. A baghouse dust collector is used to collect the dust from the ventilation air before discharge to the atmosphere. The dust from the baghouse is transferred to the reaction tanks. The crushing and screening system is shown in Figure 1.

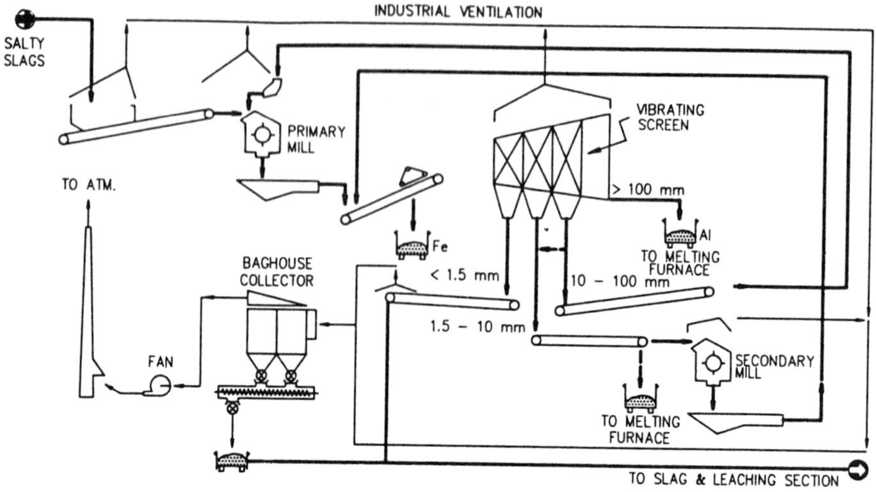

Figure 1 - Crushing and Screening System

Leaching and Filtration

The slag fines from the crushing system are conveyed to leaching reactor tanks. The crushed fines are reacted with make-up water and hot recycled water from other parts of the process. The material is leached to obtain a salt brine with a concentration of approximately 25% by weight sodium/potassium chloride. The slurry of reacted fines and the concentrated brine are pumped to a vacuum belt filter. The filtered aluminum oxide fines have a moisture content of approximately 25 to 30 percent. The filter cake is washed on the belt filter with recycled water and fresh water to achieve a chloride content less than 0.5%. This washwater is collected and recycled for further use.

Gases from the reaction tank for leaching and from the vacuum belt filter are collected in a ventilation system leading to an afterburner. A flow diagram of the leaching and filtration system is shown in Figure 2.

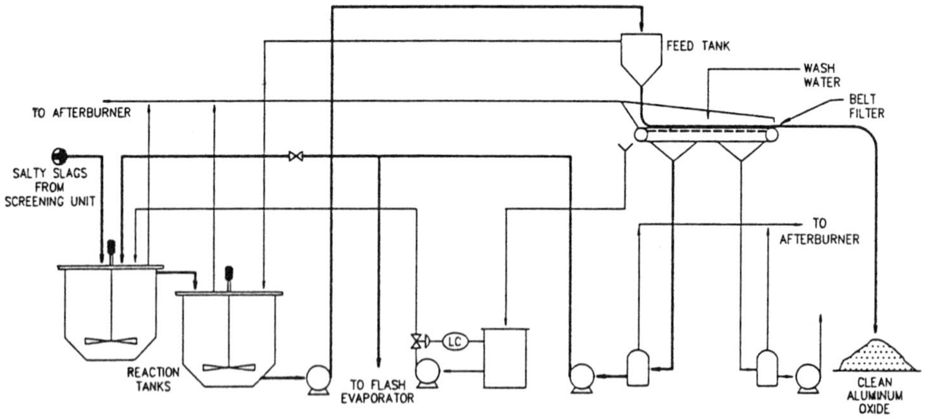

Figure 2 - Leaching and Filtration System

Brine Purification

The typical composition of the brine resulting from the leaching of salt slag is about 250 to 300 grams per liter of sodium/potassium chloride. In addition, it may contain other impurities dependent on the type and origin of the salt slag and the aluminum scrap treated. Among these impurities, ammonia, calcium sulfate and magnesium chloride create deleterious effects in the subsequent vaporization/crystallization system.

Ammonia, under the operating conditions of temperature and pressure in the vaporizer crystallizer, would be stripped from the brine according to the following reaction:

$$NH_4OH \rightarrow NH_3 \uparrow + H_2O$$

This reaction adversely impacts the pH of the magma and quickly subjects the equipment to corrosion.

Calcium sulfate, when allowed to enter the evaporator/crystallizer rapidly increases in concentration and quickly reaches a level which precipitates developing rapid scaling in the system. This ultimately causes shutdown of the system to remove the magma so that the equipment can be cleaned of the gypsum scale.

The presence of magnesium is dangerous because under certain conditions it may hydrolyze according to the following reaction:

$$MgCl_2 + 2H_20 \rightarrow Mg(OH)_2 + HCl$$

The hydrochloric acid quickly corrodes the equipment and the magnesium hydroxide tends to scale the vessels and piping.

In the Engitec Impianti brine purification section, chemicals are added to strip the ammonia and remove calcium and magnesium by precipitation and filtration in order to produce a very clean brine which can meet the specifications by the manufacturers of crystallization equipment. The ammonia is collected and burned in the afterburner.

Gas Elimination

The salt slag from aluminum scrap treatment contains impurities which when combined with water create gases according to the following reactions:

Hydrogen -

$$Al + 3\ H_2O \rightarrow Al(OH)_3 + 3/2\ H_2 \uparrow$$

Ammonia -

$$AlN + 3\ H_2O \rightarrow Al(OH)_3 + NH_3 \uparrow$$

Methane -

$$Al_4C_3 + 12\ H_2O \rightarrow 4\ Al(OH)_3 + 3CH_4 \uparrow$$

Phosphine -

$$AlP + 3\ H_2O \rightarrow Al(OH)_3 + PH_3 \uparrow$$

Hydrogen Sulfide -

$$Al_2\ S_3 + 6\ H_2O \rightarrow 2\ Al(OH)_3 + H_2S \uparrow$$

These gases are burned in the afterburner producing water, nitrogen, carbon dioxide, phosphoric acid and sulfur dioxide which pass to the flash evaporator where they are mixed with excess chemicals from the purification step. These materials are precipitated and filtered before entering the crystallizer. The gas elimination and brine purification flow diagram is shown in Figure 3.

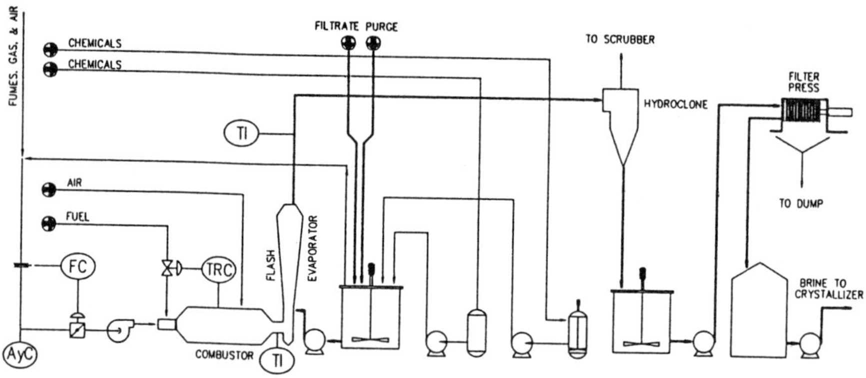

Figure 3 - Gas Elimination and Brine Purification Flow Diagram

Evaporation/Crystallization

The brine is pumped to a holding tank from which it is transferred to the flash evaporator for pre-concentration. The flash evaporator utilizes the waste heat from an afterburner unit fired with natural gas which incinerates the noxious vapors collected from all the collection points. The pre-concentrated solution is separated from the gases by a cyclone separator to a holding vessel, the gases being scrubbed before release. A portion of the slurry is recycled and a portion is sent for final evaporation. The pre-concentrated brine is filtered using a filter press and stored in a storage tank, then pumped through two heat exchangers into a draft-tube-baffle evaporator/crystallizer. The evaporator/crystallizer utilizes mechanical vapor recompression (MVR) to efficiently reuse the vapor and heat from the crystallizing process. The material is recirculated through the evaporator/crystallizer and a small stream of salt crystals is removed from the bottom of the crystallizer. The crystals are sent to a centrifuge and recovered as salt crystals. The complete evaporation/crystallization system is shown in Figure 4.

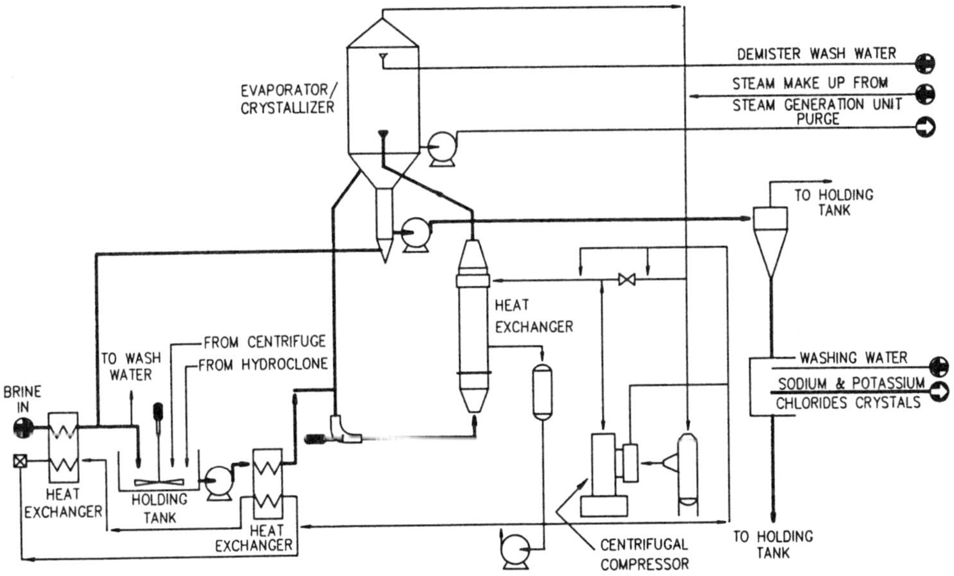

Figure 4 - Evaporation/Crystallization Flow Diagram

Plant Economics

The economic return of two standard plants is presented here. Case A plant has a capacity of 60 metric tons per day of salt slag treated while the Case B plant treats 200 metric tons per day. Costs for crushing, leaching, gas elimination, brine purification, flash evaporation (Case A) and mechanical recompression (Case B) crystallization are included.

In both cases, the composition of the salt slag is assumed to be 46% salt flux, 50% aluminum oxide fines and water reactive aluminum compounds and 4% metallic aluminum. The salt flux is assumed to have a composition of 66% NaCl, 30% KCl and 4% fluorspar.

Table I lists the annual production of products and by-products assuming 330 working days per year.

TABLE I

		CASE A	CASE B
Salt slag process - m. tons/year		19,800	66,000
Salt flux[1] - m. tons/year		9,300	29,750
Aluminum metallics[2] - m. tons/year		1,130	3,770
Aluminum oxide[3] - m. tons/year		14,320	47,680
	TOTAL	**24,750**	**81,200**
1) Moisture %		7	3
2) Metal content %		70	70
3) Chloride (as Cl_2) %		<0.5	<0.5
Moisture %		30	30

Fluorspar is removed in the aluminum oxide fines.

Capital Costs

Table II lists the capital costs (excluding building and utilities) for the milling and salt reclamation units.

TABLE II

	CASE A	CASE B
Equipment & Engineering U$ x 1,000	4,100	8,300
Civil Work & Erection U$ x 1,000	900	1,700
TOTAL	**5,000**	**10,000**
Capital Cost/m. ton Salt Slag U$ x 1,000	250	150
Building Area M^2	1,500	3,500

The equipment capital costs are considered to be provided in a turnkey project.

Operating Cost

The plant is projected to operate three shifts per day, 330 days per year. Manpower for material handling and supervision has been excluded based on the assumption that location at or near an existing plant would allow sharing of those costs.

Table III shows the costs for Case A, and Table IV shows those for Case B.

TABLE III
CASE A

Item	Unit Consumption	Cost Per Unit	Total Annual Cost	Unit Cost ($/m. ton)
Manpower	0.8 hour /m. ton	$17 /hour	$ 269,280	$13.60
Electricity	150 kwh /m. ton	0.05 /kwh	148,500	7.50
Natural Gas	170 M^3 /m. ton	0.12 /M^3	403,920	20.40
Water	1.5 M^3 /m. ton	0.26 /M^3	7,720	0.40
Maintenance	$15 /m. ton	----------	297,000	15.00
		TOTAL	**$1,126,420**	**$56.90**

TABLE IV
CASE B

Item	Unit Consumption	Cost Per Unit	Total Annual Cost	Unit Cost ($/m. ton)
Manpower	0.36 hour /m. ton	$17 /hour	$ 612,000	$ 9.30
Electricity	120 kwh /m. ton	0.05 /kwh	396,000	6.00
Natural Gas	40 M^3 /m. ton	0.12 /M^3	316,800	4.80
Chemicals	5 kg /m. ton	0.50 /kg	165,000	2.50
Water	1.5 M^3 /m. ton	0.26 /M^3	24,740	0.40
Maintenance	$10 /m. ton	----------	660,000	10.00
		TOTAL	**$2,175,540**	**$33.00**

REVENUES

Based on an average value for NaCl of $30 per metric ton and $110 per metric ton for KCl, the recovered mixture of salts in both Case A and B is about $55 per metric ton. In addition, the aluminum metallics which yields a 70% metal recovery on return to the furnace is valued at $0.80/kg. No value is considered for the aluminum oxide fines except a $40 per metric ton saved on disposal costs for the gross weight of slag treated. On-site disposal is assumed at roughly $10 per metric ton since the chlorides are now quite low.

Tables V and VI show the revenues for the Case A and Case B scenarios.

TABLE V
CASE A
(60 metric tons/day)

Item	Estimated Value	Annual Quantity (m. ton/year)	Revenue	Unit Revenue
Salt Flux	$55 /m. tons	9,300	$ 511,500	$25.80
Aluminum Metallics	0.8 /m. tons	1,130	904,000	45.70
Fines Disposal	10 /m. tons	14,320	(143,000)	(7.20)
Disposal Saved	40 /m. tons	19,800	792,000	40.00
		TOTAL REVENUE	$2,064,300	$104.30
		NET ANNUAL BALANCE		
		Revenues	$2,064,300	
		Expenses	1,126,420	
			$ 937,880	

Any increase in disposal costs of $10 per metric ton for salt slag represents about $200,000 annually in increased net balance of the project.

TABLE VI
CASE B
(200 metric tons/day)

Item	Estimated Value	Annual Quantity (m. ton/year)	Revenue	Unit Revenue
Salt Flux	$55 /m. ton	29,750	$1,636,250	$ 24.80
Aluminum Metallics	0.80 /m. ton	3,770	3,016,000	45.70
Fines Disposal	10 /m. ton	47,680	(476,800)	(7.20)
Disposal Saved	40 /m. ton	66,000	2,640,000	40.00
		TOTAL REVENUE	$6,815,450	$103.30
		NET ANNUAL BALANCE		
		Revenue	$6,815,450	
		Expenses	2,175,540	
			$4,639,910	

Every increase of $10 per metric ton for salt slag disposal increases the net balance of the project by about $660,000 per year.

THE ALCAN PLASMA DROSS TREATMENT PROCESS, A NEW SALT-FREE DROSS PROCESSING TECHNOLOGY

S. Lavoie, C. Dubé, G. Dubé
Alcan International Limited
Arvida Research and Development Centre
Jonquière, Québec, Canada, G7S 4K8

Abstract

A new technology that will revolutionize the recycling of aluminum dross by eliminating the salt fluxes used in the conventional process throughout the industry has recently been developed by Alcan with the involvement of Plasma Energy Corporation.

The high temperature process is based on a specially designed rotary furnace heated by a plasma arc gas heater. No pretreatment of the dross is required. The elimination of salt and the use of small amounts of inert gases diminish and simplify significantly the treatment of gaseous emissions. No waste material is generated by the process.

Results from three-tonne furnace operations will be presented and discussed.

Second International Symposium -
Recycling of Metals and Engineered Materials
Edited by Jan H. L. van Linden,
Donald L. Stewart, Jr., and Yogeshwar Sahai
The Minerals, Metals & Materials Society, 1990

Introduction

Dross is a by-product of all processes involving molten aluminum. The quantity of dross produced ranges from 1 to 2% in holding furnaces of primary smelters and to 3-4% in scrap melting furnaces. In most cases dross skimmed from the furnace will contain large quantities of metallic aluminum which can reach up to 80% of the weight of the material skimmed.

The classical dross processing system used to recover the aluminum is the rotary salt furnaces (RSF) process. These furnaces are oil or gas-fired and various salt fluxes are added to increase the amount of metal recovered.

The quantity of salt used is normally equal to the non-metallic content of the dross and is typically of the order of 50% of the total dross weight. The salt is necessary to protect the aluminum from the reactive atmosphere and to facilitate the agglomeration of the metallic droplets. Unfortunately, this massive use of salt has major economic and environmental shortcomings:

1) large amounts of corrosive salt vapours are emitted from the furnace and must be treated,
2) the residues from the process are toxic and expensive to dispose of or recycle,
3) the salt has to be purchased and requires energy to melt.

Different alternative processes have been proposed to replace the RSF process but none yet offered the same metal recovery efficiency or relative process simplicity.

The plasma dross treatment patented process (1) presented is the first real alternative to RSF using simple equipment and allowing efficient metal-oxide separation in dross with no addition of salt.

Comparative flow sheets for the RSF and the plasma dross treatment processes are presented in figures 1 and 2.

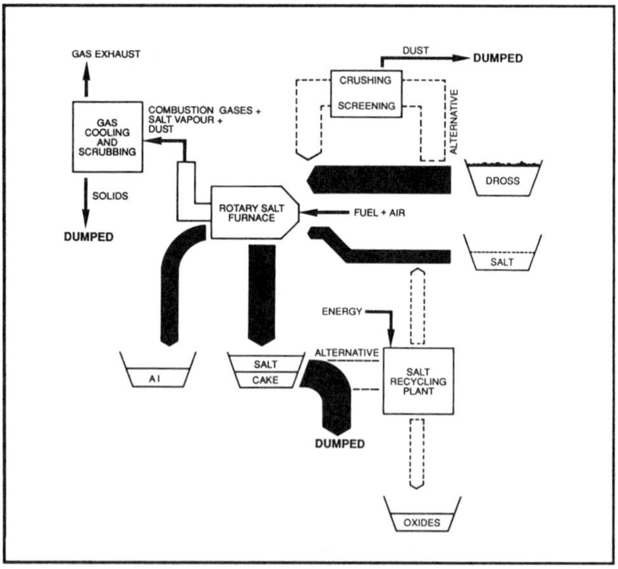

Figure 1 - Flow sheet of the rotary salt furnace process

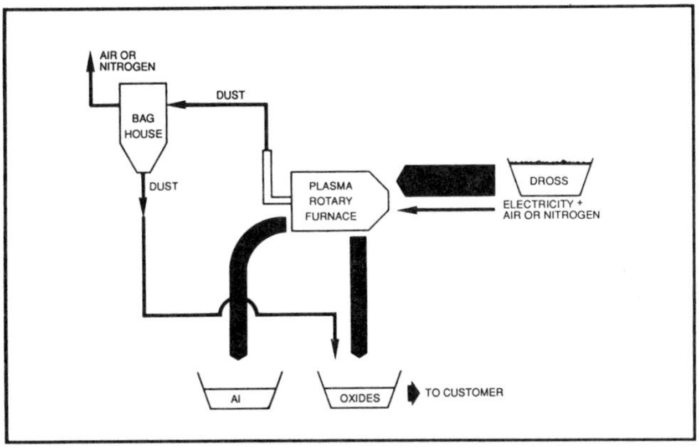

Figure 2 - Flow sheet of the plasma dross treatment process

The plasma dross treatment process

The process is based on utilization of plasma gas arc heater. Figure 3 shows a diagram of its working principle. Inside the plasma torch are two tubular electrodes placed end-to-end but separated by a small gap. During the operation a process gas, such as air or nitrogen, is injected into the small gap between the electrodes. When high-voltage power is applied, an electric arc is initiated between the electrodes. This arc heats the incoming gas at temperatures in excess of 5000 °C. At this temperature the gas is dissociated and partially ionized. This ionized gas is called a plasma. The plasma is ejected out of the torch inside the furnace and heats the dross. Because there is no combustion involved, virtually any gas can be used. Also, because of the extreme temperatures reached, the same energy can be transfered using 10 to 20 times less gas than a fossil fuel burner. Table 1 compares some characteristics of plasma and fuel flames.

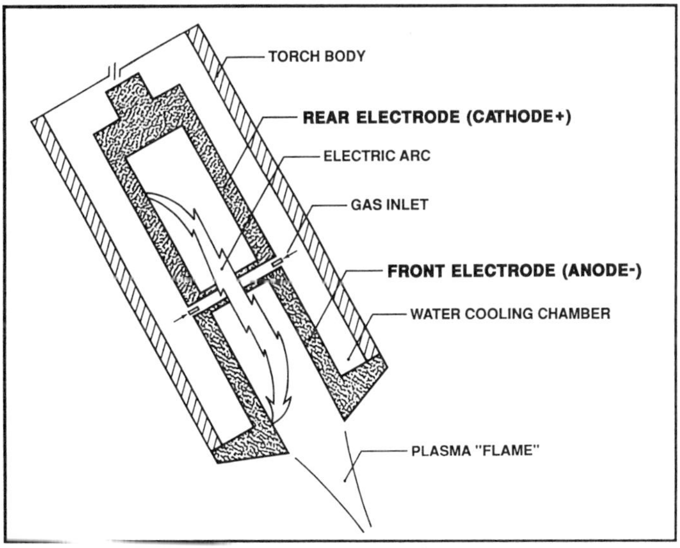

Figure 3 - Schematic of the plasma torch

Table I Comparative characteristics of plasma and fossil fuel flames

	PLASMA	FOSSIL FUEL
Temperature (°C)	4000 - 6000	1000 - 1500
Exhaust gas	N_2, Air	N_2, H_2O, CO_2, CO
Volume of exhaust gases (Nm^3/min/MW)	1.5 - 3.0	15 - 30

The plasma torch is installed in an especially designed rotary furnace. The main characteristic of this furnace as compared to the conventional RSF is its atmosphere control. The door, on which the torch is installed, is airtight and the stack is designed to maintain a slight positive pressure inside the furnace.

Aluminum in the dross is finely divided and wrapped in oxide films. The presence of humidity and organics in the combustion gases of fossil fuel burner enhance the oxidation of the fine aluminum droplets. This increased oxidation thickens the oxide wrapping and makes it more and more difficult to separate the alumimum. That is why the salt fluxes are required.

In the case of plasma heating, nitrogen and air were used successfully. These gases are much less reactive than combustion gases and the oxide film surrounding the alumimum remains thin and fragile. In the case of air, even the detrimental effect of oxygen is limited by the very small amount of gases fed into the furnace - there is simply not enough oxygen available to produce a significant oxide film thickening. The oxide film is easily broken by the mechanical stirring induced by the rotation of the furnace. Gradually all the metal agglomerates in a pool separate from the non-metallic products (NMP) and it can be tapped and recycled.

Absolutely no salt is required for all types of dross ranging from dusty dross with less than 20% metal content, to large 1 tonne blocks containing 80% metal.

Plasma dross treatment pilot plant

Alcan's work on the plasma dross treatment process is part of an integrated project aimed at

1. reducing dross formation,
2. protecting its alumimum content and
3. reducing the environmental impact of the recycling

The first two objectives are the subject of other active development projects in Alcan. New devices and operating practices which can significantly reduce the dross formation are being tested in production plants and a new way to cool the dross to avoid oxidation and dust is described in reference 2. The plasma dross treatment process is the answer to the third objective.

At the beginning of 1988 a pilot plant capable of processing industrial scale samples of dross began operating. The equipment and operation of this pilot plant is described here.

The plasma torch system is supplied by Plasma Energy Corporation (Raleigh, NC). This is a 1 MW system which could be operated either with nitrogen or air. The plasma torch is mounted on the charging door of the furnace.

The 3 tonnes capacity rotary furnace is tiltable and has a 1 meter diameter charging door. It is charged by using a semi-cylindrical spoon mounted on a fork truck. No dross pre-treatment is necessary and blocks up to one tonne were charged as well as very dusty dross.

After charging, the door is closed and the plasma turned on. The especially designed door allows for an airtight furnace operation to control the inside atmosphere. The pressure in the furnace is always maintained positive to prevent outside air infiltration. The 1 MW torch uses 1.5 to 3 normal cubic meters of gas per minute which vents through a 150 mm diameter stack.

The small volume of gases introduced and the absence of salt greatly simplify pollution control equipment. The only significant emission is the dust produced during the tumbling of the dross and entrained in the stack. This dust, mainly composed of alumina, is efficiently collected in a conventional baghouse. This simple exhaust gases treatment system contrasts with the extensive gas cooling and scrubbing equipment required with the RSF. Low concentrations of chloride and fluoride products can also be detected in the stack of the plasma furnace as they often are trace components of the dross. The chlorides are a by-product of the chlorine fluxing of furnaces and fluorides are a carry-over from electrolysis cells in smelter plant furnaces. If their concentration is not acceptable, these contaminants can be neutralized by adding lime at the inlet of the baghouse or directly into the furnace.

The furnace is continuously rotated during heating. The rotation prevents the formation of hot spots on the charge or on the refractories, improves energy efficiency and, most importantly, provides the mechanical stirring for the breakage of the oxide films and agglomeration of the aluminum.

The charge is heated to 700-800 °C. When the final temperature is reached, the metal has been separated from the oxides and is tapped from the furnace through a hole on the side.

The oxide part of the dross does not melt and it appears as a greyish powdery material that will not flow through the tap hole. This material is called the non-metallic product (NMP) and is evacuated through the charging door by forward tilting the furnace about 30° while slowly rotating it.

No mechanical scraping is normally required to empty the furnace. A thin NMP coating generally forms at the surface of the refractory. This coating is friable and does not tend to grow thicker as it is broken and formed again from batch to batch. This coating acts as a good protection for the refractory against excessive wear and metal penetration.

A hood is installed around the door to collect the dust generated during the discharging of the NMP.

Pilot plan results

A total of about 400 tonnes of dross has been processed during the course of this evaluation. The dross originated from Alcan's smelter and remelt plants in Canada, USA and Europe. The metal content ranged from about 10% to 80%. Most of the dross was cooled under argon hoods (2) prior to treatment; some of it was cooled in a rotary dross cooler and some was air cooled.

Two different gases were evaluated: nitrogen and air.

The flow sheet of the global process is shown in Figure 4.

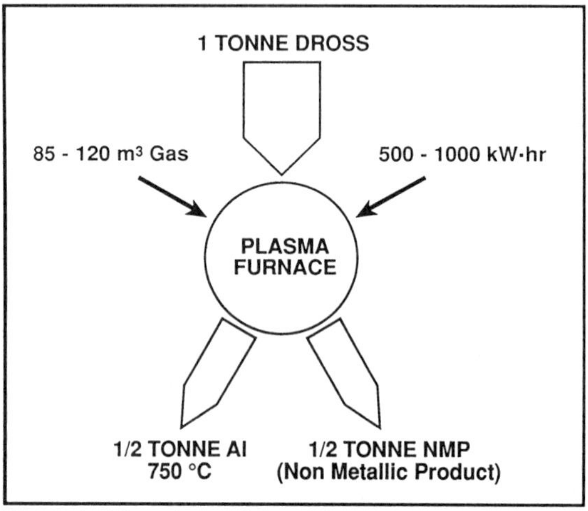

Figure 4 - Flow sheet of the global plasma dross treatment

Mass balance

The exact composition of dross is generally difficult to evaluate because of its extreme heterogeneity. Careful mass balances were necessary to measure the process efficiency.

The process efficiency is defined as the percentage of aluminum recovered over the total alumimum content of the dross (see equation 2). The difficulty is to obtain a reliable estimate of the aluminum content in the raw dross.

The aluminum mass balance of the process is expressed as:

$$Al_{IN} = Al_{CAST} + Al_{NMP} + Al_{Al_2O_3\ FORMED} + Al_{AlN\ FORMED} \quad (1)$$

Where:

Al_{IN} is the total amount of aluminum in the dross,

Al_{CAST} is the aluminum metal recovered at the end of the process,

Al_{NMP} is the metallic aluminum present in the NMP

$Al_{Al_2O_3\ FORMED}$ is the aluminum which reacted with the oxygen during the plasma treatment to produce alumina,

$Al_{AlN\ FORMED}$ is the aluminum which reacted with nitrogen during the plasma treatment to produce alumimum nitride.

The process efficiency is given by:

$$\text{Process efficiency} = \frac{Al_{CAST}}{Al_{IN}} \tag{2}$$

All the members to the right side of equation 1 are measured for all batches produced at the pilot plant. Al_{CAST} is obtained directly by weighing the aluminum tapped from the furnace. Al_{NMP} is obtained by analysis of samples of the NMP. $Al_{AlN\ FORMED}$ is calculated from:

$$Al_{AlN\ FORMED} = Al_{AlN\ NMP} - Al_{AlN\ DROSS} \tag{3}$$

Where:

$Al_{AlN\ NMP}$ is the weight of aluminum which has reacted with nitrogen and is under the form of aluminum nitride in the NMP. It is measured by chemical analysis of the NMP.

$Al_{AlN\ DROSS}$ is the weight of aluminum in the incoming dross which is under the form of aluminum nitride. It is measured by chemical analysis of samples of the dusty part of the dross.

The formation of aluminum nitride being known from equation 3, $Al_{Al_2O_3\ FORMED}$ can be deduced from the weight gained during the process. In effect, the nitrogen and oxygen which reacted with aluminum during the plasma treatment increase the total weight of the products. The fraction of the weight gain not accounted for by the aluminum nitride formation is imputed to the formation of alumina.

The oxidation and nitriding of other metallic elements are ignored. Measurements have shown, for example, that more than 80% of the magnesium content in the dross was recovered so that neglecting the formation of MgO does not introduce significant errors.

In order to get accurate measurements of weight, the furnace had to be carefully scraped at the end of each batch to remove any NMP adhering to the walls.

This evaluation technique requires careful manipulations and is prone to experimental errors. More than 400 tonnes of dross were treated representing a total of about 150 batches. This large amount of data ensures statistical significance of the results.

Table II summarizes the results of the pilot plant operation. Dross from smelter plants is produced in furnaces where little or no melting takes place. The low magnesium category groups the alloys with magnesium content ranging from 0.5 to 2%.

As can be seen in this table the process efficiency ranges between 80 to 95% and is more or less proportional to the aluminum content. This is explained by the fact that in richer dross, the chances of intimate contact between two droplets of aluminum are higher and that it is easier to drain the molten metal through the shallower bed of NMP.

Figures 5 and 6 show two typical detailed mass balances for batches with nitrogen and air plasma respectively, for magnesium free smelter dross. The weight of the NMP given in these figures includes the weight of the dust collected in the baghouse. On the average, 10.2 kg of dust per tonne of dross were collected. Table III summarizes the aluminum mass balance as given by equation 1 for the same two batches.

As shown, the process efficiencies are not significantly different. Nitrogen plasma is slightly more efficient since it is less reactive. It can be noted that some alumina is formed even with nitrogen plasma. In this case, this alumina is produced after metal tapping during discharging and cooling of the NMP.

Table II Summary of the results of the pilot plant operation

TYPE OF DROSS	QUANTITY (tonnes)	PLASMA GAS	Al RECOVERY (%)	PROCESS EFFICIENCY (%)
SMELTER Mg FREE	50 30	AIR N_2	64 ± 5 65 ± 4	92 ± 3 93 ± 3
SMELTER LOW Mg	110	AIR	45 ± 4	85 ± 3
REMELT LOW Mg	40	AIR	57 ± 3	84 ± 5
REMELT HIGH Mg	10 10	AIR N_2	41 ± 4 40 ± 4	80 ± 5 81 ± 5
RECYCLING FOUNDRY	20	AIR	30 ± 3	83 ± 6
OTHERS MIXED ALLOYS	136	AIR or N_2	9 to 66	78 to 91
TOTAL	406			

The reaction of nitrogen and oxygen with aluminum can be greatly enhanced if the dross is overheated. Beyond a given temperature, a fast thermite reaction with oxygen and nitrogen takes place which can heat up the charge and the furnace to temperatures above 1200 °C within a few minutes. Metal recovery efficiency is then greatly reduced.

Because of the cost of nitrogen, it is generally more economical to operate the plasma dross process with air. As will be seen later, air also has energy advantages.

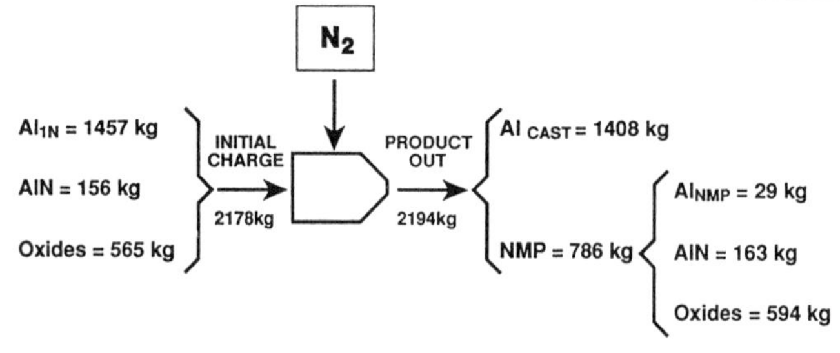

$$\% \text{ RECOVERY} = \frac{Al_{CAST}}{\text{INITIAL CHARGE}} = \frac{1408}{2178} = 64.6\%$$

$$\text{PROCESS EFFICIENCY} = \frac{Al_{CAST}}{Al_{IN}} = \frac{1408}{1457} = 96.6\%$$

Figure 5 - Typical mass balance for nitrogen plasma (magnesium free smelter dross)

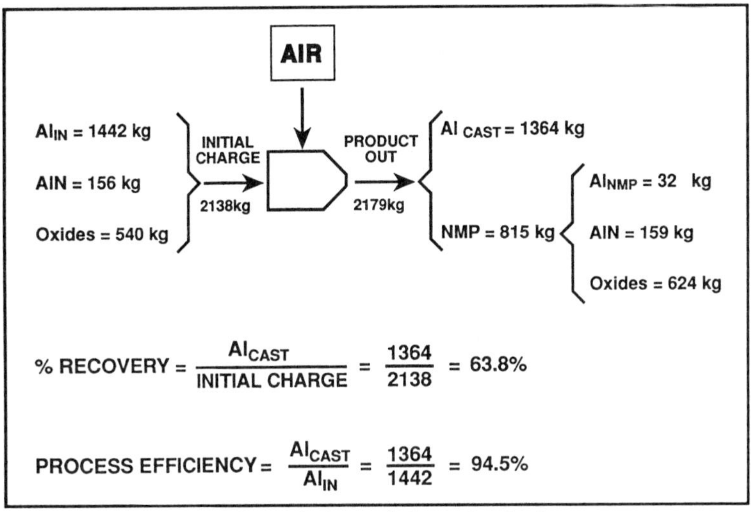

Figure 6 - Typical mass balance for air plasma (magnesium free smelter dross)

Table III Aluminum distribution as given in Equation 1 for the batches described in figures 5 and 6

	AIR PLASMA (kg)	NITROGEN PLASMA (kg)
Al_{IN}	1442 (100%)	1457 (100%)
Al_{CAST}	1364 (94.5%)	1408 (96.6%)
Al_{NMP}	32 (2.2%)	29 (1.9%)
$Al_{Al_2O_3 formed}$	45 (3.1%)	15 (1.0%)
$Al_{AlN formed}$	1 (0.2%)	5 (0.3%)

Energy balance

Figures 7 and 8 show diagrams of the energy balances for the same batches as in figures 5 and 6.

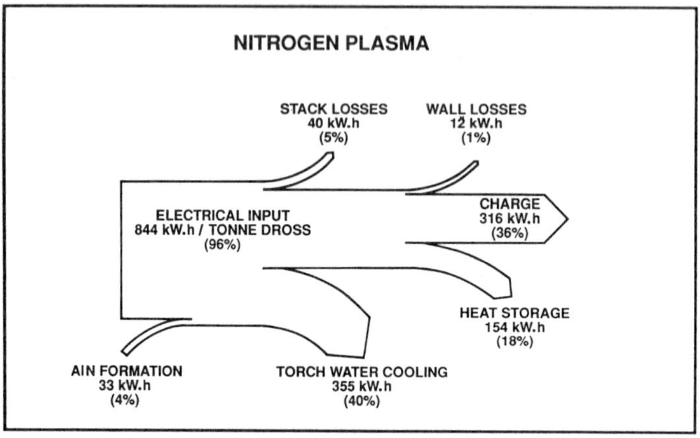

Figure 7 - Typical energy utilization with a nitrogen plasma per tonne of dross input (magnesium free smelter dross).

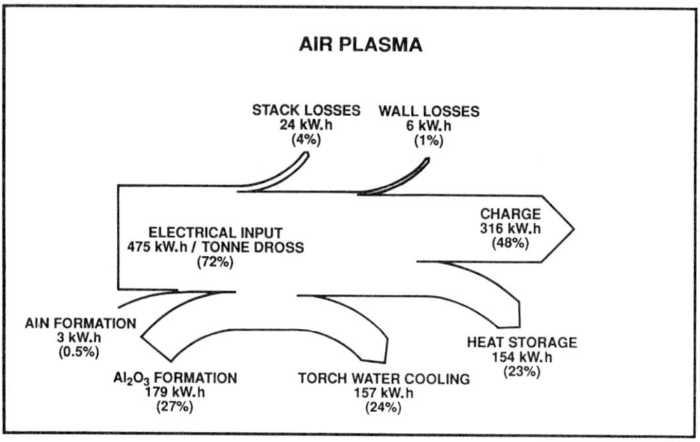

Figure 8 - Typical energy utilization with an air plasma per tonne of dross input (magnesium free smelter dross)

In both cases the net energy to the charge was about 300 kw-h per tonne of dross.

The stack losses are about one order of magnitude lower than equivalent power fossil fuel burners because the gas volumes are ten times lower.

Wall losses are the radiation / convection losses from the outside surface of the furnace surface.

Heat storage is the energy required to heat up the part of the refractory lining that cools down between two batches.

Both wall losses and heat storage were measured during extended test periods where the pilot plant operated 24 hours a day for three consecutive days.

Electrical input is the raw power input to the torch.

Torch water cooling is the energy lost in cooling the torch.

AlN and Al_2O_3 formation is the chemical energy released when nitriding and oxidizing the aluminum.

The energy efficiency of the torch is:

$$\text{Torch energy efficiency} = 1 - \frac{\text{Torch water cooling}}{\text{Total electrical input}} \tag{4}$$

In these cases the energy efficiencies were 58% for nitrogen and 67% for air. The air plasma is more efficient because the oxygen molecules are more easily dissociated than the nitrogen molecules and also because the air plasma flame is more heat conductive.

Alumina formation represents close to 20% of the total energy input when operating with air. The heat released from the formation of Al_2O_3 with the nitrogen plasma is not considered in the heat balance within the furnace as it takes place after the heating cycle is completed.

The higher efficiency and the chemical heat generation associated to the air plasma reduce the specific electric energy requirement from 844 kw-hr to 475 kw-hr per tonne of dross .

This reduction represents a significant energy cost and production time saving.

Non - Metallic Products (NMP)

The non-metallic fraction of the dross recovered at the end of the process is slightly modified by the plasma process.

As was seen in table III the aluminum nitride and oxide content increases slightly and the metallic aluminum content is reduced to 3-8%. The metallic content of the NMP is approximately inversely proportional to the metal content of the dross. The higher the metal content in the dross, the easier it is for the metal to agglomerate and drain from the NMP. Experiments have shown that more than 50% of the aluminum content of the NMP can be recovered by simple techniques like screening.

The elimination of salt reduces the toxicity of the NMP. Therefore landfilling would be made much easier. On the other hand, the absence of salt opens a much better alternative to landfilling both environmentally and economically. The main components of the NMP are oxides of aluminum, magnesium and of other metals. They also contain aluminum nitride and some metallic aluminum. Oxides from the NMP have been shown to be good grade raw material for others industries. Thorough characterization of the NMP were undertaken during the pilot plant operation and many potential applications were identified. In some cases, tests were held at potential customers' facilities to demonstrate the feasability of applications.

Conclusion

The plasma dross treatment process is a simple technology similar in its operation and equipment to the conventional rotary salt furnace process.

This new process completely eliminates the salt flux from the dross processing. This eliminates the evolution of corrosive gases from the process and, apart from the desired metallic aluminum, it generates salt free by-products which are marketable raw material for order industries.

Operating cost of the process are lower than RSF as it requires the same manpower and about the same equipment but does not require the costs associated with purchasing of the salt and landfilling or recycling of the salt cake. Capital cost is the same as the higher cost of the plasma system is offsetted by the simpler gas treatment equipment.

Metal recovery efficiency ranges from 85 to 95% which is superior to most conventional processes.

A 15 000 tonne per year plant is now being built by Alcan in Jonquière, Québec. This plant will processes all the dross produced by Alcan in the province of Québec, Canada.

This will be the first application of the plasma heating technology in an industrial environment for the aluminum industry and will provide data to confirm the findings obtained in the pilot plant. This will also be the first ever dross treatment plant featuring a closed loop, waste free, pollution free process ever developped in the world.

References

1. Canadian Patent, no 1, 255, 914.

2. Innus, A.B., "The Use of Inert Gas for Dross Cooling", Light Metals, 1986, ed. R. E. Miller (Warrendale, PA: The Metallurgical Society 1986), 777, 784

RECOVERY OF CHROMIUM AND ALUMINUM FROM F019 WASTE MATERIALS

Ranko Crnojevich, Edward I. Wiewiorowski, L. Ray Tinnin, Andrew B. Case

AMAX Metals Recovery, Inc.
Rt. 1 Box 160
Braithwaite, La. 70040

Abstract

AMAX Metals Recovery, Inc., a subsidiary of AMAX, Inc., operates a metals recycle plant near New Orleans, Louisiana for converting 25,000 tons/year of spent petroleum catalysts into four commercial products: molybdenum sulfide, alumina trihydrate, vanadium pentoxide and a nickel-cobalt concentrate. In 1989, the facility was expanded to recycle a chromium-aluminum hazardous waste material (F019) which is generated by the aluminum finishing operations. The aluminum content of this waste is conveniently coprocessed and aluminum is recovered as alumina trihydrate within the existing spent catalyst plant. This paper emphasizes several alternatives for recovering chromium from the F019 waste materials as a chromium oxide concentrate, or as a high purity salt or solutions of chromates and/or dichromates.

Second International Symposium -
Recycling of Metals and Engineered Materials
Edited by Jan H. L. van Linden,
Donald L. Stewart, Jr., and Yogeshwar Sahai
The Minerals, Metals & Materials Society, 1990

Introduction

The Resource Conservation and Recovery Act of 1976 (RCRA) established a clear direction for environmental progress in the United States and directed industries to conserve valuable resources through recovery and reuse (recycle) of waste materials.

AMAX Metals Recovery, Inc. (AMR) accepted the challenge to develop a process, which is an alternative to landfilling, for the hazardous waste material (EPA designation F019) containing chromium and aluminum generated by the aluminum metal finishing mills. The AMR plant was well suited for the task considering available facilities and expertise in metal refining, as well as the current business of recycling spent hydrotreating catalysts from the petroleum industries. The aluminum recovery facilities were already in place within the existing spent catalyst process, and several marketable-reusable chromium products are being developed.

The commercial operation for recycling F019 materials began in the last part of 1989 servicing AMAX-ALUMAX facilities and recovering chromium as a chromium oxide concentrate and aluminum as alumina trihydrate. In 1990, the effort has expanded to service other generators of the chromium-aluminum wastes as well as to upgrade and diversify the line of chromium products. The interest and response from the industry has been positive and growing. The generators of this material are electing to utilize this facility and its available technology to recycle rather than landfill the F019 waste materials.

Brief Description of The AMR Process and Facility

The AMAX Metals Recovery facility, formerly owned by Freeport Sulfur Co. came into existence in 1959, as a part of a new pioneering hydrometallurgical technology associated with the Moa Bay (Cuba) laterite project to extract nickel and cobalt from sulfide concentrates with an acid leach at elevated temperatures and pressures[1]. This operation, also known as Port Nickel plant, was successful, but short, and was closed after operating only several months because of events precipitated by the Cuban revolution.

AMAX, Inc. purchased Port Nickel in 1972 to diversify into the production of high purity metals of nickel, cobalt and copper from mattes supplied from Africa and Australia. The original facilities were considerably expanded and the commercial practice of hydrometallurgy for nickel, cobalt and copper was significantly advanced through 11 years of operation[2]. However, the nickel market went through a highly depressed stage in the early part of the eighties; and the plant was again modified in 1986 to recycle hydrotreating spent catalysts from the petroleum industry[3]. In addition, other industrial wastes including spent acid, spent caustic, Stretford solution and vanadium containing fly ash are also being accepted for recycling. From the outset, the goal of this newest operation has been to provide an alternative to landfilling and to produce commercial products without generating new solid wastes.

A partnership was formed in 1987 between subsidiaries of AMAX, Inc. and Shell Oil Company. This partnership, called CRI-MET, is the current name for the spent catalyst business.

The AMAX Metals Recovery plant is located in Braithwaite, Louisiana on the east bank of the Mississippi river, 65 miles north of the Gulf of Mexico and about 15 miles southeast of New Orleans. The plant is located on about 1100 acres of which 300 acres are occupied by the facility.

The plant is easily accessible by highways, by railroad, and quite significantly by the Mississippi river to accommodate barges and ocean-going ships.

The plant includes a great variety of equipment and support facilities designed for complete hydrometallurgical processing of metal bearing materials under extreme conditions of pressure, temperature and corrosivity. About 90% of the processing equipment and piping is constructed of stainless steel or titanium. Presently, only about 60% of the existing facilities are engaged in processing spent catalysts, thus offering considerable flexibility for future expansion and diversification.

Spent catalysts are generated by the petroleum refining industry in the process of converting crude oil into gasoline, heating oil, etc. Typical analyses of two main types of spent catalyst, "HDS" and "Resid", are shown below:

	Mo	V	Ni	Co	Al	Si	S	Oil
HDS, %	3-10	-	0.2-3	0-3	15-25	1-5	5-10	10-20
Resid, %	3-10	4-8	0.2-3	0-3	15-25	1-5	5-10	10-20

The processing scheme for spent catalyst is given in Figure 1 (page 4) to point out the coprocessing arrangement between the spent catalyst process and the chromium-aluminum process while preserving the principle of converting wastes into reusable commodities, and not generating new solid wastes.

The spent catalyst process is based on a two stage leach, one for a selective solubilization of molybdenum and vanadium, and the other for solubilization of aluminum. It converts the spent catalyst components into four products: molybdenum trisulfide, vanadium pentoxide, alumina trihydrate and a nickel-cobalt concentrate.

The spent catalyst is first fed to a ball mill, along with a recycled sodium hydroxide-sodium aluminate solution. The resulting slurry is then pumped to the first stage leach autoclave where it is leached under oxidizing conditions at elevated temperature and pressure. In this stage, sulfur is converted to sulfate, organic compounds are oxidized and molybdenum and vanadium are selectively dissolved. A liquid-solid separation of the autoclave product is performed by thickening and filtration. The solution becomes the feed for molybdenum and vanadium recovery, and the solids are forwarded to the second stage aluminum leach.

Molybdenum recovery from the first stage leach solution involves precipitation of molybdenum trisulfide in the presence of sulfuric acid. The precipitate is washed, dried and further processed at another AMAX location into an oxide or other high purity molybdenum compounds.

Vanadium is precipitated from the molybdenum free solution as hydroxide with sodium carbonate or sodium hydroxide. It is washed, centrifuged, dried, and calcined into a granular vanadium pentoxide product, and sold to manufacturers of ferrovanadium.

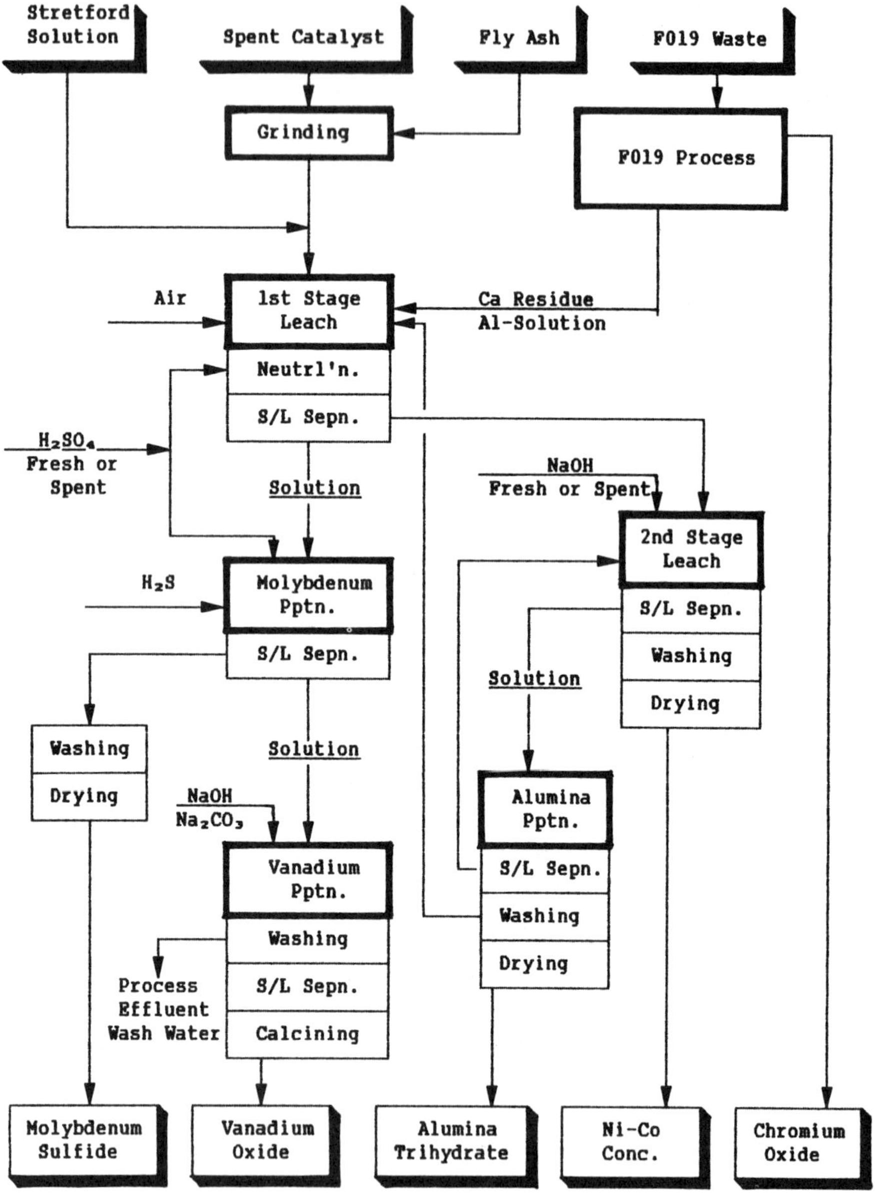

Figure 1 - Recycling of Industrial Wastes in the AMAX Metals Recovery, Inc. Facility

Solids from the first stage leach, high in alumina, are leached a second time at high temperature with strong caustic to solubilize aluminum. Following the autoclave leach, the solids are separated and washed in two stages with centrifuges. They are dried then calcined and eventually smelted at another location into a nickel-cobalt matte from which nickel and cobalt are recovered as pure metals.

The sodium aluminate solution from the second leach goes through a series of alumina precipitators. By cooling, in contact with seed material, alumina trihydrate is precipitated. It is separated from the solution and washed in three stages with centrifuges. The washed solids are dried and sold as alumina trihydrate. The sodium hydroxide-sodium aluminate solution, depleted of alumina, is recycled to the 1st and 2nd stage leach.

The process does not generate solid waste. Wash water containing Na_2SO_4 is processed through a treatment system before being discharged into the Mississippi river under permit.

Processing Schemes for the F019 Material

Manufacturers of finished aluminum products use chromic acid and other chemicals for a surface passivation or a paint-line preparation. Depending on a proprietary practice, the composition of these etchant solutions will vary greatly. In addition to chromic acid most of these solutions may contain phosphoric acid, zinc, molybdenum, HF, HNO_3 and others.

The spent etchant solutions, after being saturated with impurities, are treated for a complete chromium removal to meet stringent EPA regulations before disposal.

For chromium removal, the majority of the plants are using a similar process. The spent etchant solution is acidified to pH 2 - 4, then Cr(6) is reduced to Cr(3) with sodium bisulfite and subsequently chromium hydroxide is precipitated with NaOH, KOH, MgO, CaO or combination of these reagents.

The precipitated chromium hydroxide is normally recovered by filtration and is classified as the F019 hazardous waste. In a few installations, the F019 material is generated in a somewhat different manner whereby $NaHSO_3$ was replaced by electrolytic reduction in the "ANDCO" units. The chromium waste from the Andco units will contain much lower phosphate and much higher iron contents.

An estimated 15,000 tons of F019 waste is generated in USA annually as a wet filter cake containing 50 - 75% moisture. This material is characterized by EPA as a hazardous waste and until recently it was allowed to be landfilled. Present regulations, however, mandate immobilization or fixation of chromium before landfilling.

The approximate composition (dry basis) of the F019 material is given below:

Element	Cr	Fe	Al	Mg	Ca	Zn	P	S*
%	3-15	0.5-10	5-15	0.5-2	5-20	0.1-0.5	2-15	2-5

* Both as SO_4 and SO_3

The processing of this material presents a considerable technological challenge, and so it is understandable that recycle-recovery of this material on a commercial scale has not been available until now.

Processing F019 to Chromium Oxide Concentrate

The commercial recycle process for F019 waste, as practiced presently by AMAX Metals Recovery, is presented in Figure 2 (page 7)[4].

According to the process, the F019 material containing various proportions of chromium, aluminum, iron, calcium, magnesium, phosphorous, and other elements in minor quantities, is treated with sulfuric acid to completely dissolve aluminum and chromium while leaving calcium sulfate and some gangue material in the insoluble residue. Dissolution is carried out with sulfuric acid under conditions of a moderate acidity (pH, 1.2 - 2.0), and at an ambient or moderately elevated temperature, 70 - 150°F.

Air (oxygen) or other available mild oxidant can be contingently used during dissolution to oxidize ferrous ion to the ferric state, while retaining chromium in the trivalent state.

Following dissolution of aluminum and chromium, the pH of the reaction system, in the presence of undissolved solids, is adjusted to the pH, 2.0 - 3.0 to precipitate some iron and phosphorous as ferric hydroxide or phosphate, while avoiding coprecipitation of chromium or aluminum.

The pH adjustment is done with hydrated lime but other alkaline reagents, can be used. The slurry after the pH adjustment step undergoes a solid/liquid separation to yield a chromium-aluminum solution and an iron-phosphorous-calcium residue. The chromium-aluminum solution is further treated to precipitate and recover chromium by raising alkalinity of the solution with NaOH to the pH range of 12.5 - 13.0. Aluminum remains in solution while chromium is completely precipitated as chromium(3) hydroxide, at ambient or slightly elevated temperatures, 80 - 150°F.

Chromium hydroxide, after being properly washed, is filtered, dried and calcined at temperatures of 1000 to 1300°F to yield a chromium oxide similar in composition to a high grade chromite ore. A comparison of the compositions of the chromium oxide concentrate with a typical South African chromite ore is shown below:

Element	AMR Chromium Oxide (Calcined 1200°F)	Chromite Ore (Typ. South African)
Cr_2O_3, %	50 - 55	47.0
Fe, %	4 - 6	10.0
Al, %	3 - 5	12.0
Ca, %	4 - 6	<0.5
Mg, %	1.5 - 3	7.5
P, %	0.3 - 0.8	<0.1
Na, %	1 - 2	N.A.
Zn, %	2 - 3	N.A.
Ni, %	0.2 - 0.3	<0.1
Si, %	<0.5	2.5
C, %	<0.1	N.A.
S, %	0.1 - 0.3	N.A.

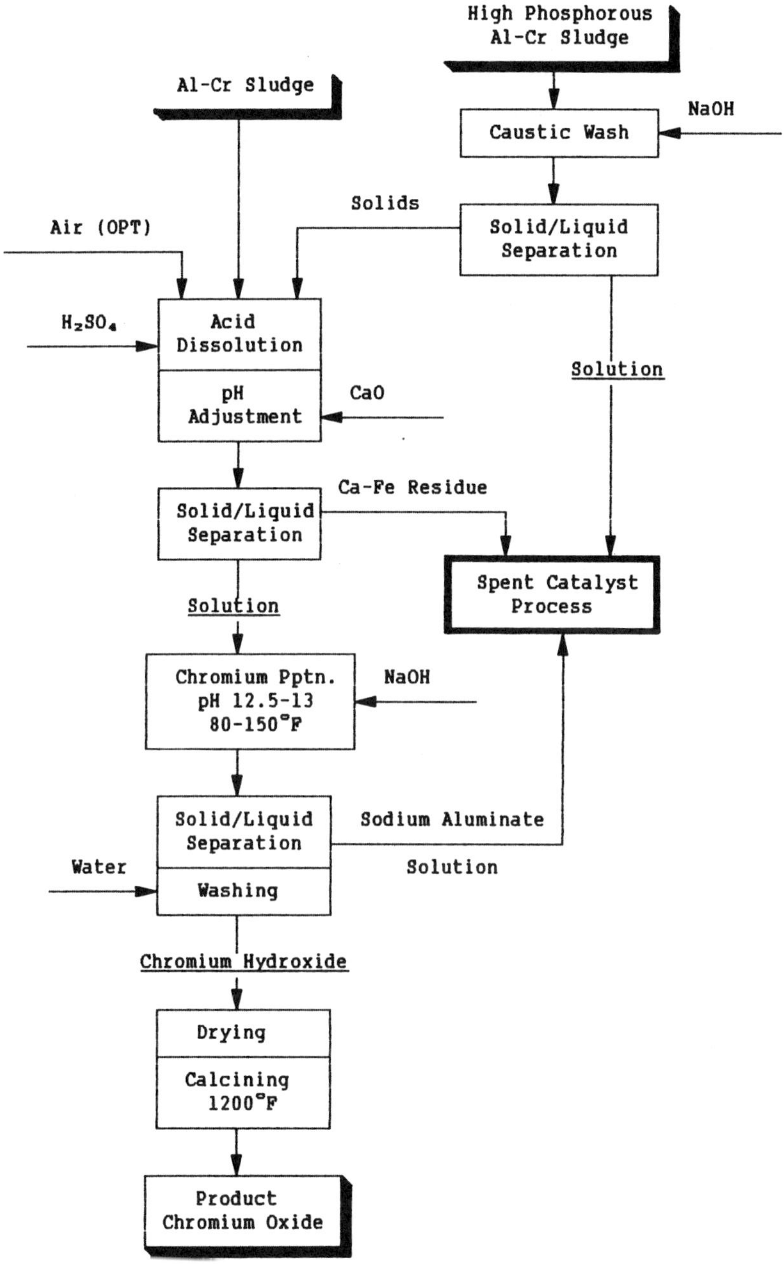

Figure 2 - Processing of Aluminum-Chromium Wastes from Etching Operations

Several comments relative to this process are appropriate at this point:

1. F019 wastes with very high phosphorous content (>12% P) have to be prewashed with a strong caustic to minimize phosphorous content in the final Cr_2O_3 product.

2. The mutually beneficial coprocessing arrangement between the spent catalyst and the F019 process is manifested in at least two areas:

   (a) The spent catalyst process uses a significant amount of lime or equivalent calcium per day, and thus the dissolution residue from the F019 process containing predominantly calcium is beneficially used in this process.

   (b) A relatively pure sodium aluminate solution, after chromium precipitation, is conveniently coprocessed in the spent catalyst process where aluminum is recovered as alumina trihydrate and regenerated caustic is reused for the leaching of spent catalysts.

3. The chromium oxide product, while having a high Cr_2O_3 content, is not suitable for all applications. The somewhat elevated phosphorous content may be objectionable to some ferro-chromium producers unless appropriately blended with a low phosphorous material. The chromium oxide product appears to be suitable as a coloring additive in glass industry, for production of fire resistant Mg-Cr bricks or as a feed material for producing a high purity chromium chemical.

Processing F019 to High Purity Chromates

Processing chromium bearing materials via a chromate route in presence of alkalies is a known practice. This basic idea was utilized in a specific manner to advantageously process the F019 material, into a high grade chromium chemicals according to the flowsheet shown in Figure 3 (page 9).

The earlier described basic process for producing chromium oxide is modified by imposing more agressive dissolution of the F019 material without a subsequent lime neutralization, to enhance a chromium recovery. Also, chromium precipitation is carried out at a lower alkalinity to enhance phosphorous removal. An unwashed chromium hydroxide is then enriched appropriately with caustic to have the Cr/Na_2O ratio of about 1 and calcined in an air atmosphere at 1200 - 1600°F.

The calcine is leached in water at pH 7.0 - 8.0 to yield a highly concentrated and pure chromate solution of a typical composition shown below:

Element	Cr	Na	Fe	Al	Mg	Zn	K	Ca	P
gpl	60-130	50-150	<0.01	0.03	0.05	<0.01	0.5-1.0	0.02	0.07

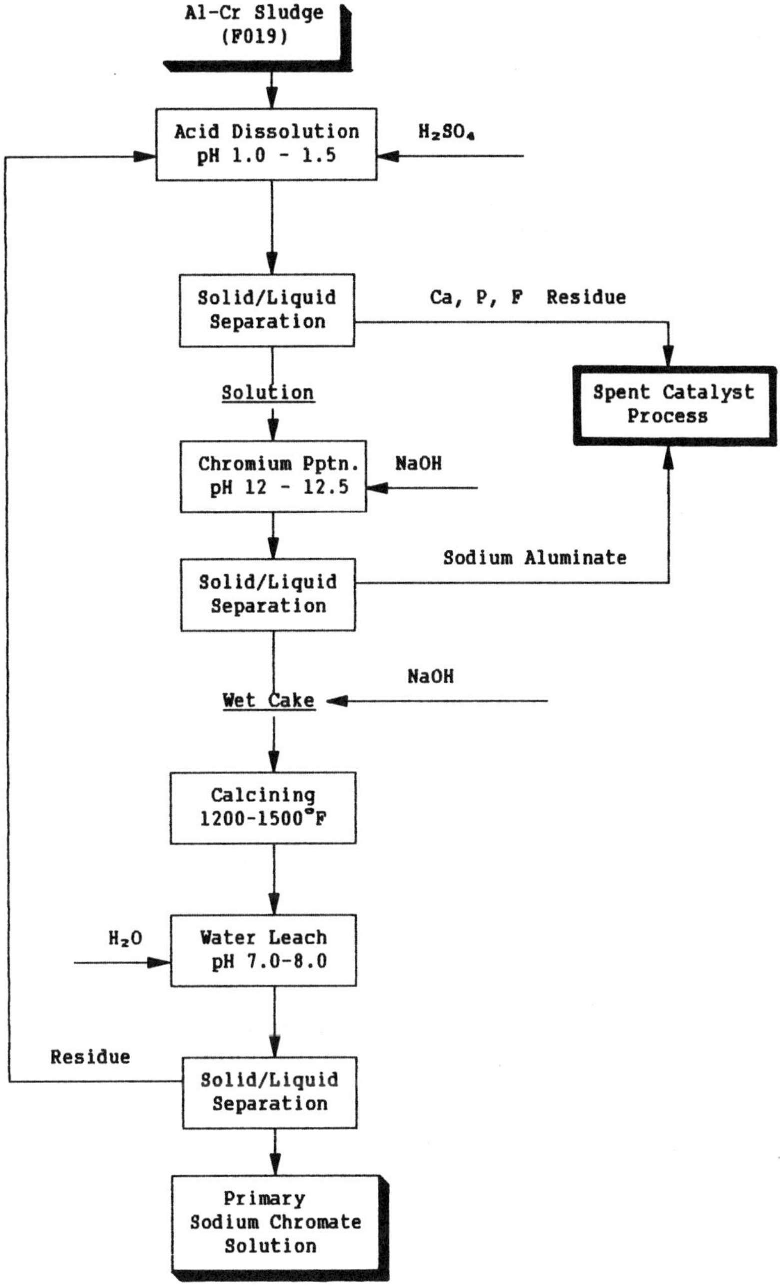

Figure 3 - Process Flowsheet for Producing High Purity Chromium Compounds from Aluminum-Chromium Waste Sludges

This process yields a solution virtually free of many originally present impurities but could contain some excess of sodium sulfate. There are several potential commercial markets for this solution for manufacture of chromic acid, pure Cr_2O_3 and various chromate salts as illustrated in Figure 4 below:

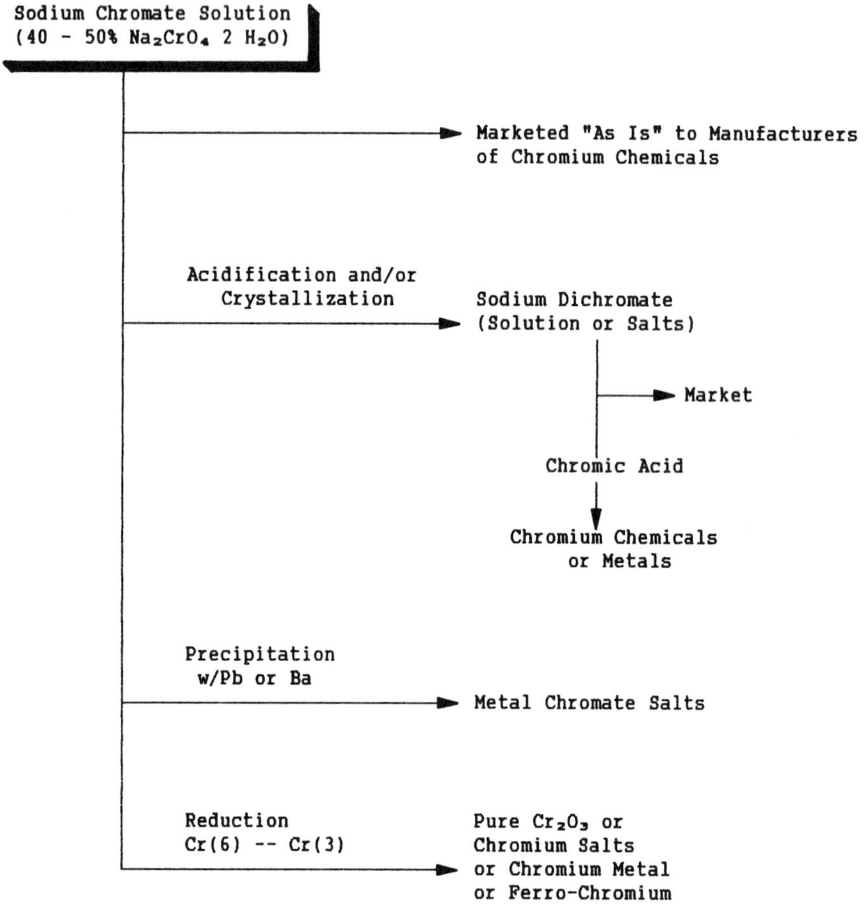

Figure 4 - Potential Markets for Sodium Chromate Solution

A direct chromate route, where the F019 material "as received" would be calcined, is also being considered. This approach is feasible, but some disadvantages have to be taken in an account:

a. A considerably larger calcining facility would have to be installed (4 - 5 times larger).

b. Blending of highly variable feed materials with alkaline reagents would require close and difficult control.

c. Increased usage of caustic reagents should be expected.

d. A lower concentration of chromium in the chromate solution with less favorable ratio of chromium to sodium makes this product less probable for direct marketing and will probably require an evaporative step and crystallization for removal of sodium sulfate.

Conclusion

Processing of the F019 hazardous waste material is a complex and expensive technology, but its commercilization at the AMAX Metals Recovery plant was made feasible in association with the existing recycling facility for spent catalyst.

The plant currently operates at the capacity of about 2000 tons of wet F019 material annually with one shift and 5 days/week operation. This capacity can be increased to 6000 tons annually by providing adequate manpower for a 7 day/week - 24 hr./day operation. Efforts continue at this facility to produce high purity chromium chemicals and to expand capacity to accommodate future needs.

AMAX Metals Recovery, Inc. is devoted to complete recycle-recovery of selected industrial wastes. This service that fulfills the environment protective intent of the RCRA Act of 1976 is now also available for the F019 material.

References

1. F. Wilson, "The Moa-Bay Port Nickel Project, New Free World Sources of Nickel and Cobalt from Cuban Ores" AIME Meeting, New York, 1958.

2. R. Crnojevich, D. H. Wilkinson, J. L. Blanco, "Contribution to the Hydrometallurgy of Nickel, Copper and Cobalt in the AMAX, Inc. Port Nickel Plant" The 112th AIME Annual Meeting, Atlanta, Georgia, 1983.

3. T. LaRue, R. Tinnin, E. Wiewiorowski, R. Crnojevich, "AMAX Port Nickel - A New Dimension in Reclaiming Spent Catalysts" The Spring 1988 AICHE Meeting, New Orleans, La.

4. R. Crnojevich, E. Wiewiorowski, A. Case, "Reclaiming of Aluminum-Chromium Waste Materials from Etching Operations" U.S. Record File 352,301, May 15, 1989.

OTHER METALS A

Session Chairman:
Paul B. Queneau

THE ANIMAL FEED INDUSTRY AS A POSSIBLE MARKET FOR "CLEAN" METAL SALTS AND COMPOUNDS FROM RECLAMATION OPERATIONS

Jon Nelson

Southeastern Minerals, Inc.
P.O. Box 1866
Bainbridge, Georgia 31717

Abstract

The animal feed industry can be a market for "clean" metal salts and compounds in the sulfate, carbonate, and oxide forms. These salts must have minimal levels of heavy metals and must be free of toxic or deleterious contaminants that could cause problems when incorporated into animal feeds. Currently, ZnO, $ZnSO_4$, MNO, $MnSO_4$, $FeSO_4$, $FeCO_3$, $CuSO_4$, CuO, $CaIO_3$, Ethylenediamine Dihydroiodide, $CoSO_4$, $CoCO_3$, and $NaSe_2O_3$ are the most commonly used trace mineral salts and compounds. These metal salts and compounds can be made up of one metal or a combination of two different metal salts in one compound. The animal feed industry is also a large consumer of MgO, $MgSO_4$, K_2SO_4, KCL, $NaHCO_3$ and various compounds of phosphorus and calcium, magnesium or ammonia.

Second International Symposium -
Recycling of Metals and Engineered Materials
Edited by Jan H. L. van Linden,
Donald L. Stewart, Jr., and Yogeshwar Sahai
The Minerals, Metals & Materials Society, 1990

Introduction and Background

The animal feed industry is a significant industrial complex that is made up on many feed mills, both large and small. A "large" feed mill will normally produce in excess of 100,000 tons of animal feed a year. These feed mills are usually owned by a feed company that sells "branded" feeds to its customers and guarantees the performance of these feeds or produces feed for its own vertically integrated operations, usually feeding animals that are owned by the company and raised by "contract" growers.

Animal feeds are usually a combination of a number of ground protein and energy ingredients such as soybean meal, cottonseed meal, corn, milo, and animal fat. In addition to these basic ingredients, the feed mill must add individual amino acids, vitamins, minerals, and trace minerals to produce a complete and balanced feed. The major or "macro" minerals that are added are usually salt, calcium carbonate and dicalcium phosphate. These macro minerals are normally purchased as individual ingredients in bulk or bagged truckload quantities. The minor minerals that are added, i.e. the trace or "mirco" minerals are usually purchased as part of a trace mineral premix that will normally contain Manganese, Zinc, Iron, Copper, Iodine, Cobalt, and Selenium and will be added to the the finished feed at a rate of one pound per ton of feed.

There is a possibility that some of the recycled or reclaimed minerals and trace mineral salts and compounds generated by industry may have an application in animal feeds. The purpose of this paper is to examine this possibility.

Mineral and Trace Mineral Ingredients Used by the Feed Industry

As you may or may not be aware, the animal feed industry in the United States and abroad is a significant consumer of various trace mineral and mineral salts and compounds. To give you an idea as to the type of mineral ingredents that are consumed by the feed industry, a listing of the ingredient definitions as published in the 1990 Association of American Feed Control Officials handbook can be found on the following page in Table I. All of these ingredients are currently acceptable for use in animal feeds, but only the ones with the asterisk following their names are used in significant enough quantity to be of interest to someone in the recycling or reclamation business.

Since the salts and compounds I have listed are used in animal feeds, they must be "clean", meaning that they must contain a minimal level of heavy metals or other mineral contaminants and must be essentially free from organic chemical residue or other toxic or deleterious agents.

Physical Characteristics of Acceptable Ingredients

In addition to being "clean", they should be of a favorable bulk density and particle size. A "favorable" bulk density can mean something different for each ingredient in question. For Zinc Oxide, it usally means a density of 100 lbs. per cubic foot or more. For Manganous Oxide, a bulk density of 80 to 90 lbs. per cubic foot is acceptable. Normally, it is preferable to have as dense a product as possible, taking into consideration the chemical nature of the mineral compound.

TABLE I

Ammonium Sulfate*	Manganese Chloride
Calcite	Manganese Citrate
Calcium Carbonate*	Manganese Gluconate
Calcium Chloride	Manganese Methionine Complex
Calcium Gluconate	Manganese Orthophosphate
Calcium Hydroxide	Manganese Phosphate
Calcium Iodate*	Manganese Sulfate*
Calcium Iodobehenate	Manganous Oxide*
Calcium Oxide	Metal Amino Chelate
Calcium Periodate	Metal Polysaccharide Complex
Calcium Sulfate*	Metal Proteinate
Cobalt Acetate	Ammonium Polyphosphate
Cobalt Carbonate*	Calcium Phosphate*
Cobalt Chloride	Diammonium Phosphate
Cobalt Choline Citrate Complex	Dicalcium Phosphate*
Cobalt Glucoheptonate	Disodium Phosphate
Cobalt Gluconate	Monoammonium Phosphate*
Cobalt Oxide	Monocalcium Phosphate*
Cobalt Sulfate*	Monosodium Phosphate*
Copper Carbonate	Phosphoric Acid*
Copper Chloride	Defluorinated Phosphate*
Copper Choline Citrate Complex	Sodium Hexametaphosphate
Copper Gulconate	Sodium Tripolyphosphate
Copper Hydroxide	Tribasic Sodium Phosphate
Copper Orthophosphate	Tricalcium Phosphate*
Copper Oxide*	Potassium Bicarbonate
Copper Sulfate*	Potassium Carbonate
Cuprous Iodide	Potassium Citrate
Diiodosalicylic Acid	Potassium Chloride*
Ethylenediamine Dihydroiodide*	Potassium Hydroxide
Ferric Ammonium Citrate	Potassium Iodate
Ferric Chloride	Potassium Iodide
Ferric Choline Citrate Complex	Potassium Sulfate*
Ferric Formate	Salt*
Ferric Phosphate	Sodium Acid Pyrophosphate
Ferric Methionine Complex	Sodium Bicarbonate*
Ferric Pyrophosphate	Sodium Carbonate
Ferric Sulfate	Sodium Iodate
Ferrous Carbonate*	Sodium Iodide
Ferrous Chloride	Sodium Molybdate*
Ferrous Fumarate	Sodium Selenate
Ferrous Gluconate	Sodium Selenite*
Ferrous Glycine Complex	Sodium Sesquicarbonate*
Ferrous Sulfate*	Sodium Sulfate*
Iron Oxide*	Sulfur*
Iron, Reduced	Thymol Iodide
Magnesium Carbonate	Zinc Acetate
Magnesium Chloride	Zinc Carbonate
Magnesium Hydroxide	Zinc Chloride
Magnesium Phosphate	Zinc Chloride Diamine Complex
Magnesium Oxide*	Zinc Methionine Complex
Magnesium Sulfate*	Zinc Oxide*
Manganese Acetate	Zinc Sulfate*
Manganese Carbonate	

* The asterisk identifies ingredients that are utilized in significant quantity by the feed industry.

Particle size is another item that can vary from ingredient to ingredient. With an ingredient like Zinc Oxide, since it is very dense, a particle size of minus 80, plus 200 mesh is preferred. Remember that the zinc oxide will be incorporated into animal feed. It is therefore important that every peck, in the case of a chicken, or every mouthfull, in the case of a pig, contains and adequate amount of ingredient particles to supply the animals biological need for that particular trace mineral or mineral element. The finer the particle size, the more particles will be dispersed in each unit of feed. Of course, some common knowledge also must govern in this area. Obviously, if Copper Sulfate were ground to a minus 80, plus 200 mesh, it would form into bagged "blocks" before it could ever be used by the feed mill. In general, most of the ingredients in question should pass through a 30 mesh screen and be retained of a 200 mesh screen. Specifications of a tighter nature than this will be dependent on the nature of each individual ingredient.

Biological Availability

In addition to an ingredient being clean, of proper bulk density and acceptable particle size, it must be biologically available to the animal once it is ingested. As an example, copper metal can be very pure, ground to an acceptable particle size and may even be of an acceptable density, unfortunately, copper metal is essentially totally unavailable to the animals biological system. We could add a significant quantity of copper metal to the feed of an animal and that animal may still die from a copper deficiency. If you will direct your attention to Table II - Manganese; perhaps this will give a you better understanding of biological availabilty values.

You will note that five studies are listed that have been done at various Land Grant Universities, testing the biological value of various Manganese compounds in baby chick and sheep diets. These studies have used Manganese Sulfate as a reference standard, with the exception of the work done by Henry et al.[4], which used Manganous Oxide. The studies also used various tissue levels of Manganese as the test parameter. As summarized in the table, the average biological values for Manganese Chloride, Manganese Carbonate, Manganese Dioxide and Manganous Oxide were 102%, 48%, 27.9% and 66.7%, respectively, when compared to the test standard, Manganese Sulfate which was designated as having a 100% relative biological value. In other words, this study tells us that one pound of Manganese from Manganese Sulfate is equal in biological value to 1½ pounds of Manganese from Manganous Oxide. In this case, since Manganese Sulfate costs twice as much per unit as Manganous Oxide, the Manganous Oxide is still the better buy. It is important that this type of cost versus biological value comparison be understood, since it can be applied to all of the significant trace mineral/mineral ingredients used by the feed industry, as long as reliable test data are available.

Types of Salts/Compounds that are Acceptable or Unacceptable

With the information in the preceding paragraphs, we now have a very basic understanding of the types of metal salts and compounds that are used by the animal feed industry, the basic physical requirements of these salts and compounds and the manner in which their value as nutrients in animal feed is determined. Perhaps we should now consider some specific types of reclaimed, recycled or by-product salts and compounds that may or may not be acceptable for use in animal feeds.

TABLE II - MANGANESE

BIOLOGICAL AVAILABILITY OF INORGANIC MANGANESE

	SOU. & BAKER[1]	BLACK et al[2]	HENRY et al[3,4]		WONG-VALLE[5]	WONG-VALLE[6]	AVE. VALUE
	Bile	Bone/Liver	Tissue	Tib./Kid.	Sheep Liv/Kid/Bone	Bone/Kidney	$MnSO_4$ Std.
$MnCl_2 \cdot H_2O$	102%	---	---	---	---	---	102%
$MnSO_4 \cdot H_2O$	100%	100%	100%	---	100%	100%	100%
MnCO3	77%	39.2%	---	---	27.8%	---	48.0%
MnO_2	29%	---	---	40%	32.9%	---	27.9%*
MnO	---	70.7%	66%	100%	57.7%	83.8% RG 80.6% FG(A) 63.6% FG(B)** 61.7% FG(C)**	66.7%

* MnO = 66.7% available relative to $MnSO_4$. 66.7% of 40% = 26.7%. This converts value in Henry *et at.*[4] to Manganese Sulfate standard.

** The purity of these two Manganous Oxides does not meet normal feed grade standards.

The first item that comes to mind is a salt or compound that is a combination of two or more different mineral elements. An example of this might be a spent catalyst. The feed industry has experience with spent catalysts and I can assure you that it does not want them. These spent catalysts are usually a combination of copper and zinc oxide, or zinc oxide that has been complexed with alumina. This complexing of the various elements renders them, not only biologically unavailable to the animal, but undesirable from a contamination standpoint. Many times, Chromium, Arsenic, Nickel, Vanadium, etc. are added to these catalysts to enhance their catalytic properties. In addition, the catalysts sometimes carry residues from the organic chemical reactions they were catalyzing. Among the chemicals found as residues might be aromatic ketones, vinyl monomers, olefins, dimethylmaleate, etc. I can assure you that none of these items are desirable in animal feeds.

If the combination compound is, as an example, a combination of two sulfates, such as Zinc Sulfate and Iron Sulfate, the probability that it can be used in animal feed increases dramatically. Since the sulfate sources of the various trace mineral ingredients are highly soluble, a compound of this type would probably be of acceptable biological value. In addition, the fact that it is a sulfate may reduce somewhat the level of possible contaminants that may be in the compound.

The fact that most of the trace elements listed in Table I that are used in significant quantities are sulfates or oxides does not necessarily mean that these are the preferred sources for use in animal feed. It does mean that they have been proven to be biologically available and are produced in adequate quantities to meet the tonnage requirements of the feed industry.

Evaluation of Salt or Compound

How can you use the information in this paper to evaluate the possibility of selling a by-product compound to the animal feed industry? Let's assume, for example, that your company is producing a Manganese Chloride it would like to sell and the production level of this compound is 5000 tons per year. Obviously, the quantity that is produced is adequate to make its use of interest to the industry, but is it biologically available? You will recall that, in Table II, we listed some work by Southern and Baker[1] demonstrating that Manganese Chloride was 102% as biologically available as the test standard Manganese Sulfate (100% available). With this information, we have now established that there is an adequate supply of the compound (5000 tons) and it can be utilized by the animal. The next step would be to establish purity standards for the compound, insure its flowability, establish a price and offer it to the animal feed industry as a replacement for some of the Manganous Oxide and Manganese Sulfate that is currently being used. This same type of scenario would apply to other salts or compounds. As a general rule, if the salt or compound is clean and is produced in significant quantity, the biological availability work can be performed, if necessary, to determine if it is of value for use in animal feed. It is also possible to use existing research work when available.

General Purity Requirements of Compounds Used by the Feed Industry

We have talked about "clean" compounds a number of times in this paper. Perhaps a listing regarding the general purity requirements of some various trace mineral ingredients might help clarify this area. These general purity requirements are as follows:

Zinc Oxide	- 72% Zn minimum. 500 ppm Pb, 250 ppm As, 80 ppm Cd and 1 ppm Hg maximums. Preferred by Walez process.
Zinc Sulfate	- 36% Zn minimum. 400 ppm Cd maximum. Usually a monohydrate salt.
Manganous Oxide	- 60% Mn minimum. 50 ppm Pb, 70 ppm As maximum. Manganese Dioxide maximum of 3%.
Manganese Sulfate	- 29.5% - 31.5% Mn minimum.
Ferrous Carbonate	- 40% Fe minimum. 3% ferric iron maxium.
Ferrous Sulfate	- 30% Fe minimum. Monohydrate salt.
Copper Oxide	- 75% Cu minimum. 95% cupric oxide minimum. 5% cuprous oxide and 1% Cu metal maximums.
Copper Sulfate	- 25.2% Cu minimum. Minus 30 mesh crystal with anticaking agent.

As you can see, various purity standards are employed for each ingredient based on the biological value of the various chemical forms of each compound. Also, each compound must contain a minimum level of heavy metals.

Current use of Reclaimed or Recycled Compounds

As far as the current use of reclaimed or recycled minerals by the feed industry is concerned, it should be pointed out that essentially 100% of the Ferrous Sulfate Monohydrate used by the industry comes from by-product material from Titanium Dioxide extraction (wet cake) of from wet Heptahydrate crystals that are extracted from the steel industry's pickling liquors. Also 100% of the Manganese Sulfate that was used by the industry up to 1988 was a by-product from the production of hydroquinone. Today, the majority of the Manganese Sulfate produced uses Manganese Carbonate bag-house dust as the feedstock for its production process. In addition, the majority of the Copper Sulfate sold to the feed industry begins as reclaimed copper metal, as does most of the Copper Oxide.

Conclusion

As you can see, the feed industry is not a stranger to reclaimed or recycled ingredients. As our world industrial complex generates additional salts and compounds that have a possible application in animal feeds, consideration should be given to this market. If you or your company need assistance in determining whether or not a product you are producing, or have access to, can be utilized by the animal feed industry, we will be happy to work with you to establish a possible application or put you in touch with other parties who may have an interest.

REFERENCES

1) **SOUTHERN, L. LEE, & BAKER, DAVID H.,** Excess Manganese Ingestion in Chick, 1983, Poultry Science 62: 642-646.

2) **BLACK, J.R., AMMERMAN, C.B., HENRY, P.R., & MILES, R.D.,** Biological Availability of Manganese Source and Effects of High Dietary Manganese on Tissue Mineral Composition of Broiler-Type Chicks, 1984 Poultry Science 63: 1999-2006.

3) **HENRY, P.R., AMMERMAN, C.B., & MILES, R.D.,** Bioavailability of Manganese Sulfate and Manganese Monoxide in Chicks as Measured by Tissue Uptake of Manganese from Conventional Dietary Levels, 1986 Poultry Science, 65: 983-986.

4) **HENRY, P.R. AMMERMAN, C.B., & MILES, R.D.,** Bioavailabilty of Manganese Monoxide and Manganese Dioxide for Broiler Chicks, Nutrition Reports International, August 1987, Vol. 36, No. 2.

5) **WONG-VALLE, J. HENRY, P.R. AMMERMAN, C.B., AND ROA, P.V.,** Estimation of the Relative Bioavailabilty of Manganese Sources for Sheep, Journal of Animal Science, 1989, 67: 2409-2414.

6) **WONG-VALLE, J., AMMERMAN, C.B., HENRY, P.R. RAO, P.V. & MILES, R.D.,** Bioavailability of Manganese from Feed Grade Manganese Oxides for Broiler Chicks, 1989, Poultry Science, 68: 1368-1373.

RECYCLING DUST IN NICKEL METALLURGY

M.N.Zevgolis* and J.F.Gaitanos**

ABSTRACT

In this work the pelletizing plant for recycling dust from laterite prereduction in rotary kilns (R/K) is studied and the critical paths are identified for a productivity increase and also for a better environmental protection. The critical paths are the capacity of the mixer and pelletizing discs and also the storaging area of green pellets. In this work the mechanical strength of pellets is also studied and as it is shown it must be at least 7-8 Kg/pellet for a pellet diameter of 8-12mm. This is achieved after 2-3 days of storaging the green pellets. In this work is also shown that strength of pellets increases as dust grain size decreases. The strength of pellets when they come from a mixture of dry dust (from polycyclones) and wet dust (from Venturi washers) of the rotary kilns No 1,2 and 3 is higher than when the pellets come from dry dust only. Aslo the pellet strength when they come from a mixture of dry dust from kilns No 1,2,3 and 4 is higher than when the pellets come from dust of kiln No 4 only and it is twice as that of pellets from dry dust of kilns No 1,2 and 3. This work is also a continuation of previous work, so that it includes additional results of two more years of industrial application. As it is shown, dust recycling has come to a level at which 85-95% of dust from the rotary kilns is utilized, i.e. dust recycling has been improved and stabilized during the last 5 years.

* Assist. Professor, Nat.Techn.Univ.of Athens, Greece
** Chief Engineer, LARCO SA

INTRODUCTION

In the metallurgical plant of General Mining and Metallurgical S.A. LARCO (G.M.& M.S.A. LARCO) at Larymna, Greece, nikeliferrous ores mined in Biotia and Euboia are smelted and refined for the production of ferronickel (FeNi) containing 18-25% Ni. The exact location of the plant as well as the mines in Greece are shown in figure 1. The metallurgical procedure is carried out in four consecutive steps as shwon in the flowsheet of figure 2 and is briefly as follows (1) (2) :

a. Crushing of the ore to the desired size (-15mm) and mixing with appropriate amounts of solid fuels. In this step pellets produced by pelletization of the dust coming from step b, below, are also added.
b. Drying and preheating of the mixture as well as partial reduction of the ore by the solid fuels in four rotary kilns (R/K).
c. Reduction smelting of the R/K product in five open bath submerged arc electric furnaces producing thus a metal phase with approximately 15% of Ni and 85% of Fe, and a slag phase.

Second International Symposium -
Recycling of Metals and Engineered Materials
Edited by Jan H. L. van Linden,
Donald L. Stewart, Jr., and Yogeshwar Sahai
The Minerals, Metals & Materials Society, 1990

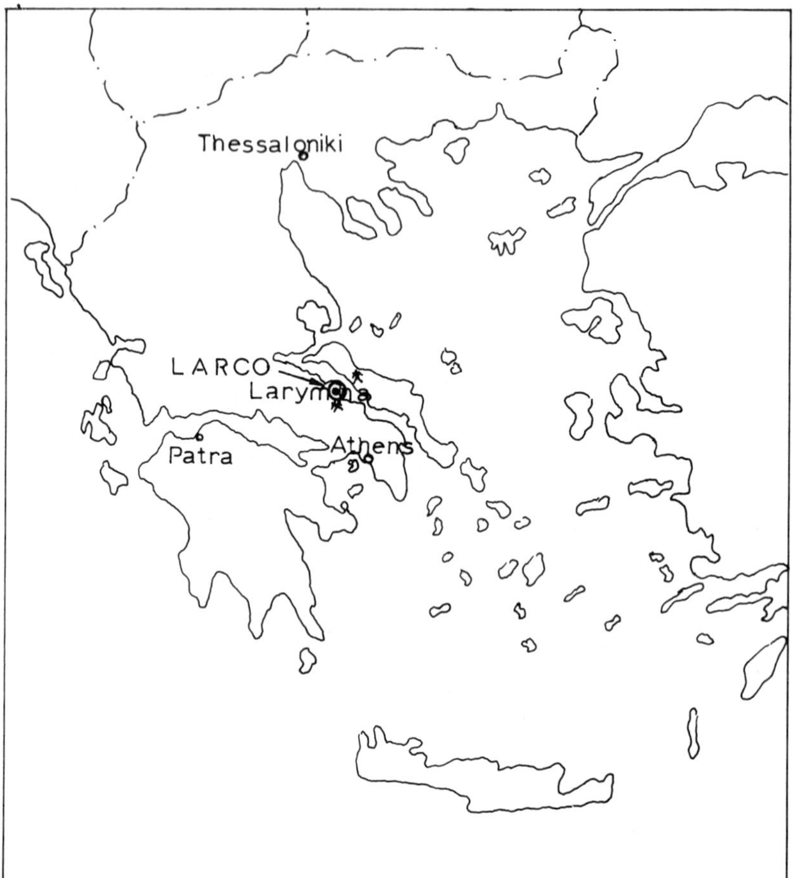

Fig 1 Map of Greece

d. Refining of the metal phase produced in an O.B.M.convertor. The final product having a nickel concentration of about 22% is granulated and delivered to stainless steel producers in Europe.

During step b of the metallurgical procedure, which is the treatment of the ore in the R/K, part of the fines existing in the feed together with those produced because of the movement of the material in the kiln and the chemical reactions taking place are carried away by the gases moving in counter flow. Dedusting of the gases is effected by a combination of dry (multicyclones) and wet (venturi scrubbers) methods for R/K 1, 2 & 3 while for R/K 4 only dry cleaning methods (electrostatic precipitators) are

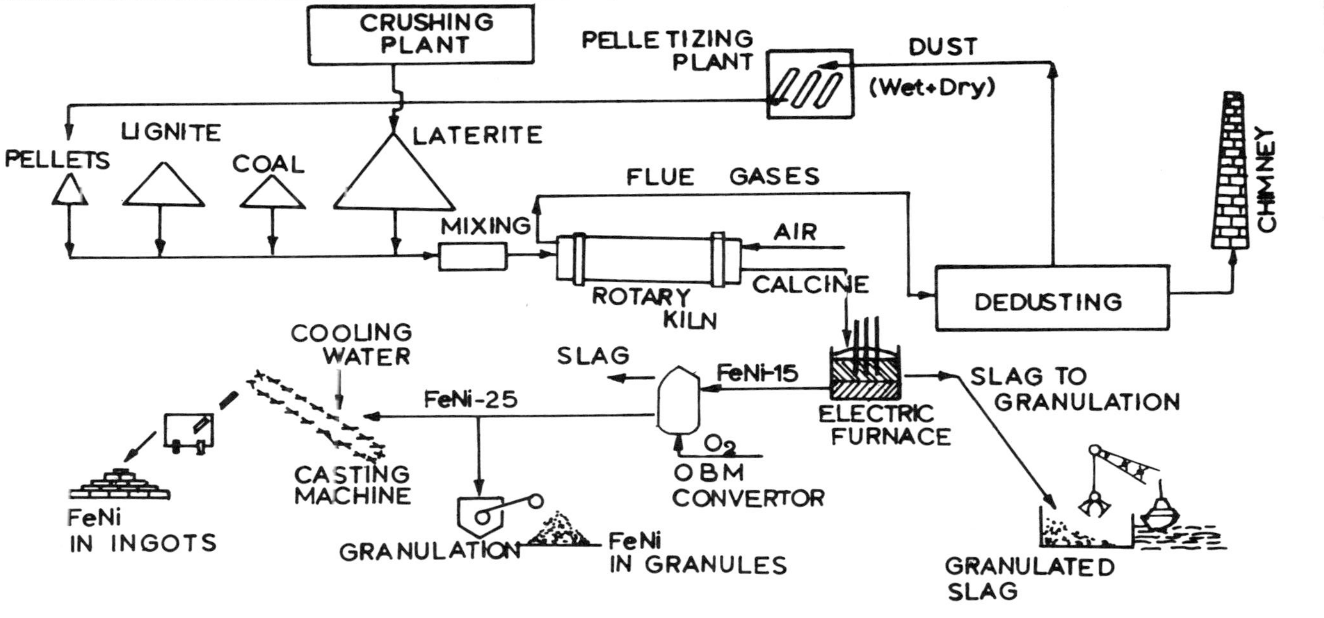

Fig. 2. Ferronickel production flowsheet

used. For the recycling of the dust thus collected a briquetting plant was erected as early as 1975 which was soon abandoned because of technical difficulties. New attempts in the direction of pelletizing, in laboratory and semi industrial scale were to follow (3). The result was an industrial scale installation operating since 1978. During the first years of operation, productivity and consequently dust recycling as shown in table I was rather low although three flying saucer type pelletizing discs with a 25 -30 tons/hour capacity were installed. Therefore part of the dry dust and all of the wet was stockpiled for exploitation in the future.

In 1985 a research programme backed by the Greek Ministry of Industry and LARCO was established with the following objectives:

a. improvement of productivity and operating conditions in the pelletizing of dust (4)(5)(6).
b. modification of known technology or creation of new one for the recycling of wet dust as well as the dust of the existing stockpile.
c. application of the findings in an industrial scale.

Therefore the research programme was to study carefully the existing pelletizing installation, find out the critical paths and give proper solutions for the increase of productivity. Moreover it was meant to study the mechanical strength of pellets in order to minimize curing area as well as decrepitation which increases the amount of dust in the R/K gases. In this paper the study on the subject and the findings are briefly outlined.

IMPORTANCE OF THE PROJECT

The amount of dust carried by the R/K gases goes upto 120 gr/Nm^3 of gases or 6.7% of laterite as shwon on table II. The amount of dust collected through wet cleaning methods range from 14.4% to 36.0% of total.dust (table II). The amount of dust stockpiled up to now (1989) is approximately 1,000,000 tons. Chemical and sieve analysis are shown in table III. One may notice the importance of nickel content ranging from 1.1% to 1.2% compared to that of the ore (1.05%) as well as that of carbon (9,0-9,5%) compared to feed mixture (6,5%) (7). Therefore collected dust consists of an excellent stockpile of raw material from the point of view of nickel and carbon content. On the other hand the recycling of dust gives an ideal solution to the environmental problem of disposing it. Disposition of dry dust is faced by moistening it by adding about 15% of water and dumping it on a preselected area. Wet dust is disposed as it is using specially designed lorries. The stockpile thus created is covered by loan to prevent carrying away by the wind till recycling is applied.

TABLE I

ROTARY KILN DUST RECYCLING

	DUST FROM DRY DEDUSTING			DUST FROM WET DEDUSTING		
YEAR (T)	WEIGHT (T)	RECYCLING OF DRY DUST (T)	%	WEIGHT (T)	RECYCLING OF WET DUST (T)	WEIGHT OF TOTAL DUST RECYCLED (%)
1977	58899	Ø	Ø	20978	Ø	Ø
1978	67911	33840	49,8	38114	Ø	31,9
1979	72386	38160	52,7	32039	Ø	36,5
1980	105338	73075	69,4	33853	Ø	52,5
1981	95386	58680	61,5	25320	Ø	48,6
1982	31883	26319	82,5	6462	Ø	68,6

TABLE II

DUST FROM THE ROTARY KILNS

	Dust From Dry Dedusting	Dust From Wet Dudusting			Total Dust (wet and dry)	
Year	Weight (T)	Weight (T)	Weight (% of laterite)	Weight (% of total dust)	Weight (T)	Weight (% of laterite)
1	47441	16658	1,33	25,99	64099	5,12
2	67911	38144	1,90	35,97	106055	5,28
3	72386	32039	1,53	30,69	104425	4,99
4	105338	33853	1,86	24,33	139191	7,66
5	95386	25230	1,95	20,98	120706	9,32
6	31883	6462	1,16	16,86	38345	6,92
7	66191	31572	2,14	32,30	97763	6,63
8	106623	22949	1,26	17,72	129572	7,13
9	104034	20420	1,09	16,50	124454	6,67
10	70210	11720	0,92	14,40	81930	6,44
11	64842	15459	1,38	19,25	80301	7,17
12	97212	18720	1,15	16,15	115932	7,12
13	114624	25020	1,29	17,91	139644	7,21
Mean Value	80314	22949	1,48	22,22	103263	6,67

TABLE III

INDICATIVE SIZE AND CHEMICAL ANALYSIS OF DUST

SIEVE ANALYSIS		CHEMICAL ANALYSIS		
Fraction (microns)	Weight (%)	Dust from Dry Dedusting (%)		Dust from Wet Dedusting (%)
+200,0	2	C	9.5	9.0
- 200,0 + 90,0	2	FeO	13.0	21.0
- 90,0 + 60,0	3	Fe_2O_3	18.0	19.0
- 60,0 + 40,0	4	Fe tot	20.0	19.0
- 40,0 + 30,0	15	Ni	1.1	1.2
- 30,0 + 28,1	2	SiO_2	35.0	35.0
- 28,1 + 21,9	10	CaO	4.6	4.6
- 21,9 + 16,5	9	MgO	4.6	5.0
- 16,5 + 11,5	9	Al_2O_3	12.0	12.5
- 11,5 + 6,7	16	Cr_2O_3	3.0	2.4
- 6,7 + 4,9	12	S	0.7	1.0
- 4,9 + 2,8	10			
- 2,8	6			

An example will provide a better understanding of the subject. Considering a 1350 tons of Ni/month production the amount of ore required is given by the equation

$$\frac{ORE}{1,055} R \frac{Ni\%}{100} = 1350$$

Where ORE = ore as it is
Ni% = nickel content on dry basis
R = nickel metallurgical recovery (0,85)
1.055 = coefficient for conversion of ore as it is to dry basis

So if Ni=1.05%, the amount of ore required will be ORE=159.580 tons/month
and the amount of dust (DST) produced will be

$$DST = \frac{6.7}{100} 159580 = 10692 \text{ tons/month}$$

therefore for an average nickel content of dust equal to 1.12% the amount of nickel gained is

$$Ni=10692 \frac{1,12}{100} 0,85 = 101,8 \text{ tons/month}$$

Metallurgical treatment of laterite would normally require 65 Kg coal/ton of ore, and this coal is imported. Consequently there is going to be a saving of coal as follows
0,065 . 10692 = 695 tons of coal/month
Also, the nickel to be extracted from the dust stockpile is

$$Ni=1.000.000\frac{1,12}{100}\ 0,85 = 9520 \text{ tons of Ni}$$

This amount of Ni has a value of

$$9520\ \frac{4,5}{0,454}\ 1000 = 94.360.000 \text{ US\$}$$

The coal to be saved will be
COAL = 1.000.000 0,065 = 65.000 tons of coal and it has a value of 65.000 45 = 2.925.000 US$.
Therefore dust recycling ends up with a considerable saving of money and natural sources, and also it protects the environment from pollution(8).

PREVIOUS WORK

Dustagglomeration has been known for a long time (9)(10)(11)(12) regarding fundamental rules and specific fields of applications. But in the case of dust coming from the treatment of Greek nickeliferrous ores it wasn't that simple because their physical and chemical parameters are more or less unique (13) and therefore existing experience was not functioning. So the "try and error" method was applied. This gives a good explanation why briquetting failed and why at the early stages pelletizing treated just 31.9% of the dust to reach 68.6% after five years of efforts (table I) with no success on wet dust until recent years.
Briefly, first attempts started as early as 1972 by feeding in the R/K briquettes produced under experimental conditions. Results were considered successfull and the erection of a briquetting plant for a 70 tons/hour capacity started in 1973. When the unit came to operation in 1975 it was proved to be inadequate for the production planed, the mechanical strength of briquettes was not satisfactory while briquetting rollers were very rapidly wearing out. As no significant improvement could be effected the idea was abandoned. Next step was to try pelletizing on a flying saucer disc with the collaboration of a German company. Experiments shown in table IV were considered successful and it was decided to run a pilot plant at LARCO's Larymna plant. So, some of the existing equipment together with a 3 m flying saucer disc were used. As it was proved, dry dust could be pelletized using common portland cement and water as binders. It was also proved that pellets produced had a sufficient mechanical strength to go through R/K without turning again into dust. So three industrial 6 m flying saucer discs were installed and pellot production started in 1978. Table I shows dust exploitation in the years after and one may clearly

see that although a considerable dust recycling had been achieved a lot of improvement was still ahead to be made.

STUDY AND CRITISISM OF THE OPERATION OF INSTALLATION

A detailed study and critisism of the pelletizing installation was carried out. Thus critical paths were pinpointed and ideas for increasing productivity were suggested and activated. In figure 3 the flowsheet of the installation is shown and one can easily distinguish the following steps in the pelletizing procedure.

a. Silos for dry dust from rotary kilns
b. Silos for cement
c. Dosing of dust and cement
d. Mixing and homogenization of raw materials, i.e. dust, cement and water
e. Belt conveyors, reversible and one direction
f. Bucket elevator
g. Silos for the mixture of raw materials
h. Dosing of the mixture
j. Pelletizing discs
k. Stocking of green pellets for induration
l. Handling and screening of indurated pellets
m. Feeding of the +4mm fraction of indurated pellets to the

A detailed discussion of these steps is given elsewhere(14). In the present work only those under the letters d, j and k are discussed because they are critical for increasing recycling.

TABLE IV
EXPERIMENTAL RESULTS OF PELLETIZING

COMPOSITION OF PELLETS (WEIGHT UNITS)			CRUSHING STRENGTH (kg/Pellet)	
Dry Dust	Binding Agent	Moisture	Green Pellets	Pellets after Induration
100	Bentonite 1	22,1	0,6	0,5
100	Fast hardening cement 10	19,5	0,5	3,2
100	Portland cement 10	22,5	0,6	1,1
100	Portland cement 5	22,6	0,6	0,4

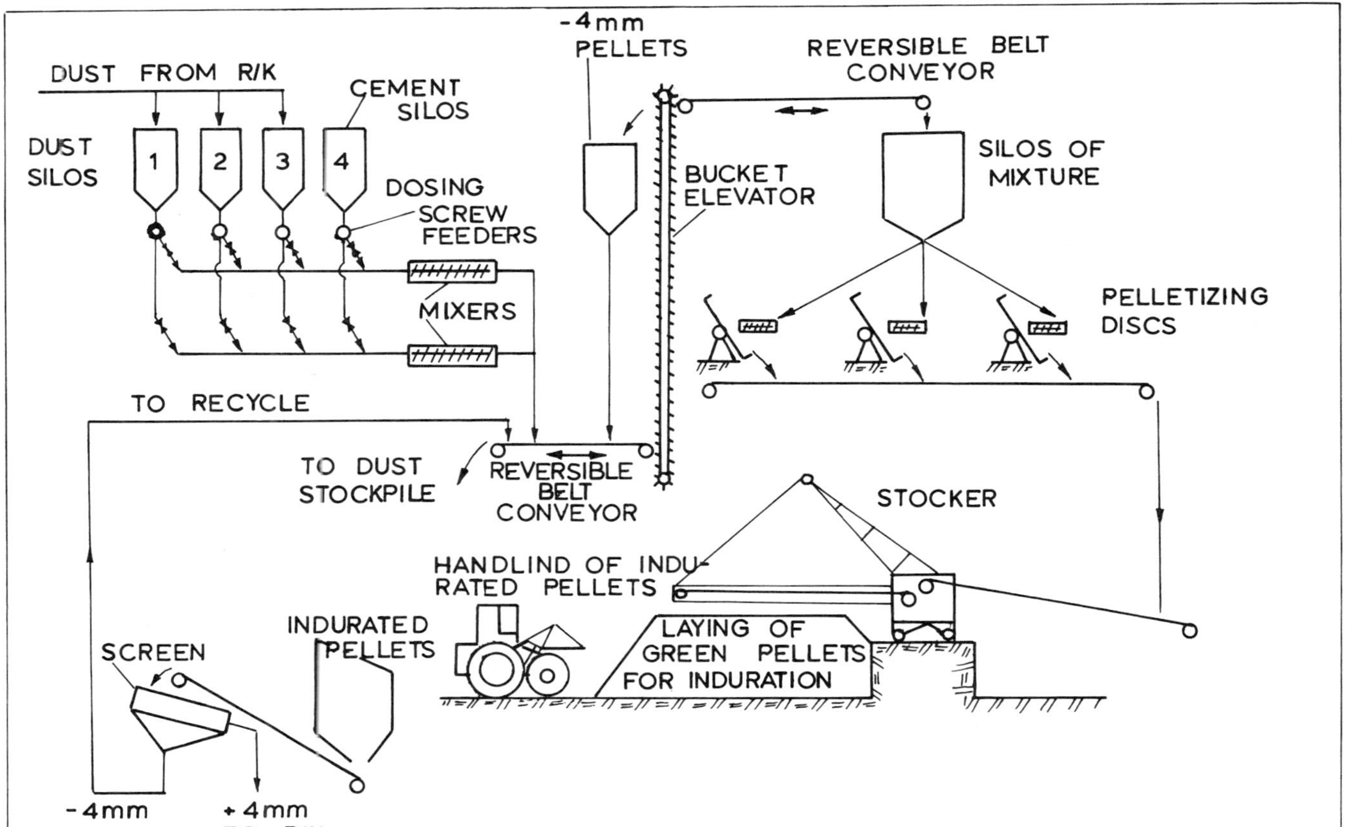

Fig. 3. Flowsheet of the pelletizing plant

Mixing and Homogenization of Raw Materials

Two Loedige blade mixers with a capacity of 50 tons each were installed for this purpose. Figure 4 shows one such mixer. Blades are fixed on the axle which rotates at a speed of 150 rpm, and they admix and homogenize the raw materials fed on one side of the mixer while ready mixture comes out on the other. Checking of the operation is done by in situ surveillance while modification of the feed rate or the proportion of the raw materials is remote controlled from the control room. An average proportion for the raw materials is 100 parts of dry dust to 20 parts of water to 10 parts of cement. According to this the daily throughput of the mixer is 350 1.30=455 tons of mixture. The time required for handling such an amount of material by a 50 ton capacity mixer is calculated to be 9 hours approximately. So at the beginning it was considered that such a mixer was enough, therefore, two were installed one to act as a reserve of the other. However it was found out that dry dust is quite a lyophobic material and so mixing with dust was not always successful. Moreover it was noticed that under certain conditions (such as low dust level in the silo, low dust temperatures, etc) dust was flowing at practically uncontrollable rates to the mixer. This didnot allow proper mixing with water and caused pollution as well as poor quality pellets. To face this problem a 100 ton mixer was decided to be installed and it proved to be satisfactory. So at present two mixers one 50 and one 100 ton capacity are in operation.

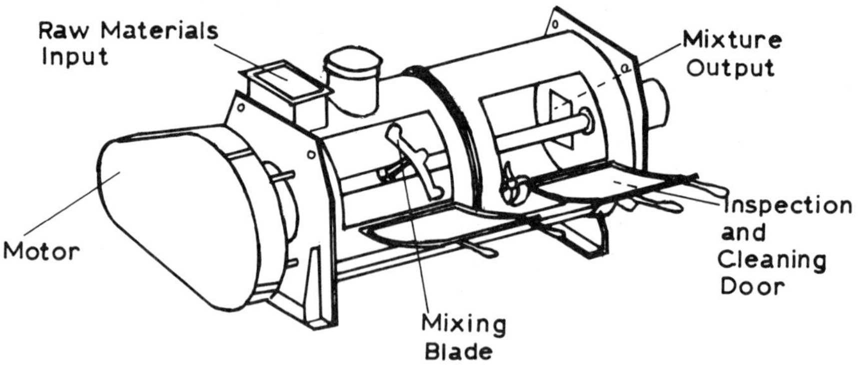

Fig. 4. Mixing and homogenization of raw materials

Considering operation time, one can realise from table V, that a 6 to 10 day period every 3 to 5 months is required for changing blades etc. Therefore another 100 ton mixer has to be installed. Such a decision would greatly increase productivity by recycling the existing dust stockpile.

Pelletizing Discs

Three flying saucer type pelletizing discs are used. They have a six meters diameter each and rotate with 9rpm at an angle of 65°. The two of them have a three step side wall and a capacity of 25 to 30 t/h, each. The third one has an one step side wall and a capacity of 70 t/h. The mixture of dust, binding material and water if fed at the center and according to the snowball technique is pelletized. Water spraying over the cascading mixture helps the procedure. Initially, all discs were of the same type. By modifying one of the discs as already mentioned and as shown in figure 5, its capacity was practically doubled while the size of pellets produced was slightly augmented. The operation of each disc is monitored by a shift worker who may act on the water spray according to the size of pellets produced and the rest conditions. As already mentioned the daily charge of the discs is about 450 tons. Therefore at a rate of 70 t/h, only 6 to 7 hours of operation are enough to cover the production required while 18 to 19 hours are required at a rate of 25 t/h. It is quite obvious therefore that a 70 t/h disc is more than enough to cover daily production with plenty of spare time for cleaning and maintenance, whereas two discs of 25 t/h each are needed.

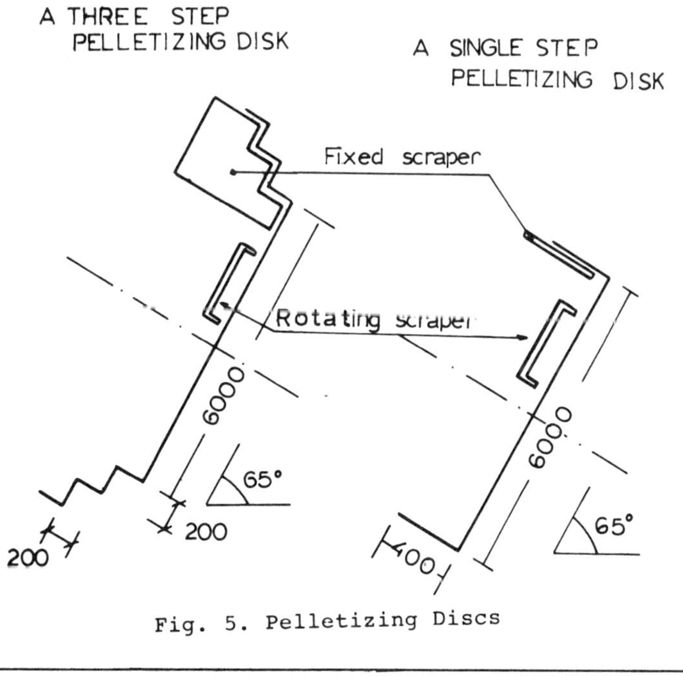

Fig. 5. Pelletizing Discs

TABLE V
OPERATIONAL TIME OF MIXERS AND PELLETIZING DISCS (%)

		MIXER I (100 t/h)	MIXER II (50t/h)	PELLETIZING DISCS	
				3 Steps	1 step
Year	Month				
1	Jan	100	-	54	56
	Feb	100	-	100	-
	Mar	100	-	98	2
	Apr	100	-	100	-
	Mai	100	-	36	64
	Jun	67	33	100	-
	Jul	100	-	75	25
	Aug	100	-	100	-
	Sept	100	-	70	30
	Oct	100	-	100	-
	Nov	88	12	71	29
	Dec	88	12	41	59
2	Jan	100	-	66	24
	Feb	100	-	100	-
	Mar	100	-	50	50
	Apr	80	20	100	-
	Mai	100	-	100	-
	Jun	100	-	100	-
	Jul	100	-	100	-
	Aug	100	-	70	30
	Sept	100	-	98	2
	Oct	-	-	-	-
	Nov	-	-	-	-
	Dec	50	-	100	-
3	Jan	100	-	100	-
	Feb	100	-	82	18
	Mar	100	-	100	-
	Apr	100	-	100	-
	Mai	100	-	100	-
	Jun	80	20	100	-
	Jul	98	2	92	8
	Aug	100	-	100	-
	Sept	100	-	99	1
	Oct	100	-	100	-
	Nov	70	30	64	36
	Dec	100	-	100	-

According to operating data for the discs as shown in table V 10 to 15% of total time is needed for maintenance. Therefore it is clear that for a better output a second disc has to be modified to a single step side one. This modification has already started.

Stocking of Green Pellets for Induration

Green pellets produced are stocked through an appropriate equipment (stocker) over a reasonably big surface area in order to obtain the required mechanical strength. It has been found out through tests that approximately 2 to 3 days are required for pellets to harden. During this time some 1500 tons of pellets are produced. Considering that the layer formed must have a maximum height of 2 m and that apparent density of pellets is 1.2 g/cm^3, a total surface area of 625 m^2 is required. At present, the space available is about 500 m^2. Therefore some extra space is needed and more than one solutions have been suggested to face the problem.

CRUSHING STRENGTH OF PELLETS

The pellets produced have to keep their shape and size while moving towards or inside the rotary kilns. The criterion for this is the amount of dust produced from the kilns. Dust should remain constant no matter what the amount of pellets is in the feed. This holds as shown in table II. For achieving this, pellets must have a given crushing strength. This is measured by a simple mechanism built for this purpose, shown in figure 6, and described hereafter. On a marked lever M a small metallic plate A, 100 x 100 mm, is fixed as shown in the figure and a standard weight F can move at will along side. On the same end with plate A the lever articulates in a way that this plate is touching another fixed plate B underneath. In order to measure the crushing strength of a pellet the later is introduced between plates A and B and the weight is moved until the pellet is crushed. Crushing strength is easy to be calculated knowing the value of the weight and its distance, as well as that of the plate, from the articulation point. To get a representative value 10 pellets of the same diameter (about 10 mm) are crushed and the mean value is considered. In table VI the effect of pellet size to the strength of pellets is given. According to tests, pellets of +13.2mm had an average crushing strength of 7.7 Kg, those between -13.2 to +6.7 mm had a crushing strength of 5.2 Kg while those between -6.7 to +3.35 mm had a crushing strength of 0.7 Kg. It was also found that pellets having a crushing strength of 7 to 8 Kg were good enough to go through the rotary kilns and keep their shape and size. To ensure this feature, 100 parts of dust were mixed with 20 parts of water and 10 parts of portland cement. Other parameters that affect the crushing strength of pellets are considered below.

TABLE VI
CRUSHING STRENGTH OF PELLETS IN TERMS OF INDURATION TIME AND PELLET DIAMETER

Samples	Pellet Diameter (m.m.)	Pellet strength (Kg/pellet) after					
		6 days	7 days	8 days	9 days	10 days	11 days
1	+13,2	n.m.	7,3	9,2	14,1	15,6	15,7
2	"	5,1	9,6	9,4	7,8	n.m.	12,5
3	"	7,3	9,2	9,4	n.m.	15,6	n.m.
4	"	8,1	n.m.	12,5	n.m.	11,5	14,1
5	"	10,3	n.m	11,5	15,6	n.m.	n.m.
Mean Value	13,2	7,7	8,7	10,4	12,5	14,2	14,5
1	-132,2+6,7	n.m.	5,8	4,6	10,8	9,5	12,2
2	"	4,4	5,1	5,7	5,7	n.m.	10,3
3	"	4,4	7,5	6,7	n.m.	11,45	n.m.
4	"	6,5	n.m.	9,2	n.m.	9,2	15,6
5	"	5,6	n.m.	8,4	n.m.	n.m.	n.m.
Mean Value	-132,2+6,7	5,2	6,1	6,9	8,3	10,0	12,6
1	-6,7+3,35	n.m.	0,7	n.m.	3,9	4.6	4,1
2	"	0,7	n.m.	0,4	2,0	n.m.	n.m.
3	"	0,5	n.m.	3,1	n.m.	5,2	n.m.
4	"	0,8	n.m.	n.m.	n.m.	n.m.	n.m.
5	"	n.m.	n.m.	0,2	1,0	n.m.	n.m.
Mean Value	-6,7+3,35	0,7	0,7	1,2	1,4	4,9	4,1

n.m. = not measured

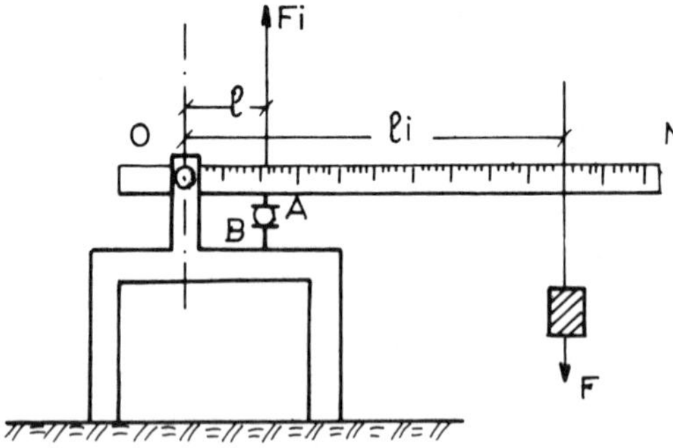

Fig. 6. Experimental set up for measuring pellet crushing strength

a. Granulometry of Dust

In table VII crushing strength data concerning dust with different granulometries are given. In figure 7 a graphic representation of these data is given. One may notice that pellets produced from dry dust coming from rotary kilns 1,2 and 3 have a smaller crushing strength than those produced from rotary kiln's 4 dry dust. This is due to the fact that dust coming from R/K 1,2 and 3 is coarser than that coming from R/K 4. This is easy to explain having in mind that multicyclones hold up the coarser particles while electroprecipitators hold up all dust particles. According to measurments (15) the mean diameter of dust particles from R/K 1,2 and 3 is 3.8 m while that for particles from R/K 4 is 3,4 m. Besides, the grain diameter for the former dust is from 1 to 380 μ while for the later is only 1.4 to 73μ.

It can also be seen that pellets produced from a mixture of dust coming from all R/K have a higher crushing strength than that of those produced from dust coming from R/K 4. The reason for this is that dusts having a greater range of granules allow less porosity to the pellets produced. It is noticed that pellets coming from dust from R/K 1,2,3 and 4 have double strength than that of those coming from R/K 1,2 and 3.

It is also noticed that crushing strength is considerably increased by adding ore fines to the dust. Initially it was planned to add these fines to the mixture for pelletizing but it was soon proven that this amount was rather small and it didnot justify the installation. So, the idea was abandonded.

b. Porosity of Pellets

The greater the range of particle size the less the porosity of pellets because small grains get in between the bigger ones and fill up the gaps. This is shown in table VIII and figure 8 where the crushing strength of pellets coming from dust with different granulometries is given.

c. Binding Material

In table IV the crushing strength of pellets with different binders, in quality and quantity, is given. It is noticed that Greek portland cement gives excellent results. It is also noticed that the crushing strength increases with the amount of binder but of course this is not the most economic solution. The results with a fast hardening cement were better than with portland cement but it caused problems in cloging the installation itself.

TABLE VII
CRUSHING STRENGTH MEASUREMENTS

TYPE OF DUST	Pellet Strength (Kg/pellet) after					
	1 day	2 days	3 days	4 days	5 days	6 days
1. Dust from R/K 1,2,3	1,9	1,9	3,9	5,2	7,3	7,3
2. " " "	1,7	1,7	1,7	2,5	3,9	3,9
3. Mean value	1,8	1,8	2,8	3,8	5,6	5,6
4. Dust from R/K 4	1,7	3,9	6,3	7,3	8,0	8,4
5. Dust from R/K 1,2,3,4	2,5	5,5	6,8	9,2	10,9	12,8
6. " " "	2,5	4,3	6,2	9,1	11,3	13,0
7. " " "	2,6	6,6	7,2	9,3	10,6	12,7
8. Mean value	2,5	5,5	6,7	9,2	10,9	12,8
9. Dust from R/K 1,2,3 and laterite	4,3	10,8	n.m.	n.m.	12,7	20,6

n.m. = not measured

d. Induration Time of Green Pellets

Induration time is of great importance when cement is used as a binder. Tables VI, VII and VIII show the increase of crushing strength with time. The fact that there was not enough free space available for green pellets to be laid and indurated, led to a thourough study of the problem. It was for this purpose that fast hardening cement was used but with poor results considering the pelletization as a whole.

TABLE VIII
CRUSHING STRENGTH OF PELLETS IN TERMS OF INDURATION TIME AND SOURCE OF DUST

Source of dust	Pellet strength (Kg/pellet) after		
	1 day	2 days	3 days
1. Dry dust from R/K 1,2,3	2,5	6,5	9,0
2. Dry and wet dust from R/K 1,2,3	4,0	7,0	12,5
3. Dry + wet dust of R/K 1,2,3 + fine pellets (-4mm)	4,0	6,0	11,0

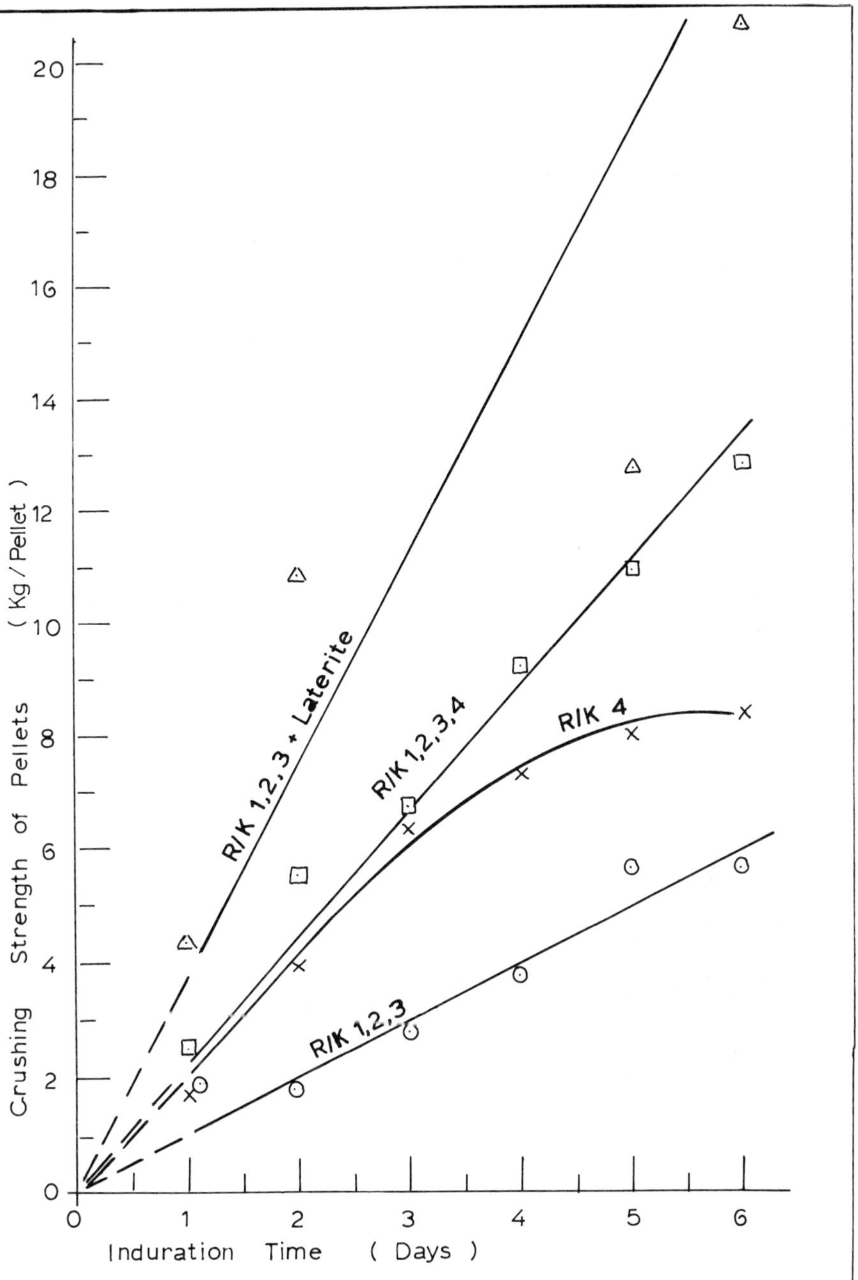

Fig. 7. Crushing strength of pellets in terms of dust composition and induration time

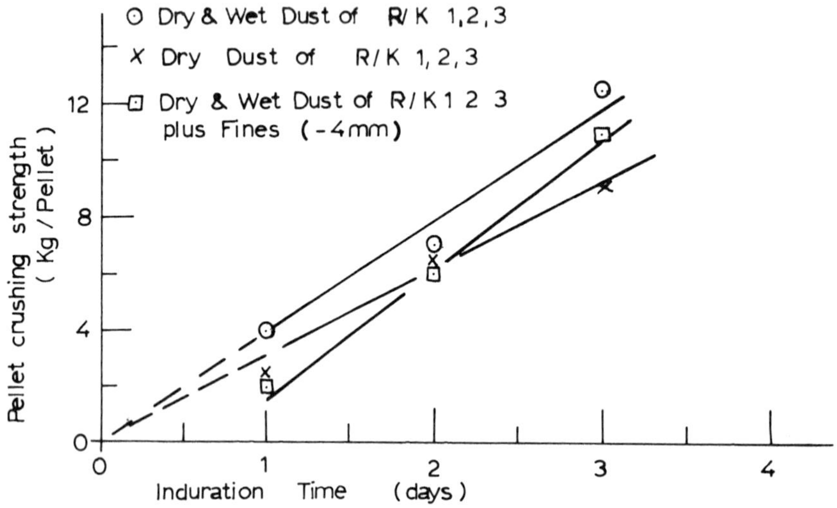

Fig. 8. Pellet strength in relation to induration time and dust composition

CONCLUSIONS

1. Critical paths for the pelletization are considered to be :
   - Mixing units, which for the given dust and productivity had to be both of at least 100 ton/hour, capacity.
   - Flying saucer discs, which had to be of 60 to 70 tons/hour, capacity.
   - Free area, for induration of green pellets.

2. The minimum crushing strength for the pellets to be handled without serious decrepitation was found to be 7 to 8 Kg/pellet.
3. Pellets coming from a mixture of dust from all four rotary kilns were found to have the highest crushing strength
4. By adding the wet dust into the mixture for pelletization, a much higher final crushing strength was obtained.
5. A 2 to 3 days induration time and a weight ratio of 100 parts of dust to 10 parts of cement, to 20 parts of water gives birth to pellets with a sufficient crushing strength no matter which the sourse of dust is.
6. A production rate of pellets of 20.000 tons/month results in a 150.000 US$ savings/month from raw materials.
7. Recycling of dust coming from the rotary kiln gases has reached a level of 85-95% during the last five years. This shows a good improvement and stability in rotary kiln dust recycling.

ACKNOWLEDGMENTS

The authors gratefully acknowledge the financial support of the General Secretariat for Research and Technology of Greece and also G.M.M.S.A. LARCO for financial support and for permission to publish the results.

REFERENCES

1. Zevgolis, M.N.: "Gas cleaning in Extractive Metallurgy as Applied by LARCO in Ferronickel Production". Mining and Metallurgical Chronicles, No 56, Mar.-Apr. 1984, pp 47-63

2. Zevgolis, M.N.: "Present State and Perspective of Environmental Protection in Extractive Metallurgy-Environmental Protection in the Greek Ferronickel Industry". Technika Chronika,Scien.Journ.of the Tech.Cham. of Greece-Sect.C, Vol 5, No 1-2,Jan-Jun. 1985, pp. 5-37

3. Tsakarisianos, A: "Tests on Green Pellet Production from Rotary Kiln Dust at SEL "Internal Report, LARCO S.A., March 1977

4. Zevgolis, M.N.: Research Proposal 5549/Jul. 1983, Greek Ministry of Research and Technology, 1983

5. Research Contract No 85053/Jul 85, The Greek Ministry of Research and Technology and LARCO SA, 1985

6. Zevgolis, M.N., Gaitanos, J.F. and Lazaridis, N.: "Recycling of Rotary Kiln Dust from Nickeliferrous Laterites". Final Report, Gen. Secretariat of Research and Technology of Greece, Jul. 1987.

7. Zevgolis, M.N. : "A Contribution to the Study of Rotary Kilns Used for Reduction of the Greek Nickeliferrous Laterites" A Thesis for Lectureship. Nat. Techn. Univ. of Athens, 1982.

8. Zevgolis, M.N. : "Pollution Control in the Mineral Industry" 1st Symposium on Mineral Resources Exploitation. The Techn. Cham. of Greece, Athens, 1978, pp 1040-1073

9. Pietsch, W.B.: "Adhesion and Agglomeration of Solids During Storage, Flow and Handling. A Survey" Journ. of Engineering for Industry, May 1969, pp 345-449

10. Ban, T.E. : "The Agglomeration of Fine Taconite Concentrate by Pelletizing" M.Sc. Thesis, Univ. of Minnesota, Jun. 1951

11. Knepper W.A. (Editor) : Agglomeration - AIME Symposium, New York, Interscience, 1962

12. Capes, C.E. (Editor). Fourth International Symposium on Agglomeration. Toronto, Canada, 1985, Co-sponsors : Iron and Steel Society, C.I.M.M., in cooperation with Society of Mining Engineers

13. Dorr. A.A. and Skretting, H.: "The Production of Ferronickel by the Rotary Kiln-Electric Furnace Process". Intern. Laterite Symposium, AIME, 1979, pp 459-490

14. Zevgolis, M.N., Gaitanos, J.F. and Lazarides, N.: "Productivity Improvement in Pelletizing Dust from Laterite Prereduction in Rotary Kilns" Symposium on Environmental Protection in Mining, Metallurgy and Materials Technology, Depart of Mining and Metallurgy, Nat. Techn. Univ. of Athens, Greece, Dec. 13-15, 1989, pp. 363-386

15. Coheur, P.: "Fines de Four Tournant" Centre de Recherche Metallurgic de Belgique (CRM). Report for LARCO, Jun. 1980

RECOVERY OF HIGH PURITY GALLIUM METAL FROM GALLIUM-ARSENIDE SCRAPS

SHIGEKI KUBO, MASAYA YUKINOBU, OSAMU YAMAMOTO, GEORGE NABESHIMA, YASUHIRO OKAJIMA

NIIHAMA RESEARCH LABORATORY, BESSHI DIVISION, SUMITOMO METAL MINING CO., LTD. 17-3 ISOURA-CHO, NIIHAMA CITY, EHIME, JAPAN 792

ABSTRACT

Recovery method of gallium metal from gallium-arsenide scraps by chlorination has been developed. The operating conditions of this process as well as comparison with other method is discussed. This chlorination process has proven to be an effective method to recover gallium metal pure enough to be recycled as a raw material to grow gallium compound crystals such as gallium-arsenide and gallium-phosphide.

INTRODUCTION

Gallium is produced as by-product in the processing of bauxite and sphalerite. Gallium belongs to the aluminum group in the periodic table and has same color as silver. History of its use as material is rather new. Gallium metal has a melting point of 29. 8 ℃ and a boiling point of 2400℃, so its temperature range existing as liquid state is wide. Vapor pressure of gallium is also extremely low even at elevated temperature. Gallium forms alloys with other metals. Chemically, when gallium is heated in ambient air or oxygen, it is not completely oxidized, but reacts easily with halogens to make compounds.[1]

Gallium was discovered in 1875 and its commercial production commenced in the 1940's. In the beginning demand for gallium was small and used only as alloy constituent. However, in these days gallium is widely applied as electronic material.

Gallium is primarily used for growing semiconductor single crystals such as gallium-arsenide and gallium-phosphide. As gallium-arsenide compound generates laser oscillation, it is applied to manufacture light emitting diode (LED) and laser diode (LD).[2] As for gallium market, world demand is reported to be 60 to 65 metric tons

Second International Symposium - Recycling of Metals and Engineered Materials
Edited by Jan H. L. van Linden,
Donald L. Stewart, Jr., and Yogeshwar Sahai
The Minerals, Metals & Materials Society, 1990

in 1988. In Japan, domestic demand is approximately 45 metric tons, which is 70 to 75% of the world demand.[3] The break-down is as follows :

Sources	Quantity in metric tons
Domestic	35
(from ore)	(7)
(from scraps)	(28)
Import	10

At the current situation, IC chips represent still smaller share for gallium market than optoelectronic devices such as LEDs and recycling of gallium from epitaxial film manufacturing process tends to increase. As can be known, reclamation of gallium from gallium-arsenide scraps is very important.

In this paper we will describe the recovery method of gallium metal from gallium-arsenide scraps at Isoura Plant. Combination of chlorination, distillation and rectification processes are applied to obtain high purity gallium matal. Major features of the process in comparison with other refining processes are also mentioned.

RECLAMATION AND PURIFICATION

We develop a simplified and cost effective process to recover high purity gallium from gallium-arsenide scraps as shown in Fig. 1 and 2.[4] Various types of scraps are crushed to the size of 2 to 5 mm. Crushed scraps are placed in a cell made of quartz, and chlorinated by introducing chlorine gas to the cell. As the reaction proceeds, mixture of chlorides is obtained in liquid state and accumulated at the bottom of the cell.

This mixture of chlorides of gallium and arsenic is transfered to the distillation column where gallium chloride and arsenic chloride are separated by distillation. The arsenic chloride separated advances to the adjacent high purity arsenic production plant as a feed.

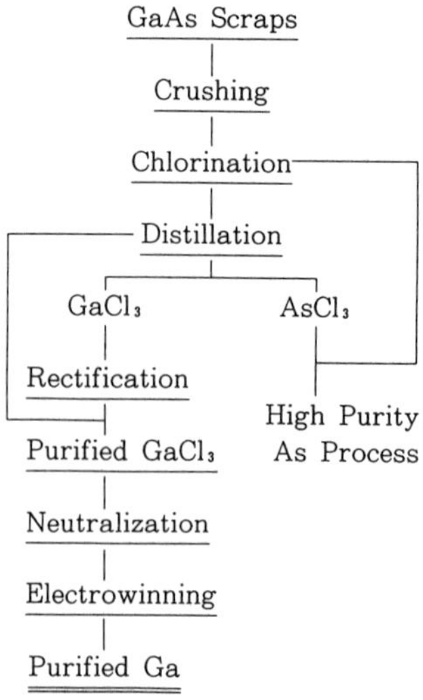

Fig 1. Flow sheet of Ga Refining Process

The distillate of gallium chloride goes to the rectifier for further purification. Then, purified gallium chloride is neutralized by addition of sodium hydroxide to form sodium gallate. This gallate solution is fed to the electrolytic cell as electrolyte, where gallium metal is electrowon. During electrowinning process gallium concentration in the electrolyte decreases, so as current efficiency. Electrowinning is continued till no gallium metal would be electrodeposited. Cathode with electrodeposited gallium is dipped in hot water to melt down gallium. Thus cast gallium has purity of 99.9999 % Minimum.

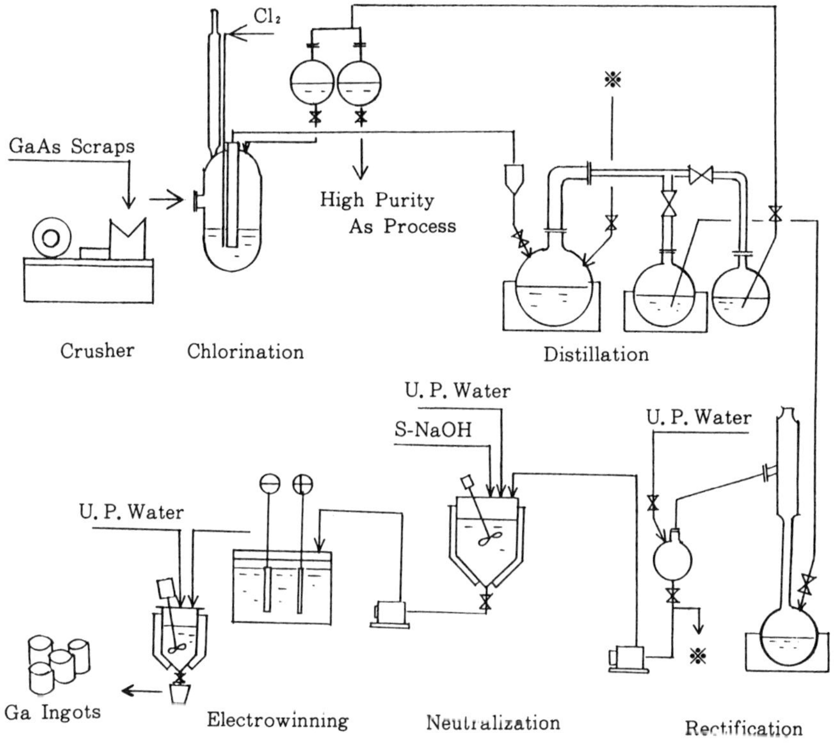

Fig 2. Flow chart of Gallium Refining Process

PREPARATION OF SCRAPS

Gallium-arsenide substrate is prepared as follows :

(1) Synthesis of compound
(2) Pulling of single crystal from compound melt
(3) Cutting to wafer from single crystal boule
(4) Polishing of wafer

(5) Epitaxial film growth on polished wafer

It is said that total yield of used metals to the final product is only approximately 10%. There are two types of scraps, which can be processed economically feasible :

(1) Soild scraps such as broken wafer and top/bottom portion of boule

(2) saw kerf : sludge generated by cutting of single crystal from boule

Scraps are also classified as doped or undoped from the point of impurities, indium is used as dopant.

Particle size of scraps determines velocity of reaction during chlorination, it is preferable to crush solid scraps to 2 to 5 mm. Dried saw kerf is charged to chlorination cell directly. Typical assay of scrap material is shown in Table I.

Table I. Typical assay of Ga-As scraps

Ga (%)	As (%)	In (ppm)	Fe (ppm)	Si (ppm)	S (ppm)
48.0	51.2	650	19	5	2

CHLORINATION

Quartz cell as shown in Fig. 3 is utilized for chlorination of scraps. 20 to 30% of arsenic chloride obtained in the following distillation column is returned to the cell not only to decrease liquidus temperature of produced chloride but also to increase reaction efficiency. Chlorination reaction is considrered to proceed according to the following reactions.

$$Ga + Cl_2 = GaCl_2 \quad (1)$$

$$GaCl_2 + 1/2\,Cl_2 = GaCl_3 \quad (2)$$

$$Ga-As + 3\,Cl_2 = GaCl_3 + AsCl_3 \quad (3)$$

In order to avoid hydrolysis of chloride, moisture in the cell should be carefully removed by passing inert gas through the assembled cell before introduction of chlorine gas. Chlorine gas is injected into the mixture of crushed scraps and recycled arsenic chloride after passing through concentrated sulfuric acid. Once introduction of chlorine gas is started, reaction proceeds by itself because of exothermic nature of chlorination process. Most of vaporized chloride is trapped by the

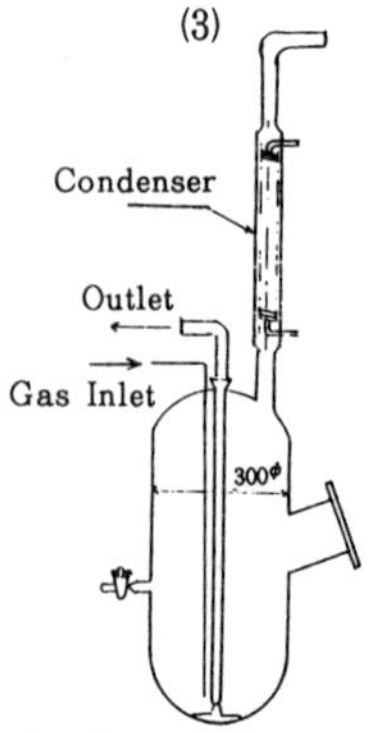

Fig 3. Chlorination Cell

condenser and returned to the cell. Melting point and boiling point of gallium and arsenic chloride are shown in Table II.

Table II. M.P and B.P of chlorides

	M.P (℃)	B.P (℃)
$GaCl_3$	77.9	201.3
$AsCl_3$	−16.2	130.4

Operating conditions of chlorination is as follows :

Blowing rate (g/hr • kg-scrap)	Chlorine efficiency (eq)
90 − 100	1.1 − 1.2

Typical impurity levels of obtained chloride mixture is shown in Table III.

Table III. Typical assay of chlorides

(ppm)

In	Fe	Si	S
4.3	18	11	4.3

DISTILLATION

As can be seen from Table II, gallium and arsenic chloride is easily separated by distillation taking advantage of difference of boiling points. Relationship between vapor pressure and temperature of various metal chlorides is shown in Fig. 4. It shows that most of metal chlorides except antimony can be separated from gallium chloride by distillation.

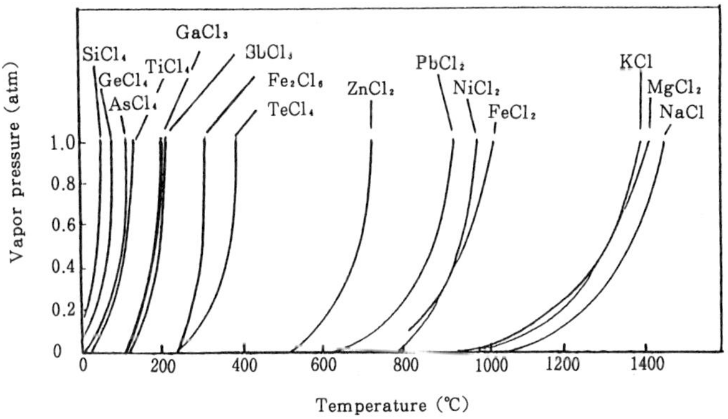

Fig 4. Vapor pressure of metal chlorides[5]

Schematic diagram of the distillation column is shown in Fig. 5, this column consists of a still and flasks for receiving distillate with condenser. Arsenic chloride comes out first when operation starts. As arsenic chloride vaporizes, the temperature of chloride in the still increases from 130° C to 190° C gradually. Distillation goes on till the volume of arsenic chloride distillate reaches to 70 to 80% of expected volume in the receiving flask. This distillate contains more volatile substances such as germanium chloride and is sent to the arsenic recovery plant. Distillation operation is resumed to boil off the balance of arsenic chloride to be recycled to the chlorination process. The end point of distillation of arsenic chloride is judged not only from the rise of chloride temperature in the still but also from precipitation of gallium chloride on the inner surface of tubings. Distillation of gallium chloride is accomplished by keeping the temperature in the still to be 200° C, and continued until the temperature starts to increase. In this operation condenser and receiving flask should be heated to 80 to 85° C to avoid solidification of gallium chloride. Distillation of gallium chloride is terminated so as to leave approximately 10 % of charged gallium chloride in the still.

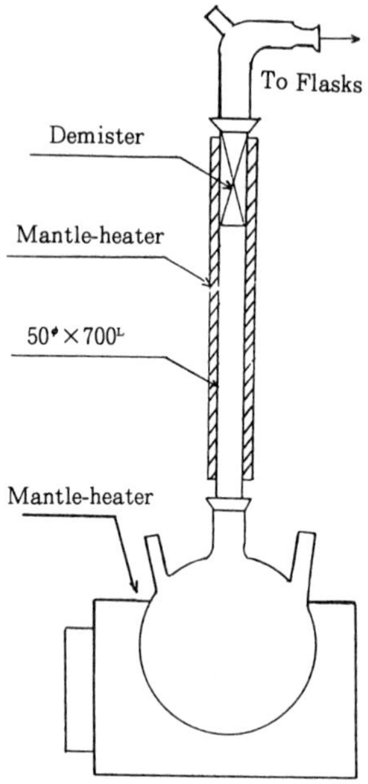

Fig 5. Distillation Column

Mean distillation rate of arsenic and gallium chloride are respectively, 4 and 6 kilograms per hour. Typical impurity levels of the distillate are shown in Table IV.

Table IV. Typical assay of distillates

(ppm)

	In	Fe	Si	S
$AsCl_3$ (first stage)	<0.2	<0.2	N. D	N. D
$AsCl_3$ (second stage)	0.3	3	N. D	N. D
$GaCl_3$	0.3	13	<0.4	<4

RECTIFICATION

Impurities such as indium and iron are concentrated in the crude gallium chloride separated by distillation. Therefore, rectification of crude chloride is necessary for removal of these elements. Schematic diagram of rectifier is shown in Fig. 6. It consists of a still, a condenser and a flask for receiving purified chloride. Crude chloride is placed to the still and the temperature in the still is kept to be 200° C.

Approximately first 5 % of distillate is returned to the distillation process because it contains small quantity of arsenic chloride. Rectification is continued until temperature in the still begins to increase. Then, usually 5 % of chloride remains in the still. Purified distillate dissolves in ultra-purified water put in the flask in advance. Rectification condition is as follows :

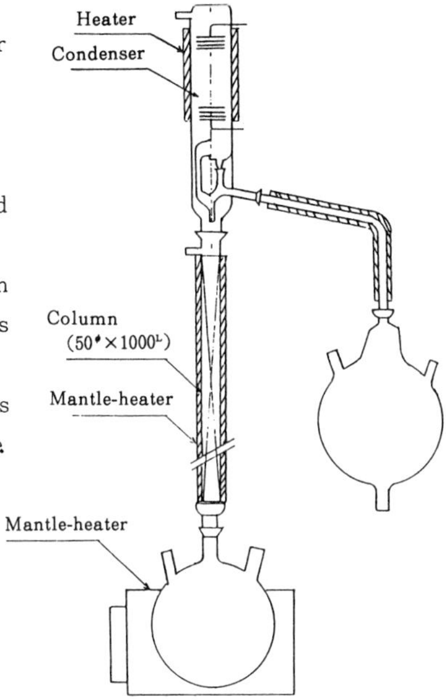

Fig 6. Rectifier

Mean distillation rate (Kg/hr)	Reflux ratio
2 – 3	3/1 – 4/1

Typical assay of purified gallium chloride after rectification is shown in Table V. As can be seen from Table V, refined chloride contains 0.2% arsenic, however, arsenic is not electrodeposited during the following electrowinning process of gallium metal.

Table V. Typical assay of refined gallium chloride

			(ppm)	(%)
In	Fe	Si	S	As
<0.2	<2	<0.4	<4	0.2

NEUTRALIZATION AND ELECTROWINNING

In order to obtain sodium gallate as electrolyte, refined gallium chloride dissolved in the flask of the rectifier is put into Teflon vessel where it is diluted with

$$GaCl_3 + 6NaOH = Na_3GaO_3 + 3NaCl + 3H_2O \quad (4)$$

ultra-purified water placed in advance, then the solution is neutralized by addition of sodium hydroxide of extra-fine grade. Gallium concentration in the gallate is controlled to 60 to 70 grams per liter. The gallate is supplied to electrolytic cell made of titanium. Gallium metal is electrowon by using titanium plate as cathode and DSE as anode. Electrode reactions are as follows :

Cathodic reaction $3\,Na^{+} + GaO_3^{3-} + 3\,H_2O + 3\,e = Ga^{\circ} + 3\,OH^{-} + 3\,NaOH$ (5)

Anodic reaction $3\,OH^{-} - 3\,e = 3/4\,O_2 + 3/2\,H_2O$ (6)

During electrolysis the cell is maintained below 20° C so that electrodeposited gallium does not melt down to the bottom of the cell. Electrowinning condition is as follows :

Electrolytic current (A)	Current density (A/dm^2)	Anode	Cathode
30 – 50	4 – 6	300×260	300×260

Electrowinning is continued till deposit on a cathode would not increase any more. Though cathode current efficiency can be maintained to 90 to 95% till gallium concentration dereases to 30 grams per liter, it deteriorates sharply below this level. When electrowinning is finished the electrolyte is neutralized by hydrochloric acid and gallium in the electrolyte is recovered as gallium hydroxide.

Electrodeposited gallium metal on a cathode is dipped into hot water to melt down. Then it is separated with water and rinsed by 1 N-HCl. Typical assay of gallium ingot thus produced by spark source mass spectrometer is shown in Table VI.

Table VI. Typical assay of products

(ppmw)

B	C	N	F	Al	Si	S	K	Fe	In	Cu	As
<0.01	0.03	0.03	0.01	<0.02	0.05	<0.02	0.06	N. D	N. D	N. D	N. D

All operations except from chlorination through neutralization are accomplished in clean room of class 10,000.[5]

CONCLUSION

As stated above major features of the process can be summarized :

(1) Processes are operated at rather low temperature.

(2) Most impurities can be removed very effectively.

(3) High purity gallium metal is obtainable.

Major impurities in gallium-arsenide scraps are indium added as dopant and carbon from crucible material, these elements can be reduced remarkably in obtained metal. Furthermore, it should be emphasized that sulfur is removed to almost trace level. This is not attainable by any other proposed processes such as thermal decomposition in vacuum, decomposition by oxidation and alkaline electrolysis. Germanium is also effectively separated, which can not to be removed by widely used segregation method for metal purification.[6]

REFERENCES

1. Renato G. Bautista, Gallium Metal Recovery. JOM, June, 1989, p. 30−31
2. Deborah A. Kramer, Gallium and Gallium Arsenide : Supply, Technology, and Uses. Bereau Mines, Infor Circular 9208, 1988, p. 1−29
3. Y. Suzuki, Gallium. Kogyoh Rare Metal, No 98, 1989, p. 39−41
4. H. Ohwa, M. Yukinobu, G. Nabeshima, M,Yasukawa, High−purity metals production of Sumitomo Metal Mining Co.,Ltd SMM). Extraction Metallurgy 89 IMM, 10−13 July, 1989, London, p. 885−898
5. K. Toyabe, C. Segawa, H. Kadoya, H. Ohwa, High Purity Arsenic Metal Production at SMM′ s Niihama Cupper Refinary. TMS−AIME, Jan 25−28, 1988, p. 351−362.
6. R. W. Bartlett, SEPARATION OF As and Ga FROM MIXED GaAs ELECTRONIC SCRAP. ARSENIC. METAL. FUNDUM. APPL. PROC. SYMP. 1988 p. 385−400

A PROCESS FOR THE TREATMENT OF FLUE DUSTS

AND OTHER WASTES CONTAINING INDIUM

Nicholas B. Gomez and John P. Hager

Department of Metallurgical and Materials Engineering
Colorado School of Mines
Golden CO 80401

Abstract

The treatment of indium-containing feed-stocks, such as flue dusts, is uneconomic if the material has a low concentration of indium and at the same time contains a significant amount of lead. Therefore, it is desirable to extract the lead prior to indium recovery. Such a process has been developed based on the selective reduction of lead sulfate, lead oxide, and lead sulfide, contained in flue dusts, with metallic sodium.

The first stage of the process entails the use of a sodium-doped lead bullion at temperatures ranging from 675-800°C. Soda ash is blended with the feed-stock and added to the sodium-doped bullion. Upon reaction, a liquid dross is formed. The liquid dross is removed, cooled, and crushed. The sodium is extracted from the powdered dross by water leaching. The resultant residue can then be processed by conventional methods for the recovery of indium. The lead reduced from the feed-stock is recovered in the molten bullion.

Indium recovery to the filter cake is typically 90-95%, while retention of lead in the filter cake is usually 5-15%. The concentration of indium in the filter cake is approximately twice that of the original feed-stock. The sodium consumption varies between 0.1 and 0.2 lbs sodium per lb of feed-stock.

Second International Symposium -
Recycling of Metals and Engineered Materials
Edited by Jan H. L. van Linden,
Donald L. Stewart, Jr., and Yogeshwar Sahai
The Minerals, Metals & Materials Society, 1990

Introduction

Indium is a soft, lustrous, silvery-white metal. When pure, it is so malleable and ductile that it is used as a gasket material for cryogenic vessels (1). Indium's wetting ability combined with its low melting point (156.6°C) makes it a suitable alloying agent for electronic solders (1). Many solder alloys containing indium have been developed with improved thermal-fatigue resistance enhancement, gold scavenging reduction, and alkaline corrosion resistance. When indium oxide is doped with tin, the result is a transparent electrical conductor that is used in liquid-crystal displays (1).

Indium is a trace metal present in some tin, lead, copper, and zinc ores. However, commercial production of indium usually involves residues, slags, and flue dusts from zinc processes. One of the more common methods for the recovery of indium involves leaching the feed-stock with sulfuric or hydrochloric acid and, after several intermediate steps, cementing the indium on zinc or aluminum sheets (2). Indium contained in lead-zinc alloys can be isolated in a high grade chloride-slag by treatment with a mixture of lead-chloride and sodium-chloride (3). Indium can then be recovered from the chloride-slag by leaching followed by sponging on aluminum or zinc. Indium can also be recovered from dilute solutions by solvent extraction using di (2-ethylhexyl) phosphoric acid or tributyl phosphate (2).

Objective

The objective was to develop a technology for the treatment of low grade, lead-containing, indium feed-stocks such as flue dusts and residues. Although such feed-stocks are amenable to conventional methods of recovery, these methods become uneconomic when low concentrations of indium are combined with a significant amount of lead (10% and above). Therefore, it is desirable to separate the lead and the indium before indium recovery. The proposed method for the separation of lead and indium involves the use of a sodium-doped lead bullion at temperatures ranging from 675-800°C. The feed-stock would be mixed with appropriate fluxes and added to the bullion. Upon reaction, a liquid dross would be formed and removed. The indium would report to the dross and any lead initially present in the feed-stock would be reduced and report to the bullion. The dross would be water leached to remove the sodium compounds in the dross. Since indium compounds are insoluble in caustic, the residue would contain the indium present in the initial feed. Using conventional methods, the resultant filter cake would then be processed for the recovery of indium

Feasibility Study

A feasibility study was conducted to determine if the desired separation of indium and lead was thermodynamically possible. A computer program was written, based upon thermodynamic principles, to model the reactor chemistry. A Gibbs Reaction Isotherm expression was written for each of the proposed reactions (presented later in this section). The computer program adjusted the extent of each reaction until ΔG was zero for all of the reactions. The elemental phase distributions were then calculated from the equilibrium extent of reaction. Thermodynamic feasibility was then established for the proposed separation of indium and lead.

Reactor Chemistry

Before the reactor chemistry could be proposed, the chemical make-up of the feed-stocks was required. The two feed-stocks characterized were Cottrell Dust Leach Residue (CDLR) and Kellogg Dust (KD). CDLR is the result of a low acid leach on a dust from a copper process. KD is a by-product of a lead process. Using X-ray diffraction and chemical assay data the compound composition of each feed-stock was determined and is presented in Table I.

The two listed materials contain more compounds than listed in Table I; however, the compounds selected comprise the principal components and those components important to commercial success, such as silver. Any other species were ignored and assumed inert to simplify the modelling process.

The reactor chemistry consists of a series of reactions that partition the various components between the bullion and dross phases. The components reporting to the dross are compounds. The components reporting to the bullion

TABLE I - COMPOUND COMPOSITION OF FEEDSTOCKS

FEED-STOCK	COTTRELL DUST LEACH RESIDUE	KELLOGG DUST
In_2O_3	0.590 %	0.230 %
$PbSO_4$	19.65 %	28.34 %
PbO	4.31 %	9.76 %
Cu_2O	11.28 %	0.08 %
$CuSO_4$	0.00 %	0.00 %
ZnO	10.08 %	8.38 %
$ZnSO_4$	0.00 %	0.00 %
As_2O_3	7.65 %	10.47 %
Ag_2O	0.075 %	0.010 %
Na_2O	20.62 %	28.57 %
OTHER	25.75 %	14.18 %
TOTAL	100.00 %	100.00 %

are dissolved elements. A reaction was written for each of the components listed in Table I; however, only the reactions involving indium and lead are listed below:

$$In_2O^d_{3(C)} + 6\ Na^b_{(l)} = 2\ In^b_{(l)} + 3\ Na_2O^d_{(C)} \qquad (1)$$

$$Q(1) = \frac{(a^b_{In})^2 * (a^d_{Na_2O})^3}{(a^d_{In_2O_3}) * (a^b_{Na})^6}$$

$$PbSO^d_{4(C)} + 2\ Na^b_{(l)} = Pb^b_{(l)} + Na_2SO^d_{4(C)} \qquad (2)$$

$$Q(2) = \frac{(a^b_{Pb}) * (a^d_{Na_2SO_4})}{(a^d_{PbSO_4}) * (a^b_{Na})^2}$$

$$PbO^d_{(C)} + 2\ Na^b_{(l)} = Pb^b_{(l)} + Na_2O^d_{(C)} \qquad (3)$$

$$Q(3) = \frac{(a^b_{Pb}) * (a^d_{Na_2O})}{(a^d_{PbO}) * (a^b_{Na})^2}$$

The superscript "d" indicates dross species, the superscript "b" denotes bullion species, and Q is the reaction quotient.

The details of the computer program are not presented in this paper. Descriptions of the solution technique and the computer program are given elsewhere (4).

Feasibility Study Results

The reactor model was applied to the two feed-stocks characterized in Table I. For each feed-stock, the computer program was executed with differing amounts of sodium per given weight of feed-stock. The elemental distribution/ recovery to the dross was plotted versus the sodium addition level. The results are presented in Figure 1.

The lead distributions reported in Figure 1 are taken on a relative basis. That is, the lead recovered in the dross is reported as a percentage of the lead initially present in the feed-stock. The elemental recovery to the bullion can be calculated by subtracting the dross recovery from 100%.

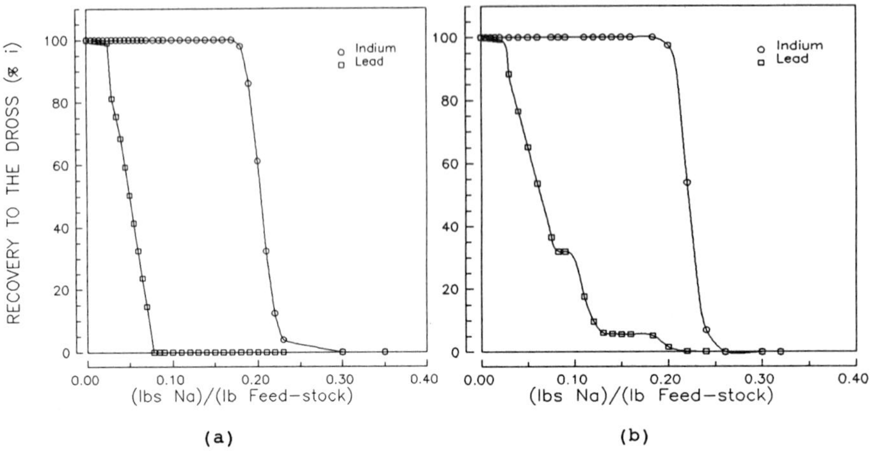

Figure 1 - Predicted recovery of lead and indium in the dross as a function of sodium addition level: a) CDLR, b) KD.

From Figure 1 it can be seen that lead is removed from the dross at lower sodium addition levels than those required to remove indium from the dross. This clearly identifies an opportunity for the separation of indium and lead present in the two feed-stocks. By adding sodium to the bullion at a level corresponding to the lead threshold value, it should be possible to selectively extract the lead to the bullion, leaving the indium in the dross.

Feasibility Study Conclusions

The results of the thermodynamic analysis indicate that the lead present in the original feed-stocks will report to the bullion with the simultaneous retention of indium in the dross. Thus, the desired separation of indium and lead present in the two feed-stocks is predicted to be feasible.

Experimental

Once the separation of indium and lead was deemed viable, an experimental plan was proposed. The experimental portion of the research program included reactor design, reactor installation, and development of an experimental procedure.

Reactor System

The proposed reactor design was derived from kettle metallurgy. Such a kettle-based technology would be readily adaptable to commercial practice. A schematic of the reactor system is given in Figure 2.

A C.A. Hones Model 172 gas fired furnace was selected and installed. The temperature range of the C.A.Hones furnace is 25-800 °C. A cast iron pot with dimensions of 10" ID and 13.5" deep was used. A lead bath of 150 lbs left a 6" free-board.

A hood assembly was fitted to the furnace to isolate the reactor environment from the workplace. Access doors were incorporated into the hood assembly for the addition and removal of material. A "T" section was mounted to the top of the hood assembly to allow passage of the off-gases to the scrubber system, and also provide access to the reactor by the stirrer.

The reactor temperature was regulated by a Honeywell Dialatrol controller. A type K thermocouple placed in the reactor was used as the input device to the Dialatrol. The Dialatrol controlled a solenoid actuated valve to turn the furnace burners on or off accordingly.

Mixing of the bullion and dross phases was accomplished by use of a stainless steel impeller and a cast iron surface agitator. These were attached

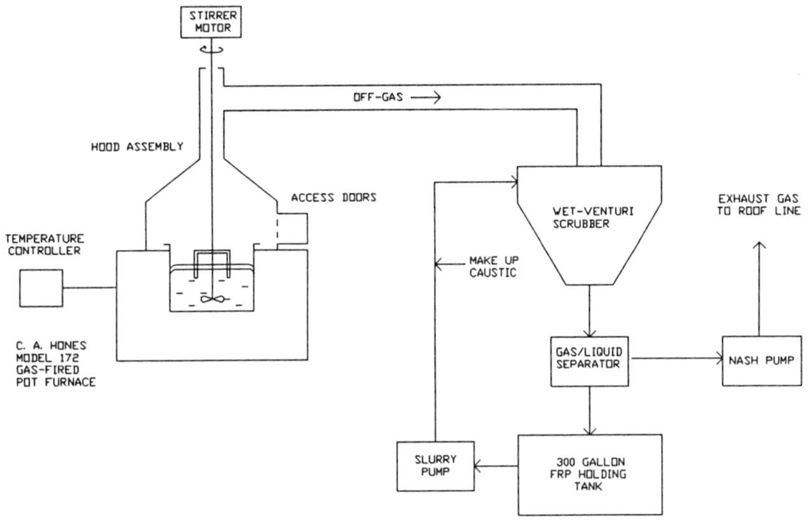

Figure 2 - Schematic of the Pb-Na reactor system

to a stainless steel shaft. The shaft was driven by a three-quarter horse power variable-speed DC motor coupled to a gear reducer.

The off-gas scrubber system consisted of a wet venturi scrubber, a gas/liquid separator, and a scrubber solution holding tank. The off-gas stream contained any solid particulates, volatile species, make-up air, and the natural-gas combustion products. The off-gas was drawn through the wet venturi scrubber by a Nash pump. The Nash pump was sized such that make-up air was drawn though the stirrer end of the "T". The flow of make-up air through the "T" insured complete isolation of the reactor environment from the laboratory. Any volatile species and solid particulates were removed from the off-gas by a stream of caustic solution that was pumped to the venturi scrubber. The caustic solution returned to the holding tank and was then recycled back to the venturi scrubber by a slurry pump. A pressure drop of 50 inches of water was chosen to insure vigorous off-gas scrubbing.

Experimental Procedure, Sodium-Lead Reactor

For each experiment, the furnace pot was charged with 135 lbs of lead (99.99%). The lead bath was then heated to the desired temperature. The stirrer was then inserted into the bath and agitation initiated.

To inject the sodium, the stirrer was turned off and a clear plastic explosion barrier was placed in front of the access doors. A pair of tongs was used to grasp a 0.25 lb chunk of metallic sodium. By reaching around the barrier and through the access doors, the sodium was rapidly immersed into the lead bath. Once the sodium was dissolved, after one second, agitation was resumed. This was repeated until the desired amount of sodium was introduced into the system. After the last sodium addition, the lead bath was sampled by casting slugs using a bullet mold.

Immediately after the sodium additions, a mixture of flux (Na_2CO_3) and feed (CDLR or KD) was placed on top of the lead bath. A typical charge weight was 5 lbs. Although the reactor was equipped with a stirrer, the mixing was found to be inadequate during the first 5-10 minutes after feed-stock addition. To overcome this deficiency, the stirrer was assisted manually to achieve satisfactory mixing.

The two phases were allowed to equilibrate for a predetermined time (usually 1 hour). Every 15 minutes after feed-stock addition, the bullion was sampled. The dross was also sampled during this time if it was molten. To sample the dross, a ladle was dipped into the molten dross layer such that no bullion was taken. The ladle was then emptied into a conical mold and allowed to cool. After the predetermined reaction time had elapsed, the dross was ladled off, cast into rectangular molds, and allowed to cool.

The addition of sodium and feed-stock, and the removal of dross were repeated the number of times prescribed by the experiment. After the last cycle, the lead bath was cooled to 400°C and ladled out and cast as ingots.

Experimental Procedure, Dross Leaching

A sample of dross was pulverized and weighed (approximately 0.110 lbs). An amount of water equal to 5 times the weight of powdered dross was placed in a beaker. A magnetic stirrer was used to agitate the water.

The dross sample was slowly introduced into the stirred water. The dissolution reaction was exothermic and if the dross was added quickly, the water boiled violently. The leach water may or may not have been heated depending upon the desired conditions for leaching.

The beaker was removed from the magnetic stirrer after one-half hour leach time and the residue was allowed to settle. The residue was separated from the liquor using a Buchner funnel. The residue was washed several times with de-ionized water. The leach liquor and water were collected and saved.

The filtrate was diluted to a known volume and submitted for chemical analysis.

The filter paper and the residue were dried in a drying oven at 120°C. Once dry, the cake and filter were weighed to obtain the residue weight. The residue was separated from the filter and ground into a powder. The residue was then submitted for chemical analysis.

RESULTS AND DISCUSSION

The experimental plan consisted of three phases. These included scoping experiments, process development experiments, and leach tests.

Scoping Runs

The objectives of the first experiment were to determine the practicality and effectiveness of doping a lead bath with sodium, and to determine the retention time of sodium in molten lead. During the experiment, the lead bath was sampled prior to sodium injection, after sodium injection, and at fifteen minute intervals thereafter for 2 hours. The experimental conditions and results are compiled in Table II.

TABLE II - RUN 1 EXPERIMENTAL CONDITIONS AND RESULTS

Bath wt (lbs)	100.88
Metallic Na (lbs)	1.80
% Na in Bullion	1.22
Dissolved Na (lbs)	1.23
Efficiency (%)	68

The amount of metallic sodium injected into the lead bath was 1.8 lbs. However, only 1.23 lbs of sodium were actually dissolved, giving an efficiency of 68% (doping efficiency for later experiments ranged from 50-70%). Because of the crudeness of the sodium injection method, this efficiency was considered acceptable. Sodium losses were attributed to oxidation and splattering. When the sodium was brought in contact with the molten lead, a portion of the sodium oxidized and formed a liquid layer (most of the remaining 32% reported to this layer). Also, small droplets of liquid metallic sodium splattered onto the side of the furnace pot. These two effects decreased the amount of metallic sodium available for dissolution.

To minimize sodium losses, a commercial application of this technology would incorporate liquid sodium injection. Sodium is usually transferred from storage tanks in the molten state. Several examples of liquid sodium use are Du Pont's tetraethyl lead process, and ASARCO's silver ore reduction and lead dross recovery processes (5) (6) (7).

As stated above, another purpose of the first experiment was to determine the stability, as a function of time, of dissolved sodium in molten lead. This is reported in Figure 3 where it can be seen that sodium loss from the bullion was gradual. After sixty minutes, 83% of the sodium initially dissolved in the bath was still in solution. After 105 minutes, 71% remained in the bath. This experiment represented the worst case scenario since the bullion was partially exposed to the atmosphere. The bullion surface was covered by a dross for all subsequent experiments. Thus, for those experiments that produced molten drosses, air oxidation of the dissolved sodium was assumed to be negligible.

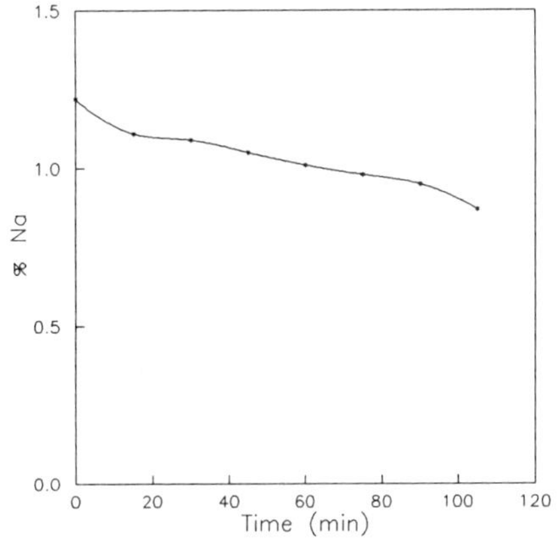

Figure 3 - Sodium retention of the bullion

The remaining scoping experiments served two purposes. First, the experiments facilitated improvements in the experimental procedure and apparatus. Second, the observations and results from these experiments allowed for the refinement of the experimental plan.

Developmental Experiments

Once the experimental procedure and apparatus were refined, several experiments were conducted to achieve the objectives stated earlier. The experimental conditions are reported in Table III. As seen in Table III, all of the drosses were molten except the Run 8 dross which was generated at 600 °C and with no soda ash.

The mass balance results for the process development experiments are reported in Table IV. Due to the nature of the experiments, a clean separation of the dross and the bullion was unattainable. This led to errors in the phase weights. Development of data reduction techniques were needed to overcome these errors. The row titled "Process Analysis" in Table IV refers to the method used to reduce the experimental data. Process Analysis 1 was based upon the measured phase weights. Process Analysis 2 equated the final bullion weight with the initial lead weight if the measured final bullion weight was less than the initial bath weight. The reactor chemistry can be represented by reactions that replace dross constituents with sodium. Since the formula weights of sodium (see Reactions 1 - 3) are less than the formula weights of the dross constituents, the final bullion weight should be greater than the initial lead bath weight. Process Analysis 3 involved the adjustment of the dross weight to achieve 100% closure on an element or elements whose initial closures were greater than 90%. This was typically a 10% increase in the weight of the dross. As seen in Table IV, all of the experimental results were not reduced using the same Process Analysis technique. The measured phase weights were used to generate the Process Analysis 1 results. However, if the

TABLE III - EXPERIMENTAL CONDITIONS AND OBSERVATIONS FOR RUNS 6-10,12

	Run 6	Run 7	Run 8	Run 9	Run 10	Run 12
Temperature (Deg C)	675	790	600	690	790	790
Lead Bath wt (lbs)	137.69	138.81	140.41	131.34	130.72	123.41
Feed-stock	CDLR	CDLR	KD	KD	KD	CDLR
Feed-stock wt (lbs)						
Cycle A	5.47	5.28	4.88	4.51	4.89	5.14
Cycle B	5.19	5.33	5.81	4.18	5.22	4.51
Cycle C	5.10	4.92	5.84	4.51	5.43	5.29
Cycle D	----	----	----	----	----	4.86
Sodium wt (lbs)						
Cycle A	4.75	4.50	4.50	4.56	4.56	2.66*
Cycle B	1.81	2.28	2.16	2.38	2.25	2.53*
Cycle C	1.88	2.28	2.25	2.34	2.25	2.75**
Cycle D	----	----	----	----	----	1.50**
Soda Ash wt%	15.00	15.00	0.00	15.00	15.00	35*/70**
Dross Descriptions	Molten	Molten	Pasty	Molten	Molten	Molten

TABLE IV - MASS BALANCE RESULTS FOR RUNS 6-10,12

	Run 6	Run 7	Run 8	Run 9	Run 10	Run 12
Feed-stock	CDLR	CDLR	KD	KD	KD	CDLR
Process Analysis	3	3	2	1	1	2
IN (lbs)						
Lead	137.69	138.80	140.41	131.31	130.72	123.41
Sodium	8.44	9.07	8.91	9.28	9.06	9.44
Feed	15.77	15.89	16.53	13.21	15.54	19.81
Soda Ash	2.36	2.38	0.00	1.98	2.33	10.49
Total	164.26	166.15	165.85	155.88	157.65	163.15
OUT (lbs)						
Bullion	137.69	138.80	140.41	131.82	136.88	123.41
Dross	22.47	24.02	14.21	13.74	16.46	30.25
Total	160.16	162.82	154.62	145.56	153.34	153.66
% Recovered	97.51	98.00	93.24	93.42	97.26	94.18
Calculated Dross Weight for 100 lbs of Feed	142.53	151.16	85.96	104.02	105.93	148.66

final bullion weight was less than the initial bath weight, the results were refined using Process Analysis 2. If, in addition, there was an element with closure greater than 90%, then the results were refined using Process Analysis 3.

As seen in Table IV, the overall mass balance closure of the experiments ranged between 93% and 98%. Dross weights were consistently between 13 lbs and 16 lbs for those experiments whose feed-stock was KD. Dross weights consistently ranged between 22 lbs and 24 lbs when the starting material was CDLR. The Run 12 drosses weighed 30 lbs. However, this experiment consisted of four cycles rather than the three cycles for the other experiments. When compared on a weight per cycle basis, the Run 12 drosses were comparable to the other CDLR drosses.

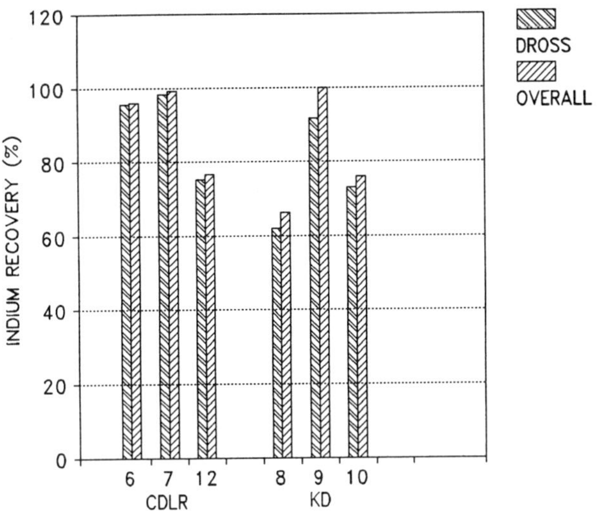

Figure 4 - Indium distributions relative to the original feed.

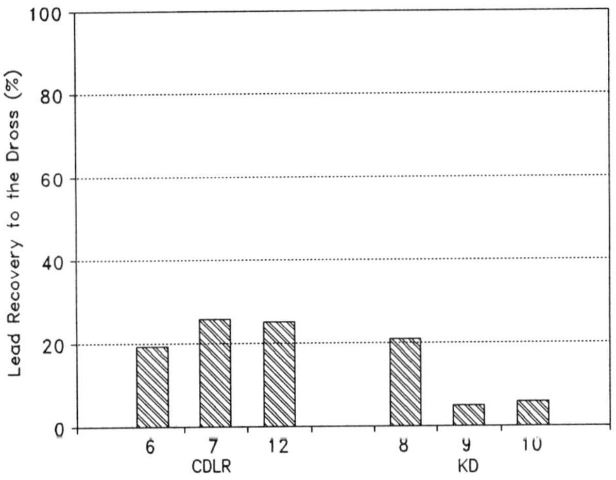

Figure 5 - Lead recoveries relative to the original feed.

Phase distributions of indium and lead initially present in the feedstocks were calculated from the phase weights listed in Table IV and the corresponding bullion and dross assays. These results are reported in Figures 4 and 5, where it can be seen that the elemental closures were sometimes less than 80%. These deviations were ascribed to analytical error.

Figure 4 illustrates the indium distribution to the dross. For the CDLR experiments, the indium recovery to the dross is greater than 90% for Runs 6 and 7, and 75% for Run 12. However, for Run 12, the overall indium recovery was only 78%. For the KD experiments, the indium recovery to the dross was 92% for Run 9 while the indium recovery was 66% and 72% for Runs 8 and 10, respectively. Regardless of the overall indium recovery, the indium distribution to the dross is at least 91% of the total indium recovery. For example, the Run 8 indium recovery was only 66.3%. However, of that 66.3%, 99% reported to the dross. Thus, it would appear that for those experiments where the indium closure was less than 80%, the lack of closure could be attributed to a low indium assay of the dross.

Figure 5 illustrates the lead recovery to the dross. Lead distribution to the bullion was not reported due to the inability to isolate the reduced feed-stock lead from that initially present in the bath. The bullion was presumed to contain that portion of the feed-stock lead that did not report to the dross. The lead recovery is based upon the lead contained in the feed. That is, of the lead initially present in the feed, only that percentage reported in Figure 5 was retained by the dross. As shown in Figure 5, the lead recoveries to the dross were quite low. This indicates that the lead initially present in the dross reported to the bullion. The simultaneous recovery of indium to the dross and reduction of lead to the bath is consistent with the initial objectives of the project.

Leach Tests

The objectives of the leach tests were removal of any soluble constituents from the dross and retention of indium in the filter cake. This would enhance the concentration of indium in the cake. Several dross samples were subjected to leach testing.

Dross leaching resulted in a cake and a liquor. The cake weights ranged from 25% to 30% of the original dross weight. The specific gravity of the leach liquors ranged from 1.05 to 1.11 g/ml.

Phase distributions of indium and lead initially present in the feed-stocks were calculated from the bullion, liquor, and filter cake weights and the corresponding bullion and dross assays. These distributions are reported in Figures 6 and 7. These recoveries are relative to the elements initially present in the feed-stock.

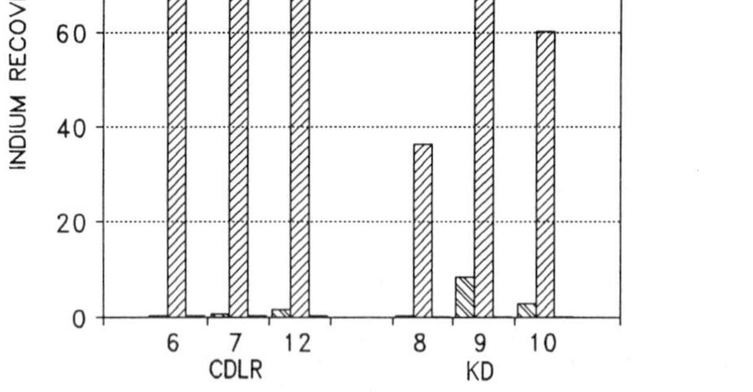

Figure 6 - Indium distributions relative to the original feed.

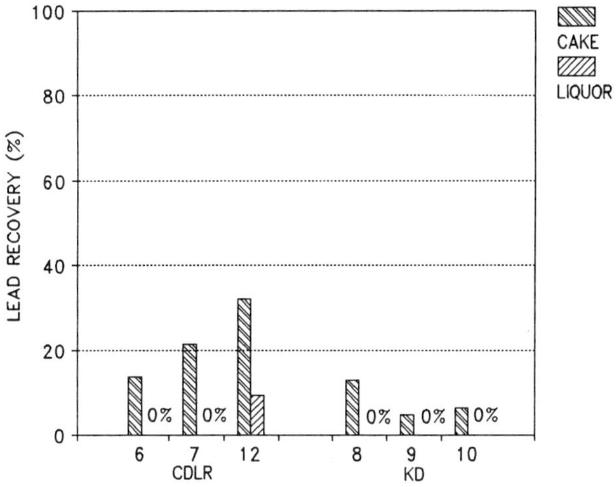

Figure 7 - Lead recoveries relative to the original feed.

Figure 6 illustrates the indium distributions to the bullion, cake, and liquor. The calculated indium recoveries to the cake were 103%, 71%, and 104% for Runs 6, 7, and 12 (CDLR experiments), respectively. The overall indium recoveries for these runs were 104,%, 72%, and 106%, respectively. For the KD experiments, the indium recoveries to the cake were 34%, 73% and 60% for Runs 8, 9, and 10, respectively. The overall closures were 37%, 82%, and 63%, respectively. Again, regardless of the overall indium recovery, indium recovery to the cake accounted for at least 90% of the total indium recovered. For example, the Run 10 overall indium recovery was only 63%. However, of that 63%, 95% reported to the cake. Also, since the experimental drosses were the materials being leached, the sum of the cake and liquor distributions should have been comparable to the distributions to the dross. Thus, it would appear that for those experiments where the overall indium closures were less than 80%, the lack of closure could be attributed to a low indium assay on the cake.

Figure 7 illustrates the recovery of lead in the cakes and liquors. As mentioned earlier, the lead recovery is relative to the lead contained in the feed. That is, of the lead initially present in the feed, the percentages reported in Figure 7 were retained by the cakes and liquors. As shown in Figure 7, the lead recoveries to the cakes were quite low. Virtually no lead reported to the liquor. Furthermore, the lead recoveries to the cakes were comparable to the dross lead recoveries reported in Figure 5. The Run 12 liquor was the only liquor that resulted in any significant lead recovery.

Reactor Model

A comparison of the actual experimental results to the predicted results (feasibility study results) was very poor. The sodium consumption was significantly higher than predicted and the elemental phase distributions were dissimilar. There are two reasons for this. First, the feasibility study did not take into account the oxidation of sodium during sodium injection. That is, there was an additional amount of sodium oxide present in the acutal system. One other explanation is in the dross activity coefficients. The computer program used activity coefficient values of unity for the dross species. This does not seem particularly likely.

The computer program was written with the ability to alter activity coefficients of the dross species. The activity coefficients were varied until the model results matched the actual results. Using the derived activity coefficients, the model could be used for predictive purposes.

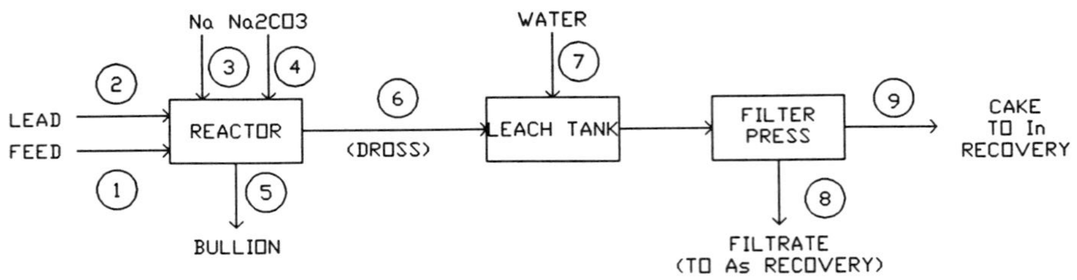

STREAM	1	2	3	4	5	6	7	8	9
SOLIDS (lbs)	100	400	12.73	70	428.8	153.9			49.7
In %	0.61				0.014	0.36			1.1
Pb %	25.4	100			98.31	2.48			7.59
Na %	0.06		100	43.39	0	28.05			0.87
Cu %	12.3				1.58	3.6			11.02
Zn %	10.2				0.014	6.59			10.21
As %	7.3				0.068	4.55			0.42
Ag %	0.091				0.018	0.01			0.03
LIQUIDS (lbs)							769.4	873.6	
H2O %							100	88.07	
In ppm								3.1	
Pb ppm								4.4	
Na %								4.89	
Cu ppm								64.4	
Zn ppm								5804	
As ppm								7781	
Ag ppm								0.05	

Figure 8 - Proposed process flowsheet.

Conclusions

Based upon the results of this research, a new sodium-based process has been developed for the separation of indium and lead present in flue dusts. The proposed process flowsheet is presented in Figure 8. The reactor design would be kettle based. The addition and removal of feed and dross can be accomplished by standard metallurgical procedures, but would be dependant upon the reactor configuration. The remaining two steps, leaching and filtration, would also be accomplished using established metallurgical technology.

The flow of material, for the process shown in Figure 8, consists of the following. First, the feed-stock is blended with soda ash. This mixture is then introduced into the reactor and agitation of the lead bath is begun. A temperature of 700 °C is recommended since this is the minimum temperature for the formation of a molten dross. Sodium is then injected into the bath. The dross-bullion mixture is allowed to react for one hour. A portion of the bullion is then sent to lead refining.

The dross is skimmed, cooled, and crushed. The crushed dross is introduced into a stirred tank and leached with water. The leach mixture is filtered. The filter cake is then processed for indium recovery. The filtrate is processed for the recovery of arsenic.

For 100 pounds of dust (combined with 70 lbs of soda ash), 400 pounds of lead bullion is doped with 12.8 pounds of sodium. This results in 153.9 lbs of dross and 428.8 lbs of bullion. Leaching of the dross produces 49.7 lbs of cake and 873.6 lbs of liquor. The compositions and flows are summarized in Figure 8.

Based upon CDLR as the starting material, the recovery of indium at the end of the leach step is approximately 90%, while 85% of the original lead in the feed-stock will be removed. Of the silver in the feed-stock, 85% reports to the bullion. This silver can be recovered during lead refining. The recovery of copper to the bullion represents 55% of the total copper contained in the feed. The remaining 45% reports to the cake and liquor (44.6% and 0.4% respectively). Of the zinc in the feed-stock, only 0.6% reports to the dross. The remaining 99.4% reports to the cake and liquor (49.4% and 50% respectively). The liquor contains the majority of the arsenic (93% of the initial arsenic). The arsenic recoveries to the cake and bullion are 3% and 4%, respectively.

The results of this research have established the technical viability of the proposed process. The specific advantage of the new process is segregation of indium to the dross with simultaneous recovery of lead to the bullion. Water leaching of the dross expedites indium recovery in the form of a filtered concentrate.

Acknowledgments

The authors would like to acknowledge Indium Corporation of America for support of this research. The assistance and guidance received from Mr. C. E. T. White and Dr. Larry Stevens are gratefully acknowledged. The authors would also like to acknowledge Mr. Michael Blenk of E. I. Du Pont for suggestions and assistance relating to the handling of sodium. One of the authors (NBG) also received support from the Bureau of Mines through a MMRRI Fellowship.

References Cited

(1) C.E.T. White, "Indium: High Technology Metal", Advanced Materials & Processes Incorporating Metal Progress, December 1989

(2) C.E.T. White, "A Review of the Extractive Metallurgy of Indium", Hydrometallurgy of Copper, Its By-products, Rarer Metallurgical Processing Dallas Symposium, 1982, 95-102

(3) Y.E. Lebedeff, "Process of Recovering Indium",U.S.Patent 2,433,770, December 1947

(4) N.B. Gomez, "Development of a Process for the Separation of Indium and Lead Present in Flue Dusts and other Wastes"(M.S. thesis, Colorado School of Mines, 1990), 5-26

(5) "Sodium: Uses and Advantages" Bulletin, Du Pont Chemical and Pigments Department, 1990

(6) M. Sittig, "Sodium: Its Manufacture, Properties, and Uses",Reinhold Publishing Corporation, New York, 1956

(7) ASARCO, "Process For Separation of Dross Elements Combining Sodium Addition to Molten Bullion Followed by Controlled Solidification of Casting", U.S. Patent 4,404,026

HIGH PURITY ZINC OXIDE FROM A

WIDE RANGE OF INDUSTRIAL RESIDUES

Faustino G. Prado
Prado & Associates, Inc.
P. O. Box 17349 Tampa, Florida 33682

ABSTRACT

Production of zinc oxide from residues of scrap recycling operations by the way of an ammonia-ammonium carbonate system has been previously reviewed. The technology is now applied to drosses, ashes, bag dusts and sludges from galvanizers and various other metallurgical operations. The spectrum of impurities considered has increased, including chloride, sulfur compounds, and metals such as calcium and magnesium. The physical and chemical problems encountered and their solutions are discussed together with the impurities control. The capability of the process to produce a uniform, free-flowing product of easy dispersion, and its ability to control particle size distribution and active surface are also presented.

Second International Symposium -
Recycling of Metals and Engineered Materials
Edited by Jan H. L. van Linden,
Donald L. Stewart, Jr., and Yogeshwar Sahai
The Minerals, Metals & Materials Society, 1990

Introduction.

Since the presentation of our paper "High Purity Zinc Oxide Production from Residues in Automobile Scrap Recycling" (1), interesting requests have been received, expanding the application of the process to a wide variety of metallurgical residues and aiming for the control of impurities we had not contemplated before. A recent publication (2) describes the results of the same process, in both pilot plant and commercial operations, for relatively pure zinc dusts (91 to 99% Zn) from vapor galvanizing.

The chemistry of the ammonia-ammonium carbonate (AAC) process for the recovery of zinc, reported in the literature as early as 1880, in its most simplified form may be figured as follow:

Solubilization in aqueous solution

$$5Zn + 5CO_2 + 20NH_3 \longrightarrow 5Zn(NH_3)_4CO_3$$

Precipitation by removal of NH_3

$$5CO_3Zn(NH_3)_4 \longrightarrow 2ZnCO_3.3Zn(OH)_2 + 3CO_2\uparrow + 20NH_3\uparrow$$

Of the two important characteristics of the process, high purity and particles size distribution, one is controlled in the solubiization step (Product Liquor Preparation) and the other at precipitation (Basic Zinc Carbonate Production).

Figure 1 is a simplified representation of the process:

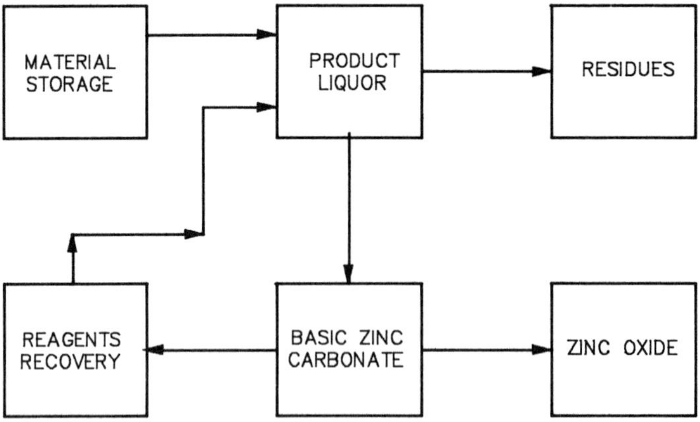

Figure 1. — Conceptual Block Diagram

The Raw Materials

In our work at the commercial plant built and operated in 1979-80, the residue treated had the following average analysis:

Zinc (total)	56.9%
Zinc (metallic)	8.7%
Lead	7.1%
Copper	9.7%
Tin	0.6%
Nickel	1.3%
Iron	1.1%
Insolubles	2.4%

Other elements, if present, were not reported and not considered. The final ZnO was only analyzed for elements indicated above.

The zinc-bearing residues already available in the market, some of them constituting a serious disposal problem, offer a wide spectrum of chemical composition and particle size distribution. They can be classified in three principal groups:

1. zinc ashes / drosses,
2. furnace dusts,
3. scrap.

Usually, zinc ashes and drosses are treated in processes such as ball-milling, for the recovery of the metal fraction, leaving a fine residue rich in impurities. This residue is characterized by a relatively high Zn content (50 to 70%), part of it in the metallic state. Generally the residue contains significant lead and tin, which, after zinc removal, should be recovered. The residues particle size distribution, before and after zinc removal, resembles a mixture of fine sand, silt and dust.

Metallurgical furnace dusts are generally extremely fine. Steel dusts are outside of the scope of this paper since, in general, they are not amenable to treatment by this process due to very low recoveries. Nevertheless, the crude zinc oxide generated by Waelz kilns falls within our scope of review. Flue dusts derived from secondary smelters, brass mills, remelters, etc., are usually not as high content in zinc (30 to 40%) and also metallic zinc as ashes and drosses. However, they do contain higher levels of copper (10 to 30%).

Zinc scrap is usually remelted for zinc recovery, generating residues for the other two categories. Alloyed zinc (brass and bronze) is also suited for the AAC process. We designed a small unit operated in 1974-75 in Los Angeles, California, for a crude separation of copper and zinc. The fed material was shreaded scrap below 50mm. Solubilization was performed in a rotary drum.

Our current work has been limited to the first two categories and to materials whose copper content is relatively low (<8%). In order to improve the chemical and physical behavior of the system, (i.e. impurities control, liquids/solids separations, and smoothness of a long term continuous operation), the design is based on balanced mixtures formulated on the availability of supplies. All the metals in the feed, with the exception of extracted zinc, are left in the residue.

When materials from group 2 with high copper content are treated, some of the zinc is alloyed and, under the conditions of operation, its recoveries are low. For these materials, in our current process scheme, an additional circuit is required to increase overall zinc extraction and to recover copper as a second residue.

Several materials from groups 1 and 2 were treated individually to confirm their behavior. Complete bench scale work was performed on several mixtures. Purity control and physical characteristics of the zinc oxide product were consistent. What follows are the chemical analyses of one of the mixtures treated, with the maximums and minimums for each controlled element present in the components of the mixture.

TABLE I
Feedstock Composition %

	individual components		mixture
	min	max	
Zn (total	34.16	69.17	53.79
Zn (metal)	0.84	38.57	7.39
Cl (chloride)	0.57	6.75	4.18
S (as SO4)	0.16	1.88	1.80
Pb	0.95	22.05	9.86
Fe	0.64	1.61	0.99
Al	0.54	2.04	0.80
Cu	0.02	5.54	2.22
Sn	0.05	5.09	1.96
Cd	0.01	0.31	0.12
Ni	0.004	0.14	0.06
Mn	0.013	0.06	0.034
Ca	0.09	0.84	0.43
Mg	0.026	0.22	0.08
Ti	0.012	0.24	0.035

Impurities Control

The impurities contemplated are:

Pb, Sn, Al, Ti, Fe, Cd, Ni, Cu, Co, Mn, Ca, Mg, Cl and S

In general, their behavior is defined by the pH and Eh (redox potential) of the aqueous system. Very important also are the ammine formation capabilites of the different cations. Published information on Zn systems are almost non-existent. Fortunately, information on the behavior of Ni in ammonia-ammonium carbonate (AAC) solutions is extensive. The work of Prof. Caron (3), the bibliographic review by Prof. Power (4) and our extensive collection of design and operating data on Ni and Cu have been very useful.

Of all the cations mentioned, only Al, Ca and Mg are more electropositive than zinc and consequently not amenable to removal by cementation by the way of metallic zinc. Al, Pb, Sn, Ti and Fe (III) are known to be insoluble in the AAC systen employed. Current knowledge indicates that the very minute amounts of such elements identified in the product liquor are in a solid phase and that their removal may be accomplished via liquid/solid separation methods.

Some Fe(II) is soluble in AAC liquors, specially in the high CO2 concentrations. Usually its conversion to Fe(III) is effective for complete removal. Howeever, when some organics (humates) are present, test work indicates that Fe(II) ions are chelated by the organic structure, keeping them soluble by protecting them against complete oxidation. This effect is voided when the CO2 decreases.

Cu, Ni, Cd, Co and Mn in their different levels of oxidation form soluble ammines. They all, including the non-chelated Fe(II), are reduced to the metallic form and rendered insoluble at the right redox potential by the presence of metallic zinc. This effect in some of the higher degrees of oxidation of Co and Mn requires longer retention time to accomplish complete removal.

Magnesium and calcium have very little solubility in the AAC liquors employed, but enough to be noticed in the produced ZnO at undesireble levels. In some of the liquors, and independently of the Ca and Mg content of the starting material, both were detected at 400 ppm each. Nevertheless, afterwards during the process, both were almost completely removed without affecting product quality.

Magnesium presents a serious scaling problem in nickel processing using AAC liquors, specially when the ores are rich in magnesium silicates. Its behaviour has been thoroughly investigated (3,4). In the early stages of leaching the solubility is high but decreases with time by reprecipitation. In our system we found a very similar pattern for both Mg and Ca, but with different parameters. The decrease in free ammonia not tied into ammine forms within the system, and the formation of solid nuclei in the early stages of basic zinc carbonate precipitation remove preferentially both Mg and Ca.

From a process point of view, the impurities control for cations is completed within the boundaries of the production of the product liquor.

Chlorides are present in almost all the residues evaluated, especially in those in group 1. The Cl solubility is almost complete in the AAC liquor, and its behaviour in Zn solutions follows very close the pattern of Cu in similar liquors (5). The tendency to form insoluble zinc (or Cu) oxychlorides is controlled by the solution pH. Since after precipitation of all the zinc as BZC all the chlorides are still in solution, a careful and controlled washing is required to achieve a low Cl content in the product ZnO.

Sulfur is also present in almost all the residues of types 1 & 2. The form in which it is received goes from sulfides to sulfates depending upon its past history, and it is largely, if not completely, dissolved. Minute amounts of sulfide have been detected in the final product. Also some sulfite is precipitated in the early stages of BZC precipitation. Proper oxidation, also convenient for cations impurities control, reduces or totally eliminates, this inclusion. The sulfate, as in similar solutions containing nickel or copper, has the tendency to form basic oxysalts. As in the case of chlorides, its precipitation is prevented by pH control. Final purity may also be achieved by the washing operation.

From the process standpoint, chlorides and sulfur sulfate control is completed within the boundaries of the basic zinc carbonate precipitation circuit.

Particle Size Distribution and Active Surface

As discussed in our previous paper (1), rate and conditions of precipitation affect the nature of the precipitate. The BZC appears as an agglomerate (Fig. 2) with a coral-like structure (Fig.3), very friable and amenable to easy dispersion. These agglomerates are made up by primary crystals of submicron size.

In early work, BZC was precipitated from low concentration solutions containing only 12 g/l zinc. Under different precipitation rates (grams/minute), the solids ranged from 100% minus 1 micron to 100% minus 10 micron. The last one, dried at 105 oC, had a Surface Area (BET) of 200 square meters per gram.

When produced on a commercial scale (140 g/l Zn in solution, 2000 kg ZnO/hour), the BZC, calcined at different temperatures and retention times, had the surface areas shown in Table II.

Table II
BET Surface Area

Temp oC	Time(hrs)	Sp. Surf. m2/g
300	16	67.5
400	1	50.8
400	16	23.8
550	0.5	25.8
750	1	10.2

Current work has reproduced and confirmed these results.

Recent work in continuous bench scale tests, has also confirmed the capacity of the system, under current design, to produce a precipitated Basic Zinc Carbonate with controlled particle size distribution and free-flowing characteristics, within the range of 10 to 100 microns.

Figure 4 shows the particle distribution curve for the BZC and Figure 5 the same material, now Zinc Oxide, after calcination at 600 oC for 30 minutes.

The BZC calcined below 750 oC produces a very friable ZnO. When suspended in an aqueous solution at 0.5% sodium hexametaphosphate or tetrasodium pyrophosphate and vigorously stirred, it is reduced to a suspension with all particulates below 1 micron.

The precipitation stage is the largest energy consumer of the process due to the fact that NH3 and part of the CO2 in the Product Liquor are removed by steam stripping. Our original consumption of 2.0 kg steam (7 bar) per kilogram of zinc oxide has been consistently corroborated. The operation and performace of this stage is independent of the raw materials processed.

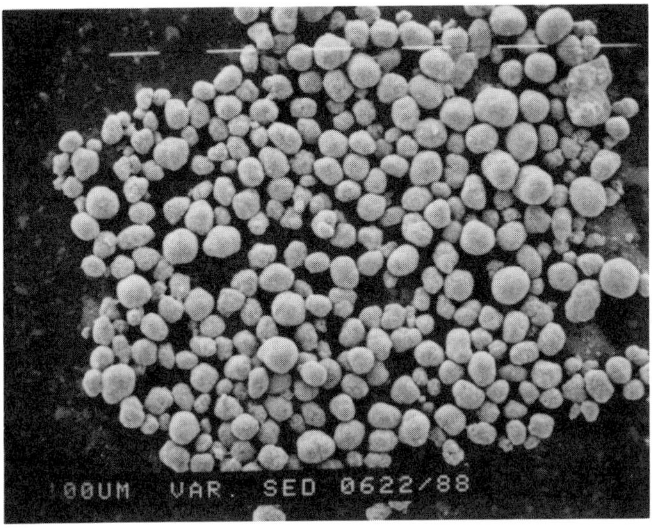

Figure 2. ZnO (X 80)

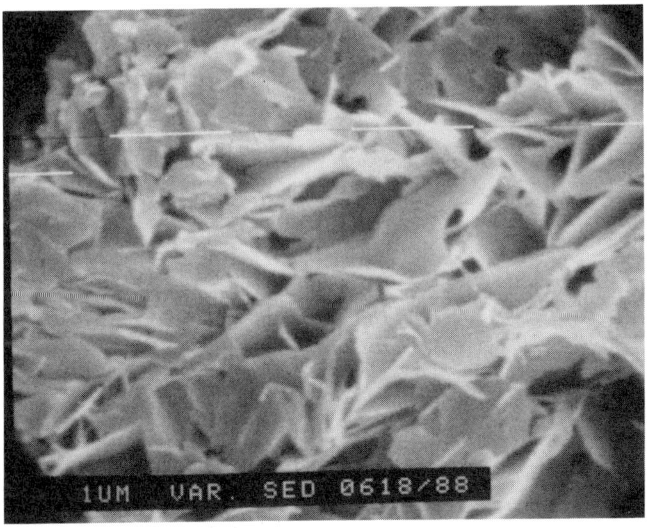

Figure 3. ZnO (X 20 000)

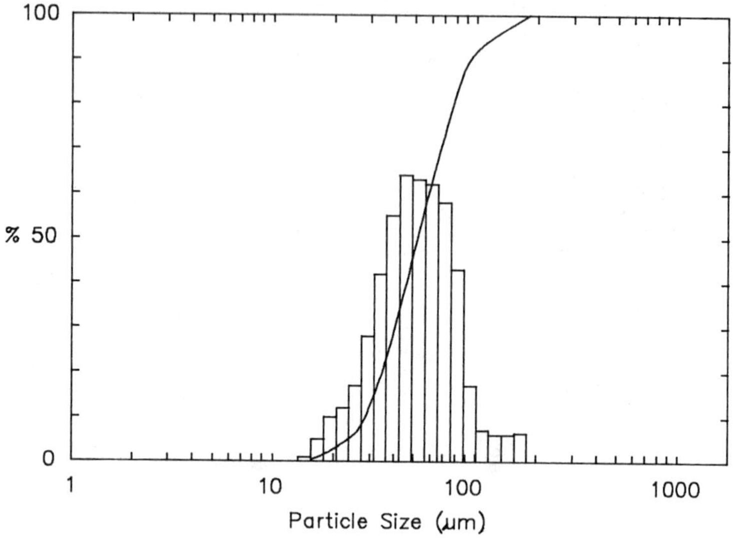

Figure 4.— BZC particle distribution

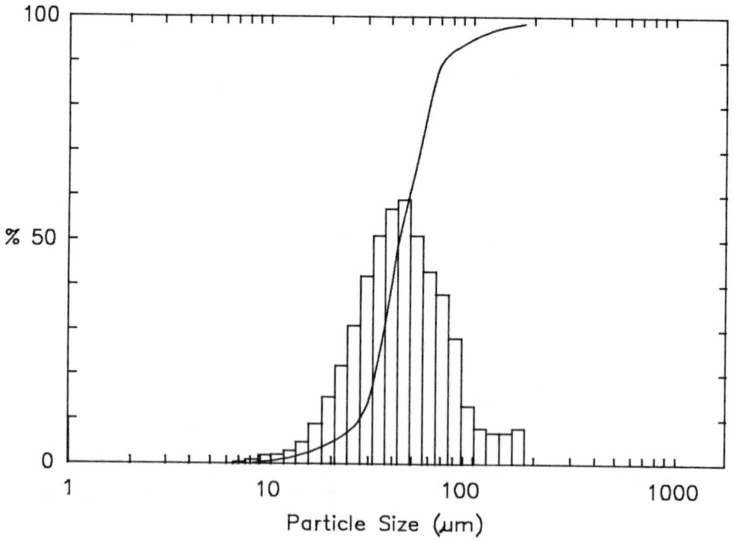

Figure 5.— ZnO particle distribution

Conclusions

The AAC (ammonia-ammonium carbonate) route for producing zinc oxide from diverse recycled residues, usually loaded with impurities, is able to produce, independently of the impurities present, a

*High Purity Product,
*Free Flowing Material.

Table III shows the impurities level achievable under normal commercial operating conditions :

Table III
Impurities Level

Pb	<0.001%
Fe	<0.001%
Al	<0.001%
Cu	<0.001%
Sn	<0.001%
Cd	<0.001%
Ni	<0.001%
Mn	<0.001%
Mg	<0.002%
Ca	<0.005%
S (SO4)	<0.005%
Cl	<0.005%

References

1. F. G. Prado, J. P. Dempsey, B. W. Wiegers, "High Purity Zinc Oxide Production from Residues in Automobile Scrap Recycling", Symposium on Recycle and Secondary Recovery of Metals, TMS, 1985, pp 183-193.

2. T. Hashimoto, T. Owaga, T. Kasai, "Development of a High Purity Zinc Carbonate Production Technology", Sumitomo Metals Vol. 40 No. 2, 1988, pp 45-54.

3. M. H.Caron, "Fundamental and Practical Factors in Ammonia Leaching of Nickel and Cobalt Ores", J. of Metals, Vol. 188, 1950, pp 67-90.

4. L. F. Power and G. H. Geiger, "The Application of the Reduction Roast - Ammoniacal Ammonium Carbonate Leach to Nickel Laterites", Minerals Sci. Engng., Vol. 9, No. 1, 1977, pp 32-51.

5. F. G. Prado, "Production of Copper Oxychloride from High Iron Copper Scrap using Combined Chlorine and Ammonia-Ammonium Carbonate Leach", 1986, Unpublished.

Acknowledgements

The author would like to thank Ir. M. Vandebriel, from REZINOX NV, Zolder, Belgium, for the pictures and graphics used in this presentation.

OTHER METALS B

Session Chairman:
Danton L. Paulson

Waste Protection in the Metal Industry

Günter Fleischer

Techninal University Berlin
Institute for Technical Environment Protection
D-1000 Berlin 12, Federal Republic of Germany

Second International Symposium -
Recycling of Metals and Engineered Materials
Edited by Jan H. L. van Linden,
Donald L. Stewart, Jr., and Yogeshwar Sahai
The Minerals, Metals & Materials Society, 1990

1. Introduction

Since the Federal Republic's Environment Programme of 1971 and Waste Management Programme of 1975, protection of waste has had top priority in waste management policies (1). Waste protection has since then become the focal point of interest for groups of all directions whereby the topic is of various relevance to them. Meant is, amongst other things, a new attitude towards consumption and production, preventative measures in the development of products and production, and the development of various methods of recycling. For an objective analysis of this important field it is necessary that the reduction of the amounts of waste in the contributive fields and their targets be clearly defined and distinguished.

Only after defining the concepts, the general objectives and the areas of waste protection in waste management, is it possible to deal with waste protection in the metal industry. The possibilities and potential of waste protection will be demonstrated by examples of causal and primary waste protection, whereby energy consumption has particular importance as a source of waste. No coherent attention has been paid to activities in the field of primary waste protection up till now, whereas recycling of metal has always been a topic in an abundance of articles. We shall describe waste protection in the metal industry with the help of a few examples.

2. Definitions

The concept "waste" is defined in the waste Act of 27.08.1986 § 1 (1). Waste in the sense of this Act are moveable objects whose owner wishes to dispose of them or whose orderly disposal is in the public interest in particular for the protection of the environment. The individual types of waste were firstly co-ordinated and given a numerical waste key number in the catalogue put out by the study group of the Länder (LAGA) (German counties).

It is important to note the the concept "production waste" can only be partially used as referred to up till now, in section §5 (1) No.3 BImSchG ("Bundesimmissionsschutzgesetz" - law for the control of the effect of gases, smoke, noise and smells etc. on neighbouring property) for firms which are required to have a permit in the sense of this law. According to this law, remaining material that arises as part of the production process may not be labelled waste if its utilisation is technically possible and can be reasonably expected. That implies that also the process of utilisation of remaining material falls into the category of waste protection.

The connotation "production waste" will be replaced in this study by the broader term "production remaining material". Remaining material in the production can either be reutilised or is waste as expressed in the waste law.

With the expression waste protection or waste prevention, measures are meant which are intended to reduce the amount of waste (4). This definition takes into consideration both groups which are actively concerned with reducing the amount of waste i.e. the production and services on the one hand and the consumer on the other. The definition, taken purely from the point of view of waste management, regards waste protection as being the sum of the measures and the scope of action which prevent waste arising from the waste producer (5). Here, as in the waste law, those involved in production and services, i.e. the group which pursues the actual prevention, are regarded as the producers of waste. A difference is made here between "quantitative waste protection" - the reduction of the total amount of waste - and "qalitative waste protection" - the reduction of the pollutants in the waste which are relevant for the environment. This is the vertical classification of waste protection. The horizontal classification of the waste protection takes place in primary waste protection and secondary waste protection (6):

- primary waste protection is the fulfillment of requirements and functions in production and services compatible with the environment with a minimal consumption of raw materials and energy.
- secondary waste protection is preventing products, the remaining material of products and used materials from arising or remaining (secondary raw materials management).

Waste protection includes also the prevention of emissions.

Secondary raw materials management is the sum of registration and recycling of products, remaining materials in production and used materials.

Recycling is the reutilisation or recycling of products or parts of products in the form of a cycle in order to protect resources (7). To be consistent this definition has to be extended to include reutilisation or recycling of remaining or used material.

3. Waste protection in waste management

Waste protection in waste management is divided into primary protection - all protection methods up to distribution to the end user - and secondary protection.

3.1 Primary waste protection in waste management

Primary waste protection is based on the principle of the fulfillment of requirements and functions by products and services in a manner harmless to the environment. The qualitative prevention component concerns the demand for ecologically harmless fulfillment, which means that possible emissions of pollutants relevant to the environment during production, use, recycling, disposal and storage of remaining materials is prevented or reduced to such an extent that the ecosystem can not be negatively influenced.

The quantitative prevention component of primary waste protection concerns the demand for fulfilling the requirements with minimal use of raw materials and energy. We point out here that various authors have indicated the connection between waste protection and use of raw materials. As examples we quote here Fußer (8) - "Alles was wir produzieren, wird früher oder später zu Müll (Everything that we produce becomes refuse sooner or later) - and Bidlingmeier (9) - Sind Rohstoffe einmal durch menschliche Tätigkeit in die Produktion gelangt, so werden sie ... eines Tages zu Abfall (If raw materials reach the production through the hands of human beings, they will one day become waste. The second statement goes further than the first but is nevertheless insufficient. The following three principles, source of waste priciples, describe the sources of waste in entirety (6):

- Every extraction of resources from the earth's crust causes waste and emissions.
- Every resource taken from the earth's crust becomes waste and emissions
- Energy consumption involves, usually, the consumption of resources.

The first waste source principle refers to a very important source of waste. In the mining industry it concerns the complete waste which arises during extraction but is unable to be used. Alone in the Federal Republic in 1982 the amount of this waste was 69 million Mg (10). The quality of the waste is, as far as the environment is concerned, not problematic but it requires the countryside for its disposal.

The second waste source principle establishes the connection between raw materials, and waste and emissions. The period of time between extraction of the resource and the arising of waste is not defined here but it should be borne in mind that the raw material does not become waste all at once but bit for bit. It is the task of primary and secondary waste protection to extend this period as far as possible.

The third waste source principle deals with energy consumption both as source of waste and energy. This connection arises from the second waste source principle. A consumption of resources takes place when using mineral oil, hard coal, brown coal, natural gas and nuclear energy as energy sources. As far as other sources such as solar energy, hydro-dynamic und wind power are concerned, the waste and and emissions quantities can be disregarded.

Primary energy consumption in the Federal Republik amounted in 1984 to 376 million Mg SKE (Steinkohleneinheit - coal equivalent) (11). The following amounts (in million Mg SKE) could be attributed to the individual energy carriers: 158 (oil), 80 (hard coal) 38 (brown coal) 60 (natural gas) 30 (nuclear energy), 7 (hydro-dynamic power) and 3 (others). Power consumption in million Mg/percent coal equivalent of the primary energy consumption, (taken from the figures of the "Gesamtverband des deutschen Steinkohlenbergbaus" - Central Association of German Hard Coal Mining (11) - and the average consumption of 330 g coal equivalent per kilowatt hour (12) produced, was distributed in 1984 as follows: 3.3/2.1 (oil), 44.1/55.1 (hard coal) 31.3/82.3 (brown coal) 10.9/18.2 (gas) 30/100 (nuclear energy) and 10/100 (hydro-dynamic power and others). In desulphurising flue gas 22.7 g gypsum was attributed per kilowatt hour to a hard coal power station and 18.7 g to a brown coal power station. Gypsum from a hard coal power station can be compared to natural gypsum and an economic utilisation would appear to be possible (14). Alongside gypsum 72 g flue ash and 18 g wet ash are produced in the same brown coal power station. If the allowed maximum amounts of 400 mg/m^3 for sulphur dioxide and 200 mg/m^3 for nitrogen oxide in flue gas are kept to, up to 1.8 g sulphur dioxide and 0.9 nitrogen oxide are emitted from a heating and power station per kilowatt hour. With such specific data, which is quoted here only as an example, it is possible to fulfil the the above-mentioned requirements after carrying out careful and comparative waste management balances.

The three waste source principles demonstrate that protection of resources in fulfilling the requirements and functions of products and services is the key to preventing waste. All those measures and actions fall into this category. which minimise, in terms of quantity and quality, the total waste and emissions caused by the product from completion of the end product to the distribution to the end user during the complete life of the product (16), and extend the lenghth of use. These measures start with the product planning and development, and the related processing planning (17). Amongst these measures are such as "recycling suitable construction", which allows as complete a recycling as possible, mininal energy consumption and low levels of waste and emissions. In this way these measures of primary waste protection - named "active recycling" by Käufer (18) - increase success appreciably in the secondary waste protection.

3.2 Secondary waste protection in waste management

Secondary waste protection and secondary raw material management can be defined as being one and the same. They consist of collecting and recycling products, remaining and used materials. Obtaining products, and remaining and used material is, indeed, the condition for recycling while the type of obtainment, such as separate collection, influences the result of recycling in a major way. Recycling concerns, when taken precisely, only the recycling of material. The notion "energy recycling" is misleading as energy cannot be recycled but only converted and this always entails loss of energy (19).

Recycling is divided into the following types of cycles (7):

- recycling during use of product,
- recycling remaining material,
- recycling used materials.

Recycling during use of the product starts in the life of the product after the distribution to the end user and is carried out by reconditioning or overhauling, whereby the form or properties of the product remain the same for reutilisation. In recycling remaining and used material the original form is dissolved and the resulting material is utilised.

Enough literature exsists about recycling so that we will go into it no further here particularly as it does not contribute further to the location of waste protection in waste management.

3.3 Overall objective of waste protection in waste mangement.

It can be generally said that in preventing waste the resources and environment are protected. The overall objective of waste prevention in waste management is that, in fulfilling the requirements and functions by products and services, resources and environment are optimally protected. Fulfillment of functions by products and services means the consumption of resources and the creation of waste and emissions. Therefore it is the goal to achieve the fulfillment with the use of the least possible resources and creation of the least possible waste and emissions for each product and each service. This principle is described by the waste protection equation in figure 1.

The quantities of raw materials and supplies, the required quantities of air and water, the amount of electrical energy consumed, which was not converted by a primary energy carrier, and the amounts of waste and emissions created for each step in the complete life of the product from the extraction of the resources from the earth's crust to the ecologically harmless disposal including recycling, are contained in this equation. This method is shown schematically in the area of validity for the waste protection equation. For "raw materials" the amounts of converted primary energy carriers and the amount of waste and emissions created during the conversion should be entered. The amount of waste and emissions that occurred during the extraction of the resources should be entered under "waste" and "emissions".

The factors b_i and b_j take into consideration the availability, the capability of regeneration and the content of pollutants relevant to the environment of each resource, whereas the factors b_n and b_m take into account the content of pollutants, toxicity and eluatebility of the waste and emissions. The factors a_1 to a_4 are used to weight the four equation components, which are different in size and unit, on the fraction line.

The quantity factors a_1 to a_4 as well as the quality factors b_i to b_m have a decisive influence on the result of assessing a product according to the waste protection equation. The determination of these factors is part of the program of precautionary waste management, waste protection, at the Technical University Berlin. The results of determining the factors will be controversial. However, as the equation is used, above all, for the comparison of products with the same function or purpose, the remaining accuracy of the factors is adjusted. The factors should be constantly updated but when they are set it is possible to obtain qualitatively evaluated information about the products.

In spite of this, this method of observation is too limited for waste management: the information about quantity must be related to the function fulfilled by the product or its serviceable life as it occurs in the waste protection equation. For the first time it is possible with this specific assessment to compare different - existing and planned - products, which have the same function or purpose, accurately and objectively select the one least harmful to the environment.

An example of this method is given for the energy consumption, one of the four elements of the waste protection equation, by Krüger et al. (21) for one-way containers for beverages. 0.33 l cans of

For different Products, which have the same Purpose or Function:

$$\frac{a_1 \Sigma b_i \text{ Raw Material}_i + a_2 \Sigma b_j \text{ Energy}_j + a_3 \Sigma b_n \text{Waste}_n + a_4 \Sigma b_m \text{ Emissions}_m}{\Sigma \text{ Serviceable Life}} = \textbf{Minimum}$$

= Measure for Ecologically Compatibility

$a_1 \ldots a_4$ = Quantity Factors
$b_i \ldots b_m$ = Quality Factors

Sphere of Application:

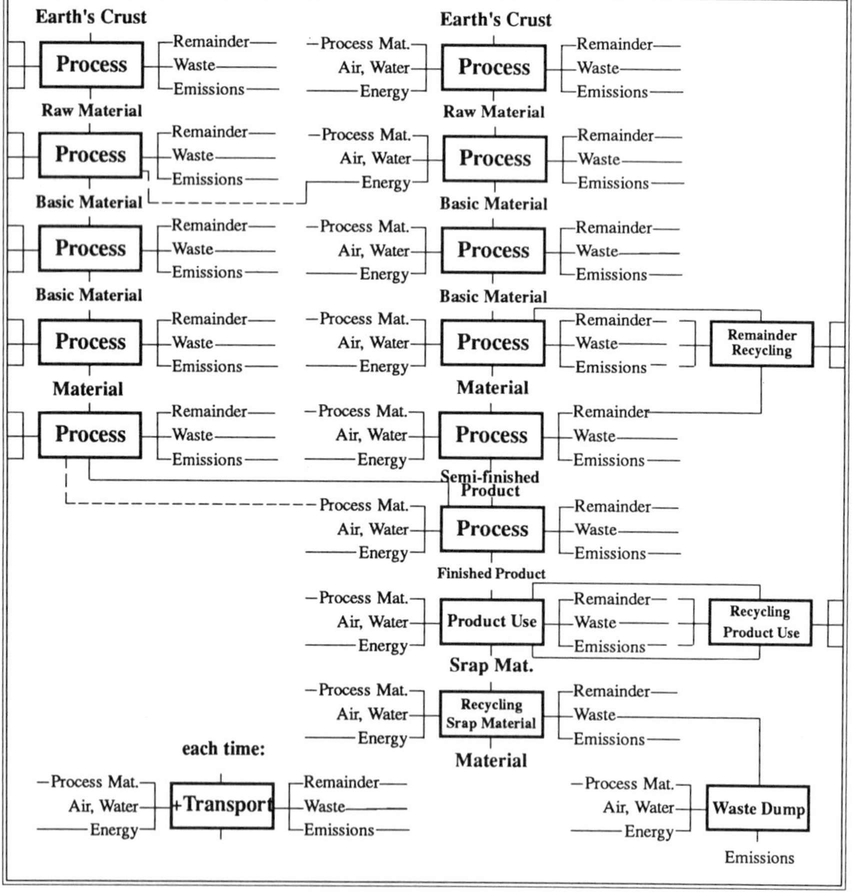

Figure 1:

Waste protection equation and its sphere of application (20) tinned sheet with an aluminium pull-off top are compared with a pure aluminium can.

Production of the steel for cans out of tinnned sheet is carried out with a blast furnace and oxygen converter. For the fabrication of cans a quality of steel is necessary which fulfills the utmost requirements of deep-drawing. The steel strip is tin-plated after hot and cold forming. Tin serves mainly as a lubricant during the following deep-drawing of the can. A coat of lacquer is applied after deep-drawing and quenching as protection against corrosion.

Table 1

Primary energy consumption in the Manufacture of cans out of tin sheet for beverages

	Body (of can)		Lid
Material	tinned sheet (0,5-2,0 g/m^2 Sn pro Seite)		Al Mg 5 Mn
Weight	29,23 g		3,88 g
		m g	E_{prim} MJ / l
1.	ore extraction and preparation	169	0,16
2.	Transport	169	0,20
3.	Sintering plant/sinter	114	0,26
4.	Blast furnace/pig iron	116	2,01
5.	Steel works/Casting slap	124,8	0,04
6.	Hot-roll/hot bath	121,1	0,37
7.	Cold-roll and tinned plate/tin strip	111,6	0,47
8.	Deep-draw/quench lacquer/can-base	88,6	1,82
9.	Limestone	37,9	0,013
10.	Lime	7,1	0,016
11.	Scrap	7,5	0,006
12.	Tin	0,16	0,017
13.	Oxygen	11,1 Nm^3	0,034
14.	Lacquer	1,5	0,021
15.	Aluminium lid	11,6	2,7
	Total		8,2

The primary aluminium for the aluminium can is obtained by dry electrolysis with Al_2O_3 which is produced from bauxite according to the Bayer process. The bars casted in billet are formed hot and cold, the body of the cans are made by deep-drawing and quenching and finally lacquered.

Table 2

Specific primary energy consumption in producing aluminium beverage cas form 100 % new metal (21)

		Body (of can)		Lid
Material		Al Mg 1 Mn 1		Al Mg 5 Mn
Weight		11,70 g		3,88 g
			m g	E_{prim} MJ / 1
1.	Bauxite extraction		224,2	0,151
2.	Transport		224,2	0,452
3.	Aluminium oxide/Al_2O_3		88,9	0,731
4.	Al-Electrolysis/primary aluminium		46,8	4,99
5.	Smelting refining, continuous cast/ingot		89,4	0,166
6.	Hot roll/hot bath		70,9	0,321
7.	Cold roll/intermediate annealing trimming/cold bath		70,9	0,10
8.	Deep drawing, iron.lacquer		46,5	2,19
9.	Lacquer		1,5	0,021
10.	Magnesium		1,0	0,41
11.	Manganese		0,5	0,06
12.	Chlorine		0,01	0,002
13.	Anode		21,5	0,931
14.	AlF_3		1,0	0,028
15.	NaOH		20,1	0,52
16.	CaO		4,1	0,014
	Total			11,1

From the single production steps it is possible to calculate the total energy required to produce a can or a litre of canned beverage. 11.1 MJ/l primary energy is required for the production of aluminium cans made of 100% new metal and 8.2 MJ/l for tinned sheet cans with aluminium lids (see Tabels 1 & 2).

The influence of recycling on the specific energy consumption will be investigated in a further step.

4. Primary waste protection in the metal industry

Intensive study in the field of primary waste protection has taken place for a lenghty period. The original goals of these activities were the saving of energy and raw materials and increasing flexibility by shortening the production chain. These goals correspond to the measures taken for primary waste prevention as laid down in the three waste source principles. Today, decreasing supplies of resources, greater ecological awareness and the resulting regulations for protecting the environment are added motivation for activity in this field.

4.1 Primary waste protection in baseic material production

Metal materials are finished construction materials with specific properties and composition. Metallic basic materials are, on the other hand, metals from the metallurgical production such as raw aluminium, crude steel or high-grade zinc. We shall present a few examples of primary waste prevention in the production of basic materials which are typical for many others. The amount of ore i.e. the amount of raw material taken from the earth, required for a basic material or metallic material depends mainly on the type of ore and its composition. The amount of energy required, on the other hand, for the same type of metal is determined by the process selected for the extraction.

4.1.1. Energy consumption

In chapter 3.1 the connection between power production and the resulting amounts of waste and emissions has already been indicated. In addition the contribution of CO_2 to the greenhouse effect through the use of fossil energy carriers must also be taken into account as a relevant factor for the environment. If an ecological and climatic catastrophe is to be avoided the CO_2 emission level of 1980 must be reduced by 35% by the year 2000 and by 90% by the year 2050 (22). Therefore the specific energy consumption, i.e. in this case the energy required for manufacturing a product or, in the sense of waste management, to fulfil a function by a product, is a major aspect for the assessment of a product in waste management. For an objective comparison of products with the same function it is necessary to consider not only the amount of energy consumed but also the type of primary energy carrier used and the the process used for generating current. An objective comparison is only possible if the data is sufficiently lucid.

The specific energy required for the production of one Mg steel is shown in figure 2 (23) dependent on the time for different countries/continents.

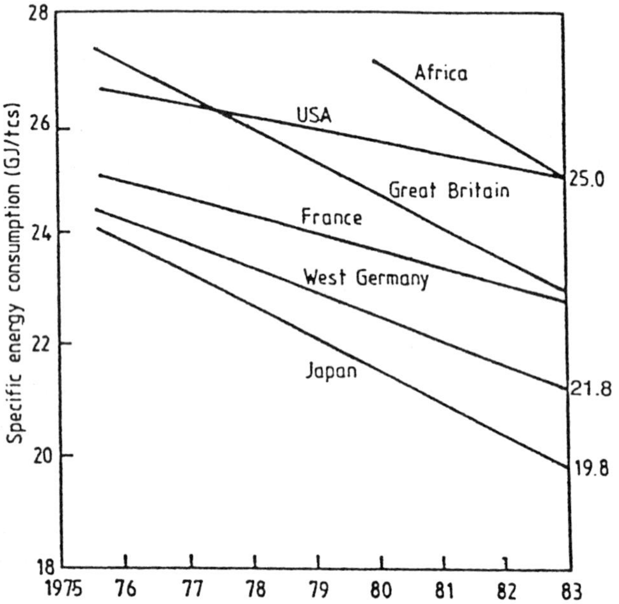

Figure 2 :

Specific energy consumption for raw steel production in GJ/Mg (23).

Alone for 1983 these figures vary from 19.8 GJ/Mg for Japan and 25.0 GJ/Mg for Africa and the USA. The same figure also shows the trend to constantly less specific energy consumption. Other writers (24) have quoted the figure for required energy at 36 GJ/Mg for the production of carbon steel so that here there is a difference between the lowest and highest estimate of 16.2 GJ/Mg for steel production. This difference, which is almost corresponds to the minimum energy requirement, demonstrates, alongside the waste protection potential, how important it is to state the origin of data in assessing different products. A possibility for further reduction of the specific energy requirement for steel production which is being studied at present is the use of the energy content of pig iron and steel work slag (23). This measure alone offers an immense potential for waste protection at an annual steel production of approx. 500 million Mg.

There is also an extreme variance in the figures for the specific energy requirement in the production of aluminium. One of the higher estimates is 300 GJ/Mg for the preparation of aluminium plus 40 GJ/Mg for the semi-finished product (24). For these figures which are from 1976 the type of current generation is not known. In a detailed study from 1989 (25) which, according to the type of product line analysis, concerns the stage from the extraction of bauxite to the production of aluminium foil, a comparison is made between generating electric current with hydro-dynamic power - hydro model - and the combustion of fossil fuel - thermal model. In the first case the total efficiency factor of power generation was 90%, in the thermal model 33%. The energy consumption for the production of the aluminium foil varies correspondingly: for generating current with hydro-dynamic power, the energy requirement is 182 GJ/Mg and when using electric power gained through primary combustion of fossil fuel 236 GJ/Mg. Thus the lowest figure in the literature about primary energy consumption for the production of semi-finished aluminium products 128 GJ/Mg, the highest 340 GJ/Mg. All these figures refer to a certain situation which must be known for the use of such data.

These variances result not only from the different form of generating electricity but also from the age of the data. In 1950 the specific energy demand of molten electrolysis for preparing aluminium was 21 kWh/kg. With furtherimprovment of this process the specific energy consumption could be reduced to approx. 13.5 kWh/kg in 1988 (26). The amount of waste and emissions could be reduced correspondingly.

4.1.2. Waste and emissions

The present-day method of steel production involves coking, sintering, blast furnace and oxygen steel works. In 1987 approx. 56% of raw steel was produced world-wide in this way. For ecological reasons as well as the cost factor, methods of production are now being sought world-wide to enable coal instead of blast furnace coke and to a large extent lump ore rather than sinter and pellets to be used, which would make coking and sintering plants superfluous. Such a so-called smelting reduction process is the Corex method (27). The first industrial production plant of this type started operation in the summer of 1988 with an annual capacity of 300,000 Mg pig iron. In figure 3 the dust, SO_2 and NO_2 emission for various smelting works with an annual capacity of 3 million Mg raw steel are shown.

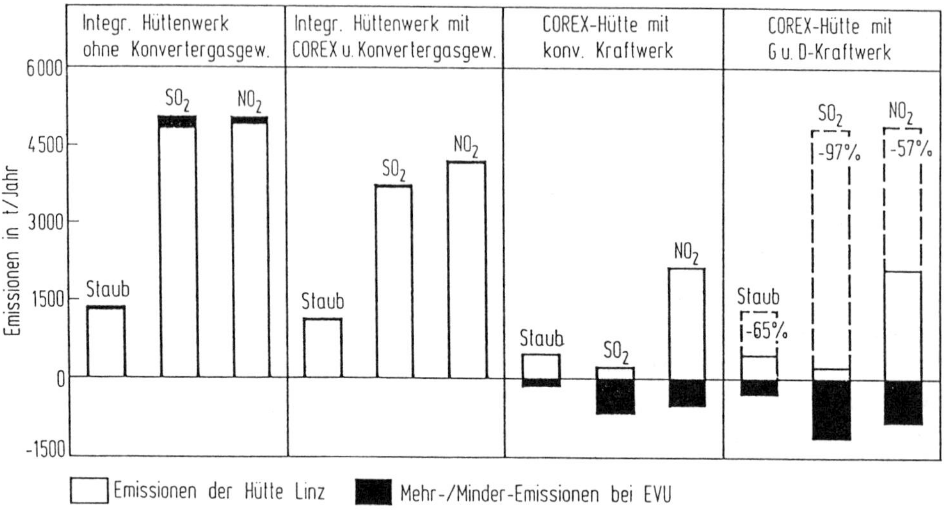

Figure 3:

Comparison of dust, SO_2 and NO_2 emission for various steel works with a raw steel production of 3 million Mg per annum

The intergrated steel work demonstrates the above-mentioned standard variant, whereas the Corex plant is shown with one conventional and one gas steam multi-purpose plant. Compared with the integrated steel work the Corex work emitts dust reduced by 65%, SO_2 by 97% and NO_2 by 57%. The conventional plant has a discharge efficiency of 37% and the gas steam multi-purpose plant 47%. Surplus power which occurs in the Corex process is directed into the grid reducing thereby emissions from the external plant. This is shown negatively in figure 3 so that it must be added to the reduction in emissions when considering the whole.

A further example of primary waste protection in the basic material industry is the development of the QSL process for producing lead. The sintering machine and the lead blast furnace in the standard process are replaced by the closed QSL reactor. Sulphide and sulphate pelleted lead concentrates as well as flue and other secondary raw materials with lead content can be processed in this reactor to a raw lead or a low-lead silicate slag. This new development without the sintering plant means that there is only one process instead of two. Through this and the fact that a clean slag which cannot be recirculated is produced, the cirulation rate of the material is reduced from 15 in the conventional method to 1 in the QSL process. This significant difference means that the QSL process requires 50% less energy and the emissions of dust 20%, SO_2 4.5% and lead 20% less than in the standard process.

Another outstanding example of primary waste prevention is the Hämatite method (29) of non-residue hydro-metallurgical zinc production. For the hydro-metallurgical zinc production the Jarosit method has been used since the end of the sixties and, since the seventies the Goethit method in addition. Since the end of the seventies the Hämatite method has been practised in Europe alongside the other two. The names of the methods come from the products of the iron precipitation. The precipitation products of the three methods as residue, the composition of the residue and the amount of residue can be found in Table 3.

Table 3

Composition and amount of concern's iron residue (29)

Element	Composition of residue Theory	 actual	Amount of residue for 100.000 jato-zinc unit in Mg
	in %		
Ammonium-Jarosite - $(NH_4)Fe_3(SO_4)_2(OH)_6$: M 481			90.000 m. 50 % H_2O
Fe	37	30	
S	13	10	
Zn	-	4	
Geothite - FeOOH: M 89			70.000 m 50 % H_2O
Fe	63	41	
S	-	3	
Zn	-	8	
Hematite - Fe_2O_3: M 160			27.000 m 10 % H_2O
Fe	70	58	
S	-	2	
Zn	-	1	

The amount of residue in the Hämatite method is clearly the least whereby the iron concentration of the residue the greatest is. The large amount of residue, 90% of produced zinc in the Jarosite method and 70% in the Goethite method have to be disposed of. On the other hand the Hämatite method is used mostly in the cement industry indicating that the method functions without residue. At present it is being continued to be developed with the goal to increase the iron content in the residue so as to utilise it in the steel industry. In the Hämatit method the operating costs are higher than in the other two. This is compensated by the saving on the disposal costs and the better yield of metal. The yield is shown in a comparison between the three methods in Table 4. As well as the increase in iron, the valuable metal in the zinc concentrate is also higher than in the other two hydro-metallurgical methods of producing zinc.

Table 4

Metal yield in sales form (in %) (29)

	Zn	Fe	Pb	Ag	Cu	Cd
Zinc concentrat analysis	52	8	1,5	200 g/Mg	0,5	0,2
Metal yield 1988						
Hematite method	>98	90	>98	>98	95	95
Jarisite method	98	-	80	80	90	90
Goethite method	97	-	80	80	80	90

4.2. Primary waste protection in the manufacturing of semi-finished products

The usual method today of manufacturing semi-finished flat products can be shown by the example of rolled product production out of copper: casting the material with the composition determined by the properties of the product to plates or slabs is done on continuous or half-continuous casting plants. The plates. already cut to length are often stored intermediately and then warmed to the hot-rolling temperature before rolling. The oxide layers and other surface errors, caused on the plates produced on the hot band, are cut off on both sides. The resulting scraps in form of shavings represents approx. 5-10% of the original weight. The next step is cold pre-rolling, intermediate annealing, pickling, cleaning, drying and final rolling to size. The final material properties are then attained by continuous annealing. This standard method, also used with few variations for other flat products, consists of a number of single steps, demanding much energy, by which residue, waste and emission evolves. By shortening the chain in the production of semi-finished products, much could be done for primary waste prevention.

Shortening the process chain can best be achieved by casting the rolling material as close as possible in time to the end of the process, as can be seen from the conventional method described above. The closer the casting is done to the final thickness of the product, the more steps can be discarded. The major problems which stand in the way of reaching this goal are achieving and maintaining the product properties and steering the casting process. Intensive work was done in the western world in 1988 in the course of 36 projects solely to achieve casting close to the final thickness for flat products out of steel (30). Energy consumption for casting steel as a finished product using the conventional method is given as being 5.3 GJ/Mg and 2.7GJ/Mg when shortening the process chain with the single-roller method (31). A distinctive comparison with the energy consumption for the production of steel plate is made in figure 4 (32). The order of the steps in the process is shown from bottom to top and the casting thickness displayed becomes less towards the right. The figure shows how much energy the individual steps require and that energy can be conserved by increasingly thinner casting and combining the casting with rolling. Not only is the energy demand reduced by having less steps, the scrap, waste and emissions are also significantly reduced.

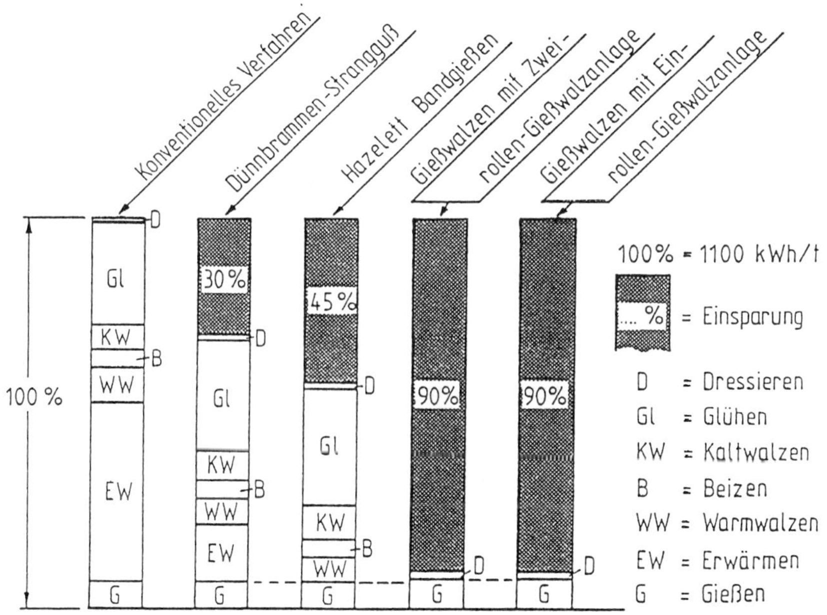

Figure 4:

Energy demand in the various methods of production and subsequent treatment of steel (32)

To produce a finished piece of steel and a finished piece of aluminium 97.5% and 99.7% respectively of the total energy is required to produce the semi-finished product (24). The remainder is required for processing the semi-finished to the finished product. This shows that significant saving of energy can only be achieved up to the point of semi-finishing and not later. This tendency can also be seen in reducing waste and emissions.

Possibilities for primary waste protection in the production of finished products in the metal industry are on the whole relatively small. On the other hand not only the primary but also the secondary waste protection can be appreciably influenced through finished product planning.

4.3. Primary waste protection through planning of finished products

As already mentioned in chapter 3.1 product planning and the necessarily associated process planning have a decisive influence on the consumption of resources and thus on the waste and emissions generated in terms of quantity and quality. According to Beitz (33) the following possibilities exist for conserving resources:

- reduction of material through optimal utilisation, longer life of the product and less waste in production.
- substitution of material for which raw materials are scarce, and thus costly, by such where the required raw materials will be available longer and are therefore cheaper.
- reduced energy consumption in the production by optimising material, fabrication, assembly and transport technology.
- reduced energy in utilisation of products by low loss processing and construction elements
- recycling of products and use of their material alongside their waste products for production of new material and renewed use of products after their first life cycle.

The success of recycling is determined during the planning phase and by a product design which foresees recycling. Generally accepted design norms have the following aims:

- to allow a high degree of recycling of the individual waste materials of the products
- to minimise the noxious proportion in old materials of the individual material types.
- to alleviate the for this purpose necessary preparation of the products to be recycled and thus reduce the costs of the preparation to a minimum.

The product planning influences to a large degree the specific resource consumption for the fulfilling of a product's function and therefore also the specific amount of waste and emissions. Beyond that the product planning, which is part of the primary waste prevention, determines the success of recycling. This influence is not dependent on one metal but is valid for all the materials with which a product is concerned.

5. Secondary waste protection in the metal industry

The secondary waste protection concerns the registration and recycling of products, product remains and used material. The recyling of metals is as old as metals themselves (35). That is the reason for the large number of publications on that topic.

For this reason we will deal not at that place at that time on recycling in the metal industry.

6. References

(1) Thomé-Kozmiensky, K. J. : Handbuch zur Planung von Abfallbehandlungsanlagen, EF-Verlag für Energie- und Umwelttechnik GmbH, Berlin 1989, S.81

(2) LAGA: Informationsschrift Abfallarten, Handbuch Müll-und Abfallbeseitigung (MuA), Ziffer 1110, 1981

(3) Jung, G.: Fachplanerische Aspekte von Entsorgungskonzepten; Informationen zur Raumentwicklung, Heft 10, 1988, S. 675 - 680

(4) Faber, M., Stephan, G. u. Michaelis, P.: Umdenken in der Abfallwirtschaft, Springer-Verlag, Berlin Heidelberg 1988, S. 30

(5) Schenkel, W. u. Kassing, D. (Hrsg.): Entsorgung 2000 -Leitfaden für Kommunen, Wirtschaft und Politik -, Verlag Bonner Energie-Report, Bonn 1988, S. 97

(6) Fleischer, G.: Aballvermeidung in der Abfallwirtschaft, in: Entsorgungspraxis Nr. 1/2, Bertelsmann Fachzeitschriftenverlag, Gütersloh (1990)

(7) VDI-Richtlinien: VDI 2243 (Entwurf), VDI-Verlag GmbH, Düsseldorf 1984

(8) Fußer, A.: Abfallvermeidung/ -verminderung, angewandte Abfallwirtschaft; Leben ohne Müll - Wunsch und Wirklichkeit? -, Sammelband zum Kongreß Abfallvermeidung/ - verminderung, 10./11.10.1988, Stuttgart 1988, S. 133 - 142

(9) Bidlingmaier, W.: Abfallvermeidung/ - verminderung durch den Verbraucher und im Haushalt - Zukünftige Rolle der Verwertungsmethoden; Leben ohne Müll - Wunsch und Wirklichkeit? -, Sammelband zum Kongreß Abfallvermeidung/-verminderung, 10./11.10.1988, Stuttgart 1988, S. 337 - 347

(10) Sutter, H.: Beseitigung, Vermeidung und Verwertung von Sonderabfällen. Abfallwirtschaft in Forschung und Praxis, Band 17, Berlin: Erich Schmidt Verlag, überarbeitete Auflage 1988

(11) Gesamtverband des deutschen Steinkohlenbergbaus: Daten und Tendenzen 1984/1985, Essen 1985

(12) Niedrig, T.: Kraftwerkstechnik im Umbruch; ENERGIE, Jahrg. 40, Nr. 7, Juli 1988, S. R 10 - R 15

(13) Bertram, J. u. Karger, R.: Entscheidungen gefordert; ENERGIE, Jahrg. 40, Nr. 7, Juli 1988, S. R 98 - R 109

(14) Backes, H.-P.: Ein hochwertiger Rohstoff; ENERGIE, Jahrg. 40, Nr. 7, Juli 1988, S. R 122 - R125

(15) Werkmann, H. J.: Nachrüstung; ENERGIE, Jahrg. 40, Nr. 7, Juli 1988, S. 14 - 20

(16) Schmincke, E.: Lebenslauf im Mittelpunkt; Müllmagazin 1/1989 S. 8 -9

(17) Fleischer, G.: Elemente der Vorsorgenden Abfallwirtschaft; AbfallwirtschaftsJournal 1 (1989), Nr. 1, S. 18 - 23

(18) Käufer, H.: Aktives Recycling von Kunstoffen; EntsorgungsPraxis 7/1986, S. 497 - 500

(19) Anonym: Diskussion zum Themenbereich Abfallvermeidung/ -verminderung durch den Handel; Leben ohne Müll - Wunsch und Wirklichkeit? -, Sammelband zum Kongreß Abfallvermeidung/ - verminderung, 10./11.10.1988, Stuttgart 1988, S. 291

(20) Fleischer, G.: Poster Abfallvermeidungs-Gleichung, auf: Postersession, 6. IRC Berlin, vom 28.11.1989 - 30.11.1989

(21) Krüger, Joachim; Reisener, Jan; Vest, Heino: Aluminium- und Weißblechgetränkedosen im Materialkreislauf; in: Fleischer, G. (Hrsg.): Abfallvermeidung in der Metallindustrie 1, EF-Verlag für Energie- und Umwelttechnik GmbH, Berlin (1989)

(22) Lesch, Karl-Heinz; Bach, Wilfrid: Reduktion des Kohlendioxids, Energie, Jahrg. 41, Nr. 5, Mai 1989, S. 30-40

(23) Gudenau, H. W.; Mahinda, N. G.; Lindner, K. H.: Survey of Different Processes used for Heat Recovery of Metallurgical Slags; Mitteilungen aus dem Institut für Eisenhüttenkunde der RWTH Aachen, Bahl, W. (Hrsg.) Bd 31, 1986

(24) Steinhilper, R.; Grundler, G.; Zöllner, S.: Recycling durch Aufarbeiten technischer Produkte in Austauscherzeugnisfertigung, Phase I und II, Forschungsbericht BMFT - FB-T 86-122, Fraunhofer-Institut für Produktionstechnik und Automatisierung (IPA), Stuttgart, Dezember 86,

(25) Fecker, I.: Herstellung von Aluminium: Ökologische Bilanz-Betrachtungen, Studie im Auftrag des Bundesamtes für Umwelt, Wald und Landschaft (BUWAL), Bern, durchgeführt von der Eidgenössischen Materialprüfungs- und Forschungsanstalt (EMPA), St. Gallen, Februar 1989

(26) Alcan informiert - extra, 1 - 1988, Alcan Deutschland GmbH, Werbung und Öffentlichkeit (Hrsg.), Nürnberg

(27) Bogdandy, Ludwig von; Nieder, Wolfgang; Schmidt, Gerald; Schröer, Uwe: Die Schmelzreduktion von Eisenerz nach dem Corex-Verfahren im kraftwirtschaftlichen Verbund, Stahl und Eisen 109 , Verlag Stahleisen GmbH, Düsseldorf, Nr. 9, 10 (1989) S. 445-452

(28) The QSL-Process - an energy saving and environmentally clean process für Leadwinning, International Symposium on Clean Technologie, 7.-18. Okt., Bundesrepublik Deutschland; Final Proceedings, Volume II: Clean Technology Practices in the Federal Republic of Germany S. U1-U22, Beitrag der Berzelius Metallhütten GmbH, Duisburg, 1985

(29) Röpenack, Adolf von: Rückstandsfreie hydrometallurgische Zinkgewinnung durch das Hämatitverfahren, Erzmetall 42 Nr. 3, VCH Verlagsgesellschaft mbH, Weinheim (1989) S. 125-129

(30) Reichelt, Wolfgang; Kapellner, Werner; Steffen, Rolf: Endabmessungsnahe Herstellung von Flachprodukten, Stahl und Eisen 108, Nr. 9, Verlag Stahleisen GmbH, Düsseldorf, (1988) S. 409-417

(31) Wood, J.V.: Direct casting of sheet by chill block techniques Materials Science and Technology, March 1988, Vol. 4

(32) Kopp, Reiner: Neuere Entwicklungen bei der Herstellung von Halbzeugen und Fetigprodukten, Stahl und Eisen 109, Nr. 4, Verlag Stahleisen GmbH, Düsseldorf (1989) S. 189-195

(33) Beitz, W.: Elemente der ressourcen- und umweltbewußten Produktgestaltung, Fortbildungszentrum Gesundheits- und Umweltschutz Berlin e. V., 179. Seminar: Ressourcen- und umweltschonende Produkte und Dienstleistungen, Berlin, 2.-3.11. 1988

(34) Meyer, Harald: Recyclingorientierte Produktgestaltung, Fortschrittbericht der VDI-Zeitschriften, Reihe 1: Konstruktionstechnik, Maschinenelemente, Nr. 98, VDI-Verlag Düsseldorf (1983)

(35) Thomé-Kozmiensky, Karl J. (Hrsg.): Recycling von Metallen, EF-Verlag für Energie- und Umwelttechnik GmbH, Berlin (1987)

THE RECYCLING OF NICKELIFERROUS WASTES FROM LATERITE PREREDUCTION IN ROTARY KILNS

M.N. Zevgolis*, J.F. Gaitanos**, J.Hajinicolaidis+, and D. Papamantellos++

ABSTRACT

In this work the results of pilot tests for recycling dust from laterite prereduction in rotary kilns are given. From these tests a screw pump conveyor was selected as the most suitable solution for wet dust recycling and the pilot work was followed by a successful industrial application. Wet dust recycling by this method is now applied successfully for more than 5 years. In order to further increase waste recycling, grinding tests were performed on pellet agglomerates and a small crushing unit is now under construction. Also, affer the necessary modifications of the pelletizing plant all of the - 4mm indurated pellet fraction is recycled. Preliminary tests were also conducted on dry dust, by injecting it in the slag of an electric reduction furnace of the plant. During these tests atmospheric pollution was noticed, so the tests should be repeated to understand better the behavior of dust in the slag.

RECYCLING OF WET DUST

As shown in table I, wet dust is approximately 25% of the total dust produced in the rotary kilns. Thus, for a Ni production of 1350 tons Ni/month, the average wet dust production of the plant is:

$$\frac{23.22}{100} \quad 10.530 = 2.340 \text{ tons/month}$$

and the amount of Ni which can be extracted from it is:

$$2.340 \; \frac{1{,}20}{100} \; 0{,}85 = 23{,}9 \text{ tons/month}$$

where 1,20 is the Ni content of wet dust and 0,85 the Ni recovery.

By achieving wet dust recycling, environmental protection as well as Ni recovery are increased[1]. Also, wet dust increases pellet crushing strength according to figure 1. For these reasons wet dust recycling was of priority, since the time when wet dedusting installation of rotary kiln gases was put in operation (1973). However, no industrial application was achieved until this research project was initiated (1985)[2,3,4,5]. The delay was due to the amount of water contained in wet dust and mostly to the first priority given to dry dust. Dry dust is approximately 75% of total dust and its recycling was also difficult[2]. Moisture in wet dust is approximately 50%. This high water content is probably due to the high Al_2O_3 (12%) and to the fineness of the dust. However, in spite its high water content, wet dust is

* Assist. Prof. Nat. Techn. Univ. of Athens, Greece
** Chief Eng., LARCO S.A.
\+ Research and Development Manager, LARCO S.A.
\++ Professor, Univ. of Patra, Greece

Second International Symposium -
Recycling of Metals and Engineered Materials
Edited by Jan H. L. van Linden,
Donald L. Stewart, Jr., and Yogeshwar Sahai
The Minerals, Metals & Materials Society, 1990

TABLE I
DUST FROM THE ROTARY KILNS

Year	Dust From Dry Dedusting	Dust From Wet Dedusting			Total Dust (wet and dry)	
	Weight (T)	Weight (T)	Weight (% of laterite)	Weight (% of total dust)	Weight (T)	Weight (% of laterite)
1	47441	16658	1,33	25,99	64099	5,12
2	67911	38144	1,90	35,97	106055	5,28
3	72386	32039	1,53	30,69	104425	4,99
4	105338	33853	1,86	24,33	139191	7,66
5	95386	25230	1,95	20,98	120706	9,32
6	31883	6462	1,16	16,86	38345	6,92
7	66191	31572	2,14	32,30	97763	6,63
8	106623	22949	1,26	17,72	129572	7,13
9	104034	20420	1,09	16,50	124454	6,67
10	70210	11720	0,92	14,40	81930	6,44
11	64842	15459	1,38	19,25	80301	7,17
12	97212	18720	1,29	16,15	115932	7,12
13	114624	25020	1,48	17,91	139644	7,21
Mean Value	80314	22949	1,48	22,22	103263	6,67

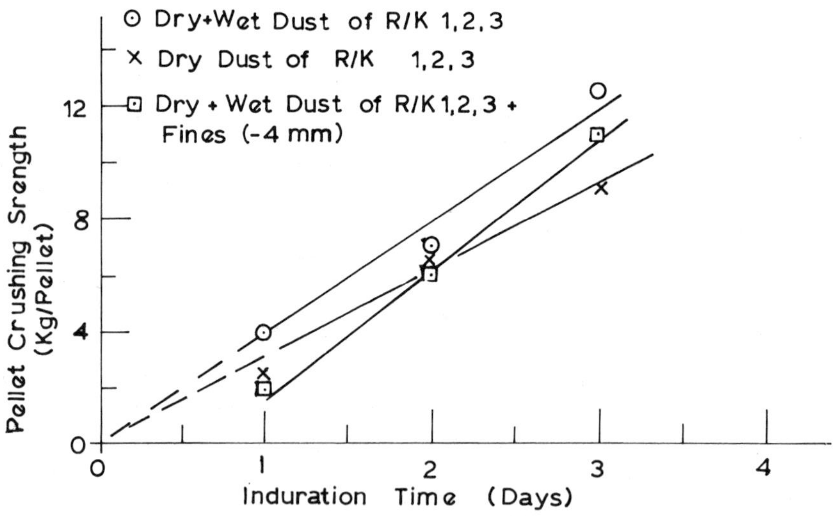

Fig.1 Pellet strength in relation to induration time and dust composition[2]

" viscous " and it does not flow easily. In fact it sticks in such a way to surfaces that its handling is difficult.

Another problem with wet dust is that its water-content is much higher than that needed to make a proper mixture for pelletizing. Pelletizing feed must have a moisture content of 15%, whereas if all dry and wet dust produced in the rotary kilns is mixed with the proper amount of cement, the moisture content of the mixture is higher than 15%, as shown below :
Let X the amount of wet dust (with 50% moisture)
and Y the amount of dry dust and cement in order to make a mixture with 15% water.

From a water mass balance it comes :

$$(X+Y)0,15 = X\ 0,50 \quad \text{and}$$

$$Y = 2,33\ X$$

For X=1 ton of wet dust, the dry dust in it is 0,5 ton and

$$Y = 2,33 \text{ tons of dry dust plus cement.}$$

Therefore, the amount of dust (without water) plus cement is

$$2,33 + 0,5 = 2,83 \text{ tons}$$

and the dust without cement is

$$\frac{100}{110} = 2,83 = 2,57 \text{ tons}$$

So, dry dust is

$$2,57 - 0,5 = 2,07 \text{ tons}$$

and the ratio

$$\frac{\text{dry dust}}{\text{wet dust}} = \frac{2,07}{0,50} = 4,14$$

According to the operating data in table 1, the ratio

$$\frac{\text{dry dust}}{\text{wet dust}} = \frac{1,00 - 0,2222}{0,2222} = 3,50$$

That is, the ratio at which dry and wet dust are produced in the rotary kilns is :

$$\frac{\text{dry dust}}{\text{wet dust}} = \frac{3,50}{1,00}$$

whereas the ratio at which they should be mixed in order to obtain 15% moisture is

$$\frac{\text{dry dust}}{\text{wet dust}} = \frac{4,14}{1,00}$$

Therefore, the moisture content of wet dust is higher then that needed for a 15% water mixture. So, if all of wet dust is mixed with all of dry dust from the kilns, the moisture content of the mixture is more them 15%. So, no additional water is needed to make a proper mixture. In this case it is difficult for dry dust and cement to get wetted from the water of wet dust, as the time spent in the mixer is not enough. Therefore, some dry dust or cement may move out of the mixer, and this causes environmental problems. It should be noticed that one of the problems in the pelletizing plant was atmospheric pollution from dust. And that most important source of dust is the mixer, either because of incomplet mixing of materials in it in the time available or because of incorrect mixing ratios of the materials.

So, if all of wet dust were to be fed in the mixer, then, more dry dust than produced in the rotary kilns should be available. Such dust has been accumulated in the past.

For wet dust recycling the following solutions were tested :

a. Chain Conveyor

This was a chain type metallic conveyor (redler) which had been forseen in the engineering design of the briquetting plant. This plant was abandoned however because of decrepitation of briquets and excessive mechanical wear of the compressing cylinders.

The chain conveyor had also to move in a zig-zag fashion and in different levels in order to get wet dust to the pelletizing plant. That is, it had to convey wet dust in different directions and hoist it to a higher than the production level. During operation a wet dust accumulation appeared at the lifting point and this gradually led to a final stop. This problem and also the excessive mechanical wear led to abandon the chain conveyor for wet dust transportation.

b. Screw Pump Conveyor

A special pump as that shown in figure 2 was also tested. It had a nominal capacity of 10 m^3/h. It was supplied by Alvailer, W. Germany. During operation, wet dust is fed in a properly formed mechanism which is rotating in a casing of special design, so that wet dust is compressed towards its exit end. From the tests it was shown that this pump could transport and also hoist easily the wet dust. However, the tests were limited in time so that the mechanical wear and reliability of the pump could not be estimated[6].

c. Screw Conveyor

A series of tests with a conventional screw conveyor, Φ400, L=4000 mm, was also conducted[6]. From the tests it was proved that wet dust transportation was satisfactory. Moreover, no serious mechanical wear, is expected according to previous experience with dry and hot dust transportation in the plant. This method might also open the way for recycling the dust accumulated during plant operation, i.e. a stockpile of one million tons.

Taking into account the results of all three solutions examined it was concluded that, although the screw pump conveyor had an unknown mechanical endurance, it provided a better environmental protection, it needed a smaller space and achieved

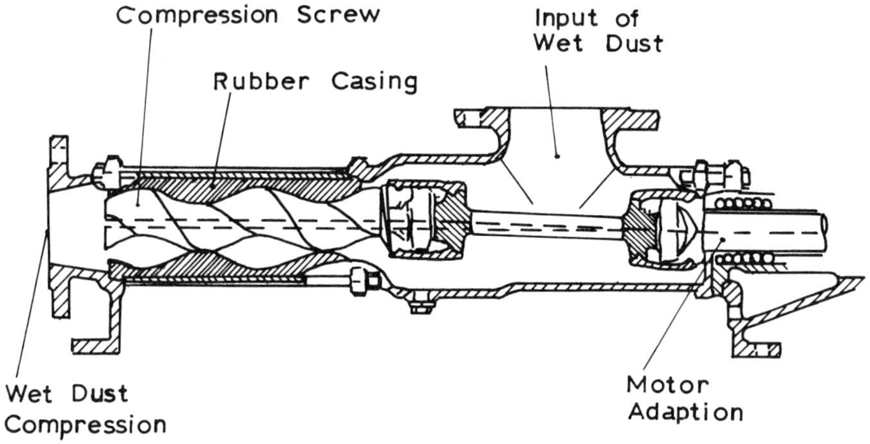

Fig.2. Screw pump for wet dust transportation

an easier dust transportation in higher vertical distances. Besides, wet dust has to be stored temporarily, as it is produced on a continuous process, whereas the mixers accepting it are working intermittently, as shown in table II. Therefore, wet dust had to be hoisted to an even higher vertical distance than that it was thought initially.

Taking into account the above discussion as well as the capital investment costs, it was decided that a screw pump would be used in industrial scale.

INDUSTRIAL OPERATION

For the industrial operation a 30 m^3/h Alvailer pump from W. Germany was used. After a short adjustment period, the screw pump worked successfully as shown in table III. As shown in the table, 71,0% of wet dust and 94,3% of total dust (wet and dry) was recycled in the fourth year of pump operation. As these are mean annual values, wet dust recycling on a monthly basis reached up to 87% and total dust up to 97,5%[7]. However, mechanical endurance of the screw pump was not satisfactory. In fact after handling 2500-3000 tons of wet dust, i.e. every 2-3 months of operation, screw and its casing are worn out and they should be replaced. So, a better construction material has to be found for the screw and its casing.

During the first three years of operation 47961 tons of dust from the wet dedusting installation of rotary kilns were

TABLE II

OPERATIONAL DATA OF VACUUM FILTERS AND MIXER

YEAR	OPERATING HOURS/MONTH	
	VACUUM FILTERS	MIXER
1980	759,2	192,0
1981	515,0	177,0
1982	266,0	161,0
1983 (Jan.-Jul.)	534,0	130,0

TABLE III

RECYCLING OF ROTARY KILN DUST

YEAR	DUST FROM DRY DEDUSTING WEIGHT (T)	DUST FROM WET DEDUSTING			WEIGHT OF TOTAL DUST RECYCLED (%)
		WEIGHT (T)	WET DUST (T)	RECYCLING (%)	
1977	58.889	20.978	-0-	-0-	-0-
1978	67.911	38.114	-0-	-0-	31,9
1979	72.386	32.039	-0-	-0-	36,5
1980	105.338	33.853	-0-	-0-	52,5
1981	95.386	25.320	-0-	-0-	48,6
1982	31.883	6.462	-0-	-0-	68,6
1983	66.191	31.572	-0-	-0-	66,7
1984	106.623	22.949	-0-	-0-	80,9
1985	104.034	20.420	3.718	18,2	86,2
1986	70.385	11.632	5.776	49,7	86,9
1987	64.842	15.459	8.371	54,1	89,6
1988	97.212	18.720	13.296	71,0	94,3
1989	114.624	25.020	16.800	67,1	85,3

recycled. They correspond to a Ni production of

$$17865\ \frac{1,20}{100}\ 0,85 = 489,2 \text{ tons}$$

TESTS WITH A PISTON PUMP

Because of the high maintenance cost of the screw pump another solution was tried for wet dust transportation. This was a a piston pump like those used in concrete transportation in civil engineering works[8]. As it was observed, wet dust was more difficult than concrete to transport, because it is more viscous, i.e. it is more difficult to flow. However, the results were such that they should be repeated.

RECYCLING THE - 4 mm INDURATED PELLETS

Indurated pellets, i.e. pellets with a crushing strength higher than 7-8 Kg/pellet, are screened so that the - 4mm fraction (the fines) is removed, in order to avoid overloading of the rotary kiln dedusting and dust handling installations. In this way rotary kiln dust remains constant as shown in table I. According to the engineering design, however, the fines could be recycled only when pellet screening and raw material mixing operations would coincide in time. Otherwise, the fines were rejected. From practical experience if came that 50% of the -4 mm fraction was rejected. Besides, the weight of fines was 15% of indurated pellets. So, the fines rejected were 7-8% of indurated pellets.

By making the appropriate modifications in the installation the total amount of - 4 mm fraction was recycled. As shown in table IV, addition of - 4 mm fraction did not affect pellet strenght.

TABLE IV

PELLET STRENGTH IN TERMS OF INDURATION TIME AND COMPOSITION

DUST COMPOSITION	PELLET STRENGTH (Kg/pellet) AFTER INDURATION OF 1 day	2 days	3 days
Dry dust of R/K No 1,2,3	2,5	6,5	9,0
Dry+Wet dust of R/K No 1,2,3	4,0	7,0	12,5
Dry+Wet dust+Fines (-4mm)	4,0	6,0	11,0

RECYCLING OF AGGLOMERATED PELLETS

According to our measurements, pellet shape and size are maintained if green pellets fall from less than one meter height, if they are stocked up to two meter thickness and if they are left to indurate for at least 48 hours. Under these conditions, some agglomeration of pellets always occurs. The amount of +100mm agglomerates is approximately 10% of indurated pellets. This fraction cannot be fed into the rotary kilns as it is. So, inde - pendent crushing tests were run at the ore and slag crushing mills of the plant. The crushed product had not much dust and could be easily fed to the rotary kilns. After that, a small crushing unit was designed and a jaw crusher was bought for crushing and recycling the +100mm agglomerates.

RECYCLING OF DUST STOCKPILE

During plant operation a dust stockpile of one million tons has been accumulated[2]. For recycling this dust the following solutions are examined.

a. Recycling Via the Existing Pelletizing Plant

According to our estimations, the pelletizing plant is almost fully occupied for pelletizing current rotary kiln dust production and only 3-6 h/day is available for recycling dust from stockpile. During this time, an additional amount of 150-200 tons pellets/day can be produced, i.e. 5000 tons/month.This corresponds to approximately 4000 tons of dust/month from the stockpile. Dust from the stockpile will be fed to the mixers by means of a belt conveyor. Transportation from the stock site to the pelletizing plant will be done by trucks at a rate of 500 tons, every three days. Close to the pelletizing plant, a place for dust temporary storage must be available. This place must be closed to avoid dust dispersion in the atmosphere.

This additional 5000 tons of pellets/month will give a pelletizing cost reduction of 20-30%.

Although this increase in pellet production is suggested, it is notconsidred as the ultimate solution because it has a very slow rate and it would take more than 20 years to recycle the total amount of dust. So, an additional solution should be found, to recycle the dust stockpile in a time span of about 5 years.

b. **Recycling Via a New Pelletizing Plant**

This plant will be approximately 1,5 Km away from the rest of the installations in order to be close to the stockpile and to avoid pollution from dust transportation. This will have however a high capital and operational cost as being away from the rest of the installations.

c. **Reduction Roasting in a Fluidized Bed Reactor**

By this solution a better utilization of carbon in the dust is achieved, especially because according to our measurements, approximately 20% of the carbon in the dust is in the volatile form. Hot calcine will be directly fed to the electric smelting furnaces.

So, a series of such tests is scheduled to be conducted.

d. **Injection of Dust in the Rotary Kiln Product**

It has been found out that green pellets cannot be fed in the electric furnace for smelting because moisture and volatiles contained give birth to huge flames interrupting normal operation.

A series of roasting tests on pellets were run in an industrial rotary kiln, 20x2,4m. The technical data of the kiln are given in table V. Again, intensive flames came out of the roasted pellets for a few minutes after they came out of the kiln. This is due to the carbon contained which had not completely burnt after a 2 hour roasting in the kiln, as shown in table VI.

Therefore, even more intensive flames are expected if dust is introduced in the calcined laterite after it comes out of the rotary kilns. If dust is injected inside laterite while it is still in the kiln, and if it (the dust) stays in the kiln until it looses all of its volatiles, i.e. more than about 2 hours, then it can be fed in the electric furnaces without operational difficulties. This requires that dust should be injected at some distance from the kiln exit. But then some of this dust will be taken in the gases and this may be decisive

TABLE V

TECHNICAL DATA OF A ROTARY KILN USED IN PELLET ROASTING

Main dimensions (DXL) m	Inlet diameter (m)	Outlet diameter (m)	Speed of rotation (RPM)		Kiln inclination (%)
			Main drive	Auxiliary drive	
2,40x20,0	0,80	1,20	1,9	0,25	3

Refractory thickness (Al_2O_3=32%) (cm)	Power of main drive, AC (KW)	Induced draft fun capacity (Nm^3/h)	Weight(steel/ refractory/ permannent load) (t)	Dedusting system	Chimney hight (m)
22,5	55	60.000	150/70/5	Dry (Dust chamber/polycyclon)	24,5

TABLE VI

EXPERIMENTAL DATA OF PELLET ROASTING IN A 2,4x20m ROTARY KILN

Feed rate (t/h)	Retention time in kiln (h)	Calcine temperature (°C)	Flue gas temperature (°C)
15	1,5 - 2,0	900	450

for the injection process. So, a proper technique has to be developed for injecting dust inside the charge at a certain distance from the kiln exit and then the dust in the gases must be determined to find out if it stays in the calcine.

e. Injection of Dust "as it is" in the Electric Smelting Furnaces

Feeding dry must of the kilns directly into electric reduction furnaces is always a challange because it is much easier than feeding it in the charge inside the kiln, and of course because it overpasses all pelletizing difficulties. It should be noted however that specific electric energy conscuption for cold dust smelting is at least two times that needed for the hot kiln calcine. However, dust is 6,7% of laterite or 6,7:0,90 =7,4% of the hot calcine. So, a series of tests was conducted in one of the electric furnaces of the plant. Dust was injected inside the slag. So, its carbon was oxidized to CO inside it. Injection rate was very low to avoid intensive CO boiling as well as electrode current flactuations.

According to the tests, some dust was taken in the flue gases, possibly because it was not injected deep enough in the slag. Since no metallurgical problems were observed these tests should be repeated to understand better dust behavior in the slag.

CONCLUSIONS

1. Pilot tests for wet dust recycling with a chain conveyor, a screw pump conveyor, a screw conveyor and a piston pump conveyor, showed that the most suitable solution was that of screw pump conveyor.
2. Recycling the dust by using a screw pump conveyor, provides a cleaner environment, requires less space and makes hoisting of the material much easier.
3. Industrial application of a screw pump conveyor for more than 5 years has been satisfactory.
4. During the fourth year of pump application 71,0% of wet dust was recycled. Mechanical wear of the screw and its casing is excessive however although not prohibitive.
5. Test work with a piston pump was satisfactory and it should be repeated to delineate its mechanical endurance.
6. Preliminary dust injection tests in the electric furnace slag were started but they should be repeated in order to better understand dust behavior in the slag.
7. After appropriate modifications in the pelletizing plant, 100% of the -4mm indurated pellets is recycled.
8. Crushing tests on pellet agglomerates were satisfactory, so a jaw crusher was bought for recycling pellet agglomerates.
9. With all improvements made, dry and wet dust recycling reached 94,3% in 1988 from 66,7% in 1983.

Acknowledgments

The authors gratefully acknowledge the financial support of the General Secretariat for Research and Technology of Greece and also G.M.M.S.A. LARCO for financial support and for permission to publish the results.

References

1. Zevgolis, M.N. : "The Problem of Handling and Recycling of Rotary Kiln Dust". Internal Report, LARCO SA, Oct. 1980.
2. Zevgolis, M.N., Gaitanos, J.F. and Lazaridis, N. : "Productivity Improvement in Pelletizing Dust from Laterite prereduction in Rotary Kilns". Symposium on Environmental Protection in Mining, Metallurgy and Materials Technology, Depart. of Mining and Metallurgy, Nat. Techn. Univ. of Athens, Athens Dec. 13-15, 1989 pp 363-386.
3. Zevgolis, M.N.:Research Proposal 5549/Jul. '83, Greek Ministry of Research and Technology, 1983.
4. Research Contract No 85053/Jul. '85, The Greek Ministry of Research and Technology and LARCO SA, 1985.
5. Zevgolis, M.N., Gaitanos, J.I. and Lazaridis N., "Recycling of Rotary Kiln Dust from Nickeliferrows Laterites", Final Report, General Secretariat of Research and Technology of Greece, Jul. 1987.
6. Zevgolis, M.N. : "Recycling of thickener Dust" Internal

Report, LARCO SA, Nov. 1983.

7. Zevgolis, M., Gaitanos, J., Hajinicolaidis J., and Papamantellos D. : "The Recycling of Wet Dust Coming from Laterite Prereduction in Rotary Kilns". Symposium on Environmental Protection in Mining, Metallurgy and Materials Technology. Depart. of Mining and Metallurgy, Nat. Techn. Un. of Athens, Athens, Dec. 13-15, 1989, pp 387-404.
8. Canlis, S : "Operational Problems of the Pelletizing Plant" Internal Report, LARCO SA, Feb. 1986.

Methodologies to Identify Scrap Metals

By W. D. Riley

Albany Research Center
U.S. Bureau of Mines
1450 SW Queen Avenue
Albany, Oregon 97321

Abstract

The Bureau of Mines has conducted research to better identify, segregate, and recycle scrap materials containing vulnerable strategic and critical materials. This paper describes some of the techniques such as X-ray and optical emission spectroscopy, thermoelectric response, and the Bureau of Mines newly developed automated spark testing technique developed for the identification and sorting of scrap metals. Results are reported that show the efficacy of these techniques on the identification of metallic scrap, consisting of stainless steels and Co- and Ni-base superalloys, in both a laboratory and a field setting.

An important phase of the work was a full scale technology transfer demonstration study conducted at a Department of Defense scrap yard located in Cherry Point, NC. Using Bureau developed technology the Department of Defense was able to improve the average return received for the sale of scrap metals from $ 0.09/kg to $ 0.95/kg.

Second International Symposium - Recycling of Metals and Engineered Materials
Edited by Jan H. L. van Linden, Donald L. Stewart, Jr., and Yogeshwar Sahai
The Minerals, Metals & Materials Society, 1990

Introduction

The United States is almost totally dependent on imports of vulnerable, strategic, and critical metals such as Cr, Co, and Mn. Among the research objectives of the Bureau's program is the conservation of these metals and the minimization of dependence on imports. A primary goal in this research program has been the reduction of downgrading and discarding of complex scrap materials containing these critical elements by improving the efficiency of recycling through the introduction and evaluation of new technologies to identify and segregate alloy scrap (1-4, 7, 12, 15-18)[3].

Historically, scrap metals have been identified by skilled sorters using a limited number of physical and chemical tests. These tests relied on object recognition, color, apparent density, magnetic properties, nature of spark properties when ground on an abrasive wheel, chemical spot tests, and spectroscopic analysis (12, 13, 20). Unfortunately, these procedures are inadequate for the identification and segregation of complex scrap alloys into recyclable fractions.

Although the advent of various refining and filtering technologies has allowed superalloy melters to use some mixed scrap to produce master alloys, ideally the mixed material should be separated into alloy classes of like chemistry (14, 17, 20, 21).

If the identification and segregation of mixed scrap can be done on a cost effective basis, recycling can be accomplished without the expense of refining, diluting or realloying. There are several ways to identify scrap metals: 1) by simple comparison to known samples as is done in spectroscopic analysis, 2) by group identification, and 3) by quantitative analysis. Each one of these identification schemes requires greater sophistication in technology and greater accuracy in elemental analysis for successful identification, respectively. It is important to consider that alloy compositions consist of ranges, so alloys can be identified from elemental analysis of lower precision and accuracy than is ordinarily expected from an analytical laboratory. The sorter can often identify an alloy by determining the amount of only a few elements in a given sample. One alloy can sometimes be picked out from a group of alloys on the basis of only one element. For example, 18-8 stainless steel scrap for targeted remelting consists of all of the 300 stainless steels (ss), with the exception of 316 ss and 317 ss molybdenum containing grades. By determining the presence or absence of 2-3 pct Mo it can be established whether or not the material is one of the Mo containing grades. This analysis, obviously, requires less inherent accuracy because the group identification can be made with greater latitude than actual chemical analysis. Comparative analysis gives the least information; "are the standard and the sample the same?" In some cases this is enough. The stainless steels again provide a good example. Assume that it has been established that an unknown is one of the 300 series 18-8 stainless steels. Semi-quantitative chemical analysis and comparison with published alloy specifications will determine the alloy number of the unknown.

Techniques such as thermoelectric response, emission spectroscopy, fluorescent X-ray spectroscopy and the newly introduced automated incandescent spark sorting technology offer potential to improve the ease and accuracy with which identification can be accomplished (1, 2, 16-19). Table I lists these methods and how well they supply the information required for identification.

---

[3]Underlined numbers in parenthesis refer to items in the list of references at the end of this chapter.

Table I.- Information supplied by commercially available techniques

	Thermoelectric Measurements	Emission Spectroscopy	X-ray Spectroscopy	Spark Testing
Alloy Name	sometimes[1]	sometimes	yes	yes
Group (SCL)	yes[2]	yes	yes	yes
Element	no	yes	yes	no
Comparison	yes	yes	yes	yes

[1] Only in certain instances, such as pure Ni, Monels, and Ti alloys.
[2] For quality control of segregated fractions.

A brief description of the principles of operation of each of the instrumental techniques is listed to acquaint the reader with the techniques. Complete discussions of these techniques can be found in (1, 2, 7, 16-19).

Thermoelectric instruments for identification of metals and alloys are based on the Seebek effect (the principal of thermocouples). The magnitude of the potential (emf) difference depends on the temperature of the junction between two metals and their composition. Thus, if a junction is formed between an unknown and a known metal at a known temperature difference (the probe is at 300° C), the unknown metal may be identified by comparing the reading with standard samples. A Technicorp model 850/950 WT Alloy Separator[4] was used to measure the thermoelectric response of the alloys listed in table II.

Under suitable excitation conditions, all metallic elements emit light of characteristic wavelengths. By observing these emissions, the material may be qualitatively or quantitatively analyzed. Typically, the steps in emission spectrochemical analysis are (1) vaporization of the sample, (2) excitation of the vapor to luminescence, (3) resolution of the resultant radiation into a spectrum, and (4) observation and analysis of the spectrum. For this particular work a Clandon Metascope was employed.

Energy dispersion X-ray spectrometry is a spectral analysis technique where secondary X-rays generated from the bombardment of a sample by α and γ emitters are measured by either a detector that can discriminate the pulse amplitudes of X-rays separated by energy, or by a balanced filter-detector combination. Because the characteristic X-rays of elements are known and the concentration of the elements is related to the intensity of the radiation, desired information on alloy type and elemental constituents can be determined X-ray Fluorescence has been used to identify stainless steels (1, 2, 16-19), superalloys (1, 2, 15-19), and copper alloys (4, 9). The three X-ray units used in this combination laboratory-field study were: (1) the Texas Nuclear 9266 Alloy Analyzer, (2) the Columbia Science Industries X-Met 840, and (3) the Kevex Analyst 6600 equipped with the X-Cite portable detector/cryostat package.

This report describes how these techniques can be used to identify and sort high alloy scrap into alloy groups as well as into specific alloys. Each of these techniques can be used individually with some success, particularly X-ray spectroscopy and the spark analysis technique. A series of laboratory-scale tests were conducted with each class of instrument (thermoelectric, optical spectroscopy, and X-ray fluorescence) to determine their usefulness as scrap identification tools. After the operational characteristics for the instruments were understood, a full-scale identification and segregation trial of the techniques was designed for demonstration at a Department of Defense scrap yard. The results of the large-scale scrap segregation and marketing test showed the practicality of these identification techniques.

[4]Reference to specific equipment does not imply endorsement by the Bureau of Mines.

Table II. - Major element concentration of alloys used in this study[1]

Alloy steel:	Major alloying elements, wt pct (nominal)							
	Fe	Ni	Cr	Mo	Co	Cu	Ti	Others
Stainless steel:								
201..........	70	5	18	n.m.	n.m.	n.m.	n.m.	Mn
301..........	73	7	17	n.m.	n.m.	n.m.	n.m.	Mn
303..........	72	9	18	n.m.	n.m.	n.m.	n.m.	Mn
304..........	69	9	19	n.m.	n.m.	n.m.	n.m.	Mn
316..........	72	10	16	2.5	n.m.	n.m.	n.m.	Mn
321..........	72	9	18	n.m.	n.m.	n.m.	5	Mn
347..........	70	9	17	n.m.	n.m.	n.m.	n.m.	Mn, Ta + Cb
Nickel-base alloys								
Monel 400....	1	67	n.m.	n.m.	n.m.	31	n.m.	Mn
Monel K400...	1	67	n.m.	n.m.	n.m.	39	n.m.	Al, Mn
Inconel 600..	8	75	16	n.m.	n.m.	n.m.	n.m.	Cu, Mn
Inconel 625..	2.5	61	22	9	n.m.	n.m.	.2	Cb, Al, Mn
Inconel X-750	7	70	15	n.m.	n.m.	n.m.	2.5	Cu, Cb, Al
Inconel 718..	19	53	19	3	1.0	n.m.	.9	Cu, Cb, Al
Hastelloy B-2.	2	65	1	28	1.0	n.m.	n.m.	Mn
Hastelloy S...	1	66	16	15	n.m.	n.m.	n.m.	Mn
Hastelloy C-276.......	6	55	16	16	2.5	n.m.	n.m.	Mn, W
Hastelloy G...	19	42	22	6.5	2.5	n.m.	n.m.	Mn, Ta, Cb
Nickel 200....	n.m.	99	n.m.	n.m.	n.m.	n.m.	n.m.	
Nickel 201....	n.m.	99	n.m.	n.m.	n.m.	n.m.	n.m.	
Nimonic 75....	3	75	19	n.m.	n.m.	n.m.	.4	Mn
Nimonic 80A...	n.m.	75	19	n.m.	n.m.	n.m.	2.4	Al, Mn
Nimonic 90....	n.m.	59	19	n.m.	16.5	n.m.	2.4	Al, Mn
Incoloy 800...	46	32	21	n.m.	n.m.	n.m.	n.m.	Mn
Incoloy 825...	30	42	21	3	n.m.	2.2	n.m.	Mn
Iron-base alloy:								
Haynes 20 MOD.	46	26	22	4	n.m.	n.m.	.4	
Cobalt-base alloys:								
Haynes 188....	3	22	22	n.m.	36	n.m.	n.m.	W, Mn
Haynes 25....	2	10	20	n.m.	51	n.m.	n.m.	W, Mn

n.m. <1 pct.

[1]In practice, alloys are specified by ranges in compositions, maximum allowed, minimum required, or balance for each element present. The nominal values in table II represent middle of the allowed range of the values used most often.

Sample Preparation

Stainless steel and Ni-, Fe-, and Co-base alloy samples, typical of those alloys prevalent in current scrap operations, were used in laboratory studies and are listed in table II. Surface impurities can interfere with all of the techniques used in this work, except automated spark testing. To overcome this problem, the testing area of each sample was prepared by grinding to clean metal with a medium grit aluminum oxide wheel. A clean sample is especially important when analyzing field test samples because many critical applications of superalloys require wear- or oxidation-resistant coatings, or fabricating or service conditions may cause the depletion of one or more elements at the surface. For example, consider table III which lists the X-ray fluorescence analysis for a scrap turbine blade before and after grinding. Because the presence of a coating is not often obvious, it is advisable to always remove the surface layer prior to analysis.

Table III.- Analysis of a turbine blade before and after grinding.

Element	Before Grinding, wt pct.	After Grinding, wt pct.
Ni.......	67.4	60.8
Cr.......	2.1	7.2
Co.......	7.8	10.6

Thermoelectric Measurements

Thermoelectric measurements, based on the Seebek effect (basis of thermocouples), allow the effect of chemistry of material to be detected as microvoltage difference across hot-cold and cold-cold junctions between a sample and two thermocouples. Table IV lists results for only nickel-base alloys, using a cobalt-base probe tip and both type 302 stainless steel and nickel 201 as standards. The selection of the best probe-standard combination is purely an empirical process. The tables reveal that there is a good first separation of the metals into broad groups of stainless steels, Monels, nickels, and superalloys. Further separation may be achieved by the judicious selection of alternate probe-standards.

Table IV - Thermoelectric response of Ni-base Superalloys (std dev 4.5 pct).

Alloy	302 ss standard	Ni 201 standard
Inconel 600.....	703	986
Inconel 625.....	656	917
Inconel 718.....	650	920
Inconel X-750...	679	952
Incoloy 800.....	615	865
Incoloy 825.....	633	883
Hastelloy C-276.	655	929
Hastelloy B-2...	728	1018
Hastelloy G.....	635	895
Hastelloy S.....	668	945
Nimonic 75......	690	989
Nimonic 80A.....	713	1020
Nimonic 90......	693	962

Emission Spectroscopic Measurements

The optical spectrograph was used for both elemental and alloy identification. At first, a series of measurements to determine specific alloying elements were conducted, such as molybdenum to separate 316/317 ss from the other 300 ss alloys and cobalt to separate Nimonic 90, 100, 105 from the rest of the Nimonic series. Figure 1 shows a typical spectra for 316 ss, the pointer in the spectra is located on the 553.30 nm Mo line. While the general patterns of all the 300 ss alloy are similar, the presence of a Mo emission line means the metal in figure 1 is 316/317 ss. During these measurements, it became obvious that certain classes of metals had similar spectral signatures and that mixed scrap could be separated into these classes based on observation of the overall spectra.

Figure 1. - Optical emission spectra of a 316 ss sample. (Pointer at the 553.30 nm Mo line.)

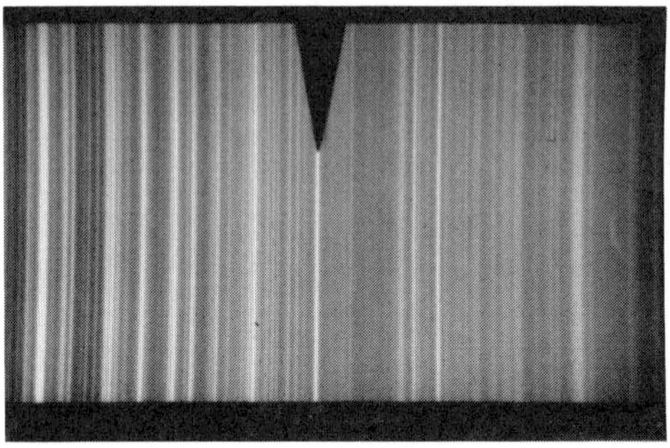

Figure 2. - Optical emission spectra of of a typical Inconel alloy.

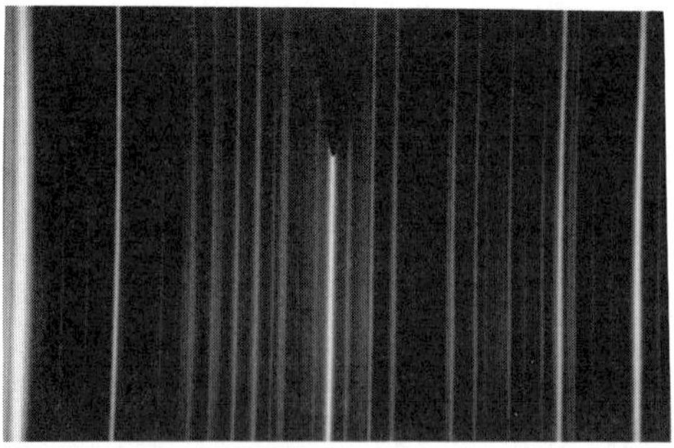

Figure 3. - Optical emission spectra of a typical Monel alloy.

All the 300 series stainless steels have a similar spectra to that shown in figure 1. The only difference is the Mo emission line in the 316/316 ss alloys. Similarities also occur in the Inconel class (fig. 2) and the Monel class, (fig. 3). With experience the operator can learn to recognize these signatures for the alloy classes.

X-ray Fluorescence Measurements

X-ray fluorescence measurements offer the best chance of positively identifying an alloy by name. All the instruments identified most of the alloys that they were programmed to identify. The advantage of some of the newer programmable units becomes apparent as new alloys enter the matrix because of the ability to easily program these instruments with the new alloy's characteristics. Preprogrammed X-ray fluorescent devices could be used to analyze and identify unknown alloys by using the elemental analysis mode and comparing it with published chemical specifications for alloys.

Combination Measurements

A combination identification method was devised that takes advantage of the high efficiencies and low cost of two of the methods studied, thermoelectric and optical emission analysis.

The thermoelectric measurements (table IV) show that it is relatively easy to identify the Fe based alloys in this array of samples. The Monels, Nickels and the high-Ni and Co alloys all report to a different fraction. In fact, with these thermoelectric measurements alone provides a separation into the fractions listed in table V.

By using the spectroscope on fraction 1 a pure "Sabot" (300 ss fraction), 200 ss fraction, and an Incoloy type alloy fraction (table VI) is obtained. The

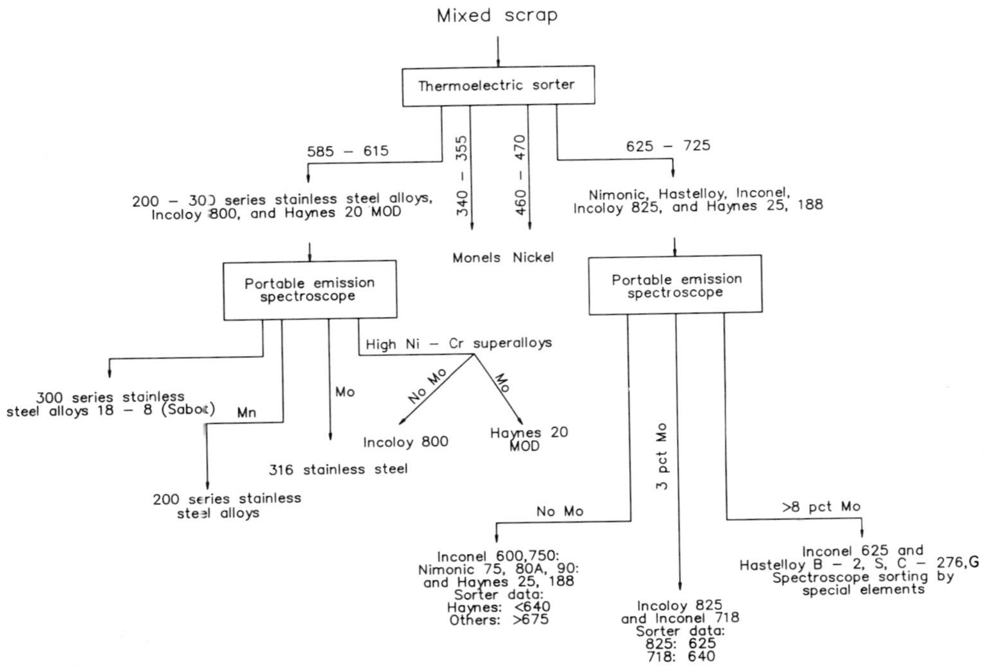

Figure 4. - Flow sheet of the combination method.

spectroscope can be applied to fraction 4 (Ni- and Co-base alloys) (fig. 4). Here again the ideal situation is to separate the alloys into classes that can be remelted together. For example, most Inconels can be recycled together and most Incoloys can be recycled together. The spectroscope allows the operator to combine all the alloys with similar spectra into marketable fractions.

Table V - Metal Fractions identified by thermoelectric response. (std dev = 4.5 pct.)

No.	Fraction Name	Thermoelectric response
1.....	200, 300 ss, Incoloys	585-615
2.....	Monels	345-355
3.....	Nickels	460-470
4.....	Ni and Co-base alloy	625-675

Large-Scale Field Tests

A series of large-scale scrap segregation and marketing tests were conducted in cooperation with the Department of Defense, Defense Reutilization and Marketing Service (DRMS), at several of its largest scrap yards in the United States. Tests were conducted at Cherry Point, NC; St. Julians Creek, VA; and Naval Air Station, North Island, San Diego, CA. The technology transfer demonstration conducted at Cherry Point, NC, fall of 1988 is described in this paper. Tests at other DRMS sites had similar results. Cherry Point, NC, was uniquely suited to Bureau of Mines technology because it is the home of a Naval Air Rework Facility so that much of the scrap generated is in the stainless and superalloy scrap category.

Scrap materials are classified by the Department of Defense, Defense Reutilization and Marketing Service (DRMS), according to a Scrap Classification List (SCL). Each type of scrap has its own category, the SCL codes serve as both a sales description and an inventory control. For high-value scrap the SCL describes a series of alloys with similar chemistry that may be melted together. For example, SCL T03 consists of the Co-base casting alloys, Haynes Stellite 23, 25, 31, 36, and L-605 (The average analysis of this SCL is 1-10 pct Ni, 12-16 pct Cr, and 50-60 pct Co). Ideally, DRMS would like to sell scrap by individual alloy, but for the purposes of this test the placement of the scrap into well identified SCL code was sufficient. The limited identification that was done was accomplished with chemical spot tests, density, and magnet tests (14, 20). These classical techniques are simply inadequate for the identification of the complex alloys that are used in modern airplane parts.

Therefore, a screening procedure was developed based on conventional identification techniques (density, spark test, and knife test) to separate the stainless steel and other high-value scrap from other categories--aluminum, carbon steel, magnesium, etc. (fig. 5). The high value scrap was then sorted visually by object recognition using samples that were identified using both X-ray spectroscopy and the combination technique. Instrumental testing was also used for quality control and the classification of odd or one-of-a-kind pieces. In addition to testing with the thermoelectric sorter and the emission spectroscope, three configurations of X-ray machines were used and analytical and identification models were optimized, by Bureau of Mines personnel, for both the X-Cite and the X-Met unit. These studies were to lead to the specification of an instrument and a segregation system for scrap metals in each of DRMS's area commands.

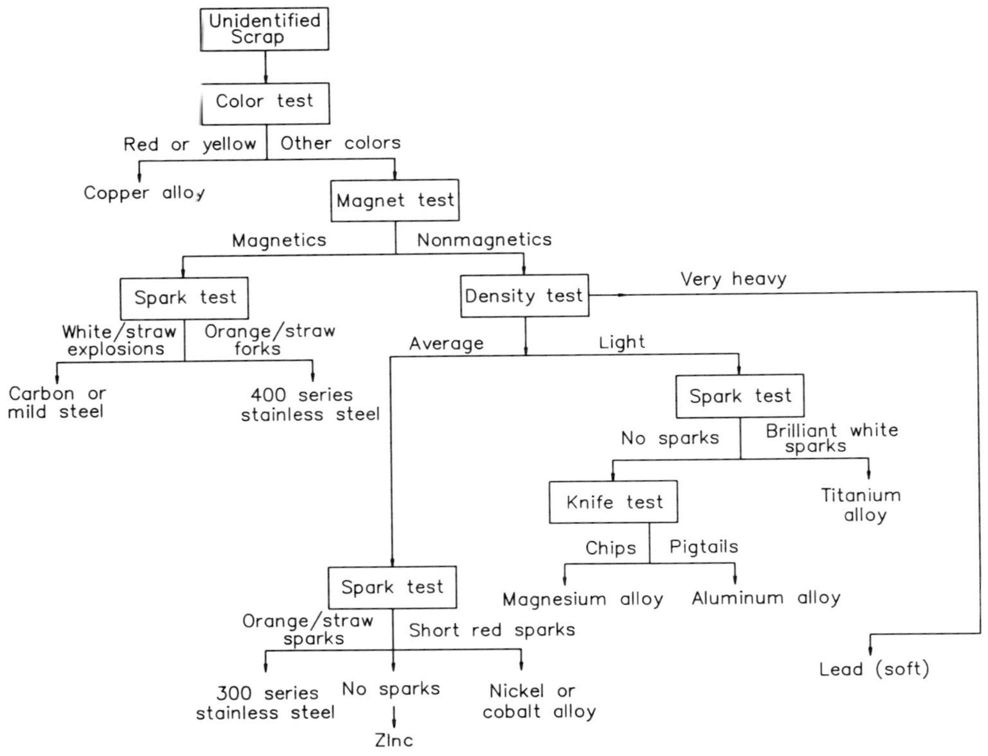

Figure 5. - Presorting procedure based on conventional identification techniques.

Testing Procedure

Scrap received at the Cherry Point scrap yard was weighed in and its classification checked. Typically the Naval Air Rework Facility labeled the scrap as mixed high-temperature scrap, no matter where the scrap was generated. The DRMS Materials Classifier would classify the material H24 (unsegregated metals) and move it to a holding area. A typical bin of mixed material might contain turbine parts, magnesium castings, titanium and aluminum alloys, and heavy and light steel. The pre-sorting operation was used to get the fractions listed in table VI. Only the nickel- and cobalt-based alloys require further processing to determine their proper SCL code. Aircraft engine scrap such as that received at the Cherry Point yard has an important advantage over other kinds of scrap. Most parts are recognizable, and in general, identical parts are made from the same alloy, or alloy class. Until DRMS acquired X-ray machines (using models developed during the tests), "sample boards" were created for classifying the material. The samples were analyzed piece-by-piece to establish their SCL codes and to determine the frequency that a particular sample appeared in a bin. The specimens ranged in size from 8 by 2.5 cm turbine blades to 10 m-long landing gear hooks weighing 350 kg. Only those samples with reasonably high generation frequencies were put on the sample boards. During a two week period approximately 500 different types of samples were identified by alloy and SCL code, 195 samples had a high enough frequency of generation (arbitrary number of 50 lbs/day Co-base alloys and 100 lbs/day Ni-base alloys) to be put on sample

Table VI - Metal fractions resulting from the flow sheet shown in figure 5

Alloy Group	Scrap Classification List Code
300 series stainless steel...	S01
400 series stainless steel...	S02
Aluminum....................	D1C/D1D
Carbon or mild steel.........	E1x
Copper......................	D2B/D4C
Lead........................	D5E
Magnesium...................	D06
Nickel- or Cobalt-based......	Txx
Titanium....................	T60
Zinc........................	D09

Table VII - Sorting categories based on minimum allowable Concentration of Ni, Cr and Co.

SCL code	Minimum Content, wt pct Ni	Cr	Co	Others
T03...	1	12	48	Mo<1
T15...	1	23	44	
T09...	18	18	38	
T25...	30	23	30	
T29...	47	18	15	Mo<1
T28...	46	23	10	
T20...	16	16	16	Ni+Co>45, Ni<39
T30...	60	7	9	
T18...	25	15	9	
T32...	41	20	1	
T05...	2	8	10	

boards. These boards proved easy to use for identifying alloys by object recognition, even for new DRMS employees. X-ray proved to be the most effective method of identifying the high value scrap. The elemental analyses provided by the X-ray machines were used to place the identified material in the SCL that provided the greatest return. A partial list of the scrap placement scheme is shown in table VII. As the X-ray analysis data was generated, the programs for the services were optimized. When possible, the programs were designed to give an alloy name and the SCL code. However with over 1500 alloys names, additional programs were written to simply give the SCL code and call the alloy name "unknown". In those cases where a unique analysis made it impossible to place the sample in an SCL, the machines displayed "unknown" and the chemical analysis.

Quality control of the sorted scrap fractions was a critical factor because mislabeled or contaminated scrap results in costly buy-backs and a loss of customer good will. Thermoelectric response was investigated as a quality control system because it offers a fast means of identification, training requirements were low, and the cost of the instrument is low. Previous research had shown that each SCL group had its own narrow range of thermoelectric response. Although there is some overlap in the standard deviations of the thermoelectric responses for the SCL codes with similar chemistries, it appeared the gross segregation errors could be identified. For example, a Co probe tip was used and the machine was arbitrarily calibrated to 600 on a piece of commercial purity Co. A bin that had been classified as 316 ss was checked, only 5 out of 123 specimens had responses that differed from the mean of the group by more than one standard deviation. These 5 specimens were later identified as one piece belonging to group S02 (400 ss--this should have been identified in the initial screening with a magnet, and four belonging to T32 (Hastelloy alloys). Approximately 900 kg of stainless steel scrap was examined and found to be sorted with a better than 95 pct accuracy. In fact, it was that found all the fractions that had been sorted by the use of the simple screening test and the sample boards were sorted with better than 95 pct accuracy. The high accuracy rating enhances the value of each sales lot. An effective rule of thumb for a misidentified piece was that a piece with a thermoelectric response that differed by more than one standard deviation (the response of known alloys) should be removed from the fraction.

Results

In five months, approximately 375,000 kg of scrap had been identified and segregated into 1 of 27 SCL codes (segregated metal accumulated during a three month period is shown in Table VIII). Historically, mixed scrap had sold at this scrap yard for prices as low as $0.09/kg. The average price for all the segregated scrap sold thus far was $0.95/kg. The segregated scrap sold for considerably more money, for example, S01; the 300 ss fraction sold for $0.29/kg, while T15, alloys with greater that 50 pct Co, sold for $14.78/kg. The rest of the prices received for the scrap fractions sold thus far is given in table IX.

Similar results were obtained at several other of the Defense Scrap Yards. The success of this program has resulted in DRMS purchasing portable X-ray equipment, which was programmed at the Bureau of Mines Albany Research Center. These machines are located at the scrap yards that generate the most high value scrap. Some of the initial data indicate that proceeds have tripled due to the marketing of adequately segregated and identified scrap produced by Bureau of Mines developed technology.

Table VIII. - Three month results of Cherry Point, NC., test.

SCL Code	Description	Principal Elements, % Ni	Cr	Co	Weight Sorted, kg	
	Stainless Steels					
S00	Stainless steels				96,400	
S01	300 Series stainless steels				5,471	
			Total Stainless		101,871	
	Ni and/or Co Based Alloys					
S3D	Misc. nickel				6,883	
T03	High cobalt alloys	1-10	12-16	50-67	302	
T04	Hastelloy C	55	16		2,209	
T05	Multimet	2-20	13-27	10-60	295	
T06	Timken 16-25-6	25	16		16	
T07	Timken 16---25-6, joined	1-25	8-16		45	
T08	Inconels	70-74	15		5,358	
T09	S-816	20	20	40	158	
T15	HS-21, 30	2-15	26-27	50-60	204	
T18	Waspalloy, Rene 41	55	19	13	2,561	
T23	Incoloys	26-30	13-20		6,928	
T25	HS-27	30	25	30	197	
T26	Invar (26% Ni, 64% Fe)	36			52	
T27	GMR-235	60	15		41	
T29	Nimonic 90	58	19	18	1,371	
T30	B-1900	64	8	10	1,045	
T32	Hastelloy X	5	22	1.5	1,373	
T60	80% Ti				3,813	
N.L.[a]	Cobalt				43	
		Total Ni and/or Co Base			32,894	
			Total High Value			134,765
	Other Scrap					
D06	Magnesium				2,565	
D1D	Aluminum				17,225	
D4C	Copper				109	
E1L	Heavy melting steel				1,407	
E1L	Painted steel				34,286	
E1R	Cast iron				51,770	
E1S	Tool steel				409	
D5E	Lead				14,655	
			Total Others		122,426	
			Total Scrap Sorted			257,191

[a]N.L., No listing

Table IX.- Results of initial sales of high-value scrap.

SCL Code	Description	Wt. Sold, kg	Avg. Price $/kg
S01	300 series stainless steels	96,000	0.29
T08	Inconels	4,600	2.67
T15	Stellites	550	14.78
T23	Incoloys	2500	0.98
T32	Hastelloy X	1370	2.29
T60	Titanium alloys	3800	4.11

Identification by the Nature of Sparks

Spark testing has been used for many years in the sorting of ferrous metals and alloys, where it has proven to be fast and reliable. Spark testing consists of observing the stream of sparks generated when a metal sample is held against a revolving abrasive wheel. Different alloying elements impart different characteristics to the spark stream--color, brightness, trajectories, and the behavior of the incandescent particles.

The main shortcomings of the traditional spark testing method are the requirement of experienced personnel and its inability to identify unknown alloys because it serves mainly as a comparison method. Its attractiveness is in its speed and the requirement of minimum sample preparation. Apart from simple precautions to prevent cross contamination, the method essentially creates its own fresh, clean surface.

The Bureau of Mines has conducted considerable research on this technique with the aim of substituting modern instrumentation for the trained observer. An instrumental technique should offer substantial improvements in sensitivity and selectivity (7, 16, 19). It was discovered that instrumental analysis of the spectral characteristics of the light emitted by the spark stream could be used for alloy identification, even in cases where traditional spark testing was unsuccessful (16, 19). Initial work was done using photodiode arrays, rapid scanning spectrometers, and fixed wavelength filters coupled with photovoltaic detectors (7, 19). The research showed that the instrument did not require highly sophisticated spectrographs to be effective.

The spectra of the spark is not a high energy spectra, like those used in atomic absorption spectroscopy. They result, in fact, from the exothermic oxidation of the fine particles dislodged by the grinding wheel. In an argon atmosphere the samples do not have characteristic sparks, only a very dull red stream appears, evidently due to the frictional heat generated by the contact of the sample on the wheel. Even Ti alloys, which are characterized by an extremely bright white spark stream in the atmosphere, have the same dull red appearance in an argon atmosphere. Microscopic examination of the grinding swarf produced in the argon atmosphere shows that the material has no rounded or burnt out particles contained in it. Hence the frictional heat, generated by grinding, is low.

The same material, when allowed to spark in air has a much brighter, more intense spectra. Microscopic examination of the swarf shows that there is globular material with rounded edges which indicates that some of the material has reached the softening point of the alloy at some point during the reaction. Obviously, the contact of the alloy and the wheel tears off fine particles from the material, frictional forces provide the initial heat and contact with the O_2 in the atmosphere allows the rapid oxidation of the material to take place. These spectral differences appear to be attributed to the characteristic exothermic energy of each alloy reacting with the O_2 in the atmosphere and forming oxide films.

When the technique proved successful, a more refined instrumentation package was constructed. The system used for most of the research is shown schematically in figure 6. The system consists of a bench grinder equipped with a medium grit Al_2O_3 grinding wheel spinning at 3,450 rpm. Light from the spark stream enters a fiber optic cable which divides into 4 separate strands leading to individual bandpass filters (peak transmissions at 542, 705, 842, and 1001 nm) followed by photodiode detectors. The photodiode detectors are connected directly to a 4-channel digitizing oscilloscope which communicates with a 16 bit computer. The computer controls sampling, storage, and data manipulation functions for the system. Table X lists the results for several alloys. The

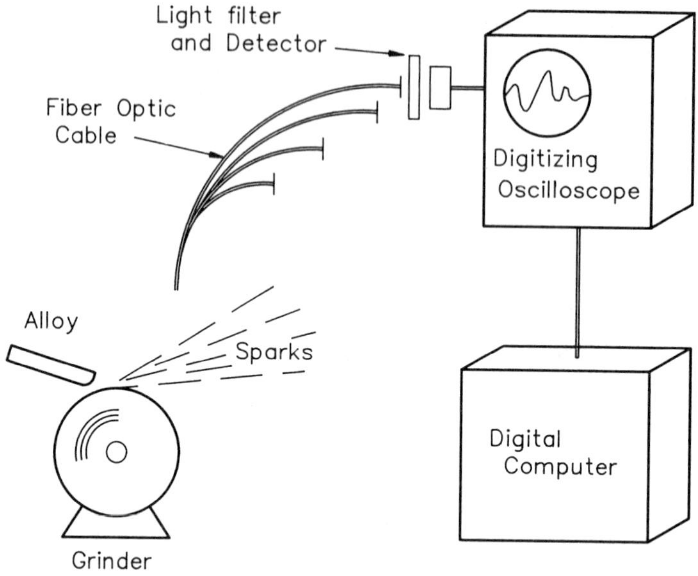

Figure 6. - Automated Spark Testing System.

Table X - Results. Time averaged values. (5 cm sec)

Metal	Four Detectors T, K	Standard Error of T, K
Udimet 500	1284	16
Haynes 188	1283	27
Inconel 600	1282	30
Inconel 625	1301	26
Hayes 25	1281	29
Hastelloy G	1283	21
Hastelloy B-2	1276	14
Hastelloy 41	1319	17
Rene 41	1339	26
B-1900	1288	20
Incoloy 800	1364	63
M2 Tool Steel, Sample 1	1388	19
M2 Tool Steel, Sample 2	1389	13
M2 Tool Steel, Sample 3	1394	15
Greek Ascoloy	1553	11
Stainless Steel 316	1614	18
Stainless Steel 430	1629	34
Stainless Steel 304	1635	25
Stainless Steel 301	1635	19
ARMCO Iron	1677	16
Stainless Steel 201	1645	17
MIL-S-17758 Mn Steel	1666	20
Medium Carbon Steel	1667	20
Ti-6 Al-4 V	2122	56

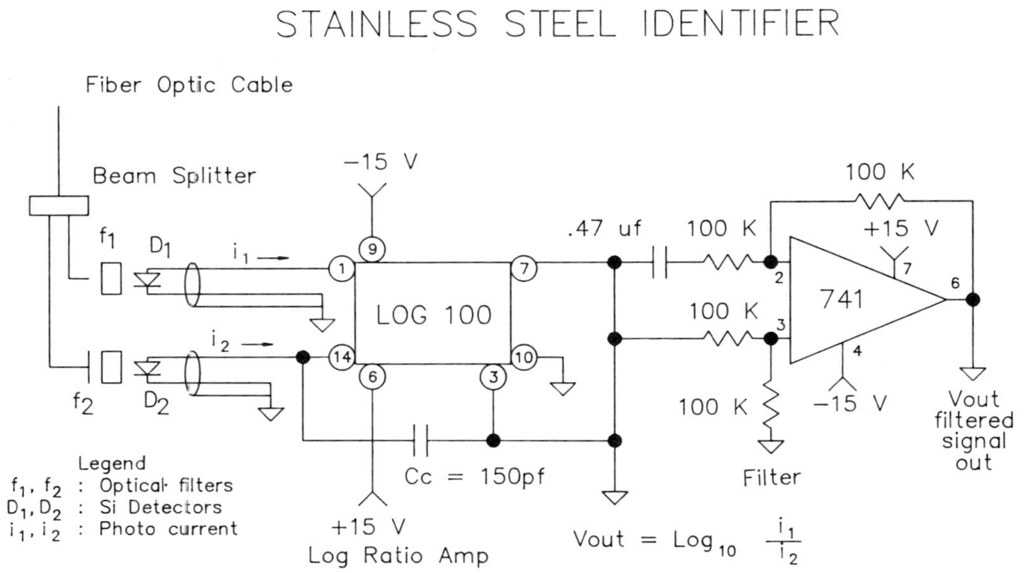

Figure 7. - Electrical schematic of the stainless steel identifier.

values are an arbitrary number based on the ratio of the intensity measurements from each photodiode. A prototype instrument has been constructed that allows for the separation of 316/317 Mo containing stainless steels from the rest of the 300 series steels resulting in a fraction that fits the Institute of Scrap Recycling Industries (ISRI) fraction "Sabot" which consists of 300 series stainless steels with less than 0.5 pct Mo. The new instrument is shown schematically in figure 7, and has been tested in laboratory experiments on 20 lb lots of mixed stainless steel scrap. In all tests the machine identified scrap with an accuracy of 95 pct.

CONCLUSIONS

Traditionally, the identification and sorting of scrap metals has been carried out by skilled sorters using various physical and chemical tests. New techniques, such as thermoelectric response, emission spectroscopy, and X-ray florescence spectroscopy, improve the accuracy and ease of rapid identification. Laboratory and field studies discussed in this paper were able to identify mixed scrap metals with better than 95 pct accuracy.

Thermoelectric response is an indirect method of identification because the reading must be compared with the response of a set of well defined, standard samples to make an adequate identification. Alloys with similar chemistries are almost impossible to identify because there is no way to isolate the effect of one individual element.

Although, it is inherently capable of quantitative analysis, emission spectroscopy is also generally used as a comparison technique. It is fairly easy to learn the skills necessary for recognizing the patterns of alloy types and for estimating the relative amounts of an element in different samples. The thermoelectric sorter and optical emission spectroscope have been combined to form an efficient, low cost system.

Portable X-ray units can identify many alloys through an integral computer and software designed to automate the identification process. In addition these units can provide an extremely accurate chemical analysis that can be compared with published chemical data for alloys or alloy classes (such as SCL's) to effect an identification.

Twenty-seven high-value alloy samples were used in the laboratory investigations of these techniques. The results of the laboratory investigations were used to establish the criteria for a field test conducted in cooperation with the Department of Defense (DOD). The success of both the laboratory and field studies is illustrated by the improvement in proceeds returned to DOD and the rapid transfer of Bureau developed technology to 15 major DOD scrap yards. The development of technology to rapidly identify and segregate high value scrap can result in more material containing critical and strategic materials being made available for recycling.

References

1. R. D. Brown, Jr., W. D. Riley, and D. M. Soboroff. Sorting Techniques for Mixed Metal Scrap. Conservation and Recycling, v. 9, no. 1, 1986, pp 73-86.

2. R. D. Brown, Jr., W. D. Riley, and C. A. Zieba. Rapid Identification of Superalloy Scrap. BuMines RI 8858, 1984, 23 pp.

3. W. J. Campbell. Energy Dispersion X-ray Analysis Using Radioactive Sources. X-ray and Electron Methods of Analysis, ed. by William J. Parrish. Plenum Press, New York, March 1968, pp.36-54.

4. W. J. Campbell and H. E. Marr, III. Identification and analysis of Copper Base Alloys by Fluorescent X-ray Spectroscopy. BuMines RI 7635, 1972, 30 pp.

5. L. R. Curwick, W. A. Peterson, and H. V. Makar. Availability of Critical Scrap Metals Containing Chromium in the United States. Superalloys and Cast Heat and Corrosion-Resistant Alloys. BuMines IC 8821, 1980, 51 pp.

6. C. L. Kusic, H. V. Makar and M. R. Mournier. Availability of Critical Scrap Metals Containing Chromium in the United States. Wrought Stainless Steels and Heat-Resisting Alloys. BuMines IC 8822, 1980, 51 pp.

7. J. M. Larrain, W. D. Riley, and R. D. Brown, Jr. Spark Testing in the Identification of Stainless Steel and Superalloy Scrap. Proceedings, 1985 International Symposium on Recycle and Secondary Materials, pp. 365-379.

8. H. E. Marr, III. Mathematical Smoothing of Digitized X-ray Spectra. BuMines IC 8553, 1972, 15 pp.

9. H. E. Marr, III. Rapid Identification of Copper Base Alloys by Energy Dispersion X-ray Analysis. BuMines RI 7878, 1974, 14 pp.

10. H. E. Marr, III and W. J. Campbell. The Processing of Energy Dispersion X-ray Data in a Timesharing Computer. Advances in X-ray analysis, v. 16, 1973, pp 206-216.

11. H. E. Marr, III and W. J. Campbell. Six Models for Interelement Correction in X-ray Analysis. Advances in X-ray Analysis, v. 19, 1975, pp 167-180.

12. A. W. Maynard and D. A. Wilson. Chemical Spot Tests for Aluminum Alloys. BuMines RI 7544, 1974, 12 pp.

13. R. Newell, R. E. Brown, D. M. Soboroff, and H. V. Makar. A Review of Methods to Identify Scrap Metals. BuMines IC 8902, 1982, 19 pp.

14. J. R. Rhodes, C. B. Hunter, D. L. Kellog, R. D. Sieberg, and T. Furuta. Application of a Computer-Coupled Radioisotope X-ray Spectrometer to Analysis of Steels. Advances in X-ray Analysis, v. 14, 1971, pp 127-138.

15. W. D. Riley and R. D. Brown, Jr. Device and Technique for Scrap Metal Spark Pattern Recognition. U. S. Pat. Appl. 623753, June 26, 1984.

16. W. D. Riley, R. D. Brown, Jr., and J. M. Larrain. Large-Scale Metals Identification and Sorting Using Instrumented Techniques. Journal of Testing and Evaluation, v.15, no. 4, 1987, pp 239-247.

17. W. D. Riley, R. E. Brown, and D. M. Soboroff. Rapid Identification and Sorting of Scrap Metals. Conserv. and Recycling, v.6. No. 4, 1983, pp 181-192.

18. W. D. Riley, B. W. Dunning, Jr., and D. M. Soboroff. Spectral Characteristics of Grinding Sparks Used for the Identification of Scrap Metals. BuMines RI 8932, 1985, 13 pp.

19. F. X. Spiegle, and E. Horowitz. Instruments for the Sorting and Identification of Scrap Metals. The Johns Hopkins University, Center for Materials Research, Baltimore, MD, 1981, 34 pp.

20. U.S. Defense Reutilization and Marketing Service (Dep. of Defense). Defense Scrap Yard Handbook. DOD 4160.21H, June 1985, 204 pp.

RECOVERY OF METAL VALUES FROM NdFeB MAGNET SCRAP

J. W. Morrison and G. R. Palmer

Salt Lake City Research Center
Bureau of Mines, U.S. Department of the Interior
729 Arapeen Drive
Salt Lake City, UT

ABSTRACT

The U.S. Department of the Interior Bureau of Mines investigated methods to separate valuable rare-earth materials from iron in NdFeB magnet scrap. A selective oxidation treatment of the scrap oxidized the rare-earth portion while leaving elemental iron. A 4:1 mixture of hydrogen to water vapor was used as the oxidizing gas with flow rate being rate controlling. Magnetic and leaching procedures attempted to separate the metallic iron and rare-earth oxides. The extremely fine grain size of the oxidized scrap prevented recovery by either technique. The best separation of rare earths from the NdFeB magnet scrap was obtained by sulfuric acid dissolution followed by precipitation of recyclable rare-earth salts. By precipitating a neodymium-sodium sulfate double salt which can be converted to a variety of useful products, many materials handling and economic disadvantages found with fluoride or oxalate direct precipitation were avoided.

Introduction

The U.S. Bureau of Mines is engaged in developing technology to treat a variety of waste processing streams that contain valuable and strategic metal values. Economic recovery from such streams would insure cost savings for industry and reduce the degree of U.S. dependence on foreign imports. As part of this effort, the separation of valuable neodymium (Nd) products from relatively worthless iron (Fe) in NdFeB magnet scrap has been studied. Since this scrap typically contains nearly 30 wt pct Nd and since the demand for Nd is increasing (1-3), economic methods for scrap treatment will have a significant impact on industrial expenditure and materials supply in the expanding area of magnet manufacture.

Permanent magnets are indispensable components of electromechanical and electronic devices such as computers, printers, motors, microwaves, and audio visual components among others. An average home contains about 40 magnets and a fully equipped car contains more than 30 (2,4,5). In 1979,

Second International Symposium -
Recycling of Metals and Engineered Materials
Edited by Jan H. L. van Linden,
Donald L. Stewart, Jr., and Yogeshwar Sahai
The Minerals, Metals & Materials Society, 1990

with the advent of the Sony "Walkman" series of miniaturized portable stereo headphones and tape players which utilized small but powerful samarium-cobalt magnets, rare-earth-alloy magnet manufacture began. Consumption of samarium oxide in the U.S. was 165 st in 1982 despite the fact that this market was totally nonexistent in 1978 (6). As the demand for permanent rare earth magnets increased, and due to unrest in Africa which affected the supply of cobalt, research was undertaken to develop other rare-earth, transition-metal compounds which could combine improved magnetic capabilities with lower raw materials costs (4).

The fortuitous discovery of a stable $Nd_2Fe_{14}B$ phase met the demand for a lower-cost, more readily-available material with improved magnetic properties. Ordinary ferrite magnets have a maximum energy product of 36 kJ/m^3 whereas samarium cobalt magnets have products over 200 kJ/m^3. NdFeB magnet material has achieved the highest energy product of all, over 250 kJ/m^3. NdFeB material is also has the best resistance to demagnetization at normal operating temperatures (7-9). The increased power of NdFeB magnets has made possible the manufacture of smaller and more efficient devices that are now used in a variety of applications.

The total market economy countries' output of Nd magnets was 250 t in 1987 and is projected to rise to 10,000 t by 1997 (10). Currently, it is estimated that 10 to 15% of magnet raw materials are being stockpiled as scrap by magnet manufacturers due to the lack of a cost-effective processing method. Scrap treatment will become an important alternative materials source because of the growing demand for individual rare-earth metals such as Nd, which was valued at $280/kg at year-end 1988 (2), and due to the projected forty-fold increase in permanent magnet production. U.S. import of rare-earth metals and alloys totalled nearly 200 t in 1988 at a cost of 2 million dollars and is expected to increase (2). It is clear that development of a viable method to process NdFeB scrap will aid domestic manufacturers economically and reduce the necessity for import of rare earth materials.

The two major methods currently being used to manufacture NdFeB magnets are rapid solidification (RS) and powder metallurgy techniques. In rapid solidification, a melted alloy is supercooled by generating melt-spun ribbon which is then crushed and treated in further processing steps such as polymer bonding or hot pressing. By this process, material with extremely fine grain size and uniformity of microstructure is obtained which has excellent magnetic properties. Typical scrap types generated by this process are slag and spillage materials from alloy melting, discarded ribbon material, and discarded finished magnets in a variety of forms. Powder metallurgy techniques involve sintering a powder which is aligned in a magnetic field. This process generates sludge from grinding stages as well as discarded sintered and unsintered (green compact) magnet pieces (1,4).

Due to availability of scrap samples, the present study concentrated on RS scrap with particular emphasis on ribbon material because, according to the manufacturer, ribbon represents the largest volume of scrap generated from this process. Analysis of RS scrap samples showed that all types have the following average composition in wt pct: 67.3 Fe, 26.3 Nd, 1.85 Pr (praseodymium), and 0.842 B. Several tests were also performed on green compact material from a powder metallurgy producer and showed this scrap to be amenable to the same treatment processes as those used for RS scrap.

Because scrap compositions are quite uniform, economical treatment methods were sought which would have the potential to efficiently separate Nd and Pr from Fe in all types of NdFeB scrap despite variations in scrap geometry and surface oxidation. The two main methods which were investigated to accomplish this were (1) selective oxidation of rare earths, and (2) acid dissolution and precipitation of rare earth salts.

Selective Oxidation

The purpose in performing selective oxidation was to oxidize the Nd in the scrap while keeping Fe in metallic form, thereby making the scrap amenable to rare-earth recovery either by magnetic separation or by dilute acid leaching of rare-earth oxides.

Previous research efforts (11) have shown that it is possible to generate a stability region diagram for the $Fe-H_2-H_2O$ system based on thermodynamic data for the various Fe oxidation-reduction reactions. For example, FeO reacts with hydrogen according to the following reaction to give elemental Fe:

$$FeO + H_2 = Fe + H_2O \quad (1)$$

The equilibrium constant for this reaction can be expressed as

$$K_{FeO} = PH_2O/PH_2 \quad (2)$$

Knowing that the combined pressures of hydrogen and water vapor are equal to 1 atm in a non-pressurized system, it is possible to express the hydrogen pressure as a function of temperature according to the following relationship:

$$PH_2 = 1/(1 + K_{FeO}) = 1/(1 + e^{-\Delta G/RT}) \quad (3)$$

By plotting the hydrogen pressure vs the temperature for each Fe oxidation reaction, and by superimposing the Nd oxidation reaction, it is possible to generate a diagram which shows the region in which metallic Fe and Nd oxide can coexist. This combination stability diagram is shown in Figure 1. For the reaction

$$1/3\ Nd_2O_3 + H_2 = 2/3\ Nd + H_2O, \quad (4)$$

thermodynamic data indicates that ΔG is on the order of +400 kcal/mol (18) for temperatures between 298° and 1,000°K. When a value of this magnitude is substituted into equation (3) above, it is clear that for this temperature range, PH_2 will effectively be equal to one. This means that Nd oxide is stable throughout the $Fe-H_2-H_2O$ diagram and that selective oxidation can be accomplished as long as the operating conditions are chosen to lie within the stability region for metallic Fe.

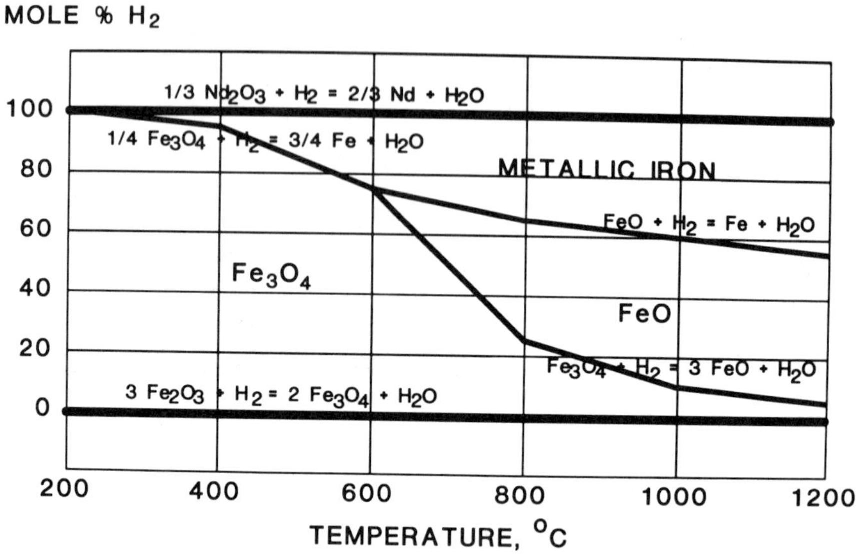

Figure 1 - Equilibrium diagram for Fe-Nd-H_2O-H_2 system.

Experimental Procedure

For this test work, the ratio of hydrogen to water vapor was chosen to be 4:1, and all tests were run at 600°C. Ribbon material ground to -400 mesh was used in all tests except one in which unground ribbon material was employed. A sample was suspended from a balance into an Astro 1000A muffle furnace through which high-purity helium gas was flowed during sample heat-up to prevent oxidation. When the desired temperature was reached, the helium was turned off and a reactant gas of the required hydrogen-water vapor ratio was flowed into the furnace. The hydrogen-water vapor ratio was controlled by flowing hydrogen gas at a known rate through a bubbler apparatus which was maintained at a temperature sufficient to generate the appropriate pressure of water vapor. The progress of the reaction was monitored by means of sample weight gain, and weight vs time data was recorded on computer through an interface with the balance.

Experimental Results

A plot of weight gained vs time for a typical test run is shown in figure 2. The amount of weight gained correlated very well with that expected for Nd oxidation alone. X-ray diffraction (XRD) data helped to confirm the fact that selective oxidation had in fact occurred. X-ray diffraction analysis of material non-selectively oxidized in an excess of oxygen, showed the presence of Fe_2O_3 and Fe_3O_4 in significant quantities along with traces of an Fe-Nd oxide phase. By comparison, analysis of the same material after selective oxidation showed that the only Fe phase present in any quantity was metallic, and the presence of Nd_2O_3 was also evident. To optimize the selective oxidation process, tests were carried out to study particle size, gas flow rate, and annealing effects on the reaction rate.

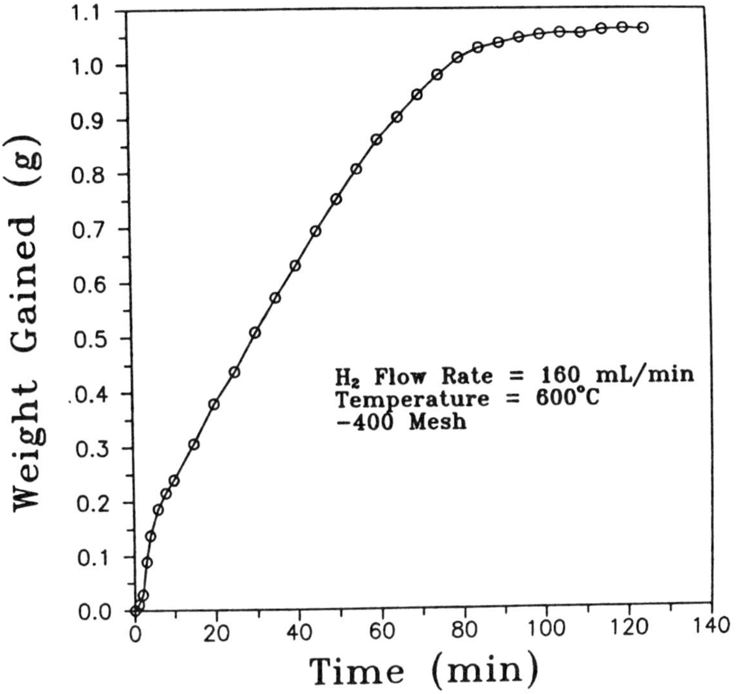

Figure 2 - Typical plot of weight gained vs time.

Particle Size Effects. The effect of particle size on the selective oxidation process was investigated by reacting ribbon material which had not been ground. Figure 3 shows plots of weight gained vs time for unground ribbon and for -400 mesh material. It is evident that particle size does not play a significant role at this gas flow rate and reaction temperature.

Gas Flow Rate. Without changing the overall ratio of hydrogen to water vapor in the reactant gas, the flow rate was varied to determine the effect of mass transport through a gas boundary layer to the reaction surface. It was found that gas flow rate significantly affected the reaction rate as shown in figure 4. A flow rate of at least 350 mL/min is required to overcome the effect of mass transfer through a gas boundary layer.

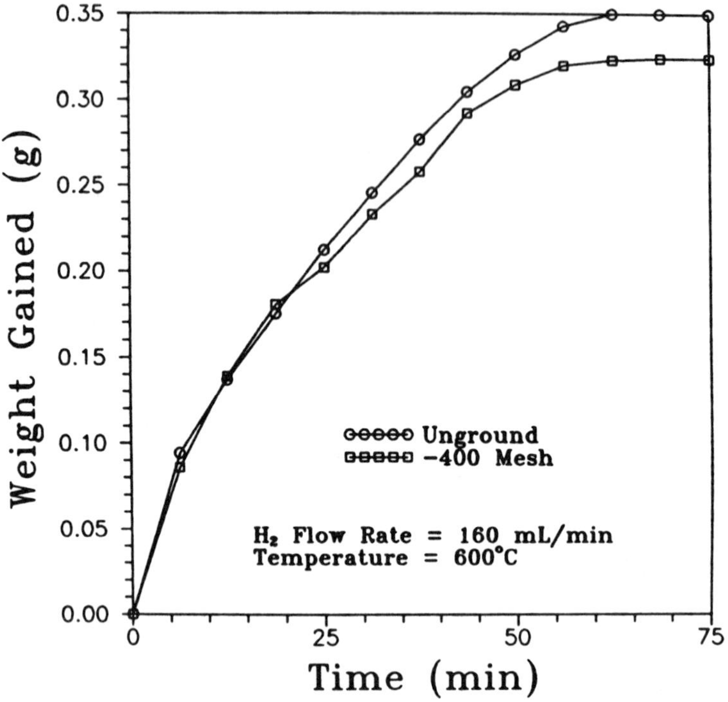

Figure 3 - Effect of particle size.

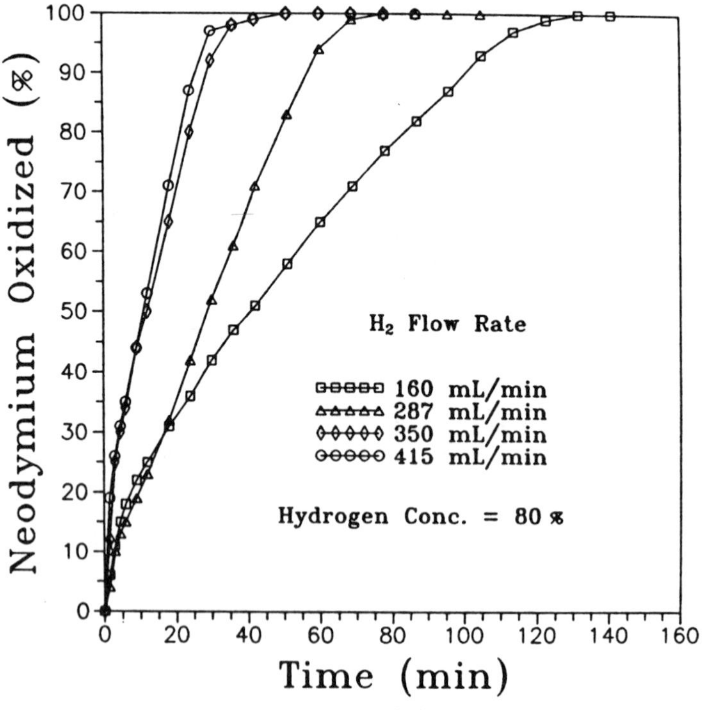

Figure 4 - Effect of gas flow rate on the rate of reaction.

Annealing Effects. Ribbon material is characterized by extremely fine equiaxed grains of the $Nd_2Fe_{14}B$ phase surrounded by a thin, intergranular, amorphous phase which is rich in Nd. A $Nd_{1.11}Fe_4B_4$ phase has also been characterized. The excess Nd compound, which may actually be a $Nd_{70}Fe_{30}$ eutectic phase, presumably segregates to the grain boundaries during solidification (8,12-15). Because the material is multi-phasic, it was thought that annealing, with its attendant grain growth, might enhance the reaction rate by aiding diffusion of the reactant gas within the sample particles. Figure 5 is a plot of the percent of Nd oxidized vs time for annealed and non-annealed material. Annealing does enhance the reaction rate slightly, but the effect is too small to be of significant value as a processing step.

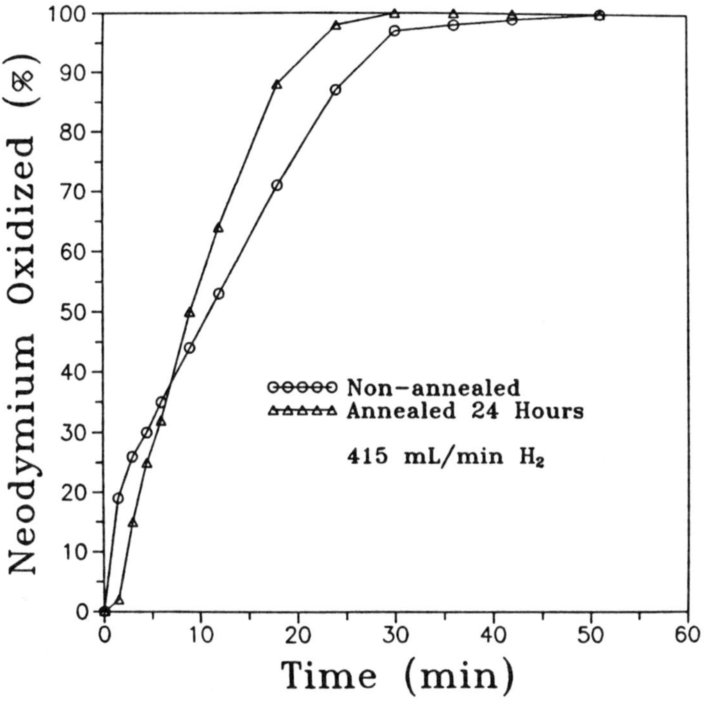

Figure 5 - Effect of annealing 24 hrs at 600°C.

Discussion

Selective oxidation was carried out with the hope that the sample morphology following reaction would lend itself to some type of separation of Nd from Fe. Because of the fine grain size of the predominant $Nd_2Fe_{14}B$ phase and because of the initial presence of the Nd-rich phase, interstitial growth of Nd oxide grains which would be amenable to a dilute acid leach was hoped for. Dilute acid leaching under a variety of conditions was employed but was not found to be effective in preferentially leaching Nd oxide out of the Fe matrix. Significant amounts of Fe also dissolved. As has already been noted, ribbon material has a complicated phase structure to begin with, and evidently the Nd oxide which forms remains intimately mixed with Fe in the lattice. Because of this, magnetic separation techniques were also found to be ineffective at achieving any separation of Fe from Nd oxide.

Selective oxidation does accomplish the oxidation of Nd alone while leaving Fe in metallic form, but separation of the two metals remained problematic. For this reason, it was decided to pursue a different processing scheme: total dissolution of the sample in stronger acid concentrations than those used for the dilute leach, followed by selective precipitation.

Acid Dissolution and Precipitation

Acid Dissolution

Although H_2SO_4 and HCl were found to be equally effective for scrap dissolution, H_2SO_4 was used in the bulk of tests due to its lower cost and because it is widely employed in industry for rare earth concentrate leaching. Precipitation of rare earth salts and double salts from sulfate systems has been extensively studied and forms the basis for many industrial practices (16,17). All types of scrap dissolved readily without heating or agitation.

Acid Requirements. Tests showed that the critical factor for scrap dissolution is the weight of H_2SO_4 added per weight of scrap dissolved. For all scrap types tested, a weight ratio of acid to scrap of 2:1 or higher is desireable both to totally dissolve the scrap and to keep the solution pH low enough to prevent Fe^{+2} oxidation to Fe^{+3}. It is known that neutral and acid solutions of Fe^{+2} ferrous ion oxidize less rapidly with increasing acidity, because Fe^{3+} is actually present in the form of hydroxo complexes, except in extremely acidic solutions (18). Fe^{+3} is undesirable in a process stream because of the formation of oxide precipitates which are contaminants. When the acid to scrap weight ratio is 2:1, the final solution pH is around one, the solutions are blue, and no Fe precipitates form over time. When the acid to scrap ratio is 1:1, the final solution pH is three, not all of the initial sample dissolves, and the greenish filtrates soon contain Fe precipitates.

In addition to determining the acid-to-scrap weight ratio required for effective dissolution, it was necessary to study the effect of acid concentration. One-hundred grams of ribbon material was completely dissolved in 1 L of 2 molar H_2SO_4 yielding a solution which analyzed 32.8 g/L Nd, 77.9 g/L Fe, and 0.9 g/L B. One-hundred-fifty grams of ribbon material was dissolved in 1 L of 3 molar H_2SO_4, but large amounts of Fe and rare earth sulfates precipitated out of solution. Although the solubility of $Nd_2(SO_4)_3$ is 80 g/L in cold water (26), and Fe sulfates are also soluble, the filtrate contained only 23 g/L Nd and 77.3 g/L Fe. It is evident that sulfate solubilities are exceeded for both Fe and Nd when the H_2SO_4 molarity is three. Operating the system at a maximum molarity of 2 is adequate to effect total sample dissolution.

Leaching Behavior of Different Scrap Types. A plot of the percent extraction of Fe and Nd vs the ratio of acid weight to sample weight for leaching of RS ribbon material is shown in figure 6. Fe appears to dissolve preferentially until an acid:scrap weight ratio of two is reached at which point the entire sample dissolves. Leaching behavior differs for hot-pressed material, however, as shown in figure 7. Nd leaches preferentially for this material, although both Fe and Nd leach

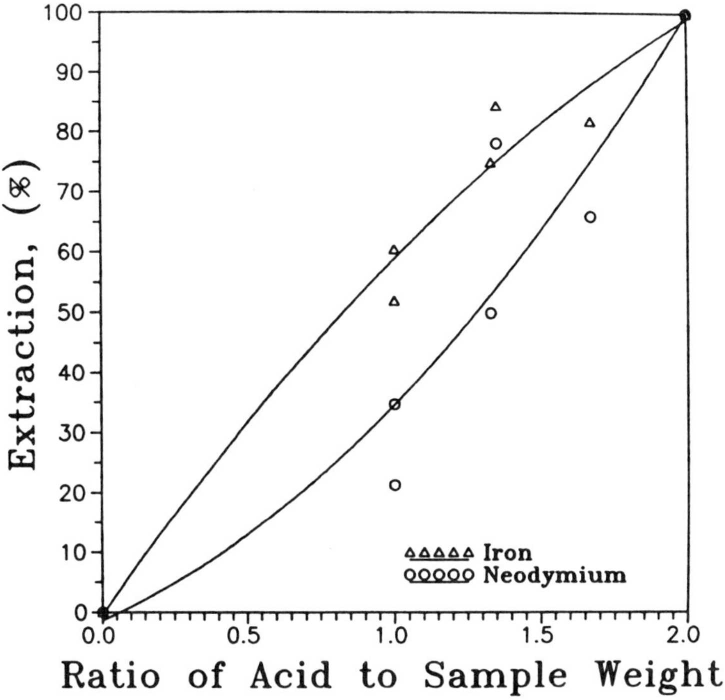

Figure 6 - Difference in rates of dissolution of Fe and Nd during leaching of ribbon material.

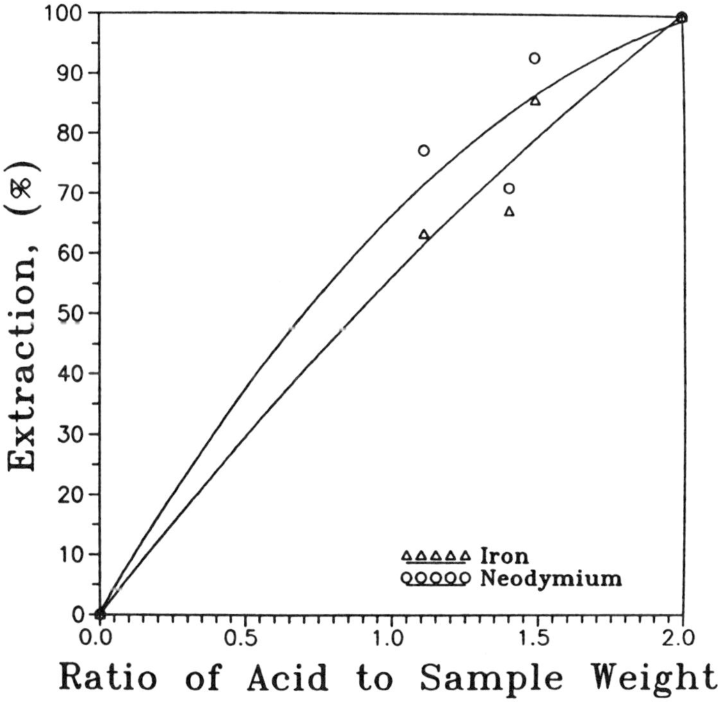

Figure 7 - Difference in rates of dissolution of Fe and Nd during leaching of hot-pressed ribbon.

more rapidly than from ribbon. Perhaps these phenomena can be explained by the effect of surface-layer oxidation which is known to occur and is obviously greater for ribbon material due to its much larger surface area per unit weight (19). If the surface layer is Fe-rich, preferential dissolution of Fe would be expected to occur. In a continuous system, a two-stage leach of ribbon material may represent the most efficient use of reagents. The first leach could be carried out at an acid-sample ratio of one to remove 60% of the Fe and about 30% of the Nd, followed by a second leach to dissolve the remainder of material, generating a concentrated Nd leach solution.

One-hundred grams of plus 10 mesh epoxy-bonded ribbon material was leached in 1 L of 2 molar H_2SO_4. This is enough acid to dissolve 100 g of any other type of scrap, but in this case, 25 g remained undissolved. Visual and SEM examination of leached pieces revealed that metal was still present in the center of each piece, so fine grinding may be required to leach this material efficiently. Other techniques such as fluxing with an organic solvent followed by magnetic removal of metal particles remain to be investigated, but it is clear that epoxy-bonded material will require an additional processing step of some kind.

Tests were carried out to determine the leaching behavior of slag, spillage, and green compact materials. All exhibited the same leaching behavior as ribbon and hot-pressed ribbon materials with regard to the amount of acid required for total dissolution.

Precipitation

Once scrap has been dissolved in acid solution, separation of rare earths from Fe can be achieved by precipitation of a rare earth salt. For acid dissolution to be a viable method for treating NdFeB scrap, a rare-earth precipitation product must be generated which can be recycled at some point into already-existing processes which use Nd salts as feed materials. Pure Nd fluorides and oxides are used in laser, glass, and ceramics manufacture, but by far the largest demand for Nd salts is in the production of Nd metal and/or alloys for magnet manufacture. At the present time two main methods exist for producing Nd metal: (1) metallothermic reduction and (2) electrolytic reduction.

In metallothermic reduction, Nd chloride, fluoride, or oxide is reduced using either calcium, sodium, or lithium metals (20). The purest and best characterized metal is produced by calcium reduction of the Nd fluoride which is generated either by precipitation from aqueous solution or by reaction of the oxide with HF gas or ammonium fluoride at low temperatures (200-500°C). Prior to calcium reduction, the NdF_3 is melted under a dynamic atmosphere of HF and argon, a process known as topping, to dehydrate it and/or remove residual oxygen which would contaminate the metal (17,21-23). Any Fe present in the fluoride is carried over into the final metal product (23).

In electrolytic reduction, an electrolytic cell is employed to reduce Nd chloride, fluoride, or oxide to Nd metal using fused-salt electrolysis. The electrolytes used are multicomponent melts of either chlorides or fluorides. The advantages of a fluoride system are that fluorides are less hygroscopic than chlorides, leading to purer final products, and fluorides can dissolve rare earth oxides directly whereas chlorides cannot. Work has also been carried out in which master alloys suitable for magnetic purposes were produced by fused-salt electrolysis of fluorides using transition-metal cathodes (22-25).

Both Nd fluoride and Nd oxide would be suitable precipitation products for recycle to current industrial operations. Fluorides can be precipitated directly from aqueous solutions, and, in fact, this is a recommended practice when large amounts of Fe are also present in solution (17). Rare earth oxides are generally obtained by roasting rare earth oxalates, and although the presence of large amounts of Fe is known to inhibit their precipitation, rare earth oxalates are one of the most important rare earth salts (17,20). Residual Fe present in the precipitates would not be problematic for either metallothermic or electrolytic reduction, particularly if the Nd metal product were intended for magnet alloy manufacture.

Fluoride and oxalate precipitation, however, have attendant problems. Fluoride precipitates tend to be gelatinous and difficult to filter, and oxalates are costly to produce and must be roasted to generate an oxide product suitable for recycle. A possible solution to the problems attendant with the direct precipitation of fluorides and oxalates is to generate an intermediate precipitate which could then be converted to either a fluoride or oxalate product depending on industrial demand. An example of such an intermediate precipitate is the double sulfate salt of Nd and Na which is used industrially for the precipitation and storage of rare-earth materials (17). The present work examined the precipitation of fluorides, oxalates, and Nd-Na double sulfate salts from dissolved magnet scrap leach liquors.

Precipitation of Neodymium Fluoride. Tests were carried out to study precipitation of NdF_3 from acid solutions of dissolved scrap using HF as the source of fluoride ion. For clarity, the term extraction refers to the percentage of an element originally present in the scrap which reports to the fluoride product. It is therefore obvious that low Fe extraction values are to be desired. Table I gives compositions of leach solutions for several scrap types along with the extraction of Fe and Nd in the fluoride product. All leach solutions were obtained by dissolving scrap in 1 molar H_2SO_4 except the last test in which 2 molar H_2SO_4 was used. The number of times the stoichiometric amount of HF used is also included in the table. It is clear that fluoride precipitation can achieve very efficient separation of Nd from Fe, although filtration was slow for all tests as expected.

The amount of HF added appears to have the greatest effect on Nd recovery. Fe extraction is influenced more by scrap type and Fe leach concentration. Figure 8 is a plot of data from a test series in which the same leach solution was divided into equal portions to which varying amounts of HF were added. It is evident that HF additions of about 1.5 times the stoichiometric amount are required for complete Nd extraction. Fe extraction was found to remain below 1% in each precipitate from this series. Adding more than four times the stoichiometric amount of HF causes Nd extraction to fall off markedly, probably due to the formation of suspensions of extremely fine particles of fluoride which are not amenable to filtration.

Table I. HF Precipitation

Scrap type	Leach liquor g/L Nd	Leach liquor g/L Fe	# times required HF stoich.	Extraction, % Nd, %	Extraction, % Fe, %	$[F^-]$ g/L final
Hot P.*	21.01	45.9	2.73	99.61	4.80	11.0
Hot P.	15.07	38.6	3.84	98.58	3.52	14.0
Ribbon	6.38	43.5	2.62	98.69	0.98	2.3
Hot P.	20.79	46.1	0.16	13.93	1.06	0.3
Ribbon	15.40	37.3	.97	59.57	.09	1.5
"	18.04	46.7	.77	66.67	.06	.3
"	"	"	1.54	98.85	.65	3.3
"	"	"	2.31	99.43	.21	6.7
"	"	"	3.09	98.89	.20	9.2
"	"	"	4.63	91.67	.21	18.8
"	"	"	5.40	65.00	.43	19.2
"	"	"	6.17	58.89	.85	25.8
Slag	18.50	48.5	1.52	99.00	2.55	2.8
"	"	"	2.28	99.50	2.05	7.3
"	15.30	37.3	.91	69.23	.34	1.0
"	"	"	1.82	93.90	2.50	5.0
"	"	"	2.72	95.64	.62	9.9
Ribbon	32.8	77.9	1.61	99.45	3.60	7.3

*Hot P. = Hot Pressed

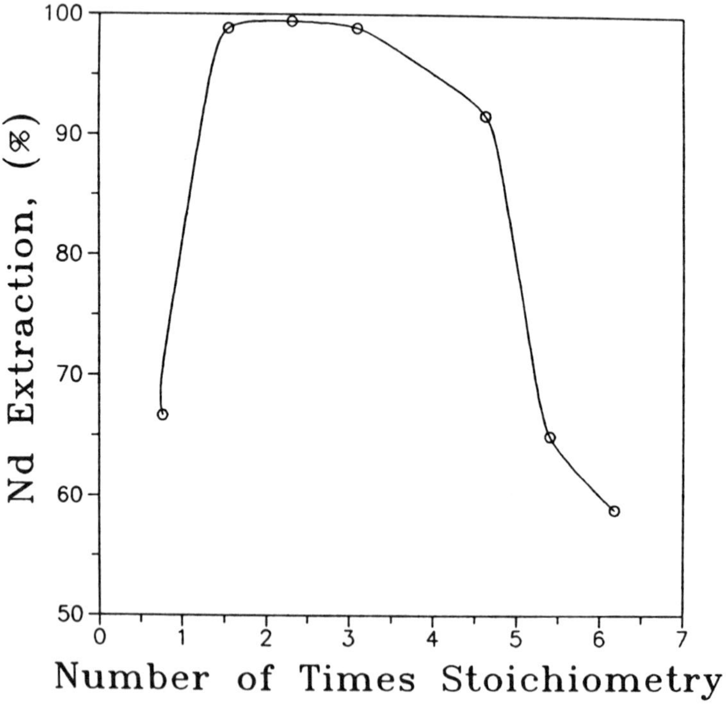

Figure 8 - Effect of number of times the stoichiometric amount of HF on Nd extraction.

The last entry in Table I represents a test carried out using 2 molar acid (hence, the higher concentration values for Nd and Fe). When the precipitate from this test was washed in hot water and then re-filtered, two effects were noted. First, the overall extraction values became 97.88% Nd and 0.18 Fe, and second, the precipitate recrystallized in a form which filtered easily. Washing was effective in removing residual Fe from the precipitate. As previously stated, fluorides typically form gelatinous precipitates which entrain residual Fe, causing slow filtration. Filtration rate remains the most serious impediment to recovery by direct fluoride precipitation from aqueous solutions (17).

It is apparent that the amount of fluoride ion remaining in solution depends on the amount of HF initially added. Even when less than the stoichiometric amount of HF is used, some fluoride ion remains in solution. Since fluoride is extremely corrosive in an operating system, it is necessary to keep its addition to a minimum. Using 1.5 times the stoichiometric amount still seems reasonable inasmuch as the Nd recovery is high while fluoride content remains around 3 g/L in most cases.

Table II shows results obtained from precipitation using several sources of fluoride ion. It can be seen that NaF achieves Nd extraction as efficiently as HF, and ammonium fluoride performs nearly as well. The other reagent tried was not as effective. It appears that using NaF as the ion source keeps the fluoride ion concentration to a minimum in the final filtrate.

Table II. Other Reagents Used to Generate NdF_3

Reagent used	Leach liquor		# x's	Extraction, %		F-
	g/L Nd	g/L Fe	Stoich.	Nd, %	Fe, %	g/L
HF	32.8	77.9	1.61	99.45	3.60	7.29
NaF	21.5	57.9	1.50	99.29	0.35	1.25
NaF	33.5	79.4	1.50	98.35	1.83	1.57
NH_4F	"	"	1.50	96.73	4.21	4.91
Na_2SiF_6	"	"	1.50	67.38	.91	9.29

Temperature effects on filtration were studied. Tests were carried out in which leach solutions were divided into two equal portions. One solution was heated to 70°C prior to precipitation, and the other was left at room temperature. Precipitation was carried out in one set of tests using HF, and NaF was used in another. Temperature did not have any appreciable effect on the rate of filtration.

Precipitation of Neodymium Oxalate. The presence of large amounts of Fe tends to interfere with oxalate precipitation. Test work indicated that when oxalic acid is added in stages, only 90% of the Nd is recovered before the precipitate becomes contaminated with Fe. Fe oxalate begins to precipitate as the concentration of Nd in solution decreases. Washing the precipitate with a solution of dilute sulfuric acid helped to remove residual Fe, leading to recovery vales for Nd and Fe of 97.76 and 0.24%, respectively. Nd oxalate can be precipitated by this method, but reagent costs may prohibit its use. Since oxalates must be converted to oxide products by roasting in a further processing step, and since the value of the oxide was only $14.50/kg at year-end 1988, it is possible that the cost of this procedure would not be justified (2).

Precipitation of Neodymium-Sodium Sulfate Double Salt. Because of the difficulty of fluoride filtration and the expense and non-selectivity of oxalate production, the precipitation of a Nd-Na sulfate double salt was investigated as an intermediate precipitate which would serve to inexpensively separate rare earths from Fe without a filtration problem. Na_2O_2 was added to acid leach solutions of magnet scrap followed by NaOH to raise the pH to 1.5. A light brown precipitate formed which filtered very easily. Although this precipitate has not been completely characterized, it is probably a form of Nd-Na sulfate double salt. Such double salts are very common although the composition is variable, and few have been adequately characterized. Double Na sulfates are listed having formulas such as $Nd_2(SO_4)_3 \cdot Na_2SO_4 \cdot 6H_2O$, $Nd_2(SO_4)_3 \cdot 3Na_2SO_4 \cdot 126H_2O$, or $NaNd(SO_4)_2 \cdot nH_2O$ where n is 0 or 1 (17,18,26). The precipitate was analyzed by ICP to contain 35.3% Nd and only 0.3% Fe representing metal recoveries of 99.5% and 0.3% for Nd and Fe, respectively. It is apparent that an effective separation was made between Nd and Fe.

The Nd-Na double salt precipitate was leached both with HF and with oxalic acid to see if pure fluoride and oxalate products could be generated. In the fluoride test, a final fluoride product was obtained which analyzed 65.4% Nd and 0.2% Fe, corresponding to overall metal recoveries of 98.2% and 0.12% for Nd and Fe, respectively. Filtration was not a problem. In the oxalate test, the final residue contained 37.2% Nd and 0.1% Fe, which corresponded to Nd and Fe recoveries of 94.6% and 0.13%. The weights of Nd and Pr in the final products indicate that nearly pure fluoride and oxalate precipitates were obtained. It is evident that forming the double salt precipitate which filters very well may be the most efficient way to remove the bulk of the Fe. Corrosion of operating equipment caused by the presence of fluoride ion would also be minimized since the Fe-rich, spent solution remaining after precipitation would contain no fluoride.

The double salt precipitate could then be treated in a variety of ways. Despite the expense of oxalate precipitation, Nd oxide is desired for some applications, and this scrap treatment method is advantageous in that it can generate either fluoride or oxide products in amounts corresponding to industrial demand.

Precipitation of Iron Jarosite. Following rare-earth precipitation, the spent leach solution which still contains as much as 77 g/L Fe, must be treated by an Fe removal process prior to disposal. Although evaporation would be one means of treating these solutions, precipitation of the Fe has the potential to conserve water and decrease the total volume of discharge from the process.

Jarosite precipitation was investigated as a method for Fe removal because it is well understood and widely used in industry. Sodium peroxide was added to the spent leach solution to serve as an oxidant and as a source of sodium ion, after which the solution was placed in a cylindrical, pyrex reactor through which air was bubbled. The temperature in the reactor was $90^{\circ}C$. A typical yellow jarosite precipitate formed which was easily removed by filtration. In addition to its filterability, a pure jarosite product is desireable because it can either be readily disposed of or converted to Fe oxide for use in paint pigments and Fe manufacture (27). It is apparent that jarosite precipitation is a viable method for Fe removal and possible recycle.

Summary

Selective oxidation with a controlled hydrogen-water vapor gas mixture successfully oxidized the Nd in NdFeB magnet scrap ribbon material while leaving Fe in metallic form. Separation of the Nd_2O_3 from the Fe phase, however, was not found to be possible with dilute acid leaching or with magnetic techniques. Apparently the oxidized phase and the metallic Fe phase are too intimately mixed to be easily separable.
Conditions for leaching magnet scrap have been effectively determined. Sulfuric acid dissolution is easily accomplished and inexpensive, requiring neither heating nor agitation. It was determined that an acid weight two times that of the scrap weight is needed for complete dissolution.

Several direct precipitation methods were examined. Pure but difficult to filter fluoride products can be generated by using 1.5 times the stoichiometric amount of fluoride ion. Both HF and NaF are efficient sources of fluoride ion. NH_4F is nearly as efficient. Pure oxalate products can also be generated from acid leach solutions but are probably too costly to be widely used.

A double salt of sodium and rare earths was produced which shows promise as an intermediate product from which pure oxalate and fluoride products can be obtained in proportions required by industry. Precipitation of the sodium-rare earth salt is economic, has no attendant filtration problem, and efficiently separates Fe from the rare-earth constituents. Fe removal from spent leach solutions can be accomplished through precipitation of a potentially useful jarosite product.

References

1. Jacqui Robbins, "Magnetic Materials Update," Materials Edge, Sept/Oct (1989), pp. 54-60.

2. James B. Hedrick and David A. Templeton, "Rare-Earth Minerals and Metals," BuMines Minerals Yearbook 1988, v. 1, 11 pp. (preprint).

3. James B. Hedrick,"Rare-Earth Metals," Sec. in BuMines Mineral Commodity Summaries 1989, pp. 128-129.

4. J. D. Livingston, "The History of Permanent-Magnet Materials, "Journal of Metals, Feb (1990), pp. 30-34.

5. I. V. Mitchell, ed., Nd-Fe Permanent Magnets: Their Present and Future Applications (New York, NY: Elsevier Applied Science Publishers, 1985).

6. Joseph G. Cannon, "Rare Earths," Eng. and Min. J., March (1983), pp. 136-137.

7. Jacqui Robbins, "Magnetic Materials Update, "Materials Edge, Sept/Oct (1988), pp. 51-52.

8. H. R. Kirchmayr, "Physical Properties of $R_2Fe_{14}B$-Based Alloys," Handbook on the Physics and Chemistry of Rare Earths, ed. K. A. Gschneidner, Jr. and L. Eyring (Amsterdam: North Holland, 1970), pp. 71-132.

9. "EEC Attraction for Rare Earth Magnet Technology,"Materials Edge, May/June(1988), p. 7.

10. Metals & Minerals Annual Review, Ed. Roger Ellis, Chris Hinde, et al., The Mining Journal Limited, 60 Worship Street, London EC2A 2HD, 1989, pp. C83-C84.

11. Robert D. Pehlke, Unit Processes of Extractive Metallurgy (New York, NY: American Elsevier Publishing Company, 1973), pp. 56-60.

12. F. H. Froes, "Rapidly Solidified Materials, "Journal of Metals, Sept (1988), pp. 21-23.

13. W. Tang, S. Zhou, and R. Wang, "An Investigation of the Nd-Rich Phases in the Nd-Fe-B System," J. Appl. Phys., 64(10)(1988), pp. 5516-5518.

14. R. K. Mishra, "Melt-spun Nd-Fe-B Magnets and the $Nd_{1+\epsilon}Fe_4B_4$ Phase," J. Appl. Phys., 64(10)(1988), pp. 5562-5564.

15. Y. Nakagawa et al., "High-Field Hysteresis Loop of Nd-Fe-B Magnet and Related Materials," IEEE Transactions on Magnetics, Mag-23(5)(1987), pp. 2530-2532.

16. S. Vijayan et al., "Rare Earths: Their Mining, Processing, and Growing Industrial Usage," Mining Eng., Jan (1989), pp. 13-18.

17. I. M. Kolthoff and P. J. Elving, eds., Treatise on Analytical Chemistry, Vol. 8(New York, NY: Interscience Publishers, 1963), pp. 1-146.

18. F. A. Cotton and G. Wilkinson, Advanced Inorganic Chemistry, 3rd Ed.(New York, NY: Interscience Publishers, 1972), p. 1067, p. 890.

19. S. S. Jaswal et al., "Electronic Structure and Surface Reactivity of $Nd_2Fe_{14}B$ and Related Compounds," J. Appl. Phys., 64(10)(1988), pp. 5577-5579.

20. "The Lanthanide Series," Comprehensive Inorganic Chemistry, Vol IV, ed. Sneed and Brasted (Princeton, NJ: D. VanNostrand Co., Inc. 1955), pp. 151-187.

21. B. J. Beaudry, P. E. Palmer, and K. A. Gschneidner, Jr., "Process Variables in the Preparation of High Purity Rare Earth Metals," Rare Earths, Proceedings of a Symposium on Rare Earths, Extraction, Preparation, and Applications,ed. R. G. Bautista and M. W. Wong, TMS, Las Vegas, Nevada 1989, pp. 187-191.

22. B. J. Beaudry and K. A. Gschneidner, Jr., "Preparation and Basic Properties of the Rare Earth Metals," Handbook on the Physics and Chemistry of Rare Earths, ed. K. A. Gschneidner, Jr. and L. Eyring (Amsterdam: North Holland,1978), pp. 173-232.

23. K. A. Gschneidner, Jr., "The Production, Quality Control, and Applications of Pure RE Metals and Alloys," Analysis and Application of Rare Earth Materials, ed. Odd B. Michelsen (Kjeller, Norway: Nato Advanced Study Institute 1972), pp. 351-358.

24. Du Senlin Tang Dingxiang, "Some Progress in Study on Preparation of Rare Earth Metals and their Alloys by Fused Salt Electrolysis," New Frontiers in Rare Earths Science and Applications, Proceedings of the International Conference on Rare Earth Development and Applications, Beijing, The People's Republic of China, Sept 1985, pp. 1117-1126.

25. Donald R. Sadoway, "Fused Salt Electrolysis of RE Metals," Rare Earths, Proceedings of a Symposium on Rare Earths, Extraction, Preparation, and Applications, ed. R. G. Bautista and M. W. Wong, TMS, Las Vegas, Nevada 1989, pp. 345-353.

26. L. Niinist and M. Lesk la, "Inorganic Complex Compounds II," Handbook on the Physics and Chemistry of Rare Earths, ed. K. A. Gschneidner, Jr. and L. Eyring (Amsterdam: North Holland, 1978), pp. 91-320.

27. V. Arregi, A. R. Gordon, and G. Steintveit, "The Jarosite Process - Past, Present, and Future," Lead-Zinc-Tin '80, ed. J. M. Cigan, T. S. Mackey, and T. J. O'Keefe, AIME, New York 1979, pp. 97-123.

Introduction

Who Is Brush Wellman?

Brush Wellman is the only fully integrated supplier of beryllium products in the free world. Primary beryllium is produced as beryllium hydroxide at the company's mineral processing mill in Delta, Utah. The principal source of feed for this mill is bertrandite ore from Brush Wellman's mines in the Topaz Spor Mountain region of Utah. These mines and the mill are capable of supplying the needs of the beryllium industry well into the next century. The Utah mill is also equipped to process beryl, a high-grade beryllium mineral, which can be purchased from a number of foreign and domestic suppliers. These beryl sources are a by-product from the production of other minerals or gems.

Beryllium hydroxide is shipped from Utah to the company's processing facilities at Elmore, Ohio, where it is converted into beryllium metal, beryllium oxide powder for beryllia ceramics and beryllium alloys, primarily beryllium copper.

Beryllium is a unique strategic material. It is one of the lightest structural materials known having a density that is one-third less than aluminum, yet it has a specific stiffness six times greater than steel. Beryllium has the highest heat absorbing capacity of any metal and is the best heat conductor, pound for pound, of the structural metals. Beryllium is transparent to X-rays, has excellent acoustical properties, and is readily machinable.

Applications for beryllium are in aircraft and satellite structures, guidance systems components, X-ray windows, audio components for home entertainment systems, optical platforms, mirrors and space telescopes.

Beryllium oxide ceramics have the highest thermal conductivity and specific heat of any oxide ceramic. At 100°C, beryllium oxide ceramics have a thermal conductivity about equal to aluminum and ten times that of aluminum oxide. Beryllium oxide ceramics are excellent electrical insulators and exhibit desirable dielectric properties. Beryllium oxide can be fabricated, using standard fabrication procedures, into strong rigid ceramics which are chemically stable. Applications for beryllium oxide ceramics include automotive electronics, microelectronic substrates, laser tubes, traveling wave tubes, aircraft engine ignitions, microwave devices and transistor bases.

High performance beryllium-containing alloys are the largest segment of Brush Wellman's beryllium business. Beryllium-nickel and beryllium-aluminum alloys are produced, but by far the most important alloy is beryllium copper. Small amounts of beryllium (0.3 to 2.0%) impart to copper, known for its formability and electrical conductivity, the additional characteristics of high strength, hardness, corrosion resistance, and wear and fatigue resistance. These unique combinations make beryllium copper the alloy of choice for electronic connectors, switches, springs, and relays. Reliability and corrosion resistance are key factors in selecting beryllium-copper alloy components for hostile environments, such as: under the hood for automotive, under the sea for cable repeater housings, and down-the-hole sensors for petroleum exploration and production.

The Resource Recovery Process and Facility

In the manufacture of beryllium metal, beryllium oxide ceramics and beryllium alloys, a number of beryllium-containing by-products and wastes are generated. Collectively, these materials contain significant amounts of beryllium and copper. The recycle of these metal values is economically as well as environmentally desirable. In fact, it has been a longstanding policy of Brush Wellman to design, manufacture, and distribute all products and dispose of all by-product materials in a safe, environmentally sound manner.

By far, the largest volume of by-products is generated from the production of beryllium copper. These by-products include spent caustic and acid pickling solutions, furnace drosses, mill scale, alloy melting skulls and ventilation collector dusts. The initial focus of our resource recovery effort was to evaluate several processing options for the recycle of beryllium and copper from alloy by-products. The process which seemed to hold the most promise was carbothermic reduction in an arc furnace.

Carbothermic reduction is used to produce a 4% beryllium-copper master alloy. The process is well known and has been practiced for more than 30 years. During this reaction, beryllium oxide is reduced to beryllium metal with carbon as a reductant. The reaction takes place in a molten bath of copper which acts to collect the beryllium and thermodynamically "drive" the reaction by keeping the activity of beryllium metal low. This reaction is performed in a carbon arc furnace. Unfortunately, a number of problems arose when plant equipment was used to recycle alloy by-products in the arc furnace. The three most critical problems were: capacity with the existing equipment, quality of the beryllium copper product and cost. After this brief plant test, carbothermic reduction as a primary method of recycle was abandoned. The arc furnace is used on a limited basis to recycle the cleaner drosses and collector dusts. However, a new method was needed to handle the more impure by-products.

The elimination of the arc furnace as a recycle option left the company with a building inventory of feed material for an unknown recycling process. The feed that was accumulating the fastest was a copper oxide/hydroxide precipitate produced from caustic treatment of spent acid pickle liquor. In reviewing the known process options, it was determined that these precipitates should be amenable for processing by electrowinning the copper from a sulfuric acid electrolyte followed by solvent extraction of the beryllium.

Electrowinning is a standard technology for the production of copper. To assist in designing a process, Bechtel Civil & Minerals Inc. of Scottsdale, Arizona was hired to design a copper electrowinning plant. The plant and process were designed based on experience in the copper industry. Typical conditions for copper electrowinning are given in Table I. These conditions were simulated in laboratory and pilot plant tests in the R & D facility of Brush Wellman.

Table I Typical Conditions for Copper Extraction

Rich Electrolyte:	45 gpL Cu
	30 gpL H_2SO_4
Lean Electrolyte:	25 gpL Cu
	70 gpL H_2SO_4
Temperature:	55 C
Anodes:	Lead Alloy
Cathodes:	Cu Starters
Current Density:	20 Amps/ft^2
Voltage:	2.5

In 1986, the first phase of an on-site resource recovery facility--an 11-cell copper electrowinning system and a beryllium solvent extraction circuit--was built to recycle beryllium and copper values from the various by-product streams generated by Brush Wellman.

The system was designed for 2,000,000 pounds of copper per year, 7 days per week operation, and 50,000 pounds of beryllium, 5 days per week operation.

The copper electrowinning operation was designed with internal recirculation of the electrolyte between the solids dissolver and the electrowinning cells (see figure 1). This gives the process a number of advantages:

(1) The amount of sulfuric acid used was minimized by recycle of the lean electrolyte.

(2) The process yield is maximized by minimizing the amount of lean electrolyte bleed.

(3) The beryllium content of the bleed is increased. This results in higher beryllium yield and minimizes the size of the beryllium extraction plant.

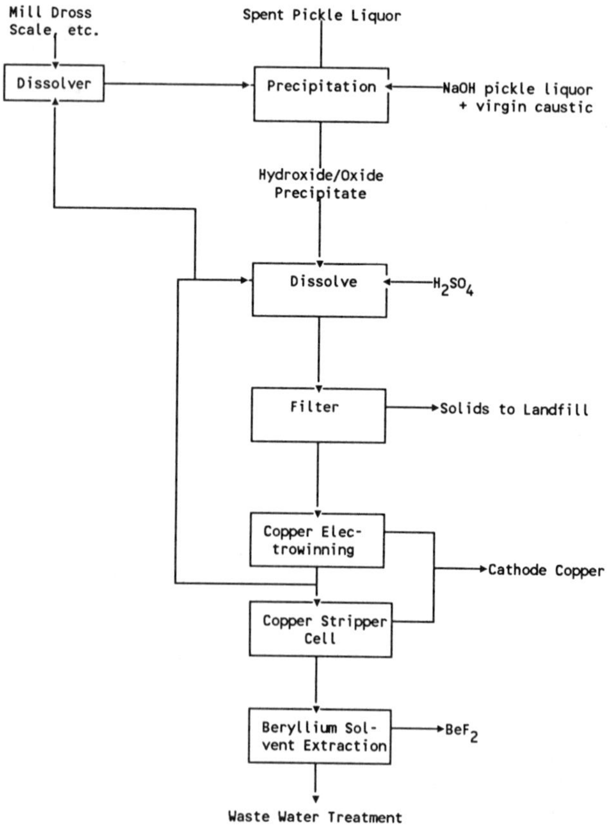

Figure 1 - Resource Recovery Flow Sheet

The resource recovery facility was further expanded in 1989 to recover copper and beryllium values from furnace drosses, mill scale, and alloy melting skulls. Provisions were also made to recover beryllium values from grinding swarf, machine swarf and etching solution wastes generated during beryllium and beryllia manufacturing operations. Metallized beryllia scrap can also be processed. The process flow sheet, which now integrates all beryllium-containing by-product and waste streams, except for ventilation collector dusts, is shown in figure 2.

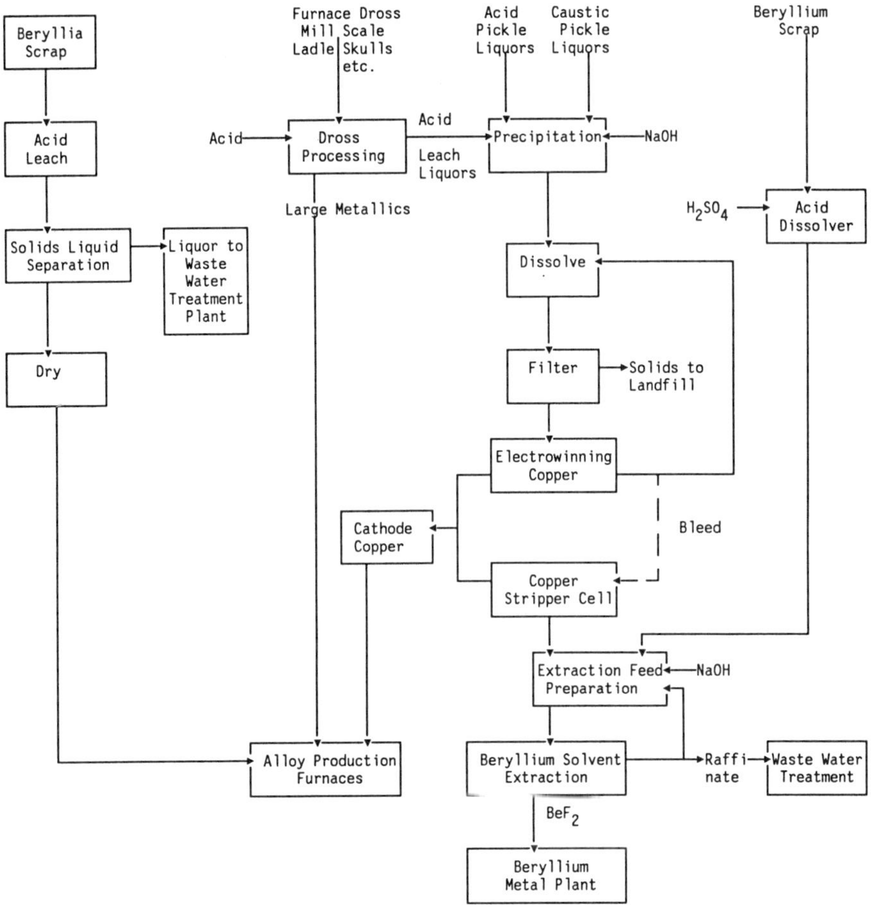

Figure 2 - Resource Recovery Flow Sheet After 1989 Expansion

Our Board of Directors recently approved a capital expansion which will allow us to process ventilation collector dusts, as well. We anticipate having this last portion in place the first quarter of 1991.

The Recycle Feedstreams

Spent acid pickle liquor is produced from the acid pickling of wrought beryllium copper. Pickling, usually done with nitric acid, improves the surface finish of the alloy by removing the outer layer of beryllium and copper oxide. The spent pickle liquor is neutralized with caustic to precipitate an impure copper oxide/hydroxide.

Furnace dross is the oxide-containing skimmings which are removed from molten alloy prior to casting. These oxides are removed so that they won't contaminate the alloy with oxide inclusions.

Mill scale is formed principally as a result of high-temperature treatment of beryllium copper alloys in an oxygen-containing atmosphere. The melting skulls are left behind in equipment such as a ladle when the molten alloy is poured. Mill scale and skulls compositionally are similar to dross.

Collector dusts are recovered from the ventilation equipment installed to protect the manufacturing workers while they handle molten beryllium-containing materials. Typical analyses of these various beryllium copper by-product feeds are contained in Table II.

Table II Analysis of Typical By-Product Feeds

A. Copper Hydroxide Precipitate from Spent Nitric Acid Pickle Liquor

Loss in drying	23.9%
Copper	51.2
Oxygen	16.0
Beryllium	0.78
Sodium	1.10
Nitrate	2.57
Silicon	0.36
Calcium	0.10
Magnesium	0.13
Iron	0.10
Chloride	0.10
All others, each	<0.10

B. Furnace Dross/Mill Scale/Skulls

Copper	80-88%
Beryllium	3-4%
Oxygen	6-12%
All Others, each	<1.0%

C. Collector Dusts

Copper	30-35%
Beryllium	10-12%
Oxygen	22-35%
Carbon	3-4%
All Others, each	<2%

Beryllia ceramics and the production of beryllium metal also generate by-products. The most important of these are grinding and machining swarf from both beryllium and beryllia processing, etching solutions from beryllium processing, and metallized scrap from ceramics.

Machine chips, saw cuttings, grinding and polishing residuals are liberated from both beryllium and beryllia during the manufacturing process. When possible, these by-products are captured without contamination and recycled directly back into the manufacturing process. This direct recycle could be as simple as a remelting process for beryllium or a dry reformation process for beryllia. It is when these by-products are contaminated with other metals or materials that they become feeds for resource recovery.

Acid etching solutions are generated when beryllium is chemically machined. Usually the etching solution is a mixture of hydrofluoric and nitric acids. Occasionally, sulfuric or hydrochloric acid is also used.

Virtually all beryllia ceramic is metallized, i.e., partially or completely coated with metal. The most common metals are molybdenum and manganese which are silk screened on as a metal paste and then fired in a furnace to effect a tenacious bond. After screening, it is common to electroplate and/or solder other metals such as nickel, copper, etc. to the metallized part. These metals preclude the direct recycle through ceramic routes, thus metallized ceramics are processed as resource recovery feeds.

These by-products, based on their characteristics, presented unique processing problems which needed special solutions to integrate them into the resource recovery facility. Also, the beryllium content of these by-products create a unique environmental challenge.

<u>How Beryllium Makes the Process Unique</u>

Beryllium and its compounds can be a health hazard when inhaled as respirable dust (below 10 μm size). The physiological reaction to beryllium is an immunological response which is similar to an acquired allergy. The majority of the people, perhaps as many as 99%, do not seem to react adversely to exposure at any level. The hazard is confined to inhalation, as there are no known ill effects resulting from other forms of contact or ingestion.

Beryllium and beryllium compounds can be safely handled with proper controls on airborne emissions. The in-plant limits established by OSHA are: (1) 2 $\mu g/m^3$ of air as an 8-hour time-weighted average exposure, and (2) 25 $\mu g/m^3$ of air maximal 30-minute exposure. The U.S. Environmental Protection Agency has set a nonoccupational ambient air standard at a monthly average concentration of 0.01 $\mu g/m^3$ of air.

These limits, although strict, are routinely maintained at Brush Wellman while processing tons of fine-powdered beryllium and beryllium compounds annually.

In order to control beryllium emissions during copper electrowinning operations, a novel design feature of the copper recovery plant was the addition of custom-built ventilation hoods above the electrowinning cells. These ventilation hoods were necessary to insure that beryllium-containing "acid mists" did not escape during the winning of copper from the beryllium-containing copper electrolyte. Acid mists are produced as a result of the oxygen gas, which is liberated from the anode while the copper is plated at the cathode. In normal operations, a layer of "ping pong" balls is floated on the top of the electrowinning cells to control this mist and produce an acceptable working environment. However, this normal provision would not be adequate to protect operators from the beryllium-containing electrolyte. To insure a safe working environment above the electrolysis cells, hoods were installed so that workers would be protected from any potential beryllium-containing acid mist even if every second hood was removed. This design was necessary to harvest the copper without shutting the entire operation down. Figure 3 shows the copper harvesting process. Ventilation hoods cover the cells on either side of the cell being harvested.

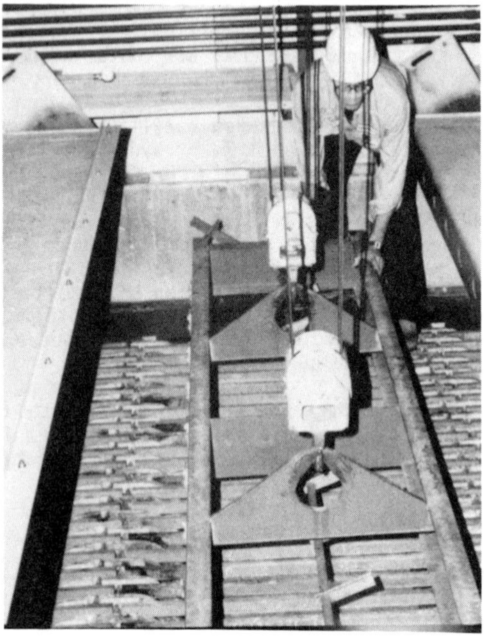

Figure 3 - Copper Harvesting Process

Shown above is the overhead crane with strongback attachment being used to lift the copper cathodes from the cell. Ventilation hoods are shown in place on either side of cell being harvested.

<u>The Challenge of Nitrate and Other Impurities</u>

The internal recirculation of electrolyte provided several advantages, but also posed several challenges as impurities accumulated in the recirculating electrolyte. One of these impurities, which we knew would be troublesome, was nitrate.

The target nitrate concentration in the electrolyte was 0.5 gpL maximum. This was established by laboratory testing to maximize the yield of copper and beryllium without sacrifice of quality products.

Excess nitrate in the acid electrolyte would cause a competing pickling reaction to occur in the electrolysis cell. This would result in a loss of cathode current efficiency. In addition, the presence of nitrate requires more costly materials of construction. Two parallel R & D programs were started to resolve the problem of the nitrate in the electrolyte. One program concentrated on removing the nitrate from the electrolyte. The second approach investigated what provisions could be made to tolerate the nitrate.

Four processes were investigated to remove nitrate. These were:

(1) Reduction of NO_3^- with a reducing agent.

(2) Distillation of nitric acid from the electrolyte solution.

(3) Calcination of copper hydroxide solids to thermally decompose the nitrate.

(4) Washing the soluble NO_3^- from the hydroxide precipitate.

Reducing agents proved to be too slow, too costly or undesirable additives to the electrolyte.

Distillation worked well but the thermoenergy required made the process too costly.

Thermal decomposition of the nitrate in the copper hydroxide precipitate by calcination was effective in removing the NO_3^-, but presented a number of health and safety problems due to airborne beryllium.

Washing the entrained nitrate from the precipitate appeared to be the most attractive removal route. The precipitated copper hydroxide does not filter well. Therefore, centrifuge dewatering was tested. Pilot testing determined that four stages of countercurrent centrifuge washing would insure an electrolyte containing less than 0.5 gpL NO_3^-. The installation and operation of this washing system would be costly.

The parallel R & D effort to design a nitrate tolerant plant proved to be the best option. The key to this design was to find a long-lived anode which would not contaminate the copper product. Lead anodes, which are the standard for copper extraction plants, are not nitrate tolerant. The conductive oxide film which forms on lead anodes spalled as a result of nitrate attack of the lead substrate. This spalled lead would contaminate the cathode copper and anode life was very poor as a result of this attack.

An anode which suited our needs was the DSA, or dimensionally stable anode, used primarily in the production of hypochlorite and caustic from sodium chloride brine. A DSA is produced by baking a precious metals/precious metals oxide coating of platinum and iridium on the surface of a titanium sheet. The precious metal coating is the active surface of the anode; the titanium forms on "inert" substrate. DSA's had been tested in electrowinning of copper and were already in use to produce high quality electrolytic copper deposits for sheet or foil applications.

For our application, we were interested in the corrosion resistance of the DSA toward nitrate. An extensive testing of the DSA, followed by engineering and economic evaluation, resulted in the selection of the DSA instead of the four-stage washing plus lead anode system.

Accelerated corrosion tests indicated an acceptable four-year life for the DSA's if the following conditions were maintained:

(1) F^- no greater than 80 ppm.

(2) H_2SO_4 no greater than 100 gpL.

(3) A minimum of 1.0 gpL Al in solution.

These conditions were maintained by carefully controlling our feeds and the process, and by adding $Al_2(SO_4)_3$ periodically to the system.

Results

The resource recovery plant was started in December 1986. Copper production has been continuous with the exception of two planned shutdowns. The beryllium extraction process started in April of 1987, along with the leaching process for beryllium and beryllia. In July of 1989, we started the dross processing part of the facility.

During the first two full years of operation, 1,700,000 pounds and 1,800,000 pounds of copper were produced. The capacity of the plant was 10-20% below expectation. Low current efficiency, due, in part, to nitrate contamination of the electrolyte is one of the contributing factors to this loss in capacity. As described before, nitrate causes a pickling reaction to occur which competes with the desired plating reaction.

Cathode copper quality is a factor which we are continuously working to improve. The quality of our product, as compared to specification for cathode copper, is shown in Table III.

Table III Typical Chemical Analyses of Recycle Copper Compared To Cathode Copper Specification

Element	Cathode Copper Specification	Typical Analysis Recycle Copper
Copper	99.95%, min.	--
Iron	100 ppm, max.	<10 ppm
Tin	70	<10
Zinc	50	<10
Chromium	10	<10
Lead	30	<10
Oxygen	600	740
Sulfur	20	80
Total of all other trace elements	0.020% max., no single element to exceed 0.010% max.	Silicon 30 ppm

The yield of recovered copper from this process is 93% which is slightly below the expected 95%. The rate of impurity bleed from the electrolyte was increased to improve product quality. This resulted in the additional 2% decrease in yield.

Beryllium extraction has operated according to design and expectations. A small decrease in yield resulted from the increased copper circuit bleed which serves as feed for the beryllium circuit. This resulted in the overall process yield decreasing from a designed 97.5% to 95%. Nominally, 50,000 pounds per year of beryllium is being recycled to beryllium products. The planned 5-day a week operation schedule has been maintained.

With the addition of dross processing in July of 1989, our capacity increased to 3,000,000 pounds of copper. In the few months that we operated in 1989, we demonstrated this increased capacity.

The recycle facility operated within OSHA limits for airborne beryllium from the day it started. Extensive air sampling proved the effectiveness of the hoods in controlling emission from the electrowinning cells. As a side benefit, there is no acid smell or mist-generated corrosion in the electrowinning building. The working environment is completely safe and decidedly more pleasant than that in a typical copper electrowinning operation. By traditional ROI calculations, there was little economic justification for this recycling effort. This investment was made because of Brush Wellman's commitment to responsibly face the challenges posed by environmental disposal issues. Alternate disposal options are expensive and pose a significant, potential long-term environmental liability. Brush Wellman's policy to dispose of all by-product materials in a safe, environmentally sound manner is the fundamental reason this plant was built. The economics of the operation have been as expected, or perhaps slightly better than expected, primarily because the price of copper has been higher than projected.

When the plant started in 1986, there were three million pounds of copper inventory as pickle liquor-derived precipitates in inventory. There is now less than one million pounds of copper remaining. The process has handled the ongoing generation of pickling by-products from the Elmore plant, while reducing this process inventory. Furnace dross processing began in July of 1989. The processing of collector dusts is scheduled to start in 1991. Our inventory of feeds will be depleted and we will need additional feeds to keep the plant operating at capacity. We are beginning the process of filling this available capacity by carefully evaluating potential feeds produced by other companies. We recently completed a market study which indicates that it is possible to maintain full capacity by offering a cost-effective alternative to disposal for other companies who process copper-based alloy products. We believe that, in time, as disposal costs and regulatory pressures increase, we can turn our resource recovery facility into a profitable business.

MIXED MATERIALS

Session Chairman:
Richard G. Sandberg

BASIC CONSIDERATIONS ON POST-USE PROCESSING

OF DURABLES FOR MATERIAL RECYCLING

Manfred Poetzschke, Reinhard Pfaff, and Hans-Peter Sattler

Metallgesellschaft Aktiengesellschaft
Reuterweg 14
D-6000 Frankfurt a. Main
F. R. Germany

Abstract

Non-ferrous metals and steel scraps from durables - consumer and capital goods - after use and service are a traditional source of the secondary metals industry. Because of the increasing use of a variety of engineered thermoplastics and non-metallics - glass, ceramic, mineral materials etc. - these metals are mostly part of metal-plastic-composite components. For the recovery of metals and plastics from these used goods improved or adapted technologies for disassembling, liberation and separation in different fractions are as indispensable as for a modified strategy of waste management. Actual demands are the prevention prior to reduction prior to reutilization prior to final disposal of waste as well as the recycling of valuable materials - metals and plastics - and their reutilization in high-level applications. Basic considerations and R & D requirements are discussed.

Second International Symposium -
Recycling of Metals and Engineered Materials
Edited by Jan H. L. van Linden,
Donald L. Stewart, Jr., and Yogeshwar Sahai
The Minerals, Metals & Materials Society, 1990

1. Introduction

In the Federal Republic of Germany, landfills and incineration plants for wastes are extremely scare. That is why an imminent waste emergency is anticipated. At the International Recycling Congress held in Berlin in 1989, already the thesis was brought up that the lack of waste treatment capacity will become the new shortage factor of our economy (1). This would then be the fourth production factor after capital, labour and land.

Waste management and waste handling strategies as the main items of environmental policy require clearly outlined terms and definitions, just like any other political activity.

How can **"waste"** and **"secondary raw material"** be integrated in the recycling process, the management of material flows becoming a component of future economic management?

Today's durables - capital and consumer goods - are of a complex nature. They may be defined as

metal - non-metal/plastic - composite components

and described as a function of

- material selection and combination
- design/mode of construction
- fabrication methods applied.

All three aspects are of paramount relevance to the environment with regard to the life cycle of the products comprising manufacture, use and disposal. From the waste management point of view, only residues or wastes burdening the environment and resources are significant in the first place (2):

Manufacture of goods:	production residues and wastes
Use of goods:	consumption and infrastructure wastes
Disposal of goods:	secondary raw materials and material for landfill dumping.

From this aspect, the service properties are of secondary importance.

Undoubtedly, the sector having the longest recycling tradition is the metal-producing industry. From the EUROPEAN ASSOCIATION OF METALS - EUROMETAUX the following definitions have been established (3) (4):

Waste:

any material or object which has no foreseeable future use and which is intended for final disposal.

Secondary raw material:

any material intended for recovery or recycling of its metal content or other recoverable components.

The BUREAUX INTERNATIONAL DE LA RECUPERATION - BIR found its own definitions for the same purpose (5):

Waste:

redundant goods, by-products or residues which have no value and must be disposed of at a cost.

Secondary raw material:
redundant goods, by-products of residues which have a value and are recycled for further use.

No matter what the agreements to be reached in Europe are going to look like, definitely durables after use or their components are not to be regarded as waste in the indentment of the waste disposal legislation. On the contrary, they are residues containing reutilizable materials which means **economic goods** from which secondary raw materials can be recovered by purposeful reprocessing.

It should be taken into account that the recycling technologies themselves are subject to scientific technical progress.

2. Material Recycling as a Waste Management Concept

The guidelines governing waste management policy in 1990 propagate the following objectives:

- Prevention prior to reduction prior to reutilization prior to final disposal of waste.

Strategic elements have been defined as follows, for instance:

- Multiple use of raw materials - **cycles of reutilizable materials**
- Quality-graded use of secondary raw materials - **reutilization at the highest possible level**
- Production and product cycles - **provision for and organization of return and utilization of products after use and service.**

For the reutilization and disposal of goods, a **sequence of priority for recycling and waste management** can be defined following the criteria of quality-graded re-use of secondary raw materials at the highest possible value level:

• Further or re-utilization as an exchange component	Product reutilization
• Recovery of inherent materials by reworking using physical processes (e.g. thermal reforming), possible value increase by additions or mixture • Chemical conversion into reutilizable basic materials (e.g. pyrolysis, hydrogenation) or use as a chemical (e.g. reductant)	Material recycling for prevention/reduction of waste
• Thermal exploitation as a fuel	Waste reutilization
• Final disposal as domestic/industrial refuse • Final disposal as hazardous waste	Landfill

Characteristic of this sequence is the topdown decrease of value addition or abandoning of a previously reached quality grade. The quantities and complexity of wastes are increasing.

If a product or component is re-used, as an exchange motor for instance, the value addition to the original component is conserved. In this case, both functions and materials are reutilized. The economic benefit is optimum, also with an eye on the production factors invested, namely "labour" and "capital".

Material recycling based on physical processes largely requires homogeneous secondary raw materials. If these are metallic, they may be supplied as feed materials to the primary or secondary smelter industry. Polymer materials if they are mostly homogeneous can be utilized as quality assured re-compounds with or without improvement by suitable additives in accordance with stipulated material specifications (6)(7)(8). In that sense, they are equivalent to virgin polymer materials. Mixed polymer scrap is restricted to specific applications; it competes with polymer mixtures prepared from domestic or industrial refuse. Reutilization of other non-metallic materials from durables has not played a significant part so far.

Material recycling calls for adequately adapted technologies (cf. Section 4). Energy is needed, but the energy originally supplied for primary production is not lost. Auxiliaries and utilities may contribute to polluting the enviroment, and residues and wastes remain. Also the latter have to be minimized.

Reutilization by purely chemical conversion may lead to economic goods which are utilizable otherwise. Examples are: contaminated aluminium scraps to be used as steel deoxidants, metal salts as leaching products or pyrolysis oils from polymer mixes.

Thermal exploitation of combustible materials - often described as "energy recycling" - ranks equal to chemical reutilization, at least at the end of a reutilization chain. It has the character of a "final solution" of material utilization prior to "non"-utilization, namely final disposal.

Chemical material conversion and also incineration often encounter lack of acceptance because of the fear of emissions and problematic residues. The licensing procedures in Europe are therefore very elaborate and time-consuming.

3. Material Flows in a Recycling Economy

In a producing economy it is essential that preferably all valuable materials should be recycled as secondary raw materials. Thus, material flows can be defined in which the term "recycling" may have different meanings (Fig. 1):

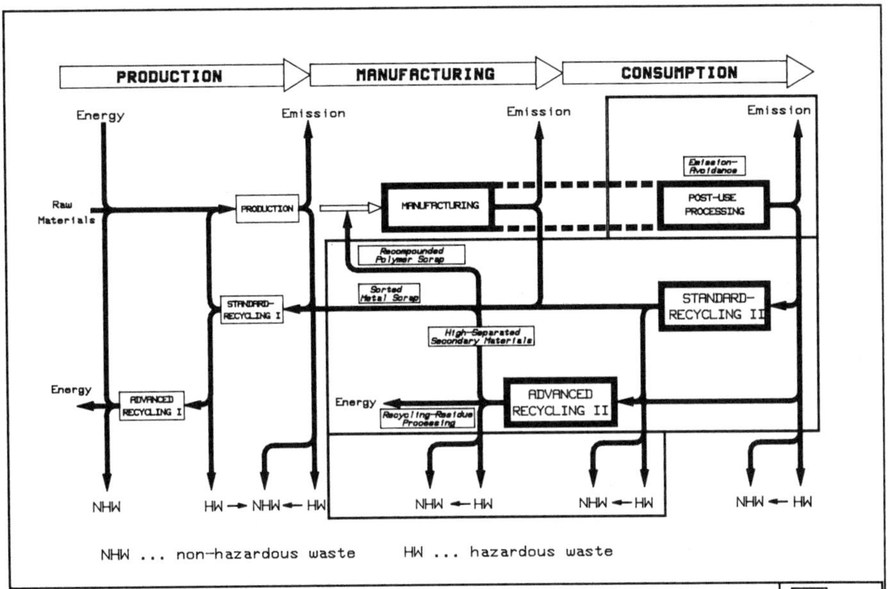

Fig.1 : Flow Diagram for the Recycling of Materials

- **Production of basic materials:** With an energy input, the so-called basic materials (metals, chemicals, polymers, etc.) are produced from raw materials as well as auxiliaries and utilities **(= primary production).**

- **Manufacturing of goods:** Consumption of basic materials to manufacture industrial and consumer goods.

- **Standard Recycling I:** Residues of basic material production and goods manufacture are recycled as secondary raw materials. This so-called **secondary production** covers metals as well as the reprocessing of spent chemicals and other usable materials.

- **Advanced Recycling I:** The residues from primary and secondary production either have to be disposed of as hazardous waste or converted into re-usable valuable materials and non-hazardous materials suitable for land-filling similar to domestic refuse. In many cases, a waste incineration unit is added for energy recovery. This young, vigorously expanding industry deals with the reprocessing of salt cake from secondary aluminium smelters, for instance, or with zinc-bearing flue dusts from electric steel plants.

- **Disposal and post-use processing of goods:** After solving logistical tasks, used durables have to be prepared for recovery of all reutilizable materials they contain. This includes the environmentally harmless disposal of used oil, coolants and lubricants, FCC-containing foams, etc. and possibly also the specific disassembling of components and structural elements which are routed to specific disposals and utilization cycles (e.g. used accumulators, cables, motors, etc.).

- **Standard Recycling II:** Collective scrap of all kinds, used cars but also large household utensils, etc. are processed according to conventional technologies for exclusive recovery of metals in the secondary production cycle (Standard Recycling I). Techniques applied cf. Section 4.

Not least on account of shortcomings in disposal and recycling control of durables as well as deficiencies in reprocessing technologies, Standard Recycling II still calls for improvement for ecological and economic reasons. As the processing is only geared to recover the metals, the technologies applied are not suitable for recovering plastics, elastomers or other non-metallic materials. Appreciable quantities of potentially hazardous waste are produced which under favourable conditions include materials that can be largely classified as domestic refuse. Adapted technologies are available for specific product groups like cables, used accumulators.

The tasks for Advanced Recycling II are derived from the recognized deficiencies and the demand of secondary smelters for clean and homogeneous scrap. Hence, except for a few initial approaches, this demand is not the state of the art but the objective of world-wide R&D-activities which seems to be within reach.

- **Advanced Recycling II:** By applying gentle liberation and separation techniques, durables - mainly metal/non-metal/plastic composite components (cf. Section 4) - are reprocessed to reutilize metals and non-metallic materials after adequate conditioning. The aim is to obtain homogeneous secondary raw materials. If they are metallic, they are recycled to secondary production. Polymer materials are returned to the goods production cycle. Residues of this process step are utilized chemically or thermally, converted to materials which can be treated like domestic refuse, suitable for landfilling, or added to other cycles of reutilizable materials as economic goods.

Conclusion: In parallel with the efforts to achieve a durable design adequate for disposal and recycling, progress in the development of recycling technologies is essential. Such technologies should focus on current and also future designs and materials. Durables suitable for recycling will not become an appreciable factor of the secondary raw materials market until after one or two decades.

4. Durables as a Secondary Raw Material Source

Durables are traditionally disposed of after use and utilized through scrap trading if they contain appreciable quantities of iron and steel as well as non-ferrous metals. In the past thirty years, however, the variety of materials used has drastically expanded. Non-metallic materials have come up; predominant are composite constructions made up of the most diverse combinations of materials. Preferred design solutions are miniature components and light-weight design. Fabrication processes utilize all rationalization options and ever more new manufacturing techniques have been developed.

Recycling of the valuable materials is a function of the following fundamental prerequisites:

1. Recyclability of the materials used
2. Separation of materials and availability of suitable techniques for this.

The two criteria mentioned are interdependent because material recycling calls for clearly assorted materials.

Recyclability of Materials

Materials can be subdivided according to material classes for instance (Table I). That they belong to one of those selected material classes is not necessarily a guarantee for their recyclability.

Table I Examples of Materials Contained in Durables

Natural materials: Minerals and their aging products		
"Regenerative" bio-materials: wood, natural fibres		
Metals:	Iron and steel	
	Stainless steels	
	Superalloys:	Ni, Co, W, Ta alloys
	Non-ferrous metals/alloys:	Mg, Al, Cu, Zn, Pb, Sn
	Precious metals	
	By-metals and special metals:	Cd, Sb, As, Bi
Inorganic materials:		Glas, ceramics, concrete, clay
Organic materials:		Thermoplastics, thermosets, elastomers
Composite fibre materials:		Sandwich composites, metal matrix-composite materials (MMS)

Metals are recyclable materials by tradition. In most cases, the purity of primary production can be restored by applying adequate secondary refining techniques.

Of the inorganic materials, glass can be recycled at a high value level as it is easy to smelt. High-grade ceramic material such as furnace brick-lining can be crushed and reprocessed by sintering.

Organic materials are offered in a great variety. Today they are the typical example of tailormade materials. On account of the slight differences in density they exhibit, they are extremely difficult to separate and, on the other hand, they cannot be recycled to form high-grade products because of chemical incompatibility in the mixture. Shortage of landfilling space and public resistance to incineration plants thus makes plastics problematic materials for waste management (9). Engineered thermoplastics are very often more expensive than metals referred to the price per weight of new materials. For this reason, it is worth making efforts for material recycling at a high application level (6)(8). In principle, in the case of organic macromolecules, degradation by processing and use has to be taken into bargain. This deterioration is of a cumulative nature and cannot always be compensated for by upgrading. After several recycling procedures they have to be earmarked for a lower-grade application or for chemical or thermal utilization. Thermosets can only be utilized chemically or thermically. Elastomers can be applied as fillers - such as crushed used tires; in most cases, however, they are utilized as auxiliary fuel, in the cement industry for instance.

Composite materials are generally considered extremely problematic for material reprocessing (2). However, one should differentiate here: so far, there have been no reprocessing techniques for fibre glass or carbon-fibre composites on a polymer basis. Metal matrix composites should be recycled by melting down to the matrix metal. Their non-metallic components would then be turned into slag.

Although, naturally, a positive answer to the question of recyclability of individual materials is a prerequisite for material recycling from durables, this alone is not sufficient. The complexity of these commodities - initially characterized as metal - non-metal/plastic composite components - obliges us to critically scrutinize the designs and production methods applied. Material combinations and miniaturizing of components, different connection techniques, non-flammable additives, coatings, etc. influence material recycling.

Preconcentration by Disassembling as a Prerequisite for Material Recycling

Techniques upstream of material recycling are required which in a first analysis separate the composite materials into useful and useless components. Fig. 2 shows a simplified flow diagram breaking the recycling operation down into individual activities.

Today, most durables are utilized mainly as secondary mixed scrap in material streams, by passing some of the levels shown. A typical example is a scrap car: It is collected by the scrap trader, partly dismantled for cannibalizing (e.g. lead accumulator, aluminium rims, coolers, various exchange parts for reuse) and subsequently, shredded. Steel and non-ferrous metals are sent to secondary smelters, the non-metallic materials are landfilled, formerly as industrial waste classified as domestic refuse, now mainly as hazardous waste (10). For other commodities product-specific disposal concepts are available, designed to avoid waste accumulation. The best known example is the lead accumulator, which once collected, is separated from the mixed scrap at an early stage and re-used almost residue free in the secondary metal smelter as lead and in the area of secondary plastic scrap as polypropylene recompound (6)(7).

In general, material recycling at a high value level is extremely difficult if bulk processes are applied to secondary mixed scrap. Hence, future waste management and utilization strategies should be geared to transform the secondary mixed scrap at an early stage into secondary metal scrap and secondary plastic scrap. This should also be done in bulk. It appears that this is the only way to enable also non-metal materials such as homogeneous thermoplastics to be purposefully recycled.

Unit Operations of Processing Technology

Unit operations of processing technology are predominant in the three secondary scrap areas:

Mechanical Processing Methods:

Liberation: (= comminution)	**Mechanical disassembling of composites** - Metal/Metal - Metal/Non-metal - Non-metal/Non-metal.	Decomposition by hand (dismantling) Scrap scissors, Torch-cutting, Shredder Impact crusher, Granulator

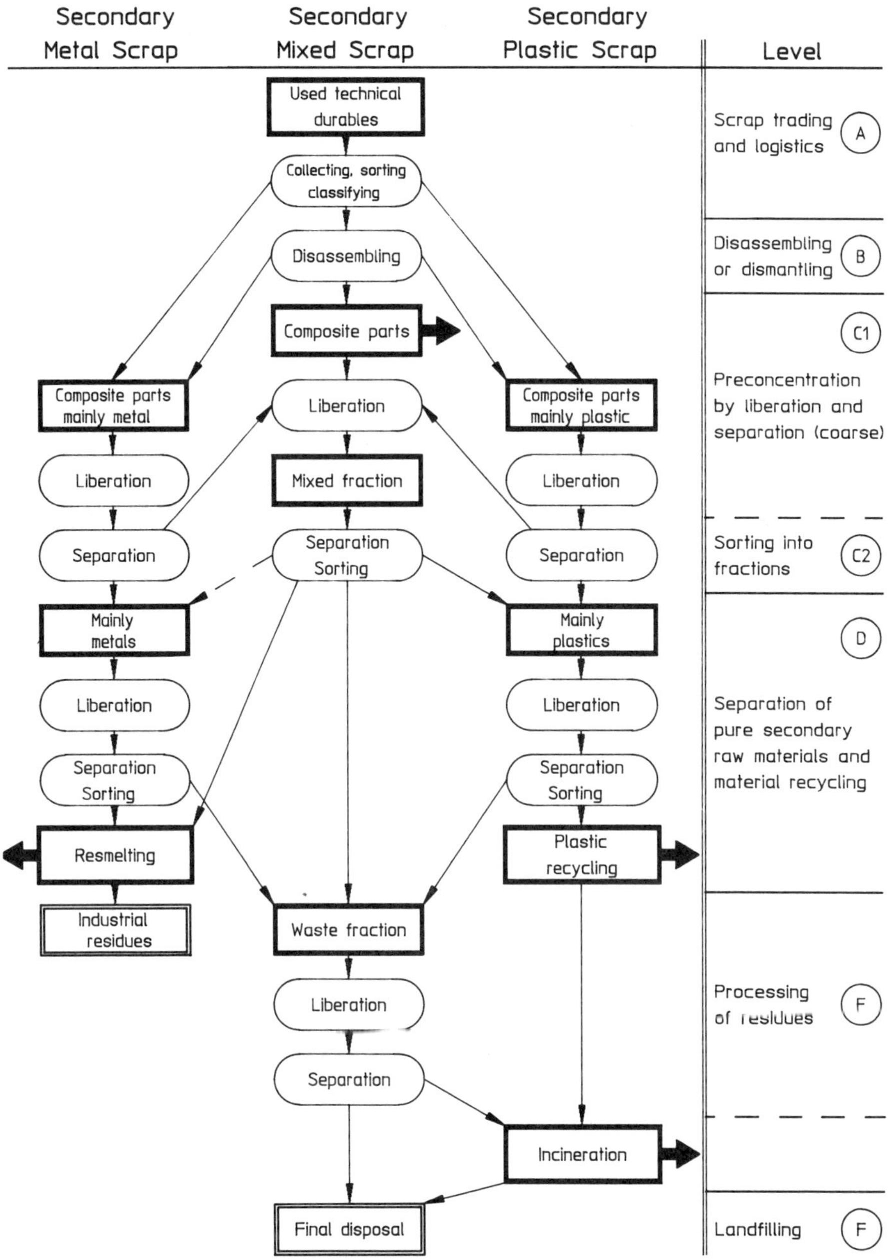

Fig.2: Simplified recycling flow diagram for durables (consumer and capital goods)

DRP/V-ZL/03.90

Separation:	**Fractionation of liberated composites**	Air separation, Screening, Rising current classification, Hydrocyclon, Heavy media separation, Magnetic separation, Eddy current separation Flotation
(Classification)	- to size range	
(Sorting)	- to density	
(Sorting)	- to other physical properties.	
Other Processing Methods:	**Division of fractions**	
	- to elementary analysis - to pure materials - to physical properties - to waste categories.	Automatic material detection, Automatic sorting techniques Melting

What is of vital importance for success or failure of material recycling is the liberation of materials which means separation of composites into mixtures of assorted material components by comminution. On account of the composite design and miniaturization of components, in this operation large amounts of fines (undersize < 15/20 mm) may accumulate. Today, this kind of scrap can only be marketed with reservations or substantial discounts on account of their deficient separation. In most cases, they require elaborate refining processing.

After liberation, the mixtures are separated into homogeneous scrap sorts. Especially important is the elimination of polymer and other combustible components. The standard waste gas cleaning equipment of secondary smelters is only conditionally suited for dealing with emissions from these undesirable substances. The smelter industry cannot remedy the shortage of utilization equipment for waste, being also subject to strict environmental regulations.

In the Federal Republic of Germany, the waste management economy has become organized as an open system on a market-economic basis. This means that chances and risks are no longer equally distributed in the recycling sequence

trade - processing - marketing of recycling materials at the highest value level - residue utilization - waste disposal:

Whilst the marketing of cannibalized components and homogeneous recycling materials may be a rather lucrative business, the residues for chemical or thermal utilization and specifically the final disposal of waste involves in part substantial costs. Evidently, the situation calls for adequate compensation mechanism.

Requirements for Research and Development

The development tasks ahead of us can be derived from the shortage or lack of techniques:

Trade and logistics:

- Establishing disposal and reprocessing routes for durables according to products or assemblies subject to similar reprocessing strategies.

Disassembling and processing:

- Development of dismantling techniques to remove assemblies
- Development of assembly-specific gentle liberation and reprocessing techniques
- Improving scrap purity by novel separation and, particularly, automated sorting techniques
- Reprocessing methods for scrap fines.

Residue utilization and final disposal:

- Improvement of chemical reprocessing methods for varying material mixes
- Improvement of incineration processes specifically with an eye on the emission of dioxins and furanes
- Immobilization technologies for converting non-combustible residues into inert material for final disposal by vitrification, for instance.

The durables industry, commanding sophisticated production techniques and elaborate distribution systems, seldom has the knowledge and experience required to recover the valuable materials contained in its spent products. To combine product design and recyclability therefore cooperations are required also aiming at further developing the recycling technologies and their repercussion on product design.

The environmental policy may oblige industry to combine product design with solutions for adequate disposal after use. Initial approaches aiming at disassembling and sorting of components can already be observed. However, this activity will only conditionally lead to new recycling techniques. The latter should rather be based on the process techniques applied in conjunction with the raw materials technologies for the production of basic materials, having to be adapted to the specific needs for material recycling.

5. Summary and Conclusions

1. At the International Recycling Congress in Berlin in 1989, the thesis was evolved that the lack of waste treatment capacities will be the new shortage of our economy, to be added as the fourth production factor after capital, labour and land. This would result in the environmental-political demand for:

   - Prevention prior to reduction to reutilization prior to final disposal of waste.

This leads us to the demands for:

- taking back and reprocessing of products after use/service
- multiple use of raw materials, i.e. establishing cycles of valuable materials and material recycling
- re-use of secondary raw material at maximum level.

2. Post-use durables are not waste but economic goods containing secondary raw materials. They may be returned to the economic cycle by utilization of components or material recycling without/with material conversion. Incineration will also reduce the waste volume.

3. For material recycling is a pure secondary resource necessary. Recyclability is a material-specific property, whilst in durables the prevailing influence is the material combination, design and manufacturing process. This imparts great importance on all material liberation and disassembling processes.

4. Presorting technologies - technologies for liberation, separation and recovery of materials - should be developed based on known raw material technologies and adapted to the specific needs of assemblies. Constant interaction with the production cycle should be safeguarded.

5. The necessary transition from a consumer-oriented throw-away society to a recycling-oriented material flow economy need not mean abstention from high technologies with tailored materials, miniaturization of components and complex designs. What is important, however, is to make the suitability for recycling as an important property for use, the main criterion for product development.

6. Real mastery implies restriction. An unlimited variety of materials and design which is superfluous, economically and ecologically unreasonable but the result of the principle of competition which dominates in the free market economy, should be critically tested against this proverb.

7. Material recycling of durables leads to components which can be sold with a profit on the one hand, and cost-intensive waste disposal on the other. A reasonable compensation of chances and risks is necessary.

8. The new strategy of recycling oriented-material economy calls for numerous innovations, but above all for the interaction of all parties concerned, i.e. producing industries and trade, waste management economy, administration and politics. Research and development will play a vital role.

References

1. K. J. Thomé-Kozmiensky and W. Schenkel, ed., Konzepte in der Abfallwirtschaft 2 (Berlin: EF-Verlag für Energie- und Umwelttechnik, 1989), 27 - 39

2. cf. (1), pages 1 - 25

3. L. Vanhecke, "The non-ferrous metals industry and the implementation of the EC directives on waste" (Paper presented at the High-level "Environment" Contact Meeting between EUROMETAUX and D. G. XI (Environment) of the CEC, Bilbao, Spain, 22./23. September 1988), 37 - 46

4. W. Blum, "The Secondary Industry and the CEC's Proposal for Directions amending Directive 75/442 on Waste and replacing Directive 78/319 on toxic and dangerous Waste", cf (3), 47 - 62

5. NN., "BIR defines waste and secondary raw material", Metal Bulletin, 22nd Febr. 1990

6. K. J. Thomé-Kozmiensky and H. Käufer, ed., Recycling von Kunststoffen 1 (Berlin: EF-Verlag für Energie- und Umwelttechnik, 1987), 165 - 183

7. R. Pfaff and D. Gneuß, "Reprocessing of Plastic Wastes", Industrial & Production Engineering, 11 (1987), 43

8. cf. (6), pages 109 - 116

9. K.-D. Sturm, "Kunststoffe bereiten große Probleme", Energie 42 (1990), 38

10. H.-P. Sattler, "Werkstoffbedingte Recyclingprobleme am Beispiel des Automobils", VDI-Berichte 694 (1988), 349

DOMESTIC BATTERY RECYCLING IN WESTERN EUROPE

J. Frenay and S. Feron

Université de Liège
Extractive Metallurgy
Rue A. Stévart, 2, B-4000 LIEGE (BELGIUM)

Abstract

A state of the art research on the recycling of domestic batteries has been carried on. All kinds of batteries were concerned : zinc-carbon, alkaline, mercury, silver oxide, nickel-cadmium... Results of this research are presented including size and evolution of the market, existing and potential recycling processes and proposals for a recycling strategy.

This research has been suported by the «Région Wallonne - Ministère des Technologies Nouvelles».

Second International Symposium -
Recycling of Metals and Engineered Materials
Edited by Jan H. L. van Linden,
Donald L. Stewart, Jr., and Yogeshwar Sahai
The Minerals, Metals & Materials Society, 1990

Introduction

The «Ministère des Technologies Nouvelles de la Région Wallonne» of Belgium incited us to study the problem of recycling used batteries. We decided together that the best way to initiate such a study was to evaluate the «market» and do state of the art research on the existing and potential recycling processes.

This paper done with the help of the above-mentioned Ministery summurarizes the results of this research.

What are used batteries

They are a large variety of domestic batteries, differing in composition and in shape.

The main parts of any battery are :
- anodes : which are metallic (for example : Zn, Fe, Cd, Pb, Li).
- cathode : graphite, copper with metallic oxides such as MnO_2 - HgO -AgO
- electrolyte : can be H_2SO_4 - NaOH - KOH - $ZnCl_2$ - NH_4Cl - organic solvant
- case : metallic or plastic, mostly in stainless steel with nickel.
- separator : plastic - cardboard
- additives : for example to avoid corrosion, such as Hg, Pb, Cd, In.

Most of the components of batteries are metallic or non metallic pieces but there are also pastes, powders, and even liquid.

Batteries exist under numerous different sizes and shapes and there is no correlation between the external aspect and the composition.

Moreover, the composition of the batteries evolves with their life. For example, in Leclanché batteries, metallic, zinc is transformed to $Zn(NH_3)_2Cl_2$ and MnO_2 to MnO(OH).

So, the used batteries form a material containing very heterogeneous parts and of complex composition.

Batteries are divided in primary and secondary cells. Main kinds are :

1. Zinc-carbon (Leclanché) : the anode is a zinc sheet and the cathode is MnO_2 powder with a graphite rod and . The electrolyte is NH_4Cl or $ZnCl_2$.

2. Alkaline-manganèse : the anode is a zinc powder with lead and mercury as an anti-corrosive agent. The cathode is MnO - C - KOH (85-10-5). The electrolyte is KOH with 6 % ZnO.

3. Mercury : the anode is zinc powder with mercury. The cathode is mercury oxide with carbon and the electrolyte is KOH + ZnO.

4. Silver oxide : the anode is zinc powder with mercury. The cathode is a compacted mixture of AgO and MnO_2 powder. The electrolyte is KOH or NaOH with ZnO.

5. Zinc-air : the anode is yet zinc powder with mercury but the cathode is activated carbon. The electrolyte is KOH.

6. Lithium : the anode is lithium and the cathode can be composed of various oxides such as MnO_2 - Bi_2O_3. The electrolyte is an organic solvent because lithium reacts with water.

7. Nickel-Cadmium secondary cells : one electrode is in Cd which transform to $Cd(OH)_2$ and the other is NiO(OH) which transform to $Ni(OH)_2$ The electrolyte is a mixture KOH + $Li(OH)_2$. Closed cells are for domestic purposes and open cells for industrial applications.

Only average compositions can be estimated for the different kinds of cells. They are given in Table 1.

How big is the «deposit»

To get a precise picture of the cells market is quite a hard job. Only a few precise and official informations exist on production or consumption of batteries.

It was estimated in 1986 represented battery market in the western world represented $ 5 billion.

Table 2 : Battery consumption in the Western World (1987)

	10^9 pieces	10^3 tonnes
Zinc - C	21.4	1453
Alk.	3.8	373
button	1	2.1
closed Ni-Cd		46
open Ni-Cd		25

From different reports and interviews we have had with professionals, we estimated the consumption for 1987 in the Western World, Western Europe and Belgium as indicated in Table 2 - 4.

Table 3 : Battery consumption in Western Europe (1987)

	10^6 pieces	10^3 tonnes
Zn - C	1575	60
Alk.	875	43
button	195	0.35
closed Ni-Cd		10
open Ni-Cd		10

Table 4 : Battery consumption in Belgium (1987)

	10^6 pieces	tonnes
Zn - C	51.8	1950
Alk.	28.8	1500
Ag	1.5	1.65
Hg	1.67	5.61
Zn-air	0.67	0.73
closed Ni-Cd		280
open Ni-Cd		230

Table 1 : Mean composition of batteries

	Zn-C	Alk	HgO	Ag_2O	Zn-Air	Li	Ni-Cd closed	Ni-Cd open
Zn	18	13	11	11	30	-	-	-
MnO_2	28	23	-	-	-	some	-	-
Hg	0.01	0.35	30	1	1	-	-	-
Ag	-	-	-	28.5	-	-	-	-
C	8	3	4.2	-	high	-	-	-
Steel	16.4	37	40	45	45	60	-	-
Ni	-	-	-	-	-	-	25	10
Cd	0.01	-	-	-	-	-	12.5	8
Pb	0.05	0.05	-	-	-	-	-	-
KOH/NaOH	-	5	7	8	5	-	10	11
$ZnCl_2/NH_4Cl$	5	-	-	-	-	-	-	-
paper/plastic	9	4	5	5	5	5	37.5	49
H_2O	10	8	3	3	3	-	20	22

From the previous tables, it is possible to calculate the estimated theoretical quantities of materials which could be recovered through a 100 % recycling of used batteries and a 100 % efficiency of the treatment processes. Values are given for Westen Europe (Table 5) and Belgium (Table 6).

Continuously for many years, the trend is to use more and more energy from portable sources. Most probably this trend will continue. Another tendancy is the use of high quality cells. So the market value has a 20 % annualy increase.

Zinc-carbone and alkaline cells form more than 90 % of the market. In developed countries the consumption of alkaline cells increases with a smaller decrease of zinc-carbon cells. However in developing countries the use of zinc-carbon cells are yet increasing.

In Belgium, an agreement was undersigned by Ministeries and battery producers in order to drastically decrease the mercury content of alkaline batteries and also to replace as far as possible mercury batteries by other kinds of batteries.

Trend exists also to use more small-size cells in place of large cells. The use of secondary cells, mainly Ni-Cd batteries, is increasing very fast at a rate of 15 to 20 % annualy. They will soon represent 10 % (in quantity) of the market.

Some evolutions must also be noted. The zinc-carbon cells could be partly replaced by the rencently developed «green power» cells, which would be less harmful for the environment. The mercury cells tend to be replaced by zinc-air cells and nickel-cadmium batteries.

How recycle

Researches were carried on in different laboratories in order to develop processes to recycle the used batteries or in some cases to treat them for a safety disposal. Some processes were tested with particular kinds of cells as for example nickel-cadmium batteries. Some processes involve the treatment of a bulk mixture.

As far as the sorting will be difficult and expensive, the best way will be to treat all the cells as a mixture. Moreover, the composition of various cells is rather similar and justify to treat the mixture of them.

Mainly for economical reasons, only few processes have been applied on an industrial scale : silver cells, nickel-cadmium batteries and mercury cells.

Sorting

Used batteries can easily be sorted by size. Unfortunately there is no correlation between size and composition of the batteries, and a partition based on shape or size is unable to give homogenous partions. Silver batteries exist only as button cells. So the collection of button cells give all the silver cells but, of course, accompagnied with a lot of other cells from which sorting is difficult.

Sorting could be obtained by visual observation, but external marks are usually discreet or even unexistant. So, hand sorting is quite difficult and expensive. An improvement should be obtained by a manufacturers agreement to clearly mark the different kinds of batteries, for example with a small colour labeling.

Breaking

Breaking, decortication or grinding is often the first step of the treatment of used batteries. It allows the liberation of the different components of the cells. But this liberation is imperfect. Particularly, some malleable metals as zinc are laminated and small solid impurities are enclosed in the metal. Imperfect liberation and particles entrapping limit the possibility of physical separations.

Table 5 : Theoretical materials to recover from batteries in Western Europe (tonnes, 1987)

	Zn-C	Alk	HgO	Ag_2O	Zn-Air	closed Ni-Cd	total tonnes	mean comp %
Zn	10800	5590	18.9	11.9	21	-	16442	15.3
MnO_2	16800	9890	-	-	-	-	26690	24.8
Hg	6	151	54.5	1.1	0.7	-	210	0.2
Ag	-	-	-	31	-	-	31	0.03
C	4800	1290	7.2	-	?	-	6097	5.7
Steel	9840	15910	68.6	48.8	31.5	3250	29149	27.1
Ni	-	-	-	-	-	2500	2500	2.3
Cd	6	-	-	-	-	1250	1256	1.2
Pb	30	22	-	-	-	-	52	0.05
KOH/NaOH	-	2150	12	8.7	31.5	1000	3201	3.0
$ZnCl_2/NH_4Cl$	3000	-	-	-	-	-	3000	2.8
paper/plastic	5400	1720	8.6	5.4	3.5	500	7638	7.1
H_2O	6000	3400	5.2	3.3	2.1	2000	11451	10.6

Table 6 : Theoretical materials to recover from batteries in Belgium (tonnes, 1987)

	Zn-C	Alk	HgO	Ag_2O	Zn-Air	closed Ni-Cd	Total tonnes	mean comp %
Zn	351	195	0.6	0.2	0.2	-	547	15.4
MnO_2	542	345	-	-	-	-	887	25.0
Hg	0.2	5.2	1.7	-	-	-	7	0.2
Ag	-	-	-	0.5	-	-	0.5	0.01
C	156	45	0.2	-	?	-	201	5.7
Steel	320	555	2.2	0.7	0.3	91	969	27.4
Ni	-	-	-	-	-	70	70	2.0
Cd	0.2	-	-	-	-	35	35	1.0
Pb	1.0	0.7	-	-	-	-	2	0.05
KOH/NaOH	-	75	0.4	0.1	-	28	104	2.9
$ZnCl_2/NH_4Cl$	98	-	-	-	-	-	98	2.8
paper/plastic	175	60	0.3	0.1	-	14	250	7.1
H_2O	195	120	0.2	-	-	56	371	10.5

Different kinds of apparatus are used to break and grind the batteries. Shredders are largely used. They work by shearing the materials. Hammer mills are also used.

Similar results are obtained by heating some kind of cells, principally the mercury cells heated to volatilize mercury. Under the increased internal pressure, these cells explode. Caution must be given to avoid damage to the furnace.

Physical processing

Ground batteries can be treated by technics used in mineral processing. These processes have the advantage to be rather cheap but they can only give impure or heterogeneous portions which must be subsequently treated by pyro- or hydrometallurgical ways.

In many processes, the first step consists in screening the ground cells to obtain two products : a coarse material containing metallic particles, papers and plastics, and a finer fraction composed of powdered compounds and sludges. A water washing allows also the separation of soluble electrolyte compounds such as soda, ammonia species and zinc salts.

The coarse fraction is then air-classified in order to separate the light materials, papers and plastics, from the heavy metallic compounds which are treated in a magnetic separator to separate the magnetic iron compounds from the non-ferrous metals. In Switzerland, Recytec developed a process to recover the non-ferrous metals by electrorefining in fused salt.

Fine material contains mercury and can be processed by pyrometallurgy with mercury distillation or by hydrometallurgy with oxidative leaching (HCl + NaClO) giving a residue containing carbon and manganese oxide and a solution containing zinc and mercury from which mercury is electrodeposed and zinc is precipitated by pH increase (TNO process, Netherland).

Pyrometallurgy

Thermal processes allow the distillation of metals having a low boiling point, mainly mercury, cadmium and zinc.

Mercury distillation from broken or unbroken cells is realized at moderate temperature, usually 600°C. Often, gases are burned at 850-950°C in a post-combustion rooom to oxidize the organic volatile compounds. Then, they are cooled close to 0°C to effectively condense the mercury. Exhaust gases are purified with active carbon filters. The major difficulty with this process is due to the presence of organic materials. These materials are volatilized in the gas phase and form with the mercury a viscous compound, called mercury butter, which can cause the clogging of the pipes, with important risks of explosion and of important damages to the furnace and its subsidiaries.

Industrial plants exist in some european countries to treat mercurial wastes and by-products. The recovered mercury can pay the treatment costs only in the case of mercury cells, thanks to their high mercury content. However, most of the plants do not accept to treat products containing organic materials. A physical separation as described above is then obligatory. For environmental reasons, the treatment of other cells and of mixture of cells necessitates the mercury recycling. This is often done by distillation.

Zinc and cadmium are volatilized at higher temperature than mercury. Depending on the kind of furnace, they can be condensed as metallic phase or recovered as oxides by oxidative cooling of the gases.

The separation nickel-iron is difficult to realize, and the presence of cadmium gives some difficulties with the distillation of mercury and zinc. For these two reasons it is better to treat distincly the nickel-cadmium batteries and all the others together.

Thanks to the cadmium price, the recycling of nickel-cadmium batteries seems to be economically realized in France (SNAM-SAVAM) and in Sweden (SAB-NIFE). Both utilize the closed furnace technology. Cadmium is distillated at 850-900°C and recovered as pure metal (99.95 %). The open furnace technology producing cadmium oxide is not to be recommanded because of unhealthy working conditions.

Zinc volatilization is a well known process in zinc metallurgy and can be applied to the zinc recovery from used batteries. However, this material is quite different from zinc concentrates and, in zinc metallurgy, pyrometallurgical processes are more and more replaced by hydrometallurgical treatment. So, except if a synergism can be obtained with a thermal treatment for mercury evolution, hydrometallurgical route would probably appear more suitable to recover zinc from the studied materials.

The recovery of nickel or manganese ferroalloys by smelting in an electric furnace can be obtained with metallic fractions from physical separations and with some by-products of thermal or hydrometallurgical treatments.

Generally speaking, the pyrometallurgical treatments are cheaper than hydrometallurgical ones. But often they must process a large tonnage to be economic, and the environmental control is more difficult. Moreover, they are not always well adapted to the treatment of complex materials.

Hydrometallurgy

Hydrometallurgical treatment of nickel-cadmium batteries has been studied in Sweden (NIFE), Great-Britain (HARWELL) and Netherland (TNO). These processes are based on leaching with $(NH_4)_2CO_3$ or HCl followed by solvent extraction, precipitation and electrowinning to recover cadmium and nickel.None of these processes has been industrially developed due to their cost, higher than in pyrometallurgy.

They are various ways to use hydrometallurgical routes directly on batteries or on by-products from a previous treatment. They are based on the leaching of metals to recover. Leaching solutions can be alkaline or acid. The subsequent pregnant solutions must undergo different steps of purification, separation and metal recovery.

Fine fraction from the physical separation of a mixture of broken cells can be leached with HCl + NaClO as mentioned before or with H_2SO_4 followed by electrowinning to recover zinc (MHO, Belgium) and in some case the manganese as dioxide (AFIF, Switzerland).

Because of the complex composition of the material to treat, sophisticated and expensives technics must be envisaged, such as solvent extraction and specific electrowinning. So, the used batteries recycling processes based on hydrometallurgy are more expensive, but they allow the recovery of better quality final products and a better control of their environmental impacts.

Conclusions

Used batteries are an heterogeneous and complex material. From this study, it appears that numerous different processes have been tested in order to recycle this material. None of them appears clearly to be the best choice. It could mean that this material is very complex and need sophisticated ways of treatment. But it can also mean that a new process could be developed from new conceptual ideas. For example, the caustic soda process recently developed to treat electric arc furnace dusts from steelmaking industries could be the basis of a new process for batteries recycling.

Anyway, the quantity of material to treat is rather low, saying 3500 tonnes for the Belgium. Valuable products to be recycled from this tonnage can be evaluated at around one million US dollar (nickel-cadmium batteries are not considered in this estimation). Depending on the process, costs of recycling can be estimated around O.6 to 1.8 $/kg. So the cost for the treatment of all the used batteries would be at least two times the value of the recovered products.Negative costs for disposal and environmental control must be added to this estimation. However, it appears quite clearly that the recycling of used batteries will be viable only with the help of public supporting.

NON-FERROUS METAL RECOVERY FROM WASTES

C. S. Brooks

Recycle Metals

41 Baldwin Lane
Glastonbury, Connecticut 06033

Abstract

Recovery of non-ferrous metals from metal finishing industry wastes awaits development of more efficient and economical technology. Principal incentives for waste metal recovery lies in reduction in disposal costs, preservation of strategic metals and compliance with hazardous waste disposal regulations. Technologies potentially applicable for separation and concentration of soluble and insoluble metal wastes are reviewed. Examples are presented of promising adaptations of ion exchange, solvent extraction and selective precipitation for recovery of nickel from actual industrial wastes.

Second International Symposium -
Recycling of Metals and Engineered Materials
Edited by Jan H. L. van Linden,
Donald L. Stewart, Jr., and Yogeshwar Sahai
The Minerals, Metals & Materials Society, 1990

Introduction

An examination is made here of the factors which facilitate creation of a more favorable environment for the recovery of nonferrous metals of economic consequence from industrial waste effluents. Principal incentives for recovery consist of reduction in the volume and toxicity of the waste effluents, the recovery of metals of economic significance and the reduction of disposal costs.

Consideration is given to the nature and volume of waste metal effluents, the identification of the metals of primary economic and strategic significance, current disposal practices, the options for recovery, the relevant regulatory framework and consideration of potential markets for recovered metals. A principal emphasis is given to a state of the art review of the technologies relevant to nonferrous waste metal recovery and promising future directions.

Sources of Metal Wastes

The sources of metal wastes are diverse in nature and in geographical distribution, as well as being impressively numerous. The volumes of all kinds of wastes generated on a nationwide scale are formidable. A national survey published in 1984, the Westat report (1), gave estimates of hazardous waste generators and treatment, storage and disposal facilities regulated under the U.S. Resource Conservation and Recovery Act (RCRA) for 1981 of more than 71 billion gallons, or 264 million tons, of hazardous waste of all sorts generated by approximately 14,000 generators. About 22% of this waste was attributed to metal related industries (SIC codes 33-37). A survey of the hazardous waste generation in a small industrial state Connecticut in 1984 (2) identified that of about 200,000 tons of all sorts of hazardous waste reported to the state's Department of Environmental Protection (DEP) for the first time in 1983 56% of the wastes are classified as heavy metals, attributed to primary metals (SIC 33), fabricated metals (SIC 34), electrical and electronic machinery (SIC 36) and transportation equipment (SIC 37).

Disposal Alternatives

The hierarchy of waste disposal priorities is commonly accepted to be the following:

- Waste volume reduction
- Waste recycling
- Detoxification treatment
  - Physical
  - Chemical
  - Biological
- Incineration
- Solidification - stabilization
- Landfill

Metal recovery from waste effluents contributes to instrumentation of the first three of these categories.

The Westat study (1) provides an overall view of current nationwide waste disposal practices of the some 14,000 generators which manage 96% of the waste on-site versus 4% off-site, with about 7% attributed to metal related industries (SIC codes 33-37). About 96% of hazardous wastes are not recycled. Of the remaining 4% which is recycled, 81% is recycled on-site versus 19% off-site. The distribution of disposal categories for the estimated volume in 1981 of 71.3 billion gallons (264 million metric tons) consists of about 48% treated in some manner, 37% stored and 15% disposed of in surface impoundments, injection wells or landfills.

Current Status of Recycling

Nonferrous metals have the second priority for recycling (3-5), after the precious metals. The nonferrous component of municipal solid waste is estimated to be about 2%, with aluminum the predominant metal. The amounts of the nonferrous metals, although minor components of municipal waste, come to about 2,000,000 tons annually, with about 500,000 tons of metals other than aluminum. The most promising potential for recycling of nonferrous metals undoubtedly lies with the metal finishing and electronics industries. The volumes of these wastes, while not impressive as new ore sources, provide the potential for yielding metal concentrations high enough to favor recovery processes. The waste generated by the mining, mineral processing and metallurgical industries in the United States are much more voluminous. They are estimated to amount to 1.8 billion tons annually (5,6) and while not present at as high concentrations as many strategic metals from the metal finishing industry, they provide high tonnage waste accumulations in many locations in western United States.

The nonferrous metals of commerce that are recycled to a significant extent as scrap metals are aluminum, copper and copper-base alloys, chromium, cobalt, cadmium, nickel, manganese, molybdenum, lead, titanium, zinc and the precious metals (7,8) although the recycling of non-precious nonferrous metals seldom achieves 50%. Aluminum, the principal nonferrous metal found in municipal waste, is currently recycled to the extennt of between 300,000 and 600,000 tons per year, but this probably represents only about 5-10% of the total available (4). Table I contains statistics on demand, imports and recycling for twelve metals of significant commercial use, in the United States. The recycling of important nonferrous metals varies widely, ranging from 6% for cadmium, 1-7% for chromium, about 6% for cobalt, approximately 24% for copper, 10-50% for lead* 25% for manganese, 17% for mercury*, 9-20% for nickel, 15% for tin, 6% for tungsten*, 6% for zinc, about 40% for titanium, 33% for tantalum and negligible percentages for niobium and vanadium* (3,8,9). Recycling precious metals is of a high order and ranges from 60-80% or better. The sources used for citing the extent of recycling indicate that the information is not highly accurate (3-10). The recycling of radioactive elements (thorium, uranium, etc.) is not public information and not considered here.

Even though the current percentages recycled are often low, the total annual tonnages for several recycled metals are not inconsequential, amounting to 1,750,000 tons of copper, 700,000 tons of lead, 350,000 tons of zinc and 1,700,000 tons of aluminum. A real potential still remains for increasing the amounts of recycling for metals such as aluminum, copper, manganese, chromium, nickel and tungsten by factors ranging from two to five (8-11).

* Not included in Table I.

Table I DEMAND, IMPORTS AND RECYCLING IN THE U.S. *

	METAL DEMAND			RECYCLED METAL	
	1979 (10^3 TONS)	FORECAST 2000 (10^3 TONS)	METAL IMPORTS PERCENT OF 1979 DEMAND	AMOUNT (10^3 TONS)	PERCENT OF 1979 DEMAND
Al	4,528.000	6,646.000	~ 90	333 - 665	5 - 10
Co	7.800	6.300	100	0.380	6
Cu	2,060.000	2,408.000	~6	~570.000	24
Cr	480.000	550.000	100	~4.5-38.5	~1 - 7
Mn	1,317.000	1,250.000	98	313.000	25
Ni	152.000	199.000	75-94	17.9-49.8	9 - 25
Nb	2.300	3.600	100	Negligible	-
Pt Group (oz) (Pt, Pd, Rh, Ru, Ir, Os)	$1,361 \times 10^3$	$2,756 \times 10^3$	99.70	$1,640 - 2,204 \times 10^3$	60 - 80
Ta	0.630	0.630	100	0.210	33
Sn	72.840	54.400	7.5	8.200	~15
Ti	516.000	618.000	63-75	247.000	40
Zn	1,437.000	1,090.000	50-76	65.400	6

* METALS WITH ANNUAL DEMAND IN EXCESS OF 1,000 TONS, EXCEPT FOR PRECIOUS METALS.

Regulatory Considerations

Enlightened management of waste effluents has been the intent of a comprehensive body of federal and state legislation designed to adequately respond to public demand for an improved environment. Three federal laws constitute the principal regulatory framework, namely,

Toxic Substance Control Act (TSCA) (12)
Resource Conservation and Recovery Act of 1976 (RCRA) (13)
Comprehensive Environmental Response, Compensation and Liability Act of 1980 (CERCLA) (14)

The intent of TSCA initially enacted in 1976 was to address the proper identification and accountability for toxic substances to protect public health. The intent of RCRA was to provide enlightened management for solid waste handling to both protect public health from toxic substances and conserve valuable natural resources. This law is the principal federal law providing incentives for recycling because the ultimate goal is to minimize, if not eliminate landfill disposal. CERCLA, the third federal law and commonly known as the Superfund, had as its intent the cleanup and protection of the public health because of the nation's earlier ignorance and neglect in properly managing its waste effluents.

There is little impetus in peacetime to build our strategic metal stock piles and current federal policy has been to drastically reduce the dollar value of the stock pile from $16,300,000,000 to $6,700,000,000* as long as there are no imminent shortages, neither Congress nor the administration has strong incentives to fund purchases, notably in a time of peace. Positive incentives for recycling can be provided if federal or state tax or other credits are provided in connection with the disposal of hazardous wastes. The administrative and regulatory incentives, as well as barriers to recycling, will not be reviewed further here.

Metal Recovery Technology

There are a number of separation and concentration processes potentially applicable to metal recovery from industrial wastes. These processes have attained a high state of development for purposes other than metal recycling. There are several published reviews of technologies which are applicable to metal finishing industry wastes (15-20), electroplating industrial practice (21,22), water treatment (23-27), recycling (28-30) and metal refining processes (31) that provide considerable detail about many of these separation-concentration processes. The author published a review of these relevant separation processes (32).

The available technologies for metal recovery are most highly developed and most favorable for efficient separation and concentration in the case of soluble metals. These technologies include electrodeposition, solvent extraction, flotation and selective precipitation. Separation technologies attractive for soluble metals in more dilute concentrations in the absence of appreciable insolubles in the colloidal size range consist of ion exchange, electrodialysis and reverse osmosis.

The purely physical separation processes such as evaporation, precipitation,

* Chemical and Engineering News 63 (28), (1985), 4

filtration, etc., while often providing the lowest costs are least likely to provide "metal by-products" with significant market value. Certain of the physical separation processes such as flotation, solvent partition by wettability modification and high gradient magnetic separations do offer promise for development into selective concentration of colloidal solids to obtain marketable by-products.

Where substantial on-site solid sludge accumulations exist, acid solubilizations can be used to obtain soluble metals amenable to separation and concentration by available highly developed processes. In the case of the latter option an inert, non-toxic, insoluble residue acceptable for land disposal becomes available. In view of the array of available separation technologies and the current absence of demonstration of applicability to metal recovery from waste sludges, further comments are in order.

It is obvious that high assays and minimum impurity content, combined with the fewest processing steps, afford the greatest promise for economical recovery. The most obvious approach for achieving high assay metals using demonstrated technology would involve processes applicable to metals in solution, such as solvent extraction, electrowinning, electrodialysis, ion exchange, reverse osomsis, etc. In view of the heavy orientation of the existing metal scrap and recycle businesses to handling metals in the elemental state, any process recovering metal will have a preferred status. On the other hand, lower cost physical separation procedures which recover metals as oxides/hydroxides or salts, such as precipitation, filtration, flocculation and evaporation, offer the least promise for recovering high assay, low impurity metals for recycle from complex metal mixtures.

Where metal wastes are generated as metal oxide/hydroxide sludges there are two principal alternatives to consider for improvement in the quality of the recovered material. One is to persuade the generator that, by accepting modification in current manufacturing procedures for handling effluents, a more valuable "by-product" versus "waste" can be obtained. This modification can be most advantageously done during installation of the initial process, rather than by retrofitting an existing process. Application of technologies permitting efficient separations from solution should be favored over processes producing insoluble sludges.

An alternative approach, primarily applicable to solid waste sludges, is to employ an acid solubilization step, followed by a second stage use of one of the demonstrated technologies applicable to metals in solution. This approach would be appropriate if the waste is generated as a metal oxide/hydroxide sludge generated from on-going precipitation/filtration effluent treatments or accumulations from previous processing.

The third alternative consists of using one of the lower cost, mostly nonselective physical separation procedures, such as flotation or solvent partition, applicable to dewatering solid sludges. This approach offers the potential for achieving high assay and minimum impurity in a single process stage for waste streams containing a single metal. It also requires more evaluation as to its applicability to metal finishing industry waste effluents than the first two alternatives, but merits development in view of the potential for obtaining improved recovery economics.

There have been several efforts to provide explicit guidelines and practical case histories to educate small generators as to how pollution control measures can provide improved economics by incorporation of recycling (33-35). There are both technical and economic disincentives to recycling, such as small waste volume, complex composition, no obvious route for significant volume reduction, costly and burdensome process modifications required, insignificant value for recovered products and lack of use or markets for recovered materials.

A flexible separation scheme (Fig. 1) for efficient recovery of nickel from metal wastes systems as diverse as electrochemical milling (ECM), nickel hydroxide sludge, waste mineral acid, a supported nickel catalyst and a waste electroless nickel has been evaluated by bench scale experiments for technical feasibility (36). Initial removal of the bulk of copper, iron and zinc contaminants is conducted at low pH in acid solution by solvent extraction. Removal of the organic complexing agents and conversion of Cr^{3+} to Cr^{6+} by auto catalysis in the presence of inorganic oxidants (H_2O_2, $NaClO_3$ or $(NH_4)_2S_2O_8$) enhances the nickel separation. Oxalate precipitation provides recoveries in the range from 99 to 93 percent for these four diverse wastes (Table II). A final sulfidation reduces the residual metals to concentrations compatible with state and EPA standards for water effluents.

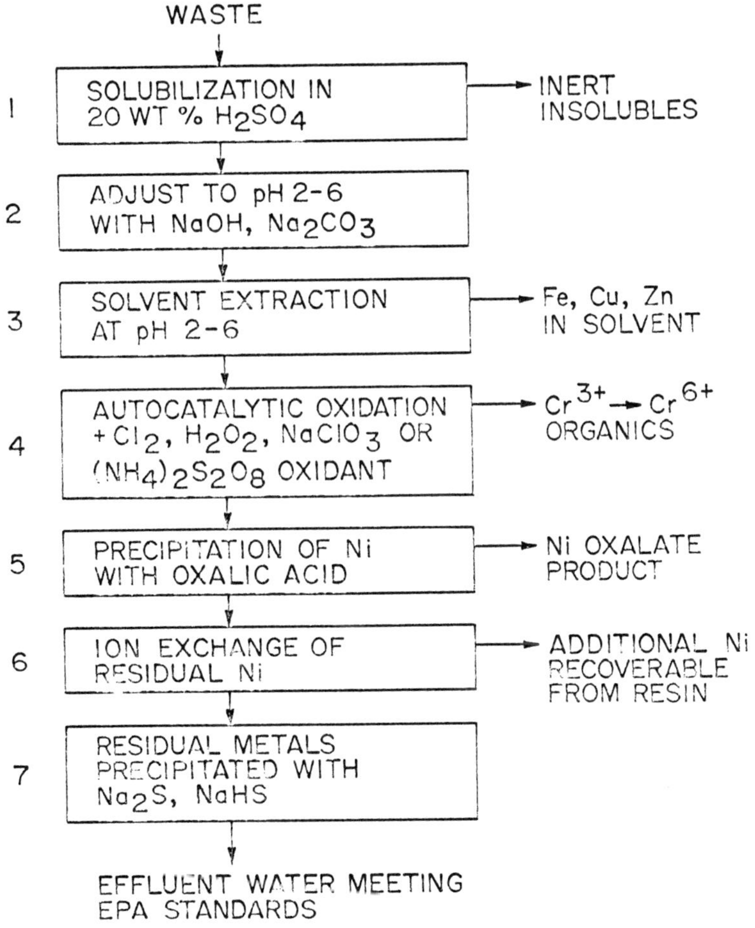

Figure 1. Separation Stages

TABLE II SUMMARY OF SEPARATION RESULTS

WASTE	WASTE NICKEL CONTENT (WT %)	SEPARATION STAGES	NICKEL WT % RECOVERED BY OXALATE PRECIPITATION	NICKEL WT % RECOVERED BY ION EXCHANGE	RESIDUAL NICKEL (PPM)
E C M	4.4	1, 2, 3, 5, 6	99.0	0.88	3.5
NICKEL HYDROXIDE	5.7	1, 2, 3, 5, 6	96.9	3.0	6.0
MINERAL ACID	3.0	2, 3, 5, 6	96.9	3.1	10.0
NICKEL CATALYST	25.0	1, 2, 5, 6	99.9	0.07	18.0
ELECTROLESS NICKEL	0.29	4, 6	92.5	2.2	217.0

Metal Recovery Economics

The principal economic factors to be considered in the recovery of metals from waste effluents consist of the following:

- Potential for assimilation into product manufactured
- Volume of metal produced and consistency of generation rate
- Assay of recoverable metal
- Nature and amounts of impurities
- Process modification costs required for recovery
- Transportation costs where applicable
- Current market price and demand for recovered metal
- Availability and geographic location of potential buyer of recovered metal
- Governmental incentives and disincentives

The framework within which recovery is accomplished ranges from the generator, through the treatment and disposal facility, or to the recycle vendor or manufacturer that can utilize recycled metals as raw materials. The most obvious economic advantages accrue to the generator for exercising the recovery options. The generator has the control on his process to design operations which minimize waste and recover materials in the most advantageous condition for recycling. For manufacturers in metal extraction and processing industries or chemical process industries, recovered metals or metal salts have the best potential for reassimilation into a manufactured product. The metal finishing and electronic industries for the most part generate metal wastes in a condition not readily assimilated into product manufacturing so that in house recycling is the preferred option.

High disposal costs, combined with rigouous enforcement of pollution emission regulations, provide the principal incentives for most generators, large or small, for the future. Although the treatment companies which collect and process wastes may also have an incentive to recycle. Off-site disposal of hazardous wastes is a big business and the major treatment companies operating nationwide, such as Chemical Waste Management, Browning Ferris, Inc./ CECOS, Zimpro, Rollins Environmental Service, SCA Chemical Services, Ensco Inc., Chem Clear Inc. and ERT, account for annual sales ranging between $2,000,000 and $150,000,000 (37). Large companies are most likely to proceed with on-site disposal solutions; the small hazardous waste generators are most dependent on off-site disposal.

The costs of disposal for hazardous wastes are subject to wide variation, dependent on the nature of the waste and geographic location, and are consistently increasing. No estimates are provided of the costs of the various disposal alternatives ranging from landfill, land treatment, incineration, chemical treatment, deep well injection, to resource recovery. High disposal costs make apparent the merits of establishing resource recovery credits in order to minimize disposal costs. The cost of transporting wastes notably designated hazardous is a very significant factor in disposal and the cost of small quantities of hazardous waste is disporportionately high. A distinct advantage for on-site recycling operations is the avoidance of transportation costs, especially if the recovered metals can be utilized on site.

Consideration of the economic incentives and disincentives for metal recycling to minimize waste disposal costs and to meet emission regulations makes it clear that the small generator must become dependent on outside technical consultant expertise appropriate state agencies or commercial sources. There have been a number of studies of disposal and recycling costs for the metal finishing industries, largely under federal (EPA) funding (27,38-52).

There are a number of commercial firms that make disposal cost assessments for various disposal alternatives for specific private industry applications such as Lancy, Metal Recovery Technology, Metpro, Eltech, Ecotech, Osmonics, Metal Removal Systems, Thibault Associates and Camp Dresser & McKee, Inc. Listings of firms providing these services can be found in an appropriate directory (53).

Successful disposal of metal finishing industry wastes by incorporation of metal resource recovery may require a combination of private, commercial and state action for creation of an infrastructure that currently does not exist. One route for creation of this infrastructure designed particularly for the off-site disposal problems of both small and large generators is through development of centralized regional waste treatment (CWT) facilities. Such facilities are used commonly in Japan and some European countries such as Germany. The need for CWT facilities in the United States is growing rapidly but such facilities are currently only in the planning stage. A CWT facility must be designed to handle the volume of metal wastes necessary to make recovery efficient and to provide economies of scale. Several analyses of the economic viability of CWT facilities have been conducted and found positive (38, 39, 43, 44). An important consideration in design of CWT facilities is provision of secure disposal by some type of land disposal of the inoccuous, irreducible residues left from destruction or conversion of the hazardous materials and recovery of resources. There are several promising processes for production on inert materials for land disposal (54). Examples that can be cited are the following processes: Solinoc, Solid Tek, Stablex, ChemFix and Poz-o-Tec.

A type of organization that plays an increasing role in recycling is the waste exchange. The number of such exchanges across the country has grown from four in 1976 at the advent of RCRA to about 36 at the present time. These exchanges are still predominantly non-profit, newsletter-type inventory listings acquainting waste generators with customers in search of raw materials, but some provide brokering and technical assistance functions. An association of the existing waste exchanges enhances their effectiveness in increasing recycling (11, 55, 56).

Strategic Metals

The strategic aspect of metal usage is a significant factor in considering which waste metals should be recovered for reuse. A modern industrial economy would be impossible without a high level utilization of metals. There are approximately 18 metals with an annual production in the United States in excess of 1,000 tons per year, with 10 metals exceeding 50,000 tons per year. In addition, there are about 14 metals of industrial significance for which there is negligible domestic production in the United States.

Entirely aside from the merits of reducing environmental impact and minimizing the costs and hazards of metal waste disposal, there are important considerations in regard to the strategic nature of many of these waste metals. Practically all metals are a precious and limited resource we can ill afford to squander. Among the 35 metals in commercial use the reserves in the United States represent less than 10% for 22 metals, based on data compiled by the U.S. Department of the Interior (57).

Several efforts have been made to assess the magnitude of the rates at which metal resources of the United States are being exhausted by current use. Two evaluations that are particularly noteworthy are those of Lovering (58) and Roper (59). Roper, in estimating the times for attaining peak production in the United States, concluded that 20 metals had passed their peak of production by 1975.

Among the some 35 metals examined, most of which are in current commerical use in the United States, 12 metals are of special importance (Table I): aluminum, cobalt, copper, chromium, manganese, niobium, nickel, platinum, tantalum, tin, titanium and zinc. All these metals are in annual use in the United States to the extent of 1,000 tons or more currently, with the exception of the precious metals and tantalum. These particular metals have in common limited U.S. reserves, extensive industrial use and strategic importance.

Future Developments

The incentives for metal recovery from metal wastes lies principally in obtaining reductions in waste disposal costs and improvement in ease of meeting pollution standards, combined with conservation of metals of economic significance and strategic value. The cumulative metal waste total volumes, even on a regional basis, do not amount to very significant fractions of the current production and use volumes for the metals of commerce and are not easily perceived as new metal sources. Furthermore, the generation sources are commonly numerous, highly diversified and geographically dispersed. The most obvious economic advantages accrue to the metal recovery option being exercised by the generator, who has the most to gain by having control over process conditions permitting recovery of "metal by-products" of marketable value from effluents with the simplest composition. Treatment companies with capability to incorporate resource recovery into existing waste metal disposal processes are second only to the generators in having the opportunity to benefit from the economic credits from recovered metals. In addition to establishing efficient separation-concentration technologies, treatment companies can play an important role in developing a collection-transportation infrastructure for concentration of the metals for transfer to potential markets.

A positive climate favoring recycling is only feasible if several conditions can be met. The first is that the corporate policy of the generator should be strongly favorable. In addition, government regulatory processes, federal and state and municipal, should provide largely positive incentives. There should be good acceptance for recycled materials, the recovered metals in active demand and the volume of recovered material significant in amount, quality and rate of generation. Also the recovered metals should meet specifications of potential buyers, and the source of recovered metals in close geographic proximity to the potential user. It is important that recovered metals can be successfully disengagned from identification as hazardous material, economical recovery technology be available and, finally, active marketing and brokerage efforts must be pursued to identify appropriate markets for recovered metals.

Successful recycling is only likely to occur in any given case if most of these factors are strongly positive.

References

1. Westat, Inc. National Survey of Hazardous Waste Generators and Treatment, Storage and Disposal Facilities Regulated under RCRA in 1981. (EPA Office of Solid Waste Contract No. 68-01-6861, 1984).

2. 1983 Connecticut Hazardous Waste Generation and Management Analysis for Connecticut Hazardous Waste Mana-ement Service. ERM-Northeast, Inc., 88 Sunnyside Boulevard, Suite 305, Plainview, NY 11803 (June 1985).

3. M.J. Spendlove, Recycling Trends in the United States: A Review. (BuMines Information Circular 8711, 1976).

4. Base Line Forecasts of Resource Recovery, 1972 to 1990. (U.S. Department of Commerce, NTIS PB 245924, March 1975).

5. H.S. Hanna and C. Rampacek, Resources Potential of Mineral and Metallurgical Wastes. (Fine Particles Processing, Vol. 2, P. Somasundaran, Ed., New York, NY, AIME, 1980),1709.

6 R.D. Hill and J.L. Auerbach, Solid Waste Disposal in the Mining Industry. (Fine Particles Processing, Vol. 2, P. Somasundaran, Ed., New York, NY, AIME, 1980), 1731.

7. Study to Identify Opportunities for Increased Solid Waste Utilization, (Natl. Assoc. of Secondary Material Industries, New York, Vol. I. EPA-SW-40D. 1-72, PB 212719, 1972).

8. Technical Options for Conservation of Metals, Case Studies of Selected Metals and Products, Office of Tech. Assessment, Congress of the United States, Washington, DC 10510, OTA 51-680 0-79-2.

9. H. Ness, Industrial Wastes, Recycling Scrap--A decade of Challenges and Frustrations, Materials Recycling, An overview of the Sixth Mineral Waste Utilization Symposium, S.A. Bortz and K.B. Higbie, Eds., (BuMines Information Circular 8826, 1980), 82.

10. M.J. Spendlove, Reclamation, Utilization, Disposal and Stabilization, BuMines Research on Resource Recovery, (U.S. Bureau of Mines 8750, 1977).

11. Proceedings of the First National Conference on Waste Exchange, Roy C. Herndon, Ed., The Florida State University, Florida Chamber of Commerce and U.S. EPA, Tallahassee, FL (March 1983).

12. Toxic Substances Control Act, 15 U.S.C.A. SS2601-2629, (Pub. L. 94-580).

13. Resource Conservation and Recovery Act of 1976, 42 U.S.C.A. SS6901-6987, (Pub. L. 94-580).

14. Comprehensive Environmental REsponse, Compensation, and Liability Act of 1980, 42 U.S.C.A. SS960-9657, (Pub. L. 96-510).

15. A State of the Art Review of Metal Finishing Waste Treatment, Battelle Memorial Institute, 505 King Avenue, Columbus, Ohio 43201, Water Pollution Control Research Series 12010 EIE (1968).

16. J.F. Zievers and C.J. Nvotny, Curtailing Pollution from Metal Finishing, Environ. Sci. & Tech. 7 (3), (1973), 209.

17. K.R. Coulter, Pollution Control and the Plating Industry, Plating 57, (1970), 1197.

18. W.P. Innes and W.H. Toller, Jr. Preliminary Steps in Waste Treatment, Plating 57, (1970), 1205.

19. M.R. Watson, Pollution Control in Metal Finishing, Noyes Data Corp., Park Ridge, MJ 07656, (1973).

20. Summary Report: Control and Treatment Technology for the Metal Finishing Industry, In-Plant Changes, (EPA 625/8-82-008, 1982).

21. A.K. Graham, Ed., Electroplating Engineering Handbook, Van Nostrand, New York (1971).

22. The Canning Handbook on Electroplating, W. Canning Ltd., Great Hampton Street, Birmingham, England (1978).

23. J.B. Berkowitz, et al., Unit Operations for Treatment of Hazardous Industrial Wastes, (NTIS PB 275054 and PB 275287, 1978).

24. R.B. Pojasek, Ed., Toxic and Hazardous Waste Disposal, Vol. 4, Ann Arbor Science Publishers, Inc., Ann Arbor, MI (1980).

25. Environmental Pollution Control Alternatives: Sludge Handling, Dewatering, and Disposal Alternatives for the Metal Finishing Industry, (EPA 625/5-82-028 1982).

26. J.G. Dean, et al., Removing Heavy Metals from Waste Water, Environ. Sci. & Tech. 6 (6), (1972), 518.

27. R.T. Coleman, et al., Sources and Treatment of Wastewater in the Nonferrous Metals Industry, (EPA-600/2-80-074; NTIS PB80-196118, 1980).

28. A.A. Cochran, et al., Development and Application of the Waste-Plus-Waste Process for Recovering Metals from Electroplating and Other Wastes, (BuMines RI 7877, 1974).

29. E.P. Crumpler, Management of Metal Finishing Sludges, EPA 530/SW/561; PB 263946, 1977).

30. S. Bhatia and R. Jump, Metal Recovery Makes Good Sense! Envir. Science & Tech. 11 (8), 1977), 752.

31. A.K. Biswas and W.G. Davenport, Extractive Metallurgy of Copper, New York, Pergamon Press, (1976).

32. C.S. Brooks, Metal Recovery from Industrial Wastes, J. Metals, 38 (7), (1986), 50.

33. D. Huisingh and V. Bailey, Eds., Making Pollution Prevention Pay--Ecology with Economy as Policy, Pergamon Press, Elmsford, NY (1982).

34. M.G. Royston, Pollution Prevention Pays, Pergamon Press, Elmsford, NY, (1979).

35. J. Kohl and B. Triplett, Managing and Minimizing Hazardous Waste Metal Sludges, Industrial Extension Service, North Carolina State University, Raleigh, NC (1984).

36. C.S. Brooks, Nickel Metal Recovery From Metal Finishing Industry Wastes Proceed., 42nd Ind. Waste Conf., Purdue U., Lewis Pub., (1987), 847.

37. F.E. Haarhoff, Managing Hazardous Waste: The Search for New Approaches. Chem. Business, May 2, (1983) 25-30.

38. M.K. Stinson, Group Treatment of Multicompany Plating Waste, Tontin, MA, Silver Project. (EPA 600/2-79-102, PB 80-102825, July 1979).

39. Environmental Pollution Control Alternatives: Centralized Waste Treatment Alternatives for the Electroplating Industry, (EPA 625/5-81-017, June 1981).

40. Clark-McGlennon Associates, Inc. and Waste Resource Associates, Managing Metal Hydroxide Sludge in Connecticut (Rpt. for U.S. EPA Region I Waste Mgy. Div. and Connecticut DEP 1982).

41. American Electroplaters Soc. and U.S. EPA, Third Conf. on Advanced Pollution Control for the Metal Finishing Industry, April 1980. (EPA 600/2-81-028, 1981).

42. EPA Fourth Conf. on Advanced Pollution Control for the Metal Finishing Industry,(EPA 600/9-82-022, NTIS PB 83-165803, 1982).

43. Pope-Reid Associates, Inc. (Technical Report: Feasibility Study--Central Recovery Facility for Minnesota Department of Energy, Planning & Development, 1982).

44. A. Gupta, et al., A Central Metal Recovery Facility, Dept. of Chem. Eng., Princeton University, NJ (Oct. 1983).

45. Environmental Regulations and Technology--The Electroplating Industry, (EPA 625/10-80-001, 1980).

46. T.H. Maugh II, Burial is Last Resort for Hazardous Wastes, Science 204, (1979) 1295.

47. M. Graff, Selling Waste Can Check Rising Disposal Costs,Chem. Eng., Oct. 4, 1982, 39-43.

48. R.A. Conway and R.D. Ross, Handbook of Industrial Waste Disposal, Van Nostrand Reinhold, New York (1980).

49. Economic Analysis of Effluent Guidelines, The Metal Finishing Industry, (EPA 230/1-74-032, 1974).

50. Wapora, Inc., Assessment of Industrial Hazardous Waste Practice, Special Machinery Manufacturing Industries, (EPA SW-141C; PB 265981, 1977).

51. Environmental Pollution Control Alternatives: Economics of Wastewater Treatment Alternatives for the Electroplating Industry, (EPA 625/5-79-016, 1979).

52. Control and Treatment Technology for the Metal Finishing Industry; In-Plant Changes (EPA 625/8-82-008, 1982).

53. Chemical Week Directory of Worldwide CPI Services, Issued Annually.

54. R.B. Pojasek, Solid-waste Disposal: Solidification, Chem. Eng., Aug.13, 1979, 141-146.

PROVEN STATE-OF-THE-ART MULTI-MATERIAL RECYCLING FACILITIES

Stephen A. Katz

Vice President, New England CRInc.

74 Salem Road

N. Billerica, MA 01862

The presentation will describe in detail the advanced, automatic methods to separate mixed glass, plastics (HDPE, PET and other resins) and metals (tin-plated steel and non-ferrous metal) from large scale recycling programs such as the state of Rhode Island's Materials Recycling Facility currently serving over 650,000 persons. This technology, in operation in over 21 plants in Europe and the United States, maximizes automatic separation methods to achieve efficient, high throughput, worker-friendly facilities recovering in excess of 200 tons of recyclable materials per day. Results of the first year of operation of Rhode Island's state-of-the-art MRF will be reviewed. Successful operating arrangements between the public and private sectors will be discussed. Innovations in design to further increase recovery will be presented. A videotape of operations will be followed by a short presentation covering the above items. A question and answer period will conclude the presentation.

Second International Symposium -
Recycling of Metals and Engineered Materials
Edited by Jan H. L. van Linden,
Donald L. Stewart, Jr., and Yogeshwar Sahai
The Minerals, Metals & Materials Society, 1990

BRIQUETTING OF NONFERROUS METAL WASTE FOR ECONOMIC RECYCLING.

CASE STUDIES.

Wolfgang Pietsch

KOPPERN EQUIPMENT, INC.
4203 Black Tree Lane
Charlotte, North Carolina 28226

Abstract

The physical properties, mainly particle size, specific surface, and bulk density, of most metal bearing waste materials are not suitable for direct recycling as part of a secondary resource recovery program. Densification into specific shapes provides a method to improve these characteristics and render particulate metal bearing wastes acceptable for reuse.

In addition to the short introduction given below, the presentation will include a pictorial review of three operating briquetting plants for particulate zinc and lead containing materials concentrated from steel mill dust as well as for brass and aluminum turnings, chafings, and borings.

**Second International Symposium -
Recycling of Metals and Engineered Materials
Edited by Jan H. L. van Linden,
Donald L. Stewart, Jr., and Yogeshwar Sahai
The Minerals, Metals & Materials Society, 1990**

	Combustion Systems	
	Air/Fuel	Air/O_2/Fuel
Firing Rate (MM BTU/hr)	9.5 - 14.0	12.0 - 14.0
Heat Cycle (hrs, tap to tap)	3-4	1-2
Average Production (lbs, 5 days/wk)	193,563	319,497
Production Increase (%)	--	65.0
Average Gas Consumption (cf/ton)	7,000	5,013
Gas Savings (%)	--	28.4
Average Baghouse Temperature (°F)	250°-270°	215°-230°
Flue Gas Volume (cfh)	132,000	67,500
Flue Gas Reduction (%)	--	49.0
Furnace Efficiency (%)	13.57	18.95

To date, there has been no adverse affect on refractory life and metal yield appears to remain unchanged. The retrofitted air/oxygen/fuel burner has proven to be a cost effective and economic method of increasing productivity from a rotary furnace while reducing flue gas volume and baghouse temperature.

ACKNOWLEDGEMENTS

The authors would like to acknowledge the assistance of Bart Bertetto and John Oldham of Air Products and Chemicals, Inc. and Serge Bermier of Quebec Aluminum Recycle for their contributions in making this program a success.

Author Index